Advances in
ORGANOMETALLIC CHEMISTRY

VOLUME 29

Advances in Organometallic Chemistry

EDITED BY

F. G. A. STONE

DEPARTMENT OF INORGANIC CHEMISTRY
THE UNIVERSITY
BRISTOL, ENGLAND

ROBERT WEST

DEPARTMENT OF CHEMISTRY
UNIVERSITY OF WISCONSIN
MADISON, WISCONSIN

VOLUME 29

ACADEMIC PRESS, INC.
Harcourt Brace Jovanovich, Publishers
San Diego   New York   Berkeley   Boston
London   Sydney   Tokyo   Toronto

ACADEMIC PRESS, INC.
San Diego, California 92101

United Kingdom Edition published by
ACADEMIC PRESS LIMITED
24-28 Oval Road, London NW1 7DX

LIBRARY OF CONGRESS CATALOG CARD NUMBER: 64-16030

ISBN   0-12-031129-1   (alk. paper)

PRINTED IN THE UNITED STATES OF AMERICA
89 90 91 92   9 8 7 6 5 4 3 2 1

Contents

Heteronuclear Cluster Chemistry of Copper, Silver, and Gold

IAN D. SALTER

Four-Electron Alkyne Ligands in Molybdenum(II) and Tungsten(II) Complexes

JOSEPH L. TEMPLETON

Department of Chemistry
The University of North Carolina at Chapel Hill
Chapel Hill, North Carolina 27599

I
INTRODUCTION[1]

A. *Scope of Review*

The past decade has witnessed renewed interest in transition metal alkyne chemistry. Early reviews of olefin and alkyne ligands are available (*1–7*), but the extensive "four-electron donor" alkyne chemistry which has blossomed since 1980 has not yet been the topic of a literature summary. Maitlis included a footnote in an early summary article, "The other π orbital ($\pi_\perp$ in our terminology) may have significant overlap . . . and cannot be ignored . . ." (*8*). Otsuka and Nakamura also mentioned the importance of $\pi_\parallel$, $\pi_\parallel^*$, *and* $\pi_\perp$ in metal acetylene complexes in their 1976 review (*9*). This article reviews the chemistry of monomeric Mo(II) and W(II) metal complexes in which the filled alkyne $\pi_\perp$ orbital (orthogonal to the MC_2 plane) is important in metal alkyne bonding (Fig. 1). Green's recent summary of conversion of molybdenum alkyne derivatives to complexes with multiple metal–carbon bonds contains abundant germane alkyne chemistry (*10*).

The term "four-electron donor" is sometimes used to describe the alkyne ligand in circumstances where alkyne $\pi_\perp$ donation supplements classic metal–olefin bonding. The utility of this scheme lies in its simplicity, and with some reluctance we shall rely on the "four-electron donor" terminology to suggest global properties of metal alkyne monomers. The general implications and specific hazards characterizing these descriptors

---

[1] Abbreviations: Ar, $p\text{-}C_6H_4R$; Cp, $\pi\text{-}C_5H_5$; Cp′, $\pi\text{-}C_5H_4Me$; Cp*, $\pi\text{-}C_5Me_5$; DCNE, *trans*-dicyanoethylene; DMAC, $MeO_2CC{\equiv}CCO_2Me$; dmpe, $Me_2PCH_2CH_2PMe_2$; dppe, $Ph_2PCH_2CH_2PPh_2$; Hacac, $CH_3C(O)CH_2C(O)CH_3$; MA, maleic anhydride; R_f, CF_3; TCNE, tetracyanoethylene; TTP, *meso*-tetra-*p*-tolylporphyrin.

1

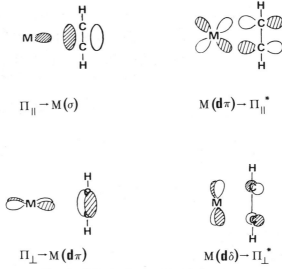

$$\Pi_\parallel \to M(\sigma) \qquad\qquad M(d\pi) \to \Pi_\parallel^*$$

$$\Pi_\perp \to M(d\pi) \qquad\qquad M(d\delta) \to \Pi_\perp^*$$

FIG. 1.   Metal–alkyne orbital interactions.

are typical of broad classification schemes in chemistry: they are often conceptually helpful but seldom specifically correct. Criteria for recognizing four-electron alkyne donation encompass stoichiometry, structure, spectra, and reactivity. Molecular orbital models provide a unifying framework for these diverse chemical properties. Group VI d^4 alkyne complexes serve admirably to illustrate the salient general features of four-electron donor alkyne chemistry.

Since the late 1970s an extensive four-electron alkyne chemistry with cyclopentadienyl Group VI derivatives has been established by the groups of Alt, Davidson, Green, and Stone, with Alt concentrating on tungsten-(II) and Green utilizing molybdenum(II) most heavily. During this same period McDonald, Newton, and co-workers have developed Group VI alkyne chemistry with ancillary dithiocarbamate ligands. The same conceptual framework undergirds all of these efforts, as well as more recent $M(CO)(RC\equiv CR)L_2X_2$ contributions.

The octahedral simplicity and ligand autonomy in $M(CO)(RC\equiv CR)$-L_2X_2 molecules make them pedagogically attractive. Members of this class of molecules were not reported until the 1980s, and the limited chemistry reported to date couples with their stoichiometry and spectral properties to encourage presentation of their chemistry in introductory sections. Accordingly this article attempts to build from $M(CO)(RC\equiv CR)L_2X_2$

to $Mo(CO)(RC\equiv CR)(S_2CNEt_2)_2$ before addressing $CpML(RC\equiv CR)X$ chemistry which provides the most extensive and diverse results in this area to date. Other important Group VI d^4 alkyne complexes, i.e., those which do not fit into one of the three major classes defined above, are treated independently.

In order to produce a tractable article, severe restrictions have been imposed on the subject matter included for review. A plethora of relevant Group VI d^4 transition metal alkyne monomers is extant for the four-electron donor topic. Important metal alkyne chemistry which does not fall under this small umbrella includes polymerization (11–13), metathesis (14), terminal alkyne to vinylidene rearrangements (15–18), metal-lotetrahedrane equilibria with monomeric metal carbynes (19,20), and applications in organic syntheses: e.g., 2 + 2 + 2 cycloadditions (21), quinone syntheses (22), and complex condensations with carbenes (23,24). Stone and group have capitalized on the isolobal analogy between the metal–carbyne triple bond in $Cp(CO)_2M\equiv CR$ (M = Mo, W) complexes and the carbon–carbon triple bond of alkynes to build an impressive array of metal clusters (25,26). Each of these major research areas is currently expanding our understanding and utilization of alkyne reagents.

B. *Early Work with M(CO)(RC≡CR)₃*

Important roles for both filled alkyne π orbitals in binding to a single Group VI metal center were recognized over 20 years ago. The synthesis [Eq. (1)] and spectral properties of $W(CO)(3\text{-hexyne})_3$ were communicated by Tate and Augl in 1963 (27). Spectroscopic methods and chemical intuition sufficed to produce an accurate description of this unusual new molecule. Although the chemical implications of the unique metal–alkyne bond in these complexes are still being uncovered, the essence of this chemistry was laid bare in the original communication. This first report (1) correctly assigned the geometry of these pseudotetrahedral tungsten derivatives (Fig. 2), (2) suggested that the alkyne was "doubly π-bonded" to the tungsten atom, and (3) invoked the effective atomic number for tungsten with two of the three alkyne ligands serving in "the unusual capacity of four-electron donors." Indeed the bulk of the work reviewed here has been spawned since 1975 by the unique ligating properties of alkynes.

$$W(CO)_3(CH_3CN)_3 + EtC\equiv CEt \xrightarrow{\Delta} W(CO)(EtC\equiv CEt)_3 \qquad (1)$$

A full report of $W(CO)(RC\equiv CR')_3$ (R = R′ = Et, Ph; R = Me, R′ = Ph) chemistry followed in 1964 (28). The strong π-acid character of the alkyne ligands was evident in the high monocarbonyl stretching fre-

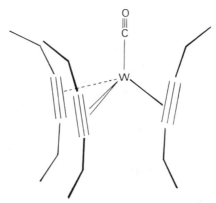

FIG. 2.   Pseudotetrahedral W(CO)(EtC≡CEt)₃ geometry.

quency (2050 ± 20 cm^{-1}) observed for these formal d^6 monomers. Note that metal configurations and oxidation states in this article were based on a neutral alkyne formalism. While the dianion formulation for alkyne ligands (RC≡CR^{2-}) can sometimes be useful for anticipating chemical reactivity patterns or molecular properties (*29*), the W(CO)(RC≡CR)₃ complexes illustrate an incongruity which discourages widespread use of this formalism. If metal electrons are distributed to make dianionic alkyne ligands, the result here is a four-coordinate d^0 W^{6+} ion (cf. WO$_4^{2-}$) with a carbonyl in the coordination sphere; surely d^0 is an unattractive metal configuration for such a demanding π-acid ligand as carbon monoxide.

Two deviations from the alkyne chemistry leading to W(CO)(RC≡CR)₃ were reported which previewed limitations in extrapolating internal alkyne results to olefins and terminal alkynes (*28*). First, the use of alkenes as reagents did not mimic results with alkynes. In the case at hand, stilbene reacted with W(CO)₅(CH₃CN) to form a six-coordinate d^6 adduct, W(CO)₅(η^2-PhCH=CHPh), rather than a tetrahedral product. Although this point concerning alkene and alkyne differences now seems self-evident, much of the early organometallic literature indiscriminately lumps olefins and alkynes together as "interchangeable" ligands. For the later transition metals it is true that there are numerous examples of analogous olefin and alkyne complexes. For early transition metals $\pi_\perp$ becomes important, usually in a constructive donor role, and olefin analogs are not accessible. Furthermore, for cases where two-electron donor alkynes would match known olefin chemistry, the filled alkyne $\pi_\perp$ orbital can cause a destructive four-electron–two-center interaction which makes alkyne complexes unstable relative to olefin analogs (*30*).

Second, reactions employing terminal alkynes as reagents did not yield tractable organometallic products (28). Rather, decomposition and polymerization resulted when HC_2H, HC_2Bu^n, or HC_2Ph were reacted with $W(CO)_n(CH_3CN)_{6-n}$ ($n = 5, 3$). These early results are explicable in light of a recent report of alkyne polymerization initiated by vinylidene complexes formed from terminal alkynes in octahedral d^6 tungsten complexes. Internal alkynes do not isomerize to vinylidenes, and polymerization does not take place (31).

A symmetry-based molecular orbital description of the unusual four-coordinate C_{3v} $W(RC\equiv CR)_3(CO)$ series of molecules was presented by King in 1968 (32). The $\pi_\perp$ orbitals of the three alkynes yield linear combinations of A_2 and E symmetry. Since there is no metal orbital of A_2 symmetry only the degenerate E combination of $\pi_\perp$ orbitals finds metal orbital mates for bonding and antibonding combinations. The three alkyne $\pi_\parallel$ orbitals serve as σ donors [$(A_1 + E)$ symmetry] as does the fourth ligand (A_1 symmetry). Thus the total metal electron count adheres to the effective atomic number rule [$W(0)(6) + 3\pi_\parallel(6) + 2\pi_\perp(4) + 1\sigma(2) = 18$ electrons].

The structure of $W(PhC\equiv CPh)_3(CO)$ reported by Bau and co-workers affirmed bonding roles for both filled alkyne π orbitals for each of the three alkyne ligands (33). The picture of three "$3\frac{1}{3}$-electron donor alkynes" provided ten bonding electrons ($A_1 + 2E$) from the three alkynes ($\pi_\parallel$ and $\pi_\perp$) to the metal center, while one alkyne $\pi_\perp$ combination (A_2) remained as a nonbonding ligand based molecular orbital.

The insight and accuracy of early work characterizing the anomalous $M(RC_2R)_3(CO)$ complexes set the stage for future developments with alkynes donating from both $\pi_\parallel$ and $\pi_\perp$ to a single metal center. The unusual stoichiometry and C_{3v} symmetry electron count rationale for these somewhat esoteric complexes inhibited extension of fundamental concepts concerning $\pi_\perp$ donation to other simple monomers, however. Although d^4 alkyne derivatives of Group V (7) of the type (π-C_5H_5)M(CO)$_2$(PhC$_2$Ph) [M = V(34), Nb (35,36)] were known prior to 1970, it remained for the 18-electron organometallic chemistry of seven-coordinate d^4 molybdenum-(II) and tungsten(II) to mature to provide a chemical backdrop which would place six-coordinate alkyne adducts in perspective. Once octahedral d^4 alkyne complexes surfaced and exhibited properties inconsistent with a two-electron donor role for the alkyne ligand, an expansive chemistry based on the simultaneous π-donor and π-acceptor properties of alkyne ligands followed.

The chemistry of Mo(II) and W(II) alkyne monomers constitutes the remainder of this article. Among the array of syntheses, structures, spectroscopic properties, and reactions, two sections deserve advertising.

Section IV describes reactions which produce four-electron donor alkyne ligands by coupling two single carbon based ligands (CR, CO, CNR). Section VIII,A describes syntheses of η^2-vinyl complexes (η^2-CR=CR$_2^-$) by nucleophilic addition to four-electron donor alkyne ligands. In the products of the coupling reactions and in the reagents of the η^2-vinyl syntheses, the role of alkyne $\pi_\perp$ donation is crucial.

In order to avoid the misleading implication that the chemistry of tetrahedral d^6 Group VI alkyne monomers has lain fallow since 1972 a few additional results are mentioned. Dithiahexyne has been incorporated into M(CO)(MeSC≡CSMe)$_3$ (M = Mo, W) products (37); substitution with dmpe yields W(dmpe)(MeSC≡CSMe)$_2$ (38). Monodentate phosphines have been substituted for carbon monoxide (39) and retained in reductive ligation reactions (40) which form W(PR$_3$)(RC≡CR)$_3$ products. The [MoCl(CF$_3$C≡CCF$_3$)$_3$]$^-$ anion exhibits an average Mo—C distance of 2.05 Å (41). Stone has combined W(CO)(RC≡CR)$_3$ with Cp(CO)$_2$-W≡CR reagents to form dimeric tungsten products (42), and Cooper has reduced W(PhC≡CPh)$_3$(CO) with loss of CO to form [W(PhC≡CPh)$_3$]$^{2-}$ (43).

II

SYNTHESES OF MONOALKYNE COMPLEXES

A. M(CO)(RC≡CR)L$_2$X$_2$ (M = Mo, W)

Syntheses of simple d^4 alkyne monomers of molybdenum and tungsten with *no* chelating or cyclopentadienyl ligands in the coordination sphere were first reported in 1982 (44). During the past several years a variety of M(CO)(RC≡CR)L$_2$X$_2$ complexes has been prepared, but reaction chemistry of the bound alkyne in these complexes has not yet been developed.

Molybdenum products with Mo(CO)(RC≡CR)L$_2$X$_2$ stoichiometries have been obtained from Mo(CO)$_3$L$_2$X$_2$ reagents (L = PEt$_3$, PPh$_3$, py) and free alkyne as in Eq. (2). Heating the reagents in 1,2-dichloroethane

$$Mo(CO)_3L_2X_2 + RC≡CR \xrightarrow{\Delta} Mo(CO)(RC≡CR)L_2X_2 + 2\ CO$$

$$L = PEt_3, PPh_3, py \quad (2)$$

for a few hours worked well for internal alkynes, while more sensitive terminal alkyne complexes were best prepared in hot methylene chloride over a period of days (45,46). Alternatively, the unsaturated Mo(CO)$_2$-(PEt$_3$)$_2$Br$_2$ reagent, formed by heating the tricarbonyl complex in solution, reacts readily with free alkyne as illustrated by the formation of $Mo(CO)(\eta^2\text{-cyclooctyne})(PEt_3)_2Br_2$ (47).

Related tungsten complexes have been prepared by diverse routes. Phosphine, phosphite, and isonitrile $W(CO)_2(RC\equiv CR)LI_2$ complexes form from $W(CO)_4LI_2$ and alkyne [Eq. (3)] (44). Parent acetylene (HC_2H) complexes were isolated as *cis*-dicarbonyl derivatives for L equal to PMe_3 or $AsMe_3$, and both *cis*- and *trans*-dicarbonyl derivatives were characterized with phenylacetylene in the coordination sphere.

$$W(CO)_4LI_2 + RC\equiv CR \rightarrow W(CO)_2(RC\equiv CR)LI_2 + 2\ CO \qquad (3)$$

Alkynes react with $[W(CO)_4Br_2]_2$ in hexane to form a dimeric product, $[W(CO)_2(RC\equiv CR)_2Br_2]_2$ (see Section III, A), which is an excellent precursor to monomeric d^4 tungsten alkyne complexes (48,49). Addition of excess nucleophile [PPh_3, $CNBu^t$, $P(OMe)_3$] cleaves the dimeric reagents and generates 2:1 adducts, $W(CO)(RC\equiv CR)L_2Br_2$, with concomitant loss of alkyne [Eq. (4)]. Tungsten(II) iodide derivatives with phosphites in the coordination sphere have been prepared from $W(CO)(RC\equiv CR)_2$-$(CH_3CN)I_2$ monomers and free phosphite [Eq. (5)] (50). Addition of internal alkynes to $MI_2(CO)_3(NCMe)_2$ in CH_2Cl_2 gives dimeric iodide bridged products, $[M(\mu\text{-}I)I(CO)(NCMe)(PhC\equiv CR)]_2$, for M = Mo or W, R = Ph or Me [Eq. (6)]. In one case a monomeric intermediate, $Mo(CO)(PhC\equiv CPh)(NCMe)_2I_2$, was isolated at low temperature (51).

$$[W(CO)(RC\equiv CR)_2Br_2]_2 + 4\ L \rightarrow 2\ W(CO)(RC\equiv CR)L_2X_2 + 2\ RC\equiv CR$$
$$L = PPh_3, P(OMe)_3, CNBu^t \qquad (4)$$

$$W(CO)(RC\equiv CR)_2(CH_3CN)I_2 + 2\ P(OR)_3 \rightarrow$$
$$W(CO)(RC\equiv CR)[P(OR)_3]_2I_2 + CH_3CN + RC\equiv CR \qquad (5)$$

$$M(CO)_3(NCMe)_2I_2 + 2\ PhC\equiv CR\rightarrow$$
$$(NCMe)I(CO)(PhC\equiv CR)M(\mu\text{-}I)_2M(CO)(PhC\equiv CR)(NCMe)I \qquad (6)$$

B. $ML(RC\equiv CR)(B-B)_2$ (B—B = Monoanionic, Bidentate Ligand)

Numerous molybdenum alkyne adducts have been prepared from the electrophilic $Mo(CO)_2(S_2CNEt_2)_2$ reagent [Eq. (7), with $RC\equiv CR' = HC_2H$, PhC_2Ph, PhC_2H, MeC_2H, MeC_2Ph, or $MeO_2CC\equiv CCO_2Me$ (DMAC)](52). Isolation of complexes containing acetylene as a ligand was uncommon through the 1970s relative to isolation of substituted alkyne complexes, but addition of acetylene gas to the reactive 16-electron mo-

$$Mo(CO)_2(S_2CNEt_2)_2 + RC\equiv CR' \rightarrow Mo(CO)(RC\equiv CR')(S_2CNEt_2)_2 + CO(g) \qquad (7)$$

lybdenum reagent $Mo(CO)_2(S_2CNEt_2)_2$ cleanly forms $Mo(CO)(HC{\equiv}CH)$-$(S_2CNEt_2)_2$. Cyclooctyne has also been successfully used as a four-electron donor ligand in this system (*47*). Alternative synthetic routes to this class of compounds avoid isolation of the air-sensitive 16-electron dicarbonyl molybdenum monomer and employ $Mo(CO)_3(S_2CNEt_2)_2$ or $Mo(CO)$-$(HC{\equiv}CH)(S_2CNEt_2)_2$ as the metal reagent (*53*). Reaction (8)(L = CO, PPh$_3$) illustrates the replacement of two two-electron σ donor ligands by a single alkyne. The parent acetylene ligand in the reagent of Eq. (9) is readily displaced by substituted alkynes, and Eq. (9) is the preferred route to heteroatom-substituted alkyne derivatives ($EtOC{\equiv}CH$, $Et_2NC{\equiv}CMe$,

$$Mo(CO)_2L(S_2CNEt_2)_2 + RC{\equiv}CR' \rightarrow$$

$$Mo(CO)(RC{\equiv}CR')(S_2CNEt_2)_2 + CO + L \qquad (8)$$

$$M(CO)(HC{\equiv}CH)(S_2CNEt_2)_2 + RC{\equiv}CR' \rightarrow$$

$$M(CO)(RC{\equiv}CR')(S_2CNEt_2)_2 + HC{\equiv}CH \qquad (9)$$

DMAC) which are difficult to purify by chromatography (*53*). The dithiophosphinate complex $Mo(CO)(HC{\equiv}CH)(S_2PPr^i_2)_2$ was reported as a four-electron donor alkyne complex in 1975 (*54,55*). Although $Mo(CO)_2$-$(S_2PPh_2)_2$ has proved elusive, the saturated $Mo(CO)_3(S_2PPh_2)_2$ complex reacts readily with phenylacetylene to form $Mo(CO)(HC{\equiv}CPh)(S_2PPh_2)_2$ (*56*).

Analogous tungsten alkyne complexes have been prepared from W-$(CO)_2L(S_2CNEt_2)_2$ [L = PPh$_3$(*57*), CO (*58*)] and alkynes as in Eq. (8). Note again that the "four-electron" nomenclature is compatible with the displacement of two two-electron ligands from these seven-coordinate d^4 metal reagents. Isolation of $W(CO)_2(S_2CNEt_2)_2$ was reported in 1982 (*59*), but it has not yet been reacted with free alkyne to yield $W(CO)$-$(RC{\equiv}CR)(S_2CNEt_2)_2$ products; presumably it would.

Isolation of mixed alkyne–olefin d^4 complexes of Mo(II) and W(II) of the type $M(\eta^2\text{-olefin})(\eta^2\text{-alkyne})(S_2CNEt_2)_2$ has been accomplished with electron-poor olefin ligands [maleic anhydride (MA), tetracyanoethylene (TCNE), and *trans*-dicyanoethylene (DCNE)] (*60*). Substitution of the carbonyl ligand in the $M(CO)(RC{\equiv}CR)(S_2CNEt_2)_2$ reagents occurred within minutes in boiling benzene for M = Mo or in boiling toluene for M = W as in Eq. (10). The higher activation energy typical of tungsten carbonyl bond dissociation relative to molybdenum analogs pervades the synthetic chemistry of the dithiocarbamate carbonyl derivatives. It appears

$$M(CO)(RC{\equiv}CR)(S_2CNEt_2)_2 + Z_2C{=}CZ_2 \rightarrow$$

$$M(Z_2C{=}CZ_2)(RC{\equiv}CR)(S_2CNEt_2)_2 + CO \qquad (10)$$

that thermal loss of carbon monoxide from $M(CO)(RC\equiv CR)(S_2CNEt_2)_2$ generates an electrophilic five-coordinate intermediate which is indiscriminately trapped by relatively poor nucleophiles if they are present in excess (CO, olefins, alkynes). In the absence of available nucleophiles a dithiocarbamate ligand in $W(CO)(PhC\equiv CPh)(S_2CNEt_2)_2$ provides additional electrons to the metal following CO loss by undergoing oxidative C—S bond cleavage to form $W(S)(PhC\equiv CPh)(S_2CNEt_2)(SCNEt_2)$ (61). An alternate route to mixed olefin–alkyne derivatives is available via sulfur atom abstraction from $W(S)(PhC\equiv CPh)(S_2CNEt_2)_2$ with PEt_3 in the presence of excess olefin [Eq. (11)]. Although only the DCNE derivative was prepared by this route (61), the synthesis was carried out at −78°C and should offer advantages for syntheses of olefin adducts which are thermally unstable.

$$W(S)(PhC\equiv CPh)(S_2CNEt_2)_2 + PEt_3 + CH(CN)=CH(CN) \rightarrow$$

$$W(DCNE)(PhC\equiv CPh)(S_2CNEt_2)_2 + S=PEt_3 \tag{11}$$

C. $CpM(RC\equiv CR)LX$ (M = Mo, W)

Cyclopentadienyl alkyne derivatives of Mo(II) and W(II) are the most populous and most extensively studied class of d^4 alkyne complexes. Both monoalkyne and bisalkyne complexes are numerous. Synthetic routes to monoalkyne derivatives can be classified as those involving addition of free alkyne to a metal complex, those replacing one alkyne of a bisalkyne precursor, simple ancillary ligand substitution of a monoalkyne metal complex, and coupling of two carbon fragments to form an alkyne ligand. We begin with incorporation of a single alkyne into monomeric metal products.

Heating seven-coordinate molybdenum reagents with internal alkynes in hexane has yielded $CpMo(CO)(RC\equiv CR)X$ products in selected cases [Eq. (12)]. The $CpMo(CO)_3Cl$ reagent was more reactive than the bromo

$$CpMo(CO)_3X + RC\equiv CR \rightarrow CpMo(CO)(RC\equiv CR)X + 2\ CO$$

$$X = Cl, Br, I, SCF_3, SC_6F_5; R = Ph \quad (62)$$

$$X = SCF_3, SC_6F_5; R = CF_3, Me \quad (63) \tag{12}$$

analog as is typical for thermal CO loss as a function of ancillary halide ligands. Photolysis was necessary to promote CO loss from $CpW(CO)_3Cl$ to form a tungsten analog, $CpW(CO)(PhC\equiv CPh)Cl$, as well as for the $CpMo(CO)_3I$ reagent (62). Photolysis of $Cp(CO)_3W(\mu\text{-SMe})W(CO)_5$ and $CF_3C\equiv CCF_3$ produces $Cp(CO)(CF_3C\equiv CCF_3)W(\mu\text{-SMe})W(CO)_5$ in 30% yield (64).

Substitution of the carbonyl ligand in $CpM(CO)(CF_3C{\equiv}CCF_3)(SC_6F_5)$ by tertiary phosphines or phosphites at ambient temperature in Et_2O provides a range of $CpML(CF_3C{\equiv}CCF_3)(SC_6F_5)$ products [Eq. (13)] (65). The isolation of a kinetic isomer of the 1 : 1 PEt_3 adduct at low temperature is particularly important. Formation of an η^2-vinyl ligand, $CF_3C{=}CCF_3$-(PEt_3), resulting from PEt_3 attack on an alkyne carbon, precedes conversion to the thermodynamic product with PEt_3 bound to the metal (65). Nucleophilic reactions to form η^2-vinyls will be discussed more fully in Section VIII,A.

$$CpM(CO)(CF_3{\equiv}CF_3)(SC_6F_5) + L \rightarrow CpML(CF_3C{\equiv}CCF_3)(SC_6F_5)$$
$$L = PR_3, P(OR)_3 \quad (13)$$

Addition of 2-butyne to $[CpMo(dppe)_2][PF_6]$ displaces one dppe chelate to yield the cationic alkyne complex, $[CpMo(dppe)(MeC{\equiv}CMe)][PF_6]$-[Eq. (14)] (66). The replacement of a bidentate phosphorus ligand with a single 2-butyne donor is a comment on the propensity of Mo^{2+} to bind alkynes.

$$[CpMo(dppe)_2][PF_6] + MeC{\equiv}CMe \rightarrow$$
$$[CpMo(MeC{\equiv}CMe)(dppe)[PF_6] + dppe \quad (14)$$

Alt and co-workers have prepared numerous alkyl (67) and acyl (68) tungsten(II) alkyne complexes. A definitive paper detailing these results was published in 1985 (69). The reaction sequence which converts CpW-$(CO)_3R'$ and free alkyne to $CpW(CO)(RC{\equiv}CR)R'$ and free CO is not simple. Low temperature photolysis ($-30°C$) of the reagents in pentane first yields *acyl* alkyne products [Eq. (15)]. These products result from trapping of the initial alkyl alkyne derivative which rapidly reacts with CO to form the observed acyl product [Eq. (16), L = CO] (69).

$$CpW(CO)_3R' + RC{\equiv}CR \xrightarrow[-30°C]{h\nu} CpW(CO)(RC{\equiv}CR)[C(O)R'] + CO \quad (15)$$

$$CpW(CO)(RC{\equiv}CR)R' + L \xrightarrow[-100°C]{} CpWL(RC{\equiv}CR)[C(O)R']$$
$$L = PMe_3, P(OMe)_3, CO \quad (16)$$

Independent experiments show that the alkyl alkyne undergoes alkyl migration to carbon monoxide at $-100°C$ where the acyl can be trapped with phosphine, phosphite, or free carbon monoxide. Although the alkyl alkyne is converted to the acyl derivative by scavenging CO under the photolytic reaction conditions, these same alkyl derivatives can be isolated following thermal disproportionation of the acyl for a few minutes in refluxing toluene. Evidently thermal loss of CO from $CpW(CO)$-$(HC{\equiv}CH)[C(O)Me]$ promotes alkyl migration from the acyl back to the metal. The acyl reagent also scavenges free CO at this temperature to form a dicarbonyl derivative containing a metallocyclic alkenyl ketone resulting from alkyne insertion into the metal acyl bond (70). The net result is 50% conversion of the acyl derivative to the corresponding alkyl [Eq. (17)].

$$2 \text{ CpW(CO)(RC} \equiv \text{CR)[C(O)R']} \xrightarrow[\Delta]{\text{PhMe}}$$

$$\text{CpW(CO)(RC} \equiv \text{CR)R'} + \overline{\text{Cp(CO)}_2\text{W[RC} = \text{CRC(O)R']}} \qquad (17)$$

Addition of the electron-rich amino alkyne, $Et_2NC \equiv CMe$, to the η^2-ketenyl tungsten complex, $(\eta^2\text{-RC} = C = O)CpW(PMe_3)(CO)$, results in trimethylphosphine substitution and opening of the η^2-ketenyl ligand to an η^1-ketenyl [Eq. (18)] (71).

$$\text{CpW(PMe}_3)(\text{CO})(\eta^2\text{-RCCO}) + Et_2NC \equiv CMe \rightarrow$$

$$\text{CpW(CO)(Et}_2NC \equiv CMe)(\eta^1\text{-RCCO}) \qquad (18)$$

In contrast to the inertness of bisalkynebisdithiocarbamate complexes, alkyne displacement from bisalkyne cyclopentadienyl derivatives is common. An extensive series of cationic $[CpMo(CO)L(RC \equiv CR)][BF_4]$ complexes has been prepared from $[CpMo(CO)(RC \equiv CR)_2][BF_4]$ reagents by substitution of one of the coordinated alkynes [Eq. (19)] (72). Reaction of the carbonyl reagent with phosphines occurs smoothly at room temperature in methylene chloride to form monoalkyne products in high yields

$$[CpMo(CO)(RC \equiv CR)_2][BF_4] + L \rightarrow$$

$$[CpMo(CO)(RC \equiv CR)L][BF_4] + RC \equiv CR$$

$$L = PEt_3, PPh_3, P(C_6H_{11})_3 \qquad (19)$$

$$[CpMo(CO)(RC \equiv CR)_2][BF_4] + 2 P(OMe)_3 \rightarrow$$

$$\{CpMo(RC \equiv CR)[P(OMe)_3]_2\}[BF_4] + CO + RC \equiv CR \qquad (20)$$

(72). Addition of trimethylphosphite under similar conditions causes displacement of both an alkyne and the carbonyl ligand [Eq. (20)] (73). The resultant $\{CpMo(RC \equiv CR)[P(OMe)_3]_2\}^+$ cation has been a vehicle for important conversions of C_2 ligands as reviewed by Green (10). Clearly, the π acidity of $P(OMe)_3$ and the net positive charge are important in promoting loss of the lone carbonyl ligand to yield the bisphosphite cation. Analogous bisphosphine cations have not been prepared directly from $[CpMo(CO)(RC \equiv CR)L]^+$ reagents, but the availability of $[CpMo-(CH_3CN)(RC \equiv CR)_2]^+$ provided a route to $[CpMo(RC \equiv CR)L_2]^+$ for $L = PEt_3$, PMe_3, $PMePh_2$, and chelating phosphines [dppe $(Ph_2PCH_2CH_2-PPh_2)$, dmpe $(Me_2PCH_2CH_2PMe_2)$, and $Ph_2PCH = CHPPh_2$] [Eq. (21)] (72).

$$[CpMo(CH_3CN)(RC{\equiv}CR)_2][BF_4] + 2\,PR_3 \rightarrow$$

$$[CpMo(RC{\equiv}CR)(PR_3)_2][BF_4] + RC{\equiv}CR + CH_3CN \qquad (21)$$

Displacement of trimethylphosphite from $\{CpMo(MeC{\equiv}CR)[P-(OMe)_3]_2\}^+$ with anionic sulfur nucleophiles produces neutral monoalkyne molybdenum derivatives [Eq. (22)] (74). Methoxide has also been incorporated into monoalkyne cyclopentadienyl derivatives [Eq. (23)] (75).

$$\{CpMo(MeC{\equiv}CMe)[P(OMe)_3]_2\}^+ + SR^- \rightarrow$$

$$CpMo(MeC{\equiv}CMe)[P(OMe)_3]\,(SR) + P(OMe)_3 \qquad (22)$$

$$[CpM(CO)(MeC{\equiv}CMe)_2]^+ + OMe^- \rightarrow$$

$$CpM(CO)(MeC{\equiv}CMe)(OMe) + MeC{\equiv}CMe \qquad M = Mo,\,W \qquad (23)$$

Cationic terminal and internal monoalkyne complexes have also been prepared by low temperature alkyne addition to Beck's electrophilic $[CpMo(CO)_2L][BF_4]$ reagents where BF_4^- is weakly bound to the 16-electron metal center [Eq. (24)] (76).

$$[CpMo(CO)_2L][BF_4] + RC{\equiv}CR \rightarrow [CpMo(CO)(RC{\equiv}CR)L][BF_4] + CO \quad (24)$$

A cationic acetylene carbene complex has been prepared by alkylation of an acyl oxygen (77). The alkyne ligand remains intact and retains the spectral properties of a four-electron donor on conversion of the neutral acyl to the cationic heteroatom carbene [Eq. (25)]. Mayr has entered

$$CpW(CO)(HC{\equiv}CH)[C(O)R] + [Et_3O][BF_4] \rightarrow$$

$$\{CpW(CO)(HC{\equiv}CH)[C(OEt)R]\}[BF_4] \qquad (25)$$

carbene alkyne chemistry by combining $(W{=}CHPh)Cl_2(CO)(PMe_3)_2$ with base and diphenylacetylene to form $(W{=}CHPh)Cl_2(PhC{\equiv}CPh)(PMe_3)_2$ (78). Geoffroy and co-workers have also prepared carbene alkyne Group VI complexes. The unstable d^6 W(0) systems they developed, including $W(CO)_4(PhC{\equiv}CPh)[C(OMe)Ph]$, have no vacant metal $d\pi$ orbital to allow constructive donation from $\pi_\perp$ (79).

D. Mo(II) and W(II) Alkyne Complexes with Other Stoichiometries

Five-coordinate d^4 alkyne derivatives are rare. Addition of acetylene gas to $Mo(CNBu^t)_4(SBu^t)_2$ in toluene at 30°C forms $Mo(HC{\equiv}CH)(CNBu^t)_2$-$(SBu^t)_2$ as a single acetylene displaces two bulky isonitrile ligands. Slightly higher temperatures (50°C) are required to promote formation of

$$Mo(CNBu^t)_4(SBu^t)_2 + RC\equiv CR \rightarrow Mo(RC\equiv CR)(CNBu^t)_2(SBu^t)_2 \quad (26)$$

$PhC\equiv CH$ and $PhC\equiv CPh$ analogs [Eq. (26)] (80). Five coordination is also found in the porphyrin alkyne complex, $Mo(TTP)(PhC\equiv CPh)$ (TTP= mesotetra-p-tolylporphyrin) (81). Reduction of a toluene solution of $Mo(TTP)Cl_2$ with lithium aluminum hydride in the presence of excess diphenylacetylene produced the violet $Mo(TPP)(PhC\equiv CPh)$ adduct [Eq. (27)].

$$Mo(TTP)Cl_2 + PhC\equiv CPh + LiAlH_4 \rightarrow Mo(TTP)(PhC\equiv CPh) \quad (27)$$

For comparative purposes the syntheses and properties of formally seven-coordinate d^4 biscyclopentadienyl alkyne complexes will be presented. The absence of a vacant $d\pi$ orbital places these d^4 complexes unambiguously in the two-electron alkyne category, i.e., $N = 2$ applies where N is defined as the formal electron donor number for the alkyne ligand. As in the porphyrin case a reductive ligation route provided access to these molecules. Sodium amalgam reduction of Cp_2MoCl_2 in the presence of alkyne yields $Cp_2Mo(RC\equiv CR)$ [Eq. (28)] (82). Note that, in

$$Cp_2MoCl_2 + RC\equiv CR + Na/Hg \rightarrow Cp_2Mo(RC\equiv CR) \quad (28)$$

accord with a two-electron alkyne donor role, ethylene can be utilized successfully in an analogous preparation to form $Cp_2Mo(H_2C\equiv CH_2)$. The molybdenum diphenylacetylene derivative was first prepared by thermolysis of the biscyclopentadienyldihydride complex Cp_2MoH_2 in toluene with alkyne present (83). Photolysis of $Cp_2Mo(CO)$ in the presence of alkynes is another viable route to these 18-electron compounds (84). Similar results are available with tungsten reagents although the yields are somewhat lower (85).

III

SYNTHESES OF BISALKYNE COMPLEXES

Bisalkyne derivatives have been crucial to the development of a comprehensive model for $\pi_\perp$ donation in d^4 monomers. The presence of two equivalent alkynes in the coordination sphere allows unambiguous interpretation of certain bonding properties, and in particular a formal donor number of three applies for each alkyne ($N = 3$). Bisalkyne complexes have been exploited to prepare monoalkyne monomers as well as for alkyne coupling reactions and ligand based transformations.

A. $M(RC{\equiv}CR)_2L_2X_2$ Complexes

Few monomeric d^4 bisalkyne complexes with only monodentate ligands in the coordination sphere have been reported. The only molybdenum(II) complex in this category is $Mo(PhC{\equiv}CR)_2(CO)(NCMe)I_2$ (R = Ph, Me), which was included in a reaction scheme illustrating products accessible from cleavage of the iodide bridges in dimeric $[Mo(PhC{\equiv}CR)(\mu\text{-I})(I)(CO)(CNMe)]_2$ reagents (51). Efforts to convert $Mo(RC{\equiv}CR)(CO)L_2X_2$ complexes to bisalkyne derivatives were not successful (46).

Hexafluorobutyne reacts with $[WBr_2(CO)_4]_2$ to form an unusual five-co-ordinate complex, $W(CF_3C{\equiv}CCF_3)_2(CO)Br_2$, with ν_{CO} at 2172 cm^{-1}, above the frequency of free carbon monoxide (2143 cm^{-1}) or that of H_3BCO (2165 cm^{-1}) (86). Further data characterizing this material will be informative; it is known to react with $P(OMe)_3$, but it does not form an η^2-vinyl product (see Section VIII,A).

Addition of internal alkynes to $[WBr_2(CO)_4]_2$ in hexane at room temper-ature produces a dimeric product with two alkynes bound to each tungsten [Eq. (29)] (48). The structure of the dimer is believed to be a conlateral bi-octahedron with two bridging bromide ligands which can be cleaved with added ligand to form monomeric alkyne derivatives. A 1:1 reagent ratio with added isonitrile produces a bisalkyne monomer [Eq. (30)] while 2 equiv of CNBu^t per tungsten result in substitution of one alkyne to form $W(MeC{\equiv}CMe)(CO)(CNBu^t)_2Br_2$ (48).

$$[WBr_2(CO)_4]_2 + 4\ RC{\equiv}CR \rightarrow$$

$$(RC{\equiv}CR)_2(CO)BrW(\mu\text{-Br})_2W(CO)Br(RC{\equiv}CR)_2 \qquad (29)$$

$$[W(MeC{\equiv}CMe)_2(CO)Br_2]_2 + CNBu^t \rightarrow W(MeC{\equiv}CMe)_2(CO)(CNBu^t)Br_2 \quad (30)$$

The reaction of $WI_2(CO)_3(NCMe)_2$ with PhC_2R (R = Ph or Me) yields dimeric $[WI_2(CO)(NCMe)(PhC{\equiv}CR)]_2$ complexes which are susceptible to further alkyne addition to form $M(PhC{\equiv}CR)_2(CO)(NCMe)I_2$ (51). Monoalkyne derivatives have been prepared from $W(RC{\equiv}CR)_2(CO)(NCMe)I_2$ (50).

B. $M(RC{\equiv}CR)_2(S_2CNEt_2)_2$ (M = Mo, W)

Molybdenum bisalkyne products form directly when $Mo(CO)_2(S_2CNEt_2)_2$ and excess alkyne are heated in methylene chloride for 18 hours (52), or, more quickly, in refluxing benzene or toluene [Eq. (31)] (87). The parent acetylene complex, $Mo(HC{\equiv}CH)_2(S_2CNMe_2)_2$, has

$$Mo(CO)_2(S_2CNEt_2)_2 + 2\ RC{\equiv}CR \xrightarrow{\Delta} Mo(RC{\equiv}CR)_2(S_2CNEt_2)_2 + 2\ CO \quad (31)$$

proved elusive. Small amounts of this bisalkyne have been prepared from $Mo(CO)(HC\equiv CH)(S_2CNMe_2)_2$ in autoclaves pressurized with $HC\equiv CH$ (52,87). Substitution of the carbonyl ligand in $Mo(CO)$-$(DMAC)(S_2CNEt_2)_2$ is facile relative to analogous alkyne complexes, and the preferred route to $Mo(DMAC)_2(S_2CNEt_2)_2$ is to allow DMAC and $Mo(CO)(HC\equiv CH)(S_2CNEt_2)_2$ to react in CH_2Cl_2 at room temperature for 2 hours.

Molybdenum bisalkyne complexes form more readily in the pyrrole-N-carbodithioate ligand system ("pyrroledithiocarbamate") than in the corresponding dialkyldithiocarbamate systems (88). The pyrrole nitrogen is reluctant to share electron density with the attached CS_2 moiety since the aromatic stabilization of the five-membered NC_4 ring is lost in resonance form **ii**. As a result of decreased electron donation from the

dithiocarbamate ligands the monocarbonyl CO stretching frequencies of $Mo(CO)(RC\equiv CR)(S_2CNC_4H_4)_2$ are typically 10–20 cm^{-1} higher in energy than in S_2CNMe_2 analogs, and the carbonyl is sufficiently labile that $Mo(RC\equiv CR)_2(S_2CNC_4H_4)_2$ complexes form spontaneously at room temperature (88). These bisalkyne products form cleanly from either Mo-$(CO)_3(S_2CNC_4H_4)_2$ or $[Et_4N]Mo(CO)_2I(S_2CNC_4H_4)_2]$ and excess alkyne; the anionic iodide complex is more easily prepared and less prone to decomposition than the carbonyl metal reagents $Mo(CO)_n(S_2CNC_4H_4)_2$ ($n = 2, 3$) (89). Even the parent acetylene complex, $Mo(HC\equiv CH)_2$-$(S_2CNC_4H_4)_2$, has been prepared in 10% yield with pyrroledithiocarbamate ligands in the coordination sphere.

Several mixed bisalkyne molybdenum complexes have been prepared [Eq. (32)] (87). Alkyne exchange is known to occur in the $Mo(CO)$-$(RC\equiv CR)(S_2CNMe_2)_2$ reagents (90), and this complicates isolation of pure mixed bisalkyne products.

$$Mo(CO)(PhC\equiv CH)(S_2CNMe_2)_2 + EtC\equiv CEt \xrightarrow[\Delta]{PhH}$$

$$Mo(PhC\equiv CH)(EtC\equiv CEt)(S_2CNMe_2)_2 \qquad (32)$$

The only report of tungsten bisalkyne bisdithiocarbamate complexes to date is a footnote which describes the formation of $W(PhC\equiv CPh)_2$-$(S_2CNEt_2)_2$ from $W(CO)(PhC\equiv CPh)(S_2CNEt_2)_2$ and $PhC\equiv CPh$ in boiling methanol (91). The abundance of bisalkyne molybdenum complexes in the literature relative to tungsten is another reflection of the comparative ease of thermal CO substitution for molybdenum, while harsher conditions are required to promote tungsten carbonyl substitution reactions.

C. CpM(RC≡CR)₂X Complexes

Bisalkyne cyclopentadienyl derivatives of molybdenum and tungsten have been prominent in the development of Group VI alkyne chemistry in the 1980s. Unlike the dithiocarbamate systems where addition of a second alkyne produces a substitutionally inert bisalkyne derivative, interconversion of monoalkyne and bisalkyne cyclopentadienyl complexes is facile and synthetically useful.

Six-coordinate bisalkyne products are available from $CpM(CO)_3Cl$ (M = Mo, W) for some internal alkynes [Eq. (33)] (*92*). The molybdenum reagent yields bisalkyne products for $CF_3C_2CF_3$, PhC_2Ph, PhC_2Me, and

$$CpM(CO)_3Cl + 2 RC≡CR \rightarrow CpM(RC≡CR)_2Cl + 3 CO(g) \qquad (33)$$

$MeC≡CMe$ in refluxing hexane (*93*). Removal of CO from the system is important in the case of hexafluorobutyne in order to avoid formation of a cyclopentadienone derivative. The bromide and iodide molybdenum derivatives with 2-butyne and $CF_3C_2CF_3$ ligands were also prepared although the iodide gave low yields (*62*). The reaction to form $CpMo(ClCH_2C≡CCH_2Cl)_2Cl$ proceeds nicely in hexane, while CH_2Cl_2 is a better solvent for obtaining bisalkyne products from $CpMo(CO)_3Cl$ and $HOCH_2C≡CCH_2OH$ (*94*). Tungsten carbonyl reagents require higher temperatures for reaction as usual, and no clean products were accessible with 2-butyne for tungsten (*93*).

Mixed bisalkyne complexes were prepared from monoalkyne precursors in several cases by room temperature reactions [Eq. (34)] (*93*). Higher temperatures produced bisalkynes via replacement of the original alkyne ligand.

$$CpM(CO)(PhC≡CPh)Cl + CF_3C≡CCF_3 \rightarrow$$
$$CpM(PhC_2Ph)(CF_3C_2CF_3)Cl + CO \qquad (34)$$

Other neutral cyclopentadienyl alkyne complexes have been prepared by chloride substitution reactions with sulfur nucleophiles; pentafluorophenylthiolate products have been reported [Eq. (35)] (*95*). These reactions are sensitive to the identity of the sulfur nucleophile (*96*), and attack at an alkyne carbon to yield η^2-vinyl ligands is common (*97*).

$$CpM(CF_3C≡CCF_3)_2Cl + Tl(SC_6F_5) \rightarrow CpM(CF_3C≡CCF_3)_2(SC_6F_5) + TlCl \qquad (35)$$

Another route to neutral bisalkyne complexes is from the trifluoro-methylacyl precursor which deinserts carbon monoxide to yield trifluoro-methyl molybdenum products. Photolysis of $CpMo(CO)_3[C(O)CF_3]$ in the presence of $CF_3C{\equiv}CCF_3$ forms $CpMo(CF_3C{\equiv}CCF_3)_2(CF_3)$; CpMo-$(DMAC)_2CF_3$ is formed without photolysis (98). Addition of hexafluoro-butyne to $CpMo(CO)(MeC{\equiv}CMe)(CF_3)$ forms the mixed bisalkyne via CO substitution.

Cationic bisalkynes are an important independent category of cyclo-pentadienyl derivatives. These complexes have been accessed by oxida-tive cleavage of the $[CpMo(CO)_3]_2$ dimer in the presence of free alkyne. For indenyl derivatives a stepwise procedure oxidizing $[(\eta^5\text{-}C_9H_7)Mo(CO)_3]_2$ with silver ion in acetonitrile allows isolation of $[(\eta^5\text{-}C_9H_7)Mo(CO)_2(MeCN)_2][BF_4]$ which reacts with alkynes to form $[(\eta^5\text{-}C_9H_7)Mo\text{-}(RC{\equiv}CR)_2(CO)][BF_4]$ products [Eq. (36)] (73). For $\pi\text{-}C_5H_5$ derivatives the acetonitrile complex is difficult to isolate, but $[CpMo(RC{\equiv}CR)_2\text{-}(CO)][BF_4]$ products form directly when $AgBF_4$ is added to a methylene chloride solution of $[CpMo(CO)_3]_2$ with alkyne present [Eq. (37)].

$$[(\eta^5\text{-}C_9H_7)Mo(CO)_2(MeCN)_2]^+ + 2\ RC{\equiv}CR\ \rightarrow$$

$$[(\eta^5\text{-}C_9H_7)Mo(RC{\equiv}CR)_2(CO)]^+ + 2\ MeCN + CO$$

$$RC{\equiv}CR = BuC_2H,\ MeC_2Me,\ PhC_2Ph,\ HC_2H,\ MeC_2H \qquad (36)$$

$$[CpMo(CO)_3]_2 + 2\ AgBF_4 + 4\ RC{\equiv}CR\ \rightarrow$$

$$2\ [CpM(RC{\equiv}CR)_2(CO)][BF_4] + 2\ Ag(0) + 4\ CO(g) \qquad (37)$$

Replacement of the lone carbonyl ligand in $[CpMo(RC{\equiv}CR)_2(CO)]\text{-}[BF_4]$ was difficult, and no product was isolated from photolysis or heat-ing of an acetonitrile solution of $[CpMo(MeC{\equiv}CMe)_2(CO)][BF_4]$. Detection of free 2-butyne suggested that alkyne dissociation was occu-ring, and, indeed, isolation of $[CpMo(MeC{\equiv}CMe)_2(MeCN)][BF_4]$ with a labile acetonitrile ligand was accomplished once free $MeC{\equiv}CMe$ was added to the MeCN reaction solution prior to thermolysis (72).

Oxidative cleavage of $[CpMo(CO)_3]_2$ is not the only route to cationic bisalkyne products. The electrophilic $[CpMo(CO)_3][BF_4]$ reagent prepared by hydride abstraction from $CpMo(CO)_3H$ with $[Ph_3C][BF_4]$ (99) readily adds alkyne to form monoalkyne and bisalkyne products depending on the reaction conditions and the alkyne substituents [Eq. (38)]. Bisalkyne products were best prepared from $\{CpMo(CO)_2[P(OPh)_3]\}[BF_4]$ re-agents (76).

$$[CpMo(CO)_3][BF_4] + PhC{\equiv}CPh\ \rightarrow\ [CpMo(PhC{\equiv}CPh)_2(CO)][BF_4] + 2\ CO \qquad (38)$$

Bergman and Watson have prepared $[CpM(RC\equiv CR)_2(CO)][PF_6]$ (M = Mo, W) by protonation of $CpM(CO)_3Me$ in the presence of 2-butyne in acetonitrile or more cleanly by CF_3CO_2H addition in benzene to form $CpM(CO)_3(O_2CCF_3)$ [Eq. (39)] which then reacts with alkyne in CH_3CN to form bisalkyne products [Eq. (40)]. Both 2-butyne and 2-heptyne derivatives have been prepared in this manner (75).

$$CpM(CO)_3Me + CF_3CO_2H \xrightarrow{\text{PhH}} CpM(CO)_3(O_2CCF_3) + MeH \quad (39)$$

$$CpM(CO)_3(O_2CCF_3) + RC_2R \xrightarrow[\text{2. NaPF}_6,\ \text{MeOH}]{\text{1. MeCN}} [CpM(RC\equiv CR)_2(CO)][PF_6] \quad (40)$$

IV

ALKYNE FORMATION FROM COUPLING C₁ LIGANDS

A. RC≡COR' Formation

In 1979 Fischer and Friedrich reported formation of a hydroxytolylacetylene ligand from photolysis of $Cl(OC)_4W\equiv CC_6H_4Me$ with acetylacetone (Hacac) [Eq. (41)] (100). The hydroxyalkyne ligand present in this six-coordinate d^4 tungsten product clearly resulted from coupling of the original carbyne ligand with a carbonyl, with proton addition required at some point to form the observed $HOC\equiv CAr$ unit.

$$Cl(CO)_4W\equiv CAr + Hacac \xrightarrow{h\nu} Cl(acac)W(CO)_2(HOC\equiv CAr) + CO \quad (41)$$

Another example of a coupling reaction to form an alkyne ligand was reported in 1982 by Schrock and co-workers (101). Addition of $AlCl_3$ to $W(\equiv CH)Cl(PMe_3)_4$ under a CO atmosphere in chlorobenzene generated $W(CO)(\eta^2\text{-}HC\equiv COAlCl_3)(PMe_3)_3Cl$ [Eq. (42)]. This carbon–carbon bond forming reaction couples the two fundamental C_1 units, CH and CO, and ultimately forms a six-coordinate d^4 alkyne derivative.

$$W(CH)Cl(PMe_3)_4 + 2\ CO + AlCl_3 \rightarrow$$

$$W(CO)(HC\equiv COAlCl_3)(PMe_3)_3Cl + PMe_3 \quad (42)$$

Stepwise formation of hydroxy and alkoxy alkynes from carbonyl–carbyne coupling reactions has been studied by Kreissl and co-workers over the past decade (102). These coupling reactions typically form ketenyl complexes prior to electrophilic addition at oxygen to yield alkyne products. Carbyne–carbonyl conversion to ketenyl complexes which undergo electrophilic addition at oxygen to form alkyne derivatives is now a well-established reaction pathway.

Addition of PMe_3 to the saturated carbyne reagent $Cp(CO)_2W\equiv CAr$ (Ar = p-C_6H_4Me) at $-40°C$ in Et_2O produces ketenyl products via CO—CR coupling (103). The product can be either an η^1- or an η^2-ketenyl complex depending on the position of equilibrium (43) as determined by the PMe_3 concentration (104). Addition of electrophiles to the η^2-ketenyl oxygen produces coordinated $ArC\equiv COR$ alkynes [Eq. (44)]. Kreissl and

$$CpW(PMe_3)_2(CO)(\eta^1\text{-}ArC\!\!=\!\!CO) \rightleftharpoons$$

$$CpW(PMe_3)(CO)(\eta^2\text{-}ArC\!\!=\!\!CO) + PMe_3 \qquad (43)$$

$$CpW(PMe_3)(CO)(\eta^2\text{-}ArCCO) + [Me_3O][BF_4] \rightarrow$$

$$[CpW(CO)(ArC\equiv COMe)(PMe_3)][BF_4] + Me_2O \qquad (44)$$

co-workers have synthesized both neutral (105) and cationic (106) η^2-alkyne cyclopentadienyl derivatives by Lewis acid addition to η^2-ketenyl precursors. The carbonyl ligand in the cationic alkyne complexes is susceptible to substitution by PMe_3 or CN^- (107). In 1983, Stone and co-workers also generated cationic alkyne complexes by addition of acid or alkylating agents to $CpW(CO)(PMe_3)(\eta^2\text{-}ArCCO)$ (108). They reported the structure of the hydroxyalkyne adduct as well as reaction products of $Cp(CO)_2W\equiv CAr$ plus acid to form carbene complexes.

Angelici and co-workers have recently reported that PEt_3 adds to $[HB(pz)_3](CO)_2W\equiv CSMe$ (pz = $C_3H_3N_2$) to form an η^2-ketenyl product which can be methylated with $MeOSO_2F$ at room temperature to yield a cationic complex with an unusual alkyne ligand, $MeOC\equiv CSMe$ [Eqs. (45) and (46)] (109).

$$[HB(pz)_3](CO)_2W\equiv CSMe + PEt_3 \rightarrow$$

$$[HB(pz)_3](CO)(PEt_3)W(\eta^2\text{-}MeSCCO) \qquad (45)$$

$$[HB(pz)_3](CO)(PEt_3)W(\eta^2\text{-}MeSCCO) + MeOSO_2F \rightarrow$$

$$\{[HB(pz)_3](CO)(PEt_3)W(\eta^2\text{-}MeSC\equiv COMe)\}[SO_3F] \qquad (46)$$

Carbonylation of a tungsten carbyne ligand has also been photochemically induced in the cyclopentadienyl system. Photolysis of $Cp(CO)_2$-$W\equiv CAr$ with PPh_3 in Et_2O produces the η^2-ketenyl product much more rapidly than thermal reaction (110). Coupling reactions of carbynes and carbonyls have also been promoted by addition of chelating ligands to tungsten carbyne reagents. Addition of NaS_2CNEt_2 to $(W\equiv CPh)Cl$-$(CO)_2(py)_2$ forms $[W(CO)(\eta^2\text{-}PhCCO)(S_2CNEt_2)_2]$ in 96% yield (111).

Anionic η^2-ketenyl complexes were first reported in 1983 from the reaction of (W≡CPh)(CO)$_2$Br(phen) with KCN to form [W(CO)(η^2-PhCCO)-(CN)$_2$(phen)]$^-$ (112).

Reaction of W(≡CR)(CO)$_2$(py)$_2$Cl with pyrrole-2-carboxaldmeth-ylimine (C$_6$H$_8$N$_2$) in the presence of base (KOH) yields an anionic tungsten ketenyl product, [W(CO)(η^2-RCCO)(C$_6$H$_7$N$_2$)$_2$]$^-$. This complex containing two anionic, bidentate Schiff base ligands can be methylated at the ketenyl oxygen to form a neutral d^4 alkyne complex, W(CO)-(RC≡COMe)(C$_6$H$_7$N$_2$)$_2$ (R = Me, Ph) (113). A similar pattern of reactivity has been observed with the unsaturated [(dppe)(CO)$_2$W≡CCH$_2$Ph]$^+$ carbyne reagent and [S$_2$CNEt$_2$]$^-$ (114). A neutral ketenyl product forms, (dppe)(S$_2$CNEt$_2$)W(CO)(OCCCH$_2$Ph), which can be protonated with HBF$_4$ or alkylated with [Me$_3$O][BF$_4$] to yield cationic alkyne products, [(dppe)(S$_2$CNEt$_2$)W(CO)(MeOC≡CCH$_2$Ph)][BF$_4$]. Reversibility of the carbyne–carbonyl coupling step has been demonstrated by rapid incorporation of ^{13}CO into both the terminal carbonyl and the ketenyl carbonyl position of W(CO)(η^2-OCCCH$_2$Ph)(S$_2$CNEt$_2$)(dppe) at room temperature under a ^{13}CO atmosphere (114).

Mayr has lucidly discussed the factors which promote coupling of carbon monoxide and carbyne ligands (113). He notes that a common feature of ketenyl precursors is a strong donor trans to the carbyne ligand. The well-known trans-halotetra(carbonyl)carbyne tungsten complexes resist thermal ketenyl formation in neat PMe$_3$ even though photochemical coupling has been achieved. In addition to cyclopentadienyl ligands trans to the carbyne other trans ligands which have produced coupled products include CN$^-$, dithiocarbamates, and Schiff bases. Mayr suggests that negative charge localization on the carbyne favors carbonylation of the carbyne. Geoffroy and Rheingold have proposed that the photoexcited state of Cp(CO)$_2$W(≡CAr) responsible for carbonylation is a metal-to-carbyne charge transfer state that promotes carbon monoxide migration to the electron-rich carbyne carbon (110).

It may be that all oxygenated alkyne syntheses to date occur by formation of a ketenyl intermediate prior to Lewis acid addition to give the observed alkyne products. For both the Fischer (100) and the Schrock (101) coupling reactions one can propose rapid stepwise ketenyl to alkyne conversions based on analogy with Kreissl's work (105). The coupling of carbyne and carbonyl ligands to form an η^2-ketenyl ligand decreases the metal electron count by two, hence the need for a trapping ligand. Addition of acac$^-$ or CO to the electron-deficient ketenyl intermediate could be followed by Lewis acid addition to the oxygen, either H$^+$ or AlCl$_3$ for the Fischer and Schrock examples, respectively, to give the observed alkyne products.

B. *RNHC≡CNHR Formation*

Reductive coupling of two isonitrile ligands to form RNHC≡CNHR bound to Mo(II) was communicated by Lippard and co-workers in 1977 (*115*). Details and extensions of these reductive coupling reactions which add two electrons and two protons, equivalent to net addition of H_2, to the $[Mo(CNR)_6X]^+$ reagents (M = Mo, W) have been published during the 1980s. Successful extrapolation from isonitrile coupling to the isoelectronic carbonyl coupling reaction was reported in 1986 in a related tantalum d^4 system as $(dmpe)_2ClTa(CO)_2$ was converted to $(dmpe)_2ClTa(Me_3$-$SiOC≡COSiMe_3)$ (*116*).

The use of zinc as a reducing agent with water as a proton source in THF solution produces yields of up to 90% of monomeric $[Mo(RNHC≡CNHR)(CNR)_4X]^+$ products [Eq. (47)], with R = Bu^t most thoroughly studied (*117*). One can image ligand based reduction to form $[RNC≡CNR]^{2-}$ followed by protonation at nitrogen to yield the observed

$$[Mo(CNR)_6X]^+ + Zn + H_2O \xrightarrow{\text{THF}} [Mo(RNHC≡CNHR)(CNR)_4X]^+ \quad (47)$$

diaminoalkyne ligand, but it is doubtful that the reaction mechanism bears any resemblance to this mnemonic scheme. Although other reductants can be used successfully, zinc is the reductant of choice. A proton donor is required, and reducing agents stronger than zinc generate H_2 rather than coupling isonitriles.

The isonitrile coupling reactions are mechanistically complex. Recent work with $[Mo(CNR)_4(bpy)Cl]^+$ illustrates this point (*118*). Treatment of the cationic bypyridine complex with Zn, H_2O, and $ZnCl_2$ in THF led to isolation of two complexes containing the coupled ligand RNHC≡CNHR, but neither corresponded to the simple product of direct coupling with retention of the other ligands. Rather, the five ancillary ligands in the two d^4 products were three isonitriles and one bpy (i.e., Cl$^-$ replacement by CNR to give a dication) and four isonitriles and a chloride (i.e., bpy replacement by two isonitriles). No reaction takes place unless both a reductant and a proton source are present. The coupled ligand has been oxidatively removed as the oxamide, RHNC(O)C(O)NHR, with retention of the carbon–carbon bond (*117*).

Lippard has noted that more electron-rich metal reagents, as measured by the ease of electrochemical oxidation, undergo reductive coupling more readily (*119*). Of course the coupling reaction is a ligand based reduction which formally adds one electron and one proton to each of two isonitrile ligands to form RNHC≡CNHR while maintaining the original metal oxidation state. Other operational factors which favor coupling have also been identified. Molecular orbital calculations indicate that an acute angle between the two ligands to be coupled is important (120); certainly the seven-coordinate $[M(CNR)_6X]^+$ reagents meet this criterion. In addition to an electron-rich metal center with a high coordination number, proper alignment of the two ligands is also necessary, and Lewis acid assistance may result from binding the heteroatom of the linear precursor ligand.

Addition of a Lewis acid to either CNR or CO prior to coupling is consistent with the pattern of carbyne–carbonyl couplings reported by Kreissl (105). Although available data does not directly test this mechanistic hypothesis, protonation of an isonitrile would generate an aminocarbyne ligand, a reaction known in other systems (121). This is an attractive first step in the isonitrile coupling reaction since it directly utilizes available metal electron density by tying up metal lone pairs in the newly formed

$$\ddot{M}\!\leftarrow\!C\!\equiv\!NR + H^+ \;\rightleftharpoons\; M\!\equiv\!C\!-\!\ddot{N}\!\!\begin{smallmatrix}\diagup H^+ \\ \diagdown R\end{smallmatrix}$$

metal carbyne π bonds. This would account for the importance of electron–rich metal centers in coupling reactions. Of course neither coupling nor reduction has taken place with the simple addition of a Lewis acid to the linear ligand β position, but now a carbyne and an isonitrile are present which could conceivably couple. Unlike carbonyl carbyne coupled products to date which have been trapped by added ligand, the electron-deficient coupled product here adds two electrons to regain saturation. The $(RNC)M\!\equiv\!CNRH$ fragment would seem to be an attractive starting point for coupling, reduction, and protonation.

Indeed, in 1988 Lippard and co-workers isolated metal carbynes from tantalum carbonyl and rhenium isocyanide reagents (121a). Addition of 1 equiv of Pr^i_3SiCl to $[(dmpe)_2(CO)_2Ta]^-$ allows the isolation of $[(dmpe)_2-(CO)Ta\!\equiv\!COSiPr^i_3]$, which has been structurally characterized. This complex contains a cis-carbyne carbonyl fragment and yields an alkyne ligand when treated with R_3SiX in accord with previous results (116). This chemistry merges nicely with the extensive studies of Group VI carbyne–carbonyl coupling reactions by Kreissl to suggest that a unifying mechanistic theme pervades these diverse coupling reactions.

C. PhC≡CMe Formation

Another class of coupling reactions consists of a single example recently reported by McDermott and Mayr (122). Conversion of W(≡CPh)-

$$
\begin{array}{c}
\text{Me} \diagdown \; {}^{O}\!\!\diagup \\
\text{C} \diagdown \; \text{CO} \\
\text{Br}-\text{W}\!\!\equiv\!\!\text{CPh} \\
\text{L} \diagup \!\! | \diagdown \!\! \text{CO}
\end{array}
\;\Bigg]^{-}
\quad
\xrightarrow[\;-78°C\;]{\overset{\text{O}\;\text{O}}{\overset{\|\;\|}{\text{BrC}-\text{CBr}}}}
\quad
\begin{array}{c}
\overset{\overset{\text{O}\,\text{O}}{\|\,\|}}{} \\
\text{Me} \diagdown \; \text{OCCBr} \\
\text{C} \diagdown \; \text{CO} \\
\text{Br}-\text{W}\!\!\equiv\!\!\text{CPh} \;+\; \text{Br}^{-} \\
\text{L} \diagup \!\! | \diagdown \!\! \text{CO}
\end{array}
\qquad (48)
$$

$CO)_3(PPh_3)Br$ to an anionic acyl complex can be realized with methyllithium addition, effectively initiating the classic Fischer carbene preparation. Addition of oxalyl bromide, still at $-78°C$, would be expected to generate a biscarbyne precursor complex as shown in Eq. (48). Addition of excess triphenylphosphine and warming to room temperature produce a deep blue solution which yields $W(CO)(PhC\equiv CMe)(PPh_3)_2Br_2$, previously characterized by Davidson and Vasapollo (48). Presumably the carbene carbyne intermediate loses bromooxalate to form an electron-deficient biscarbyne. Coupling of the carbyne ligands may assist in displacing $CO_2 + CO + Br^-$, or it may occur rapidly at the biscarbyne stage [Eq. (49)]. The formation of an alkyne from two carbyne fragments for a d^4 configuration is in agreement with theoretical predictions of Hoffman, Wilker, and Eisenstein (123,124). More recently, the inherent reactivity of monomeric biscarbyne derivatives has been discussed in qualitative terms based on extended Huckel calculations of model octahedral complexes (125).

$$
\begin{array}{c}
\overset{\overset{\text{O}\,\text{O}}{\|\,\|}}{} \\
\text{Me} \diagdown \; \text{OCCBr} \\
\text{C} \diagdown \; \text{CO} \\
\text{Br}-\text{W}\!\!\equiv\!\!\text{CPh} \;+\; \text{L} \\
\text{L} \diagup \!\! | \diagdown \!\! \text{CO}
\end{array}
\quad \longrightarrow \quad
\begin{array}{c}
\overset{\text{O}}{\overset{\|}{}} \\
\text{C} \;\; \text{L} \\
\text{Br}-\text{W}\!\leftarrow\!\! \overset{\text{CMe}}{\underset{\text{CPh}}{\|\|}} \;+2\text{CO} + \text{CO}_2 \\
\text{L} \diagup \!\! | \diagdown \!\! \text{Br}
\end{array}
\qquad (49)
$$

V

STRUCTURES OF Mo(II) AND W(II) ALKYNE COMPLEXES

Metal–ligand bond lengths, ligand orientations, intraligand bond lengths, and selected angles assist in revealing important metal–alkyne interactions. Metal–carbon bond lengths are particularly useful for assessing the importance of alkyne $\pi_\perp$ donation. A beautiful illustration of the relationship between metal–carbon distances for alkyne ligands and $\pi_\perp$ donation is provided by a pair of cobalt(I) complexes, $[Co(PhC\equiv CPh)(PMe_3)_3]^+$ (with $N = 4$) and $[Co(PhC\equiv CPh)(PMe_3)_3(CH_3CN)]^+$ (with

$N = 2$). The average Co—C(alkyne) distance is 0.13 Å longer for the five-coordinate complex with a formal two-electron donor diphenyl-acetylene (*126,127*). More than half of the four-electron donor alkyne ligands in Mo(II) and W(II) monomers exhibit metal–carbon bond lengths in the narrow range of 2.02 to 2.04 Å, and more than 90% are included in the 2.03 ± 0.03 Å range (Table I). Among the four-electron donor examples only CpMo(CO)(MeC≡CMe)(SC$_6$H$_4$SPh) has a metal–carbon bond exceeding 2.06 Å (2.03 and 2.09 Å for this complex) (*74*), and the role of thiolate π donation competing effectively with alkyne $\pi_\perp$ donation may account for the deviation. Competition for the vacant $d\pi$ orbital will decrease alkyne $\pi_\perp$ donation relative to cases with no other π-donor ligands in the coordination sphere. Given the diversity of alkynes and auxiliary ligands in Table I the consistency of four-electron donor M—C distances is noteworthy.

The structures of Cp$_2$Mo(PhC≡CPh) (*81*) and Cp$_2'$Mo(CF$_3$C≡CF$_3$) (*128*) have been included in Table I as standards for two-electron donor alkyne ligands in d^4 monomers. The observed metal–carbon distances of 2.14 (R = Ph) and 2.13 Å (R = CF$_3$) exceed the four-electron average by 0.10 Å, a substantial distance difference.

A range of carbon–carbon bond lengths from 1.27 to 1.33 Å encompasses all of the alkyne adducts except for structures reported by Lippard which contain diaminoalkyne ligands (*129*); extensive π delocalization in the R$_2$N—C≡C—NR$_2$ ligand system may account for these long C≡C distances of 1.35–1.40 Å. Both two-electron and four-electron donor alkynes exhibit bound C≡C distances at the short end of the range (≤1.30 Å); the inadequacy of carbon–carbon bond distances as a probe of metal–alkyne bonding was recognized by McDonald and co-workers prior to 1980 (*57*).

Deviation of the bound alkynes from linearity is another germane structural parameter. The C≡C—R angles in Table I range from 130 to 150° with 140 ± 5° encompassing the majority of examples. Again, the diaminoalkyne ligands (*129*) are geometrically unique as they exhibit angles of 128–130°, i.e., they deviate roughly 50° from linearity compared to the more common 40° distortion characterizing other bound alkynes.

A. *M(CO)(RC≡CR)L$_2$X$_2$*

Both Mo(CO)(PhC≡CH)(PEt$_3$)$_2$Br$_2$ (*46*) and W(CO)(HC≡COAlCl$_3$)-(PMe$_3$)$_3$Cl (*130*) contain the *cis*-M(CO)(RC≡CR) fragment common to all octahedral d^4 L$_4$M(CO)(RC≡CR) complexes characterized to date. The Mo—C(alkyne) distances of 1.98 and 1.99 Å and the W—C-(alkyne) distances of 2.03 and 2.01 Å in these complexes are typical of

TABLE I

STRUCTURAL DATA FOR MO(II) AND W(II) ALKYNE COMPLEXES

Complex	M—C(alkyne), Å	C≡C, Å	<C≡C—R, °	Ref.
Monoalkyne complexes				
Mo(CO)(PhC≡CH)(PEt$_3$)$_2$Br$_2$	1.98, 1.99	1.27	134	46
W(CO)(HC≡COAlCl$_3$)(PMe$_3$)$_3$Cl	2.01, 2.03	1.32	137, 144	130
W(=CHPh)(PhC≡CPh)(PMe$_3$)$_2$Cl$_2$	2.04, 2.06	1.33	—	78
W(CO)(HC≡CH)(S$_2$CNEt$_2$)$_2$	2.02, 2.04	1.29	130, 136	57
W(CO)(PhC≡COMe)(C$_6$H$_7$N$_2$)$_2$	2.02, 2.03	1.31	143, 144	113
W(MA)(PhC≡CH)(S$_2$CNEt$_2$)$_2$	2.01, 2.01	1.32	137	60
W(CO)$_2$(HOC≡CC$_6$H$_4$Me)(acac)Cl	2.04, 2.04	1.30	138, 141	100
{W(CO)(CF$_3$C≡CCF$_3$) [C$_2$(CF$_3$)$_2$P(OEt)$_2$] (μ-Br)}$_2$	2.00, 2.01	1.34	136, 138	212
CpMo(CO)(CF$_3$C≡CCF$_3$)(SC$_6$F$_5$)	2.02, 2.05	1.30	137, 138	135
CpMo(CO)(MeC≡CMe)(SC$_6$H$_4$SPh)	2.03, 2.09	1.27	142, 145	74
[CpMo(CO)(PhC≡CPh)(PPh$_3$)] [BF$_4$]	2.03, 2.04	1.29	135, 141	76
[CpW(CO)(HOC≡CC$_6$H$_4$Me) (PMe$_3$)] [BF$_4$]	2.00, 2.06	1.37	133, 135	108
Cp(CO)(CF$_3$C≡CCF$_3$)W(μ-SMe) W(CO)$_5$	2.03, 2.06	1.30	137, 140	64
[(η^5-C$_9$H$_7$)Mo(CO) (MeC≡CMe)(PEt$_3$)] [BF$_4$]	2.03, 2.06	1.29	140, 142	72
CpW(CO) (Et$_2$NC≡CMe)[η^1-C(CO)C$_6$H$_4$Me]	2.03, 2.04	1.34	137, 141	71
CpMo(MeC≡CMe) [P(OMe)$_3$]SC$_6$H$_4$NO$_2$)	2.02, 2.06	1.31	137, 140	74
[CpW(MeOC≡CC$_6$H$_4$Me)(PMe$_3$)$_2$] [BF$_4$]	1.99, 2.01	1.34	139, 143	107
[η^5-C$_9$H$_7$)Mo(MeC≡CMe) (PMe$_3$)$_2$][BF$_4$]	2.00, 2.03	1.31	134, 139	72
[CpMo(PPh$_2$C$_6$H$_4$CH=CH$_2$) (MeC≡CMe)] [BF$_4$]	2.00, 2.04	1.29	137, 142	136
[Mo(Bu^{t}HNC≡CNHBu^t) (CNBu^t)$_4$I] [PF$_6$]	2.03	1.36	130	129
[Mo(Bu^{t}HNC≡CNHBu^t) (CNBu^t)$_4$Br]·$\frac{1}{2}$ZnBr$_4$	2.03	1.35	129	129
[Mo(Bu^{t}HNC≡CNHBu^t) (CNBu^t)$_4$(CN)][PF$_6$]	2.05	1.40	128	129
[Mo(Bu^{t}HNC≡CNHBu^t) (CNBu^t)$_3$(bpy)[PF$_6$]$_2$	2.04	1.37	129	118
Mo(PhC≡CPh)(TPP)	1.97, 1.98	1.32	136, 145	81
Mo(HC≡CH)CNBu^t)$_2$(SBu^t)$_2$	2.04, 2.05	1.28	148, 152	141
Mo(PhC≡CPh)(CNBu^t)$_2$(SBu^t)$_2$	2.05, 2.06	1.28	139, 140	141
Cp$_2$Mo(PhC≡CPh)	2.14, 2.14	1.27	—	81
(η^5-C$_5$H$_4$Me)$_2$Mo(CF$_3$C≡CCF$_3$)	2.13, 2.13	1.28	138, 139	128
Bisalkyne complexes				
Mo(MeC≡CMe)$_2$(S$_2$CNC$_4$H$_4$)$_2$	2.02, 2.06	1.24	146	88
	2.04, 2.09	1.25		
CpW(CF$_3$C≡CCF$_3$)$_2$Cl	2.05, 2.06	1.23	139	92
	2.07, 2.07	1.28		
[CpMo(CO)(MeC≡CMe)$_2$][BF$_4$]	2.06, 2.08	1.27	146	137
	2.12, 2.14	1.28		
[CpMo(CH$_3$CN)(MeC≡CMe)$_2$][BF$_4$]	2.06, 2.07	1.27	145	137
	2.06, 2.07	1.28		
[Cp(CO)$_3$W(μ-η^1,η^2-C≡CPh)W(CO) (HC≡CPh)Cp] [BF$_4$]	2.06, 2.07	1.30	138	138
	2.10, 2.21	1.30	147	

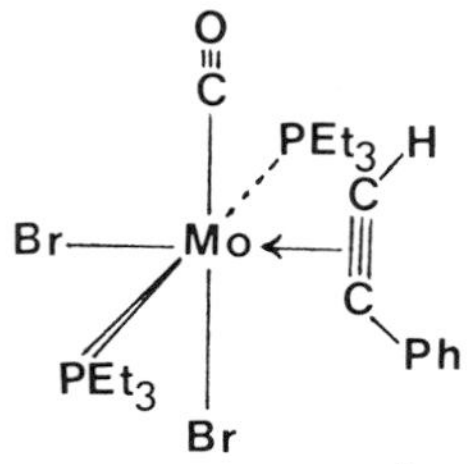

FIG. 3.   Octahedral Mo(CO)(PhC≡CH)(PEt₃)₂Br₂ geometry.

tightly bound four-electron donor alkynes. The ancillary ligands in Mo-(CO)(PhC≡CH)(PEt₃)₂Br₂ adopt a trans-phosphine–cis-bromide arrangement as shown in Fig. 3.

The Lewis acid–oxygen adduct which serves as a ligand in W(CO)-(HC≡COAlCl₃)(PMe₃)₃Cl can profitably be viewed as an alkyne bound to tungsten(II) (101). The C₂ unit, with a C—C bond length of 1.32 Å, is indeed cis to CO and lies parallel to the W—CO axis. The three PMe₃ ligands are meridionally disposed with the lone chloride trans to alkyne. Note that overall the complex is neutral with chloride and four-coordinate aluminum carrying formal negative charges to counterbalance the W²⁺ formal charge.

In addition to short metal–carbon bonds indicative of M—C multiple bond character, the strength of the metal–alkyne bond is also reflected in systematic distortions of the cis ligands away from the alkyne. This feature is most informative in L₄M(CO)(RC≡CR) complexes with no chelates to dictate metal–ligand angles. The fact that cis ligands in the metal–alkyne plane are more than 90° from the alkyne midpoint can be rationalized on steric grounds. Considering each alkyne carbon to be a single donor converts the geometrical description from octahedral to pentagonal bipyramidal with steric hindrance clearly greatest in the pentagonal belt of ligands including C₂. While electronic factors almost certainly impact on the L—M—L angles around this plane we defer this analysis and consider the two axial ligands of the pentagonal bipyramidal formulation. These ligands should not be crowded by the cis alkyne, which lies perpendicular to the L$_{ax}$—M—L$_{ax}$ axis, and in either an idealized octahedral or idealized seven-coordinate formalism these two ligands adopt an angle of 180° relative to one other.

In the same way that terminal oxo or nitrene units create a pyramidal distortion in M(O)L₅ (131) and M(NR)L₅ (132) complexes, alkynes also tend to pull the metal out of the plane of the four adjacent ligands. In W(CO)(HC≡COAlCl₃)L₃Cl the two PMe₃ ligands cis to the alkyne form an L—W—L angle of 161° (130), and in the molybdenum structure the

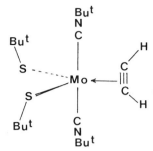

FIG. 4.   Trigonal bipyramidal Mo(HC≡CH)(CNBu^t)$_2$(SB^t)$_2$ geometry.

comparable L—Mo—L angle is 164° (for the trans-PEt$_3$ pair) (46). Again, these are orthogonal to the C_2 axis so steric factors are minimal. In a qualitative sense these angle distortions provide more metal orbitals for binding the alkyne ligand at the expense of the other ligands in an effort to more completely accommodate $\pi_\parallel$, $\pi_\parallel^*$ and $\pi_\perp$ alkyne–metal bonding. One can consider a square pyramidal molecule to represent the limit of this process. In this context we include here the Mo(II) porphyrin structure with the shortest Mo—C bond lengths yet reported for an alkyne adduct of Mo(II), 1.97 and 1.98 Å (81). This molecule counts only 16 electrons even when including four-electron donation from the diphenylacetylene ligand.

The only other examples of five-coordinate molybdenum(II) alkyne complexes are Mo(RC≡CR)(SBu^t)$_2$(CNBu^t)$_2$ (R = H, Ph) (133). These unusual molecules adopt a trigonal bipyramidal structure with the π-acid isonitriles in axial positions while the alkyne occupies an equatorial position and is aligned along the axial direction (Fig. 4). In other words the alkyne points along the π-acid axis as usual in d^4 monomers. The presence of two thiolates in the equatorial belt provides additional π-donor electrons for this unsaturated d^4 complex as described in the molecular orbital section which follows. The Mo—C bond lengths of 2.04–2.06 Å are on the high end of the four-electron donor range and probably reflect π-donor competition with thiolate lone pair electrons.

Another important six-coordinate alkyne derivative with monodentate ligands which has been structurally characterized is W(=CHPh)-(PhC≡CPh)(PMe$_3$)$_2$Cl$_2$ (78). Here a single-faced π-acid carbene ligand is cis to the alkyne rather than a linearly ligating carbonyl π-acid ligand. The plane of the metal benzylidene fragment is orthogonal to the trans-P(1)—W—P(2) axis as anticipated based on optimal $d\pi$ orbital utilization (Fig. 5). The structure is quite similar to that of W(CO)(HC≡COAlCl$_3$)(PMe$_3$)$_3$Cl, and again the two phosphines cis to the alkyne are swept back away from the alkyne [∠P(1)—W—P(2) = 157°].

Fig. 5.   W(=CHPh)(PhC≡CPh)(PMe₃)₂Cl₂ geometry.

Several structures of coupled isonitrile alkyne complexes are available. Variations among the cation geometries are small for [Mo-(Bu^{t}NHC≡CNHBu^t)(CNBu^t)₄X]$^+$ with X = Br, I, or CN (*129*). The cationic bipyridine complex also displays similar diaminoalkyne parameters (*118*). Alkyne Mo—C distances of 2.03–2.05 Å are found with bend back angles of 128–130°. The C≡C bond lengths of 1.35–1.40 Å are uniformly longer than those of other d^4 alkyne adducts, presumably reflecting the delocalized π system involving both planar nitrogens, the alkyne carbons, and the d^4 Mo^{2+} ion. Lippard has commented on the striking similarity of the [Mo(RHNC≡CNHR)(CNR)₄X]$^+$ product geometry to that of the [Mo(CNR)₆X]$^+$ reagent geometry (Fig. 6) (*115*).

B. M(L)(RC≡CR)(B—B)₂

The structure of W(CO)(HC≡CH)(S₂CNEt₂)₂ reported in 1978 solidified the utility of the four-electron alkyne donor concept (Fig. 7) (*57*).

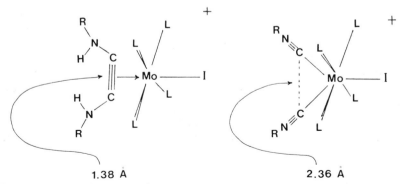

Fig. 6.   Graphic comparison of coupled RHNC≡CNHR alkyne product geometry with [Mo(CNR)₆I]$^+$ reagent geometry.

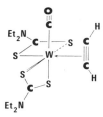

FIG. 7. Octahedral W(CO)(HC≡CH)(S₂CNEt₂)₂ geometry.

Two geometric features affirmed the hypothesis that both $\pi_\parallel$ and $\pi_\perp$ provide electron density for binding acetylene to tungsten. First, the short tungsten–carbon distances to acetylene (2.03 Å) indicate multiple bond character. Second, alignment of the alkyne parallel to the W—C≡O axis uniquely positions $\pi_\perp$ to overlap the lone vacant metal $d\pi$ orbital. Recognition that the C≡C bond length is *not* a sensitive indicator of alkyne $\pi_\perp$ donation was also an important general point noted by McDonald and co-workers (57). Secondary features such as W—S bond distances reflect the trans influence of the carbonyl and alkyne ligands as well as variable π bonding from the dithiocarbamates to the tungsten as a function of metal $d\pi$ orbital occupancies (134).

Comparison of the structures of W(CO)(HC≡CH)(S₂CNEt₂)₂ (57) and W(MA)(PhC≡CH)(S₂CNEt₂)₂ (60) indicates that major changes accompany replacement of carbon monoxide with a single-faced π-acid olefin ligand. The tungsten-to-alkyne carbon bond distances change only slightly (Table I), but the orientation of the alkyne in the mixed olefin alkyne complex is rotated 90° relative to the carbonyl case. The $d\pi$ orbital ordering is shuffled accordingly as described in Section VI. The two unsaturated C₂ ligands, maleic anhydride and phenylacetylene, are cis to one another and parallel (Fig. 8). The disparity between average metal–carbon bond distances to the alkyne (2.01 Å) and to the olefin (2.25 Å) underscores substantial bonding differences between these two unsaturated organic fragments serving as ligands to the d^4 tungsten center (60).

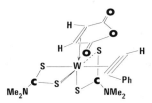

FIG. 8. Octahedral W(MA)(HC≡CPh)(S₂CNMe₂)₂ geometry.

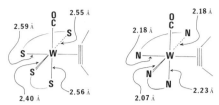

FIG. 9. W—L bond lengths in $W(CO)(HC{\equiv}CH)(S_2CNEt_2)_2$ and $W(CO)(MeOC{\equiv}CPh)(C_6H_7N_2)_2$.

Structures of other alkyne complexes containing chelating ligands have been reported for products derived from carbyne–carbonyl coupling reactions. The Schiff base pyrrole-2-carboxaldehydemethylimine has been used to generate anionic η^2-ketenyl products, and the methylated alkyne analog then belongs to the class of bisbidentate carbonyl alkyne d^4 monomers. The structures of $W(CO)(MeOC{\equiv}CPh)(C_6H_7N_2)_2$ (*113*) and $W(CO)(HC{\equiv}CH)(S_2CNEt_2)_2$ (*57*) are very similar. As in alkyne complexes with only monodentate ligands, the two donor atoms cis to the alkyne ligand are distorted away from the alkyne. The relevant N—W—N angle is 153° while the analogous S—W—S angle is 150°. Some of this distortion from linearity may be due to chelate bite angles, but clearly the influence of tungsten–alkyne multiple bonding also drives this distortion.

Differences in the W—N bond lengths follow the pattern established by the W—S distances in $W(CO)(HC{\equiv}CH)(S_2CNEt_2)_2$. The W—L bond trans to CO is long, while the W—L bonds in the N_3 or S_3 plane are intermediate for the chelate that lies in the plane and short for the one end of the chelate tied to the position trans to CO (Fig. 9).

The $W(CO)_2(HOC{\equiv}CC_6H_4Me)(acac)Cl$ product of carbyne–carbonyl coupling is one of few examples of a structurally characterized dicarbonyl alkyne d^4 monomer (*100*). Details of the geometry of the tungsten dicarbonyl alkyne fragment were not published in the original communication, and further work with two π acids adjacent to an alkyne ligand is required to assess general geometric properties of such moieties.

C. CpML(RC≡CR)X

The metal–alkyne bonding parameters of molecules of the type $CpML(RC{\equiv}CR)X$ are similar to those of other classes of four-electron donor alkynes in Table I. Both cationic and neutral derivatives with cyclopentadienyl ($Cp = \eta^5\text{-}C_5H_5$) or indenyl ($\eta^5\text{-}C_9H_7$) illustrate the structural regularities of these molecules. The preponderance of molybdenum derivatives relative to tungsten reflects more extensive and diverse synthetic chemistry with molybdenum to date.

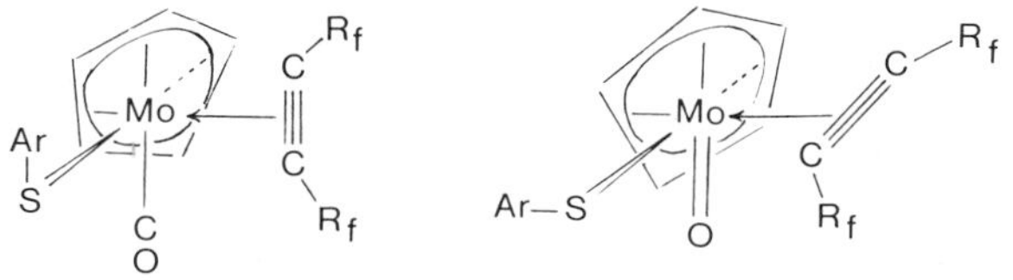

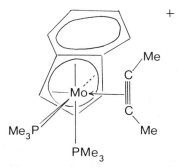

Fig. 10. Alkyne orientations in d^4 CpMo(CO)(CF$_3$C≡CCF$_3$)(SC$_6$F$_5$) and d^2 CpMo(O) (CF$_3$C≡CCF$_3$)(SC$_6$F$_5$).

Alkyne orientation in these complexes depends on the two cis ligands. An early comparative structural study of CpMo(CO)(CF$_3$C≡CCF$_3$)- (SC$_6$F$_5$) and CpMo(O)(CF$_3$C≡CCF$_3$)(SC$_6$F$_5$) which revealed alkyne orientations differing by 90° for the d^4 and d^2 configurations (Fig. 10) was important in elucidation of electronic factors dictating alkyne rotational preferences (135). All of the d^4 CpM(CO)(RC≡CR)L examples in Table I have the alkyne aligned parallel to the single M—CO axis.

The 2-butyne ligand in CpMo[P(OMe)$_3$](MeC≡CMe)(SC$_6$H$_4$NO$_2$) lies along the Mo—P axis (74), effectively preferring to be parallel to the M—L π-acid axis as in the monocarbonyl cases. This necessarily orients the alkyne perpendicular to the cis Mo—S axis. Given that the thiolate ligand has lone pair electrons available for π donation to the metal, the orthogonal relationship between C$_2$ and M—S is desirable so that both of these π-donor ligands confront the lone vacant $d\pi$ orbital.

The only complex in Table I with identical ligands in the two positions cis to the alkyne is the [(η^5-C$_9$H$_7$)MoL$_2$(MeC≡CMe]$^+$ cation with L = PMe$_3$ (72). Here the 2-butyne is parallel to one Mo—L vector and perpendicular to one Mo—L vector (Fig. 11). The π-ligand properties of the PMe$_3$ ligands are probably not sufficiently dominant to create a substantial alkyne preference between the $d\pi$ orbital combinations which are available.

Fig. 11. [(η^5-C$_9$H$_7$)Mo(MeC≡CMe)(PMe$_3$)$_2$]$^+$ geometry.

Kreissl has pointed out that the η^1-ketenyl ligand in $CpW(CO)(Et_2$-$NC{\equiv}CMe)[\eta^1\text{-}C(CO)C_6H_4Me]$ acts as a multiple electron donor with resonance form **ii** below an important contributor to the description of this ligand (71). The impact of this π-donor role for the η^1-ketenyl ligand on the cis-alkyne geometry is minimal as the aminoalkyne ligand structural parameters fall in the range typical of four-electron donor alkynes.

i   ii

The first olefin–alkyne Group VI d^4 complex to be structurally characterized contained a coordinated vinyl ligand which was part of a styrene attached to a metal-bound phosphine (136). This chelating Ph_2P-$(o\text{-}C_6H_4CH{=}CH_2)$ ligand displayed metal–carbon bond lengths of 2.27 Å from Mo(II) to the olefin, whereas the 2-butyne ligand in the coordination sphere displayed metal–carbon bond lengths (2.02 Å) 0.25 Å shorter than those to the alkene. This bond length differential for the unsaturated C_2 ligands in $\{CpMo[Ph_2P(o\text{-}C_6H_4CH{=}CH_2)](MeC{\equiv}CMe)\}^+$ (135), also evident in $W(\eta^2\text{-}MA)(PhC{\equiv}CPh)(S_2CNEt_2)_2$ (60), drives home the inadequacy of equating alkynes and olefins as ligands in d^4 monomers.

D. *Structures of Bisalkyne Complexes*

Although relatively few bisalkyne complexes have been structurally characterized, the pattern of cis alkynes lying parallel to each other is firmly established. Metal–carbon distances of 2.06 and 2.07 Å predominate in bisalkyne complexes (Table I). Shorter distances are rare, and two substantially longer lengths of 2.12 and 2.14 Å for $[CpMo(MeC{\equiv}CMe)_2$-$(CO)][BF_4]$ (137) are notable. Only a small difference separates typical M—C distances for $N = 4$ (2.03 ± 0.03 Å and $N = 3$ (2.07 ± 0.02 Å), and the two classes overlap in the long $N = 4$ and short $N = 3$ region. Carbon–carbon bond lengths of 1.27 ± 0.03 Å for $n = 3$ lie in the shorter half of the $N = 4$ values, but again these C—C distances are not reliable for guiding bonding considerations.

The structure of $CpW(CF_3C{\equiv}CCF_3)_2Cl$ reported in 1974 exhibited the key geometric features now assumed to characterize all d^4 bisalkyne complexes (92). The two cis alkynes lie approximately parallel to the adjacent M—L axis, here W—Cl (Fig. 12). Average values of W—C (2.06 Å), C≡C (1.27 Å), and R—C≡C angles (139°) (93) are representative of

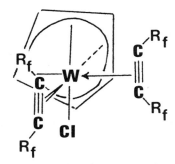

FIG. 12.   Octahedral CpW(CF$_3$C≡CCF$_3$)$_2$Cl geometry.

other monomeric bisalkyne structural parameters. More recently, the structures of two closely related cations have been reported (137). The pseudooctahedral description with parallel cis alkynes accurately describes [CpMoMeC≡CMe)$_2$L]$^+$ for both L = CO and L = MeCN. Although the structural similarities are many, a significant difference is found in the Mo—C(alkyne) distances adjacent to L which average 2.06 Å for L = MeCN and 2.13 Å for L = CO. The only other cyclopentadienyl derivative included in the bisalkyne structural data is an unusual tungsten acetylide which is π bound to a CpW(HC≡CPh)(CO) fragment (138). The net result can effectively be considered a mixed bisalkyne monomer.

The geometry of Mo(MeC≡CMe)$_2$(S$_2$CNC$_4$H$_4$)$_2$ is octahedral (88). The two alkynes are cis and parallel (Fig. 13), reminiscent of the olefin–alkyne structure found for W(MA)(PhC≡CH)(S$_2$CNEt$_2$)$_2$. The cis-bent C≡C—Me angle of 146° is typical of alkyne ligands. The L—M—L angle

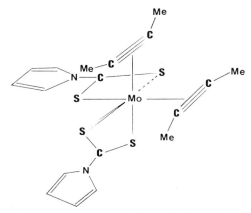

FIG. 13.   Octahedral Mo(MeC≡CMe)$_2$(S$_2$CNC$_4$H$_4$)$_2$ geometry.

between the two alkyne midpoints is 99°, larger than the idealized octahedral right angle. This is a common feature of cis π-donor ligands [c.f. $L_4MoO_2^{2+}$ complexes with O—Mo—O angles near 105° (*139*)]. Metal–carbon bond distances to the two alkynes average 2.05 Å for Mo-$(MeC\equiv CMe)_2(S_2CNC4H_4)_2$, intermediate between four-electron donor distances (approaching 2.0 Å) and two-electron donor cases (approaching 2.15 Å). Other examples of intermediate alkyne $\pi_\perp$ donation result from competition with cis π-donor ligands such as oxide (*140*) or sulfide (*134*) and exhibit metal–carbon bond lengths of 2.10–2.15 Å.

VI

MOLECULAR ORBITAL DESCRIPTION OF d^4 ALKYNE MONOMERS

The seminal role of analogous alkyne complexes differing only in oxo for carbonyl ligand exchange deserves emphasis. Simultaneous publication of the informative structures of $CpW(CO)(CF_3C\equiv CCF_3)(SC_6F_5)$ and $CpW(O)(CF_3C\equiv CCF_3)(SC_6F_5)$ provided a firm geometric basis for an accurate molecular orbital description of the alkyne ligand in these complexes (*135*). Qualitative molecular orbital presentations based on octahedral models have been reinforced by calculations on various d^4 monomers, but the essence of alkyne binding through $\pi_\parallel^*$ acceptance and $\pi_\perp$ donation as reflected in orientational preferences has remained intact.

A theoretical comparison of three diverse model compounds designed to mimic the bonding in structurally characterized diphenylacetylene adducts has probed $\pi_\perp$ donation (*141*). Identifying free $PhC\equiv CPh$ as **1** allows use of the mnemonic labels **2**, **3**, and **4** for the alkyne adducts Cp_2Mo-$(PhC\equiv CPh)$, $Mo(SR)_2(CNR)_2(PhC\equiv CPh)$, and $Mo(TTP)(PhC\equiv CPh)$, respectively. The $C\equiv C$ distances increase slightly from **2** to **4** (1.27, 1.28, and 1.32 Å), but more striking are the decreasing Mo—C bond lengths: 2.14, 2.05, and 1.97 Å. Bonding due to $\pi_\parallel$ donation to a vacant metal σ acceptor orbital and $\pi_\parallel^*$ acceptance from a filled metal $d\pi$ orbital, analogous to classic Dewar–Chatt–Duncanson metal–olefin bonding (*142–144*), is evident in the calculations. While these two interactions are important in setting the equilibrium geometry, the interaction of the filled alkyne $\pi_\perp$ orbital with a vacant metal $d\pi$ orbital tunes the bonding. The energy of the first available $d\pi$ orbital of appropriate symmetry to serve as an acceptor for $\pi_\perp$ decreases systematically and more nearly approaches the energy of $\pi_\perp$ as one proceeds from **2** to **3** to **4**. This accounts for the increased electron flow from $\pi_\perp$ to Mo(II) which is calculated to be 0.15, 0.22, and 0.25 e for **2**, **3**, and **4**, respectively. In simplistic effective atomic number terms a two-electron donor alkyne in **2** saturates the metal while even four-electron alkyne donation in **4** produces an electron count of only 16 for molybdenum. Complex **3** cannot be unambiguously classified due to simultaneous π donation from the two thiolates and the alkyne competing for two available vacant metal acceptor orbitals; $N = 3$ for the alkyne in **3** is a reasonable first approximation.

These calculations confirm the expectation that "four-electron donation" is in detail a misnomer, but of course "two-electron donation" is not a rigorous term for any ligand either. The calculations indicate that $\pi_\perp$ donation into a low-lying $d\pi$ orbital can indeed be important, but as the $d\pi$ acceptor orbital is driven to higher energy by other ligands or by increasing electron density at the metal this $\pi_\perp$-to-$d\pi$ donation is attenuated. For 16-electron fragments which form alkyne adducts a two-electron donor description with $N = 2$ is appropriate as $\pi_\parallel$ utilizes the lone vacant metal orbital for σ^* formation, and no low-lying metal orbital is available to accept $\pi_\perp$ electron donation. For a 14-electron fragment two vacant metal orbitals will exist, and these may be used to form M—L σ^* and π^* combinations with $\pi_\parallel$ and $\pi_\perp$, respectively. Of course a continuum of $d\pi$ orbital extensions and energies are conceivable, and the extent of constructive $\pi_\perp$ donation will vary accordingly.

Detailed extended Huckel calculations treating $CpM(CO)_2L$ complexes, including L = alkyne, were presented in 1979 by Schilling, Hoffmann, and Lichtenberger (*145*). This thorough treatment nicely rationalized orientational features of numerous ligands in these pseudooctahedral cyclopentadienyl derivatives. Substitution of one carbonyl ligand eliminates the symmetry plane and has a dramatic impact on $d\pi$ orbital combinations and orientational preferences as described in a companion paper by Schilling, Hoffmann, and Faller (*146*). In fact monocarbonyl alkyne derivatives have been more thoroughly investigated than any other class of d^4 alkyne complex, and we now turn to the alkyne alignment which is easily understood in the context of $d\pi$ orbital occupancies as dictated by a single cis carbonyl ligand.

The allocation of electrons in diamagnetic d^4 $L_4M(CO)(RC{\equiv}CR)$ complexes requires a two below one splitting of the octahedral t_{2g} metal $d\pi$ orbitals. Davidson, Sharp, and co-workers rationalized the restricted rotation and observed geometries of $CpMo(CO)(RC{\equiv}CR)(SR)$ in terms of alkyne–$d\pi$ interactions (*63*), and McDonald, Newton, and co-workers accurately described the $M(CO)(RC{\equiv}CR)(S_2CNEt_2)_2$ series by considering metal $d\pi$ interactions with both the carbonyl and alkyne ligands (*57*). These same orbital concepts also account for the diamagnetism and stability of $M(CO)(RC{\equiv}CR)L_2X_2$ derivatives (*46,48*). Extended Huckel molecular orbital (EHMO) calculations are in complete accord with these qualitative analyses and conclusions (*147*).

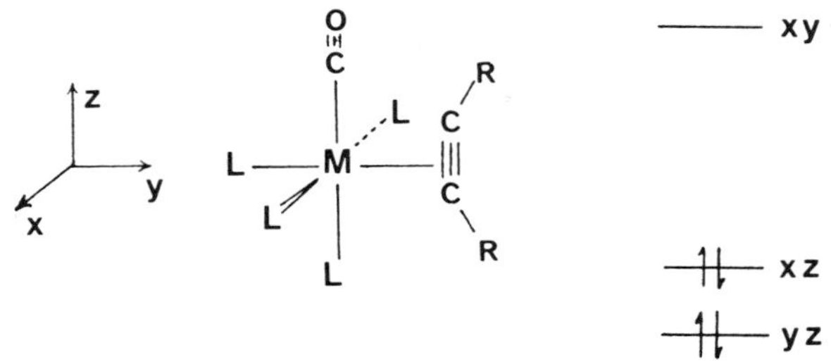

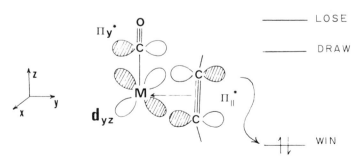

FIG. 14.   General $d\pi$ orbital splitting pattern for d^4 L$_4$M(CO)(RC≡CR) complexes.

An adequate model for *cis*-L$_4$M(CO)(RC≡CR) can be constructed by placing the four L ligands in adjacent octahedral sites and neglecting their π effects. Placing the π-acid CO along the z axis selectively stabilizes d_{xz} and d_{yz} among the original nest of three $d\pi$ metal orbitals (Fig. 14). This relegates the d^4 electrons to d_{xz} and d_{yz}, with d_{xy} remaining vacant. Placement of the alkyne along y completes the σ-bonding framework regardless of the alkyne orientation since donation from $\pi_\parallel$ to the sixth metal σ acceptor orbital is independent of rotation. Alkyne $\pi_\parallel^*$ and $\pi_\perp$ orbitals encounter two $d\pi$ orbitals directed along y, both linear combinations of d_{xy} and d_{yz}. In order to optimize *both* back donation into $\pi_\parallel^*$ *and* forward donation from $\pi_\perp$ the alkyne must be oriented parallel to the M—CO axis, as observed in L$_4$M(CO)(RC≡CR) complexes. The result is overlap of $\pi_\perp$ with the vacant d_{xy} orbital and overlap of $\pi_\parallel^*$ with the filled d_{yz} orbital, both constructive metal–alkyne bonding interactions.

A secondary feature of the *cis*-M(CO)(RC≡CR) molecular orbital description is the three-center–two-electron bond built from carbonyl π_y^*, alkyne $\pi_\parallel^*$, and metal d_{yz} (Fig. 15) (*45*). EHMO calculations locate the completely in-phase bonding combination, dominated by d_{yz}, approx-

FIG. 15.   Three-center–two-electron bond in L$_4$M(CO)(RC≡CR) complexes.

imately 0.4 eV below the HOMO which consists of d_{xz} stabilized by CO π_x^*. The three-center bond is redundant in terms of π-acid stabilization of d_{yz}, and indeed replacement of the carbonyl ligand with a single-faced π-acid olefin or carbene to create complexes with three simple two-center–two-electron $d\pi$–ligand π bonds has been achieved. No d_π orbital degeneracies remain in $M(CO)(RC\equiv CR)L_4$ as the $d_{xy}-\pi$ metal-alkyne $\pi_\perp$ antibonding combination constitutes the LUMO, 1.12 eV above the HOMO. Efficient utilization of each $d\pi$ orbital in a constructive metal–ligand π bond is evident here.

The potential impact of $\pi_\perp$ on monomeric metal alkyne chemistry is reflected in overlap calculations for $\pi_\parallel$, $\pi_\parallel^*$, $\pi_\perp$, and $\pi_\perp^*$ with metal orbitals in σ, π, π, and δ M—L bond formation (147). As expected, σ overlap (0.42) is the largest and δ overlap (0.04) is negligible, but the equality of $\pi_\parallel^*$ and $\pi_\perp$ overlap with metal $d\pi$ orbitals (0.25) suggests that both these interactions deserve attention in rationalizing metal alkyne chemistry. Since $\pi_\parallel^*$ is analogous to the olefin π^* of the Dewar–Chatt–Duncanson bonding model (142–144), $\pi_\perp$ donation may be equally important in describing metal alkyne bonding. The $\pi_\perp$ terminology should not obscure the atomic orbital origin of metal alkyne multiple bond character. Carbon $2p$–metal $d\pi$ overlap is commonly invoked for metal–ligand π bond formation in metal carbonyls, carbenes, and carbynes. The metal–carbon distances in many Group VI metal–alkyne complexes, near 2.0 Å, qualify for similar descriptions.

Having belabored the molecular orbital description of d^4 $L_4M(CO)$-$(RC\equiv CR)$ complexes, the advantages and drawbacks of the four-electron terminology are evident. Counting $\pi_\perp$ donation, recognizing that it is M—L π in character rather than M—L σ, leads to compliance with the 18-electron rule for these complexes. Given the M—L π^* nature of the LUMO here we can expect systematic deviations from spectral and chemical properties of classic 18-electron complexes where an M—L σ^* orbital constitutes the LUMO. Recalling general properties associated with alkenes compared to alkanes in organic chemistry is not completely amiss here in anticipating properties of "16-electron monomers stabilized by alkyne $\pi_\perp$ donation" relative to classic 18-electron monomers. Alternative descriptions utilizing bent bonds avoid the four-electron nomenclature completely and account for observed metal–alkyne geometries quite nicely (148).

Octahedral olefin–alkyne d^4 complexes are characterized by a one-to-one match of each of the three metal $d\pi$ orbitals with a ligand π function as mentioned above. Three constructive two-center–two-electron metal–ligand π bonds result. Extended Huckel calculations on $W(H_2C=CH_2)$-$(HC\equiv CH)(S_2CNH_2)_2$ produce the $d\pi$ level ordering shown in Fig. 16

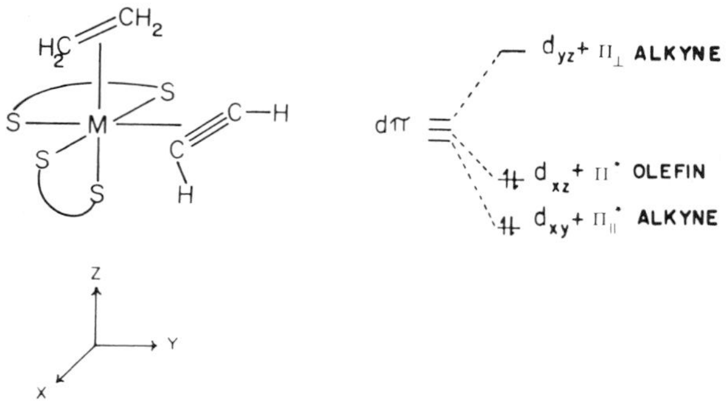

FIG. 16.   Octahedral d^4 cis-(η^2-olefin)(η^2-alkyne)ML$_4$ $d\pi$ orbital splitting pattern.

(60). The ground state geometry produces a LUMO consisting of d_{yz} destabilized by $\pi_\perp$ donation; the alkyne is a four-electron donor. Vacant π^* orbitals of the two single-faced π-acid C$_2$ ligands stabilize d_{xy} ($\pi_\parallel^*$) and d_{xz} (π^*) which house the four metal electrons.

In the idealized ethylene–acetylene model complex the HOMO1 is the olefin stabilized d_{xz} while the HOMO2 orbital, d_{xy}, reflects alkyne $\pi_\parallel^*$ overlap. The M—C alkyne distances employed in the calculation increase overlap responsible for the alkyne–metal π interactions relative to the olefin which is further from the metal and overlaps less (60). The $d\pi$ bonding contribution of the single-faced π-acid olefin is to stabilize the lone filled $d\pi$ orbital which is independent of the alkyne. This role is compatible with the successful incorporation of electron-poor olefins cis to the alkyne in these d^4 monomers. It may well be that the HOMO1 and HOMO2 orbitals in isolated complexes are reversed relative to the model complex as a result of electron-withdrawing substituents present on the olefins.

Steric factors probably prohibit simultaneous rotation of the olefin and alkyne C$_2$ units which would crowd all four metal-bound carbons into the same plane. Separate rotation of each unsaturated ligand was explored theoretically using the EHMO method. Rotation of the olefin destroys the one-to-one correspondence of metal–ligand π interactions. Overlap of the filled d_{xz} orbital with olefin π^* is turned off as the alkene rotates 90°, creating a large calculated barrier for olefin rotation (75 kcal/mol). Alkyne rotation quickly reveals an important point: the absence of three-center bonds involving $d\pi$ orbitals allows the alkyne to effectively *define* the linear combinations of d_{xy} and d_{yz} which serve as $d\pi$ donor and $d\pi$ acceptor orbitals for $\pi_\parallel^*$ and $\pi_\perp$, respectively. Thus there should be a small electronic barrier to alkyne rotation (the Huckel calculation with fixed metal ligand positions produced a barrier of 12.5 kcal/mol). The theoretical conclusion that these two adjacent unsaturated C_2 ligands will have entirely different rotational energy profiles is consistent with dynamic NMR results (see Section VII,A). The essence of the analysis is that an electronic barrier should exist for olefin rotation but not for alkyne rotation (60).

Mayr has described the geometry and bonding in a cis-carbene alkyne tungsten derivative, $W(CHPh)(RC{\equiv}CR)L_2X_2$, in terms of the single-faced π-acid benzylidene (:CHPh) imposing no electronic constraints on the orientation of the adjacent alkyne (78). The observed orientation of the benzylidene plane relative to the alkyne is orthogonal to that of the C_2 olefin unit in the maleic anhydride derivative, $W(MA)(RC{\equiv}CR)$-$(S_2CNEt_2)_2$. This orthogonality, a common feature of carbene/olefin analogs, is consistent with the single-faced π-acceptor properties of a neutral CR_2 fragment (vacant carbon $2p$ as a π-acceptor orbital) and a neutral $R_2C{=}CR_2$ ligand (vacant π^* as a π-acceptor orbital).

Alt has prepared a cationic cis-carbene alkyne derivative, $\{CpW(CO)$-$(HC{\equiv}CH)[C(OEt)Me]\}^+$ (77), but the presence of both an alkyne ligand and a carbonyl ligand in the d^4 coordination sphere as well as a heteroatom on the carbene complicates orientational predictions relative to $W(CHPh)$-$(RC{\equiv}CR)L_2X_2$. No structural data are yet available for this cationic carbene alkyne complex.

Dicarbonyl derivatives, $W(CO)_2(RC{\equiv}CR)LI_2$, depart from the simple molecular orbital analysis which so nicely rationalizes the omnipresent cis mono-carbonyl alkyne geometry. Although trans-dicarbonyl fragments are relatively rare, this reflects the dominance of d^6 octahedral complexes where π-acid ligands indeed attempt to avoid a trans disposition. Octahedral d^4 dicarbonyl complexes are in fact ideal for mutual trans carbonyl placement, much like d^2 MO_2 complexes are trans while the more common d^0 MO_2 fragments are cis (131). Trans CO ligands stabilize both d_{xz} and d_{yz} to create the two-below-one $d\pi$ orbital splitting which is required for diamagnetic d^4 octahedra. This leaves d_{xy} vacant, and alignment of the alkyne parallel to the OC—M—CO axis then finds the vacant d_{xy} for $\pi_\perp$ donation and the filled d_{yz} for $\pi_\parallel^*$ acceptance (Fig. 17). The net result is simply replacement of the ligand trans to CO in $L_4M(CO)(CR{\equiv}CR)$ with a second CO ligand which further stabilizes the two occupied $d\pi$ orbitals of the d^4 metal center.

For cis-carbonyl derivatives the alkyne orientation is not deduced so simply, and no structural data are yet available for the d^4 cis-$M(CO)_2$-$(RC{\equiv}CR)LX_2$ series. Nonetheless conclusions based on studies of other cis-dicarbonyl units in the coordination sphere with π-donor ligands provide some guidance. The first point to make is that the alkyne will prefer to be cis to both carbonyls in order to avoid an orbital conflict between $\pi_\perp$

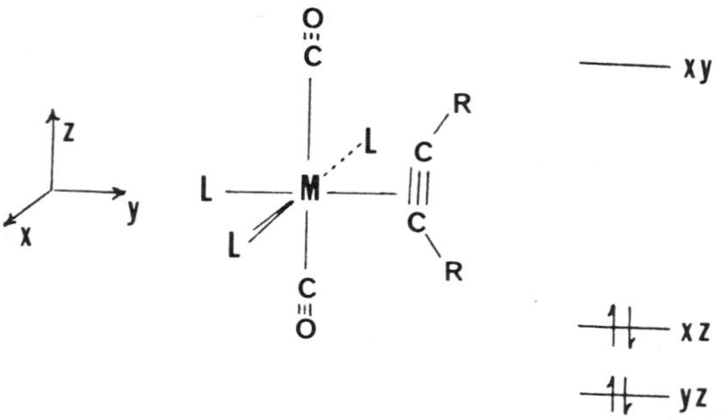

FIG. 17.   *trans*-$(OC)_2M(RC{\equiv}CR)L_3$ $d\pi$ orbital splitting pattern.

donation and CO π^* acceptance competing for a shared metal $d\pi$ orbital (*125*). Second, the alkyne orientation will depend on which two $d\pi$ orbitals are occupied, and this in turn hinges primarily on whether the OC—M—CO angle is obtuse or acute (*149,150*).

In contrast to *cis*-ML(CO) fragments where adjacent single-faced π ligands choose between lying parallel to or orthogonal to the M—CO axis, the choices for a single-faced π ligand adjacent to a *cis*-$M(CO)_2$ fragment are relative to the bisecting symmetry axis, not along one or the other M—CO axes. The $d\pi$ orbital in the $M(CO)_2$ plane is stabilized by both π-acid ligands, therefore low in energy, therefore filled. For a OC—M—CO angle of 90° the remaining two $d\pi$ orbitals, vertical relative to the M-$(CO)_2$ plane, are degenerate and above the filled $d_{x^2-y^2}$ orbital (Fig. 18). Few $M(CO)_2L_2X_2$ complexes have a triplet ground state [L = py is an interesting exception (151)], and the *cis*-$M(CO)_2$ angle dictates which vertical $d\pi$ orbital is preferentially stabilized and will therefore house the second metal based electron pair.

If the $M(CO)_2$ angle is acute, the interior vertical $d\pi$ orbital (d_{yz} in the coordinate system of Fig. 18) will be stabilized and occupied. A cis alkyne would then be in the plane bisecting the $M(CO)_2$ unit so that $\pi_\parallel^*$ would see the filled d_{yz} orbital while $\pi_\perp$ would encounter the vacant d_{xz} orbital. Alternatively, an obtuse $M(CO)_2$ angle would reverse the roles of d_{yz} and d_{xz} and lead to a predicted alkyne orientation orthogonal to the $M(CO)_2$ bisector plane (Fig. 18). Note that these arguments are valid only for d^4 monomers and completely irrelevant for d^6 dicarbonyl alkyne derivatives such as $W(CO)_2(ZC{\equiv}CZ)_2(dppe)$ (*30*).

The *cis*-$W(CO)_2(RC{\equiv}CR)LI_2$ monomers reported by Umland and Vahrenkamp (*44*) exhibit two carbonyl IR absorptions as anticipated for *cis*-$M(CO)_2$ moieties. The lower frequency CO absorption is more intense for L = PMe_3 or $AsMe_3$ with $PhC{\equiv}CH$ or $HC{\equiv}CH$ as the alkyne,

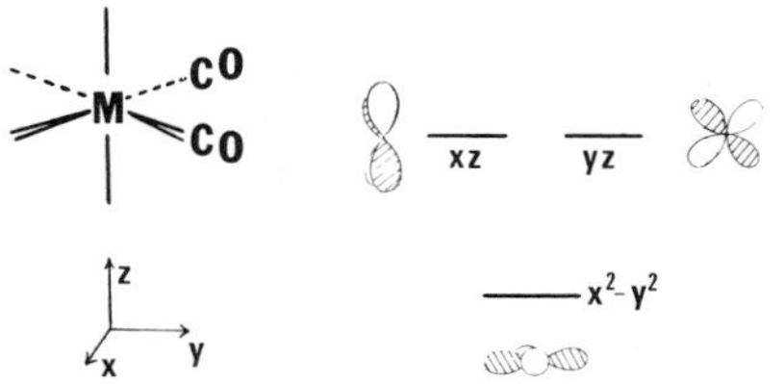

FIG. 18.   *cis*-$(OC)_2ML_4$ $d\pi$ orbital splitting pattern for a OC—M—CO angle of 90°.

suggesting that the *cis*-$M(CO)_2$ angle is obtuse (*152*). By inference alkyne orientation **ii** seems likely (Fig. 19). Since two discrete proton signals are observed for the HC≡CH ligand, isomer **II** is indicated rather than the more symmetric possibility isomer **I** (Fig. 19). If the ground state alkyne orientation is in fact opposite to that predicted here either arrangement of the ancillary LI_2 ligands creates distinct environments for the two ends of the alkyne ligand.

An exemplary molecular orbital description of $Mo(RC≡CR)(SR)_2$-$(CNR)_2$ (*133*) is relevant to the above discussion of trans-dicarbonyl geometries. Here the two isonitriles are linearly ligating π-acid ligands, albeit weaker than carbonyls. Although the coordination geometry is trigonal bipyramidal rather than octahedral, placing the isonitriles along the

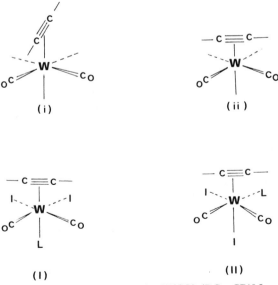

FIG. 19.   Possible isomers for $W(CO)_2(RC≡CR)LI_2$.

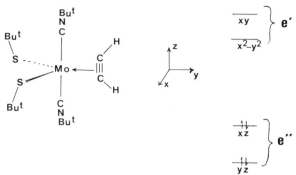

FIG. 20. Orbital splitting pattern for the four $d\pi$ orbitals of trigonal bipyramidal d^4 Mo(HC≡CH)(SR)$_2$(CNR)$_2$.

$\pm z$ axes selectively stabilizes d_{xz} and d_{yz} to house the four electrons of the Mo(II) center. The alkyne then lies parallel to this axis in order to enjoy backbonding into the $\pi_\parallel^*$ orbital while confronting a pair of vacant metal orbitals of idealized e' symmetry (largely d_{xy} and $d_{x^2-y^2}$) in the xy plane to serve as acceptors for $\pi_\perp$ donation (Fig. 20). The two thiolates are also equatorial ligands, and they have their R groups in anti-upright orientations so that both sulfur atoms can also donate electron density from filled lone pair p orbitals to the metal e' orbitals in the xy plane (133). This competition among the three equatorial π-donor ligands, two thiolates, and one alkyne for donation to two metal $d\pi$ orbitals complicates assigning an alkyne donor number N, but the ratio of three π donors for only two $d\pi$ orbitals is reminiscent of M(CO)(RC≡CR)$_3$ bonding where $N = 3\frac{1}{3}$ applies due to equivalence of the three alkyne ligands.

Octahedral d^4 cis-ML$_4$(RC≡CR)$_2$ complexes provide an unambiguous example of "three-electron" donor alkynes due to symmetry guidelines imposed by the C_{2v} point group (147). Of course no odd-electron chemistry is implied here, but rather it is donation of a total of six electrons from the two alkyne ligands which leads to conformity with the 18-electron rule. More insight is provided by recognition of the simple but pervasive three-center ligand $d\pi$ bonding scheme which has important implications for other less symmetrical complexes involving two π-donor ligands competing for a single metal $d\pi$ acceptor orbital. The effective donation of three electrons from each alkyne in Mo(RC≡CR)$_2$(S$_2$CNEt$_2$)$_2$ as presented by McDonald and co-workers in 1975 (52) has been subsequently reinforced by both experimental and theoretical work (137).

Although the two chelates in M(RC≡CR)$_2$(S$_2$CNEt$_2$)$_2$ complexes reduce the symmetry to C_2 from the C_{2v} properties of the isolated cis-M-

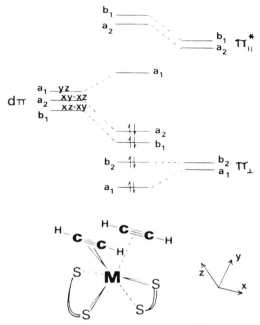

FIG. 21.   Metal $d\pi$–alkyne π interactions in octahedral d^4 M(HC≡CH)$_2$L$_4$ monomers.

(RC≡CR)$_2$ fragment, the salient metal-alkyne bonding scheme is unchanged: the out-of-phase $\pi_\perp$ ligand orbital combination has no metal orbital mate (Fig. 21). This M—L nonbonding π combination constitutes the central orbital of a standard three-center–four-electron scheme originating from the two filled alkyne $\pi_\perp$ orbitals and the metal d_{yz}. Strongly bonding and antibonding molecular orbitals result from plus and minus combinations of the d_{yz} orbital with the in-phase $\pi_\perp$ ligand orbital combination (147).

Both combinations of alkyne $\pi_\parallel^*$ orbitals find filled $d\pi$ orbital symmetry matches in these d^4 complexes. Extended Huckel calculations on Mo(HC≡CH)$_2$(S$_2$CNH$_2$)$_2$ indicate a large HOMO–LUMO gap of 1.62 eV. These octahedral complexes have proved to be quite robust and resist exchange and substitution reactions in accord with a substantial frontier orbital energy gap (153).

In summary, the molecular orbital description of Mo(RC≡CR)$_2$L$_4$ includes a nonbonding $\pi_\perp$ combination of b_2 symmetry, alkyne donation of two electrons each from $\pi_\parallel$ to metal σ orbitals, and one filled bonding orbital from overlap of both $\pi_\perp$ orbitals with the lone vacant $d\pi$ orbital.

The result is six alkyne electrons in two σ and one π metal–ligand bonding orbitals, i.e., two three-electron donor alkynes ($N = 3$) are present in the coordination sphere.

VII

PHYSICAL PROPERTIES OF d^4 ALKYNE COMPLEXES

A. NMR Spectra

Mo(CO) (HC≡CH) (S$_2$CNEt$_2$)$_2$ is characterized by a unique ^{1}H-NMR singlet at 12.3 ppm assigned to the metal bound HC$_2$H ligand (52). This low-field ^{1}H chemical shift and the chemical properties of this six-coordinate d^4 acetylene complex were considered a reflection of effective four-electron donation from the alkyne to the d^4 metal center. A similar proton chemical shift in Mo(CO)(HC$_2$H)(S$_2$PPr$_2^i$)$_2$, also prepared by McDonald and coworkers, had earlier been compared to ^{1}H shifts of the cyclopropenium ion ($\delta = 11.2$ ppm) where two electrons are delocalized over three carbon π orbitals (54). This analogy highlights delocalization of the alkyne $\pi_\perp$ electron pair to a vacant metal $d\pi$ orbital in the transition metal MC$_2$ triangle which resembles C$_3$H$_3^+$. The characteristic low-field ^{1}H chemical shift of the acetylenic protons in Mo(CO)(HC≡CH)-(S$_2$PPr$_2^i$)$_2$ serves as a definitive and convenient spectroscopic probe for four-electron donation from terminal alkynes in other monomers (154). Data for related terminal alkyne complexes are collected in Table II with most ^{1}H values near 13 ppm for W(RC≡CH) derivatives and near 12 ppm for Mo(RC≡CH) derivatives. Since ^{1}H resonances in this region are rare for organic molecules it is often possible to monitor the formation or disappearance of terminal four-electron donor complexes even when numerous ^{1}H signals complicate the high-field portion of the spectrum.

The global utility of this ^{1}H-NMR alkyne probe is decreased by the scarcity of terminal alkyne adducts relative to the abundance of internal alkyne adducts. Diphenylacetylene and dimethylacetylenedicarboxylate (DMAC) are two particularly popular alkyne ligands which have no acetylenic proton to monitor. An empirical correlation between alkyne $\pi_\perp$ donation and ^{13}C chemical shift for the bound alkyne carbons has been recognized (155) which partially fills this spectroscopic need. A plot of alkyne ^{13}C chemical shifts, which span over 100 ppm (Tables II and III), versus N, the number of electrons donated per alkyne to fulfill the effective atomic number guideline, reveals both the advantages and the limitations

TABLE II

NMR and IR Data for Mo(II) and W(II) Terminal Alkyne Complexes

Complex	^{1}H, ppm	^{13}C, ppm ($^1J_{CH}$, Hz)	$\nu(CO),^a$ cm^{-1}	Ref.
Mo(CO)(PhC≡CH)(PEt$_3$)$_2$Br$_2$	13.0	225, 225 (204)	1950	46
Mo(CO)(n-BuC≡CH)(PEt$_3$)$_2$Br$_2$	13.7		1944	46
Mo(CO)(n-BuC≡CH)(dppe)Cl$_2$			1981	46
Mo(CO)(n-BuC≡CH)(dppe)Br$_2$	10.8	218, 194 (215)	1985	46
	10.7 (isomer)	219, 199 (isomer)		
Mo(CO)(PhC≡CH)(dppe)Br$_2$	10.9	211, 201	1995	46
	10.9 (isomer)	212, 203 (isomer)		
cis-(OC)$_2$W(PhC≡CH)(PMe$_3$)I$_2$	12.5		1993, 2059	44
trans-(OC)$_2$W(PhC≡CH)(PMe$_3$)I$_2$	13.0		2011	44
cis-(OC)$_2$W(PhC≡CH)(AsMe$_3$)I$_2$	12.4		1995, 2054	44
trans-(OC)$_2$W(PhC≡CH)(AsMe$_3$)I$_2$	12.9		2005	44
cis-(OC)$_2$W(PhC≡CH)(CNBu^t)I$_2$	13.3		2013, 2068	44
trans-(OC)$_2$W(PhC≡CH)(CNBu^t)I$_2$	13.2		2015	44
cis-(OC)$_2$W(HC≡CH)(PMe$_3$)I$_2$	12.6, 12.8		2006, 2064	44
cis-(OC)$_2$W(HC≡CH)(AsMe$_3$)I$_2$	12.5, 12.7		2007, 2063	44
W(CO)(HC≡COAlMe$_3$)(PMe$_3$)$_3$Cl	12.8	226, 196 (200)	1960	101
W(CO)(HC≡COAlCl$_3$)(PMe$_3$)$_3$Cl		231, 194 (203)	1970	101
W(=CHPh)(PhC≡CH)(PMe$_3$)$_2$Cl$_2$	12.9	218, 211 (201)		78
Mo(CO)(HC≡CH)(S$_2$PPr^i)$_2$	12.3		1950	54
Mo(CO)(PhC≡CH)S$_2$PPr^i)$_2$	12.5	210 (218)	1931	56
Mo(CO)(HC≡CH)(S$_2$CNEt$_2$)$_2$	12.3		1960	52,57
W(CO)(HC≡CH)(S$_2$CNEt$_2$)$_2$	12.5, 13.5	206 (210), 207 (210)	1925	57,58
W(CO)(HC≡CH)(S$_2$CNMe$_2$)$_2$	12.6, 13.6		1925	58
Mo(CO)(PhC≡CH)(S$_2$CNEt$_2$)$_2$	12.6	209, 205 (215)	1934	52,58
W(CO)(PhC≡CH)(S$_2$CNEt$_2$)$_2$	13.3	206	1924	58
W(TCNE)(PhC≡CH)(S$_2$CNEt$_2$)$_2$	13.3			60
W(DCNE)(HC≡CH)(S$_2$CNEt$_2$)$_2$	11.6			60
W(DCNE)(PhC≡CH)(S$_2$CNEt$_2$)$_2$	13.4			60
W(MA)(PhC≡CH)(S$_2$CNEt$_2$)$_2$	13.3	223, 224		60
W(MA)(HC≡CH)(S$_2$CNEt$_2$)$_2$	13.2			60
W(MA)(MeC≡CH)(S$_2$CNEt$_2$)$_2$	13.2			60
[CpMo(CO)(Bu^{t}C≡CH) (PPh$_3$)] [BF$_4$]	12.1	207	1971	72
[CpMo(CO)(Pr^{i}C≡CH) (PPh$_3$)][BF$_4$]	12.5		1990	72
[CpMo(CO)(Bu^{t}C≡CH) (PMe$_3$)$_2$][BF$_4$]	11.6			72
CpMo(CO)(Bu^{t}C≡CH) [P(OMe)$_3$](SMe)	10.2	211, 175		74
CpMo(Bu^{t}C≡CH)(S$_2$CNMe$_2$)	13.9			74
CpMo(SC$_6$H$_4$SPh)(Bu^{t}C≡CH)	10.2			74
[CpMo(CO)(PhC≡CH)(PPh$_3$)] [BF$_4$]	13.1		1990	76
[CpMo(CO)(HC≡CH)(PPh$_3$)] [BF$_4$]	12.9, 12.0		1977	76

(continued)

TABLE II (continued)

Complex	^{1}H, ppm	^{13}C, ppm ($^1J_{CH}$, Hz)	$\nu(CO)$,[a] cm^{-1}	Ref.
$\{(\eta^5\text{-}C_9H_7)Mo(MeC{\equiv}CH)$ $[P(OMe)_3]_2\}[BF_4]$	10.6			73
$\{CpMo(Bu^tC{\equiv}CH)[P(OMe)_3]_2\}[BF_4]$	11.6			73
$[CpMo(Ph_2PC_6H_4CH{=}CH_2)$ $(Bu^tC{\equiv}CH)][BF_4]$	10.2			136
$CpW(CO)(HC{\equiv}CH)Me$	12.0, 12.6	192 (212), 187 (204)	1925 1710*	67
$CpW(CO)(HC{\equiv}CH)Et$	11.8, 12.4	191 (212), 186 (205)	1925	69
$CpW(CO)(HC{\equiv}CH)Pr^n$	11.9, 12.4	191 (212), 186 (205)	1922	69
$CpW(CO)(HC{\equiv}CH)Bu^n$	11.9, 12.4	191 (212), 186 (205)	1922	69
$(\eta^5\text{-}C_5H_4Me)W(CO)(HC{\equiv}CH)Me$	11.8, 12.5	192 (212), 187 (205)	1924	69
$(\eta^5\text{-}C_5Me_5)W(CO)(HC{\equiv}CH)Me$	11.2, 12.3	192 (209), 189 (202)	1906	69
$(\eta^5\text{-}C_5Me_5)W(CO)(HC{\equiv}CH)Pr^n$	11.2, 12.2	191 (209), 189 (201)	1905	69
$CpW(CO)(HC{\equiv}CH)Ph$	12.2, 12.5	191 (215)	1925	69
$(\eta^5\text{-}C_5Me_5)W(CO)(HC{\equiv}CH)Ph$	11.5, 12.1	190	1915	69
$CpW(CO)(MeC{\equiv}CH)Me$	11.5 12.0 (isomer)	201, 184 199, 187 (isomer)	1921	69
$CpW(CO)(PhC{\equiv}CH)Et$	12.7 12.0 (isomer)	198, 190 202, 185 (isomer)	1924	69
$CpW(CO)(PhC{\equiv}CH)Pr^n$	12.6 12.0 (isomer)	198, 190 202, 186 (isomer)	1924	69
$(\eta^5\text{-}C_9H_7)W(CO)(HC{\equiv}CH)Me$	12.0, 12.6		1925	161
$(\eta^5\text{-}C_9H_7)W(CO)(MeC{\equiv}CH)Me$	11.3 12.1 (isomer)		1914	161
$CpW(CO)(HC{\equiv}CH)[C(O)Me]$	12.1, 12.6	188 (215), 186 (209)	1938 1702*	67,69
$CpW(CO)(HC{\equiv}CH)[C(O)Et]$	12.1, 12.6	187 (214), 186 (209)	1955	69
$CpW(CO)(HC{\equiv}CH)[C(O)Pr^n]$	12.1, 12.5	187 (215), 186 (209)	1955	69
$CpW(CO)(HC{\equiv}CH)[C(O)Bu^n]$	12.1, 12.6	187 (214), 186 (209)	1959	69
$(\eta^5\text{-}C_5H_4Me)W(CO)$ $(HC{\equiv}CH)[C(O)Me]$	12.0, 12.6	187 (214), 187 (208)	1940	69
$(\eta^5\text{-}C_5H_4Me)W(CO)$ $(HC{\equiv}CH)[C(O)Pr^n]$	11.5, 12.3	186 (211), 188 (205)	1940	69
$CpW(CO)(PhC{\equiv}CH)[C(O)Et]$	12.8 12.3 (isomer)	199, 188 198, 186	1959	69
$CpW(CO)(PhC{\equiv}CH)[C(O)Pr^n]$	12.8 12.3 (isomer)	199, 187 197, 185	1954	69
$(\eta^5\text{-}C_5H_4Me)W(CO)$ $(HC{\equiv}CH)[C(O)Pr^n]$	11.6, 12.4	201, 197	1950	69
$CpW(PMe_3)(HC{\equiv}CH)[C(O)Ph]$	11.8, 12.5	200 (195), 196 (193)		69
$CpW(PMe_3)(HC{\equiv}CH)[C(O)Me]$	11.9, 12.2	195 (193), 191 (190)	1710*	67

(continued)

TABLE II (*continued*)

Complex	^{1}H, ppm	^{13}C, ppm ($^1J_{CH}$, Hz)	$\nu(CO),^a$ cm^{-1}	Ref.
CpW[P(OMe)$_3$](HC≡CH) [C(O)Me]	11.8, 12.2	191 (202), 186 (205)		67
[CpW(=C(OEt)Me) (CO)(HC≡CH)][BF$_4$]	12.6, 12.8	190, 187	1962	77
CpW(CO)(HC≡CH)(NO)	7.8, 8.5	106 (223), 98 (230)	1985 1723*	160
CpW(CO)(HC≡CC(O)Me)(NO)	9.3	121, 113 (221)	2008 1700*	160
Mo(HC≡CH)(SBu^t)$_2$(CNBu^t)$_2$	10.4	172 (215)	1565*	133
Mo(PhC≡CH)(SBu^t)$_2$(CNBu^t)$_2$	10.4	184, 172 (211)	1672*	133
[CpMo(HC≡CH)(η^6-C$_6$H$_6$)][PF$_6$]	7.8		1689*	66
Cp$_2$Mo(HC≡CH)	7.7	118 (200)	1613*	82,84
Cp$_2$Mo(MeC≡CH)	7.1	126, 106 (198)	1736*	82

^{a} Compounds for which the ν(C≡C) frequency has been reported are included with an asterisk to indicate this value.

TABLE III

NMR AND IR DATA FOR Mo(II) AND W(II) INTERNAL MONOALKYNE COMPLEXES

Complex	^{13}C, ppm	$\nu(CO),^a$ cm^{-1}	Ref.
Mo(CO)(MeC≡CMe)(PEt$_3$)$_2$Br$_2$	237, 229	1939 1645*	46
Mo(CO)(PhC≡CPh)(PEt$_3$)$_2$Br$_2$		1951	46
Mo(CO)(EtC≡CEt)(PEt$_3$)$_2$Br$_2$		1939	46
Mo(CO)(π-cyclooctyne)(PEt$_3$)$_2$Br$_2$	240, 232	1920	47
Mo(CO)(MeC≡CMe)(PEt$_3$)$_2$Cl$_2$		1943 1645*	46
Mo(CO)(MeC≡CMe)(PPh$_3$)$_2$Cl$_2$	240	1956 1670*	46
Mo(CO)(EtC≡CEt)(PPh$_3$)$_2$Cl$_2$		1946	46
Mo(CO)(MeC≡CMe)(PPh$_3$)$_2$Br$_2$		1950	46
Mo(CO)(MeC≡CMe)(py)$_2$Cl$_2$		1937	46
Mo(CO)(MeC≡CMe)(dppe)Cl$_2$		1970	46
Mo(CO)(EtC≡CEt)(dppe)Cl$_2$		1977 1671*	46
Mo(CO)(EtC≡CEt)(dppe)Br$_2$		1985 1625*	46
Mo(CO)(PhC≡CPh)(dppe)Br$_2$		1990	46
cis-(CNBu^t)$_2$W(CO)(MeC≡CMe)Br$_2$	203	2002	49

(*continued*)

TABLE III (continued)

Complex	^{13}C, ppm	$\nu(CO),^a$ cm^{-1}	Ref.
cis-(CNBu^t)$_2$W(CO)(PhC≡CMe)Br$_2$	207, 202	2010	49
[W(CO)(PhC≡CPh)(MeCN)I$_2$]$_2$		2075, 2000 1675*	51
[Mo(CO)(PhC≡CPh)(MeCN)I$_2$]$_2$		2050, 2005 1690*	51
[Mo(CO)(PhC≡CMe)(MeCN)I$_2$]$_2$		2045, 1990 1620*	51
Mo(CO)(PhC≡CPh)(MeCN)$_2$I$_2$		1978	51
W(CO)$_2$(HOC≡CC$_6$H$_4$Me)(acac)Cl		2060, 1960 1675*	100
(W=CHPh)(PhC≡CPh)(PMe$_3$)$_2$Cl$_2$	223		78
[W(CNBu^t)$_4$(RHNC≡CNHR)I]I	182		117
[Mo(CNBu^t)$_4$(Bu^{t}HNC≡CNHBu^t)I]I	193		117
[Mo(CNBu^t)$_4$(Bu^{t}HNC≡CNHBu^t)Br] · $\frac{1}{2}$ZnBr$_4$	196		117
[Mo(CNBu^t)$_4$(Bu^{t}HNC≡CNHBu^t)Cl]Cl	199		117
[Mo(CNBu^t)$_4$(Bu^{t}HNC≡CNHBu^t)(CN)][PF$_6$]	200		117
[Mo(CNBu^t)$_4$(Bu^{t}HNC≡CNHBu^t)$_5$][BPh$_4$]$_2$	203		117
{[HB(pz)$_3$](CO)(PEt$_3$)W(MeOC≡CSMe)}[FSO$_3$]	231, 198	1959	109
Mo(CO)(PhC≡CPh)(S$_2$CNEt$_2$)$_2$		1932	53
Mo(CO)(Me$_3$SiC≡CSiMe$_3$)(S$_2$CNEt$_2$)$_2$		1910	53
Mo(CO)(MeC≡CMe)(S$_2$CNEt$_2$)$_2$		1920	53
Mo(CO)(EtC≡CEt)(S$_2$CNEt$_2$)$_2$		1908	53
Mo(CO)(π-cyclooctyne)(S$_2$CNEt$_2$)$_2$	217	1900	47
Mo(CO)(π-cyclooctyne)(S$_2$CNMe$_2$)$_2$	218	1896	47
Mo(CO)(DMAC)(S$_2$CNMe$_2$)$_2$		1975	53
Mo(CO)(Et$_2$NC≡CMe)(S$_2$CNMe$_2$)$_2$		1913	53
W(CO)(PhC≡CPh)(S$_2$CNEt$_2$)$_2$		1927	53
W(CO)(HOCH$_2$C≡CCH$_2$OH)(S$_2$CNEt$_2$)$_2$		1930	53
W(CO)(Et$_2$NC≡CMe)(S$_2$CNEt$_2$)$_2$		1904	53
W(CO)(MeC≡CMe)(S$_2$CNEt$_2$)$_2$		1910	53
W(CO)(EtC≡CEt)(S$_2$CNEt$_2$)$_2$		1910	53
W(CO)(DMAC)(S$_2$CNEt$_2$)$_2$		1956	53
W(CO)(π-cyclooctyne)(S$_2$CNEt$_2$)$_2$	214	1881	47
W(CO)(π-cyclooctyne)(S$_2$CNMe$_2$)$_2$	215	1878	47
W(CO)(Ph$_2$PC≡CPPh$_2$)(S$_2$CNEt$_2$)$_2$		1920	53
W(CO)(PhC≡COMe)(C$_6$H$_7$N$_2$)$_2$	232, 182	1927 1691*	113
W(CO)(MeC≡COMe)(C$_6$H$_7$N$_2$)$_2$		1917	113
CpMo(CO)(PhC≡CMe)Cl		1944	93
[CpMo(CO)(MeC≡CMe)(PEt$_3$)][BF$_4$]	230	1968 1682*	72,76
[CpMo(CO)(MeC≡CMe)(PPh$_3$)][BF$_4$]	234	1973 1679*	72,76
{CpMo(CO)(MeC≡CMe)[P(C$_6$H$_{11}$)$_3$]}[BF$_4$]		1959	72

(continued)

TABLE III (*continued*)

Complex	^{13}C, ppm	$\nu(CO)$,^{a} cm^{-1}	Ref.
[CpMo(CO)(PhC≡CPh)(PEt$_3$)][BF$_4$]		1974 1688*	72,76
[CpMo(CO)(PhC≡CPh)(PPh$_3$)][BF$_4$]	232	1981 1645*	72,76
[CpMo(CO)(EtC≡CMe)(PEt$_3$)][BF$_4$]		1930	72
[CpMo(CO)(EtC≡CMe)(PPh$_3$)][BF$_4$]		1930	72
[CpMo(CO)(MeC≡CPh)(PPh$_3$)][BF$_4$]		1940	72
[(η^5-C$_9$H$_7$)Mo(CO)(MeC≡CMe)(PEt$_3$)][BF$_4$]		1957	72
[(η^5-C$_9$H$_7$)Mo(CO)(MeC≡CMe)(PPh$_3$)] [BF$_4$]	213	1966	72
[CpMo(PMePh$_2$)$_2$(MeC≡CMe)][BF$_4$]	229		72
[CpMo(Ph$_2$PCH=CHPPh$_2$)(MeC≡CMe)][BF$_4$]	216, 201		72
[CpMo(CO)(MeC$_6$H$_4$C≡CC$_6$H$_4$Me)(PEt$_3$)][BF$_4$]		1976	72
[CpMo(CO)(MeC$_6$H$_4$C≡CC$_6$H$_4$Me)(PPh$_3$)][BF$_4$]		1981	72
CpMo[P(OMe)$_3$](MeC≡CMe)(SMe)	196, 185		74
CpMo[P(OMe)$_3$](PhC≡CPh)(SMe)	196, 194		74
CpMo[P(OMe)$_3$](MeC≡CMe)(SePh)	202, 193		74
CpMo[P(OMe)$_3$](MeC≡CMe)(SC$_6$H$_4$NH$_2$)	197, 188		74
CpMo[P(OMe)$_3$](MeC≡CMe)(SC$_6$H$_4$OMe)	198, 189		74
CpMo[P(OMe)$_3$](MeC≡CMe)(SC$_6$H$_4$Me)	199, 190		74
CpMo[P(OMe)$_3$](MeC≡CMe)(SC$_6$H$_5$)	200, 190		74
CpMo[P(OMe)$_3$](MeC≡CMe)(SC$_6$H$_4$NO$_2$)	205, 196		74
CpMo(CO)(MeC≡CMe)(OMe)	176, 169	1880	75
CpMo(CO)(MeC≡CMe)(SMe)		1936, 1920	74
CpMo(CO)(MeC≡CMe)(S$_2$CNMe$_2$)	219		74
CpMo[$\overline{SC_6H_4SPh}$(—o)](MeC≡CMe)	209		74
CpMo(CO)[SC$_6$H$_4$SPh(—o)](MeC≡CMe)	179, 181	1924	74
{CpMo(CO)(PhC≡CPh)[P(OPh)$_3$]}[BF$_4$]		2011 1652*	76
CpMo(CO)(MeC≡CMe)(σ-CMe=CMe$_2$)	188, 186	1912	183
CpW(CO)(PhC≡CMe)Cl		1921	93
Cp'W(CO)(MeC≡CMe)[C(O)Me]	193, 192	1940	69
CpW(CO)(MeC≡CMe)[C(O)Et]	194, 193	1940	69
CpW(CO)(MeC≡CMe)(OMe)		1890	75
{CpMo($\overline{Ph_2PC_6H_4CH}$=CH$_2$)(MeC≡CMe)}[BF$_4$]	231, 216		136
W(CO)(EtC≡CEt)(μ-CHAr)(C$_2$B$_9$H$_8$Me$_2$) Mo(CO)$_3$(η^5-C$_9$H$_7$)	178, 172		158
CpW(CO)(CF$_3$C≡CCF$_3$)(μ-SMe)W(CO)$_5$		1730*	64
CpW(CO)(CF$_3$C≡CCF$_3$)(SMe)		1999 1720*	190
CpW(CO)(CF$_3$C≡CCF$_3$)(SPh)		1999 1718*	190
CpW(CO)(CF$_3$C≡CCF$_3$)(SC$_6$H$_4$Me)		1998 1717*	190

(*continued*)

TABLE III (*continued*)

Complex	^{13}C, ppm	ν(CO),^{a} cm^{-1}	Ref.
CpW(CO)(CF$_3$C≡CCF$_3$)(SC$_6$F$_5$)		2001	*190*
		1720*	
CpMo(CF$_3$C≡CCF$_3$)(PEt$_3$)(SC$_6$F$_5$)		1675*	*65*
CpW(CF$_3$C≡CCF$_3$)(PEt$_3$)(SC$_6$F$_5$)		1650*	*65*
CpMo(CF$_3$C≡CCF$_3$)[P(OMe)$_3$](SC$_6$F$_5$)		1690*	*65*
CpW(CF$_3$C≡CCF$_3$)[P(OMe)$_3$](SC$_6$F$_5$)		1665*	*65*
CpMo(CF$_3$C≡CCF$_3$)(PMe$_2$Ph)(SC$_6$F$_5$)		1675*	*65*
CpW(CF$_3$C≡CCF$_3$)(PMe$_2$Ph)(SC$_6$F$_5$)		1640*	*65*
CpMo(CF$_3$C≡CCF$_3$)(PMePh$_2$)(SC$_6$F$_5$)		1678*	*65*
CpMo(CF$_3$C≡CCF$_3$)(PPh$_3$)(SC$_6$F$_5$)		1678*	*65*
CpMo(CO)(PhC≡CPh)CF$_3$		1962	*98*
CpMo(CO)(MeC≡CMe)CF$_3$		1968	*98*
[CpW(CO)(PMe$_3$)(MeOC≡CMe)][BF$_4$]	227, 198	1975	*105*
CpW(CO)(PMe$_3$)(Et$_3$AlOC≡CC$_6$H$_4$Me)	223, 195	1951	*105*
[CpW(CO)(MeOC≡CC$_6$H$_4$Me)(PMe$_3$)][BF$_4$]	231, 195	1975	*105*
[CpW(CO)(EtOC≡CC$_6$H$_4$Me)(PMe$_3$)][BF$_4$]	229, 194	1976	*105*
CpW(CO)(Cl$_3$BOC≡CC$_6$H$_4$Me)(PMe$_3$)		1970	*105*
[CpW(PMe$_3$)$_2$(EtOC≡CC$_6$H$_4$Me)][BF$_4$]	227, 192	1975	*107*
[CpW(PMe$_3$)$_2$(MeOC≡CC$_6$H$_4$Me)][BF$_4$]	226, 200	1970	*107*
[CpW(MeOC≡CMe)(PMe$_3$)$_2$][BF$_4$]	228, 201		*107*
[CpW(CO)(HOC≡CC$_6$H$_4$Me)(PMe$_3$)$_3$][BF$_4$]	227, 192	1975	*108*
[CpW(CO)(HOC≡CMe)(PMe$_3$)][BF$_4$]	223, 197	1970	*108*
[CpW(CO)(MeOC≡CMe)(PMe$_3$)][SO$_3$CF$_3$]	227, 198	1975	*108*
Cp$_2$Mo(PhC≡CPh)		1771*	*85*
Cp$_2$Mo(CH$_3$C≡CCH$_3$)	115	1830*	*85,155*
Mo(PhC≡CPh)(SBu^t)$_2$(CNBu^t)$_2$	183	1755*	*133*
Cp$_2$W(MeC≡CMe)		1750*	*85*
Cp$_2$Mo(CF$_3$C≡CCF$_3$)		1782*	*85*
(η^5-C$_5$H$_4$Me)$_2$Mo(CF$_3$C≡CCF$_3$)		1778*	*128*

^{a} Compounds for which the ν(C≡C) frequency has been reported are included with an asterisk to indicate this value.

of such a classification scheme (Fig. 22). Given that N is restricted to discrete values $(2, 3, 3\frac{1}{3}$, and 4) by definition [N = (total number of electrons donated by alkynes)/(number of alkyne ligands)], the correlation between N and the continuum of possible ^{13}C chemical shifts is surprisingly useful. The N values derived from simple 18-electron guidelines are compatible with more sophisticated molecular orbital descriptions involving alkyne $\pi_\perp$ combinations in each case. In spite of the theoretical void behind this correlation, these ^{13}C guideposts have frequently been cited in describing molecular properties of alkyne complexes.

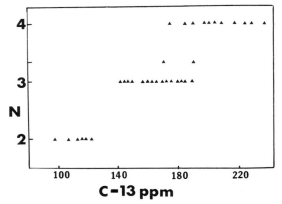

FIG. 22.   Alkyne carbon ^{13}C chemical shift (ppm) versus formal alkyne electron donation number, N.

In cases where an alkyne is bound to a single metal in a dinuclear complex the alkyne carbon ^{13}C chemical shift can provide an indirect assessment of metal–metal bond order. Examples to date are consistent with alkyne $\pi_\perp$ donation dominating at the expense of metal–metal multiple bonding: $W_2(CH_2SiMe_3)_4(\mu\text{-CSiMe}_3)_2(HC{\equiv}CH)$, δ (^{13}C) = 217 ppm (156), and $[Cp(MeC{\equiv}CMe)W(\mu\text{-CHAr})(\mu\text{-CO})Pt(PMe_3)_2]^+$, $\delta(^{13}C)$ = 191 ppm (157). Chemical shifts of 172 and 178 ppm for the 3-hexyne ligand in $(EtC{\equiv}CEt)W[\mu\text{-}\sigma,\ \eta^3\text{-CHAr}(C_2B_9H_8Me_2)](\mu\text{-CO})Mo(CO)$-$(\eta^5\text{-C}_9H_7)$ are compatible with N equals 3 to 4 as reflected in W—C bond distances of 2.08 and 2.12 Å. This unusual mixed molybdenum–tungsten dimer reflects insertion of a carbyne CAr group into a cage B—H bond (158).

Single bond J_{CH} coupling constants for bound alkynes also offer insight into the bonding of terminal alkynes to metals. For $M(CO)(HC{\equiv}CR)$-$(S_2CNEt_2)_2$ complexes $^1J_{CH}$ values well below the free alkyne value of 250 Hz but still above 200 Hz are observed (58). Since $^1J_{CH}$ is related to the fraction of s character in the C—H bond it addresses hybridization directly and thus crudely quantifies perturbation of the alkyne from sp hybridization on binding to the metal.

Dynamic NMR studies of $W(CO)(HC{\equiv}CH)(S_2CNEt_2)_2$ and related complexes (58) indicate that the barrier to alkyne rotation is around 11–12 kcal/mol (Table IV). Discordant metal–ligand π interactions anticipated as the alkyne rotates away from the ground state orientation are evident in EHMO calculations (Fig. 23). When the alkyne is orthogonal to the M—CO axis the d_{yz} orbital is stabilized by CO $\pi_y{}^*$ but destabilized by alkyne $\pi_\perp$ while d_{xy} becomes the lowest lying $d\pi$ orbital due to

JOSEPH L. TEMPLETON

TABLE IV

Alkyne Rotational Barriers in Mo(II) and W(II) Monoalkyne Complexes

Complex	$\Delta G^{\ddagger}$, kcal/mol	Ref.
$Mo(CO)(MeC\equiv CMe)(PEt_3)_2Br_2$	13.0	46
$Mo(CO)(MeC\equiv CMe)(PEt_3)_2Cl_2$	11.4	46
$Mo(CO)(MeC\equiv CMe)(py)_2Cl_2$	9.3	46
$Mo(CO)(MeC\equiv CMe)(PPh_3)_2Cl_2$	9.6	46
$W(CO)(MeC\equiv CMe)(PPh_3)_2Cl_2$	9.6	46
$Mo(CO)(\pi\text{-cyclooctyne})(PEt_3)_2Br_2$	13.0	47
$W(CO)(\pi\text{-cyclooctyne})(PEt_3)_2Br_2$	12.7	47
$cis\text{-}(CNBu^t)_2W(CO)(MeC\equiv CMe)Br_2$	11.8	49
$trans\text{-}(PPh_3)_2W(CO)(MeC\equiv CMe)Br_2$	10.1	49
$(W=CHPh)(PhC\equiv CH)(PMe_3)_2Cl_2$	12.1	78
$W(CO)(HC\equiv CH)(S_2CNMe_2)_2$	11.9	58
$W(CO)(HC\equiv CH)(S_2CNEt_2)_2$	11.7	58
$W(CO)(MeC\equiv CMe)(S_2CNEt_2)_2$	11.1	58
$W(CO)(\pi\text{-cyclooctyne})(S_2CNEt_2)_2$	10.3	47
$W(CO)(\pi\text{-cyclooctyne})(S_2CNMe_2)_2$	10.9	47
$W(CO)(Ph_2PC\equiv CPPh_2)(S_2CNEt_2)_2$	11.4	58
$Mo(CO)(\pi\text{-cyclooctyne})(S_2CNMe_2)_2$	10.5	47
$Mo(CO)(\pi\text{-cyclooctyne})(S_2CNEt_2)_2$	10.8	47
$[CpMo(CO)(MeC\equiv CMe)(PEt_3)][BF_4]$	15.0	72
$[CpMo(CO)(MeC\equiv CMe)(PPh_3)][BF_4]$	12.9	72
$\{CpMo(CO)(MeC\equiv CMe)[P(C_6H_{11})_3]\}[BF_4]$	15.4	72
$[(\eta^5\text{-}C_9H_7)Mo(CO)(MeC\equiv CMe)(PEt_3)][BF_4]$	14.9	72
$[(\eta^5\text{-}C_9H_7)Mo(CO)(MeC\equiv CMe)(PPh_3)][BF_4]$	12.8	72
$[CpMo(Ph_2PCH\equiv CHPPh_2)(MeC\equiv CMe)][BF_4]$	15.8	72
$[CpMo(dppe)(MeC\equiv CMe)][BF_4]$	14.4	66,72
$[CpMo(dmpe)(MeC\equiv CMe)][BF_4]$	11.2	72
$[CpMo(PMePh_2)_2(MeC\equiv CMe)][BF_4]$	9.1	72
$[CpMo(CO)(MeC_6H_4C\equiv CC_6H_4Me)(PEt_3)][BF_4]$	13.1	72
$\{CpMo[PPh_2(o\text{-}C_6H_4CH\equiv CH_2)](MeC\equiv CMe)\}[BF_4]$	9.6	136
$CpMo(CO)(MeC\equiv CMe)(\sigma\text{-}CMe\equiv CMe_2)$	19.9	183
$CpMo[P(OMe)_3](MeC\equiv CMe)(SePh)$	13.2	74
$CpMo[P(OMe)_3](MeC\equiv CMe)(SC_6H_4NH_2)$	15.6	74
$CpMo[P(OMe)_3](MeC\equiv CMe)(SC_6H_4OMe)$	14.5	74
$CpMo[P(OMe)_3](MeC\equiv CMe)(SC_6H_4Me)$	14.3	74
$CpMo[P(OMe)_3](MeC\equiv CMe)(SC_6H_5)$	14.0	74
$CpMo[P(OMe)_3](MeC\equiv CMe)(SC_6H_5NO_2)$	12.5	74
$CpW(CO)(HC\equiv CH)Me$	18.2	67
$CpW(CO)(HC\equiv CH)Et$	18.8	69
$CpW[P(OMe)_3](HC\equiv CH)[C(O)Me]$	17.6	67
$CpW(PMe_3)(HC\equiv CH)[C(O)Me]$	16.8	67
$(\eta^5\text{-}C_9H_7)W(CO)(HC\equiv CH)Me$	18.4	161
$(\eta^5\text{-}C_9H_7)W(CO)(HC\equiv CMe)Me$	18.5	161
$CpW(CO)(HC\equiv CH)Pr^n$	18.5	69
$CpW(CO)(HC\equiv CH)Bu^n$	18.2	69
$Cp'W(CO)(HC\equiv CH)Me$	17.9	69
$Cp^*W(CO)(HC\equiv CH)Me$	17.9	69
$Cp^*W(CO)(HC\equiv CH)Pr^n$	18.2	69
$CpW(CO)(HC\equiv CH)Ph$	17.5	69
$Cp^*W(CO)(HC\equiv CH)Ph$	17.2	69
$CpW(CO)(MeC\equiv CH)Me$	16.9	69
$CpW(CO)(MeC\equiv CMe)Me$	18.0	69
$Cp'W(CO)(MeC\equiv CMe)Me$	17.4	69
$CpW(CO)(PhC\equiv CH)Et$	17.7	69
$CpW(CO)(PhC\equiv CH)Pr^n$	18.6	69

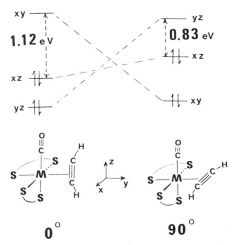

FIG. 23.   Electronic factors influencing $d\pi$ orbital energies and alkyne rotational barriers.

overlap with $\pi_\parallel^*$. The net result is a calculated activation barrier of 10.5 kcal/mol for the idealized $Mo(CO)(HC\equiv CH)(S_2CNH_2)_2$ model complex, with virtually all of the electronic factors localized in metal $d\pi$–ligand π energy changes (147).

Terminal alkynes in $M(\eta^2\text{-olefin})(RC\equiv CH)(S_2CNEt_2)_2$ (M = Mo, W) complexes display low-field signals (11.5–13.5 ppm) indicative of four-electron donation from the alkyne (60). Alkyne carbons in W(MA)-$(PhC\equiv CH)(S_2CNEt_2)_2$ appear near 223 ppm in the ^{13}C-NMR spectrum, somewhat below the 206 ppm chemical shift of the alkyne in the carbonyl analog, $W(CO)(PhC\equiv CH)(S_2CNEt_2)_2$. The parent acetylene ligand, $HC\equiv CH$, produces only a singlet in the ^{1}H spectrum of both the DCNE and MA tungsten derivatives even at low temperature ($-99°C$). The two ends of the alkyne are not equivalent by symmetry, and facile alkyne rotation is suggested by these results.

The trans-substituted DCNE ligand can form two diastereomers on binding to the $M(RC\equiv CR)(S_2CNEt_2)_2$ fragment, as either CN or H can lie above the equatorial chelate ring (Fig. 24). Indeed, two isomers are evident in the room temperature ^{1}H-NMR spectrum of $Mo(DCNE)$-$(HC\equiv CH)(S_2CNMe_2)_2$ (60). Two pairs of doublets can be assigned to the static trans DCNE protons in the two isomers, so olefin rotation is slow. The presence of only two acetylene proton signals suggests that the ends of the alkyne are averaged on the NMR time scale in each isomer. Isomer interconversion is not evident at 77°C (this would require flipping the olefin faces since olefin rotation alone is inadequate to interconvert the isomers).

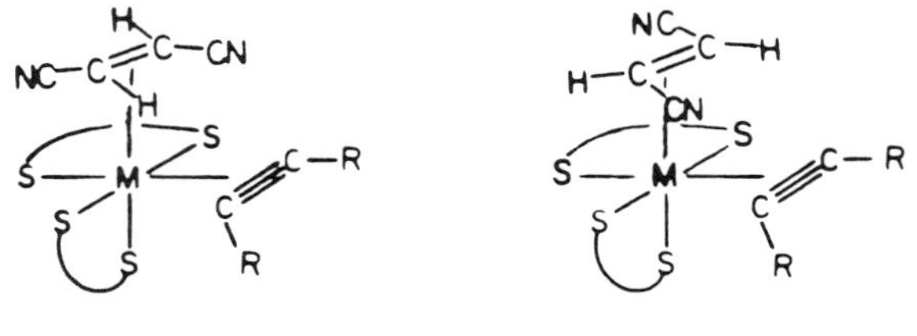

FIG. 24.   Isomers of $M[\eta^2\text{-}trans\text{-}H(NC)C=CH(CN)](RC\equiv CR)(S_2CNEt_2)_2$.

Furthermore, no exchange of the two independent olefin proton sites in either isomer was observed, indicating that olefin rotation is not ocurring at 77°C. Thus, the experimental results are consistent with molecular orbital expectations: the barrier to olefin rotation is large while the barrier to alkyne rotation is small.

Acetylenic proton chemical shifts for $M(CO)(RC\equiv CH)L_2X_2$ (M = Mo, W) (46) and $W(CO)_2(RC\equiv CH)LI_2$ (44) fall in the 12–14 ppm range typical of four-electron donor alkynes (Table II). Complexes containing the dppe chelate exhibit higher field acetylenic proton signals (10–11 ppm) for both Mo and W. The dppe chelate will force a cis-L_2 geometry on the metal which may alter electronic factors to account for decreased $\pi_\perp$ donation and to cause comparatively high ^1H chemical shifts. A more likely explanation, given that other spectral properties (including ^{13}C alkyne chemical shifts) are not unique, is that the terminal alkyne proton resides in the shielding cone of one of the dppe phenyl groups and is moved up roughly 1 ppm as a result. Carbon-13 chemical shifts for these complexes are in the 200–240 ppm range typical of four-electron donor alkyne ligands. Single bond J_{CH} values of 215, 204, and 203 Hz have been reported for $Mo(CO)(HC\equiv CBu^n)(dppe)Br_2$, $Mo(CO)(PhC\equiv CH)(PEt_3)_2Br_2$ (46), and $W(CO)(HC\equiv COAlCl_3)(PMe_3)_3Cl$ (101), respectively.

Alkyne rotational barriers that have been determined for several complexes fall between 9 and 13 kcal/mol (Table IV). Activation energies for analogous molybdenum and tungsten pairs differ only slightly. The activation energy is not simply steric in origin since the $\Delta G^\ddagger$ trend for Mo(CO)-$(MeC\equiv CMe)L_2Cl_2$ is py < PPh_3 < PEt_3 as a function of L with the largest L associated with a barrier of intermediate magnitude (46). Given that both $\pi_\parallel^*$ acceptance and $\pi_\perp$ donation influence the barrier, as do steric factors, interpretation of the relatively small activation energy differences requires caution.

Stereochemical rigidity characterizes the alkyne ligand in some complexes, as in cis-$W(CO)_2(HC\equiv CH)LI_2$ (L = PMe_3, $AsMe_3$) where two independent alkyne proton signals are observed (*44*). A single rigid isomer with I and L trans to the two cis-CO ligands, respectively, could give the observed ^{1}H spectrum. Alternatively, two isomers with rotating alkynes could give the observed ^{1}H spectrum. In one isomer both I ligands could be trans to carbonyls, in the other one I could be trans to alkyne (Fig. 19).

Barriers of 11.8 and 10.1 kcal/mol are found for *cis*-WL$_2$(CO)-(MeC≡CMe)Br$_2$ with L = CNBu^t and PPh$_3$, respectively. The coalescence temperature of −14°C for L = P(OMe)$_3$ (*48*) allows an estimate of $\Delta G^\ddagger$ of 12–13 kcal/mol for the cis-P(OMe)$_3$ isomer. The trans-P(OMe)$_3$ isomer exhibits no ^{1}H-NMR temperature dependence which, barring accidental degeneracies, suggests facile alkyne rotation down to −80°C. A lower barrier in the trans-P(OMe)$_3$ derivative relative to the trans-PEt$_3$ derivative is consistent with the greater π acidity of the phosphite ligands decreasing the distinction between the $d\pi$ orbitals directed toward the alkyne. Rotation of the alkyne by 90° is more favorable for P(OMe)$_3$ since the d_{xy} orbital can be stabilized by the two phosphites as well as by $\pi_\parallel^*$.

The number of cyclopentadienyl complexes with terminal alkyne ligands is large enough to look for ^{1}H chemical shift trends among different subgroups. Cationic molybdenum derivatives of the type [CpMo(RC≡CH)-LL′]$^+$ (*72*) display resonances between 11 and 13 ppm. Although only four neutral complexes containing thiolate ligands and terminals alkynes (*74*) appear in Table II, their ^{1}H signals vary between 10 and 14 ppm, a larger spread of values than among the cationic complexes. The extensive series of tungsten complexes studied by Alt includes CpW(CO)-(RC≡CH)(R′) and CpW(CO)(RC≡CH)[C(O)R′] with acetylenic proton shifts of 11–13 ppm (*69*). Incorporation of a nitrosyl ligand moves the terminal alkyne proton upfield [CpW(CO)(HC≡CH)(NO): 8.5, 7.8 ppm], compatible with an NO$^+$ formalism producing an octahedral d^6 W(0) monomer with no vacant $d\pi$ orbital available to accommodate $\pi_\perp$ donation (*159,160*). Another example of a saturated Cp derivative is [CpMo-(HC≡CH)η^6-C$_6$H$_6$)]$^+$, formally a seven-coordinate d^4 monomer, which exhibits a singlet at 7.8 ppm for the bound acetylene (*66*). Alkyne carbon-13 signals for cyclopentadienyl derivatives with $N = 4$ appear in the 200 ppm region as expected. Cationic molybdenum alkynes typically resonate between 200 and 235 ppm, while alkyne carbons in neutral complexes with a thiolate ligand appear at somewhat higher fields (175–210 ppm).

Tungsten derivatives, both CpW(CO)(RC≡CH)(R′) and CpW(CO)-(RC≡CH)[C(O)R′], consistently exhibit ^{13}-C signals between 185 and 200 ppm with single bond $^1J_{CH}$ values of 190–205 Hz for alkyl derivatives and 205–215 Hz for acyl derivatives (*69*). Incorporation of methoxide in place of the alkyl or acyl ligand shifts the alkyne carbons slightly upfield to 169 and 176 ppm (*75*), presumably reflecting competition between alkyne

$\pi_\perp$ donation and OMe^- π donation. Again, comparison with the nitrosyl complex is relevant as the alkyne carbons in $CpW(CO)(HC{\equiv}CH)(NO)$ resonate near 100 ppm (106 and 98 ppm), in accord with expectations for d^6 octahedral alkyne derivatives (160).

Rotational barriers of numerous monoalkyne complexes have been measured, with 2-butyne a particularly popular probe. Barriers for cationic $[CpMo(CO)(MeC{\equiv}CMe)L]^+$ complexes range from 12.8 to 15.4 kcal/mol with the triphenylphosphine derivative exhibiting a barrier at the low energy end and trialkylphosphine derivatives at the high energy end (72). This is consistent with other data which suggest that the alkyne rotational barrier is sensitive to differences in π acidity of the two cis ligands L and L' in $CpM(RC{\equiv}CR)LL'$ complexes. The two $d\pi$ orbitals directed toward the alkyne are differentiated by their interactions with L and L'. As L and L' diverge in their π-acid or π-base properties the energies and extensions of the two $d\pi$ orbital combinations available to the alkyne increasingly differ. One would anticipate a large alkyne rotational barrier for very different L and L' ligands, and a small barrier for similar L and L' ligands.

More subtle effects determine alkyne rotational preferences in symmetric $CpM(RC{\equiv}CR)L_2$ complexes with identical L and L' ligands as described by Hoffmann and co-workers (145). The extensive molecular orbital discussion of $W(CO)_2(RC{\equiv}CR)LX_2$ complexes in Section VI described the factors which remove the dicarbonyl complexes from the simple guidelines appropriate for analysis of monocarbonyl complexes. The experimental barrier of 9.1 kcal/mol measured for $[CpMo(MeC{\equiv}CMe)-(PMePh_2)_2]^+$ is lower than those of related L = CO, L' = PR_3 complexes (72), but it is certainly not negligible as might first be expected for L = L'. Steric factors may account for most of the activation energy required to rotate the 2-butyne ligand in this complex.

Barriers of 12.5-15.5 kcal/mol for neutral $CpMo(CO)(MeC{\equiv}CMe)-(SR)$ complexes are quite similar to rotational barriers in cationic complexes (74). Given the π acidity of CO and the π basicity of SR^-, these barriers are surprisingly small. Sulfur donor ligands tend to be electronically flexible, and the soft thiolate may facilitate alkyne rotation by simultaneous rotation of the thiolate substituent.

For cyclopentadienyl tungsten(II) derivatives barriers of 18-19 kcal/mol characterize alkyne rotation in both the alkyl and acyl cases (67,69,161). Extended Huckel calculations indicate that steric factors play a role in these complexes in addition to the standard $d\pi$ orbital electronic factors (147). Smaller barriers attend replacement of CO in $CpW(CO)-(HC{\equiv}CH)X$ complexes with either $P(OMe)_3$ or PMe_3. No alkyne rotation was observable by NMR for the saturated nitrosyl analog, suggesting an activation energy greater than 21 kcal/mol (*160*).

Bisalkyne d^4 monomers, with $N = 3$ by symmetry, exhibit proton and carbon chemical shifts at higher fields than those of monoalkynes with $N = 4$. The proton chemical shift of 10.45 ppm for Mo(PhC≡CH)$_2$-(S$_2$CNEt$_2$)$_2$ (*52*) falls nicely between the four-electron donor Mo(CO)-(PhC≡CH)(S$_2$CNEt$_2$)$_2$ case (12.6 ppm) and the two-electron donor (π-C$_5$H$_5$)$_2$Mo(HC≡CH) case [7.68 ppm (Table II)]. Additional data for bisalkyne complexes, including pyrrole-N-carbodithioate derivatives, support a correlation of ^{1}H chemical shifts with alkyne $\pi_\perp$ donation, with three-electron donors typically near 10.0 ± 0.5 ppm. Similar ^{1}H values are found for cyclopentadienyl bisalkyne complexes with terminal alkyne ligands. Chemical shifts between 8.5 and 10.5 ppm characterize all the neutral and cationic bisalkynes listed in Table V except for [CpMo-(RC≡CH)$_2$(MeCN)]$^+$ where one isomer has δ near 11 ppm for the acetylenic proton (*72*).

Relatively few ^{13}C-NMR data have been reported for bisalkyne complexes. The range of ^{13}C alkyne chemical shifts is roughly 170-190 ppm for Mo(RC≡CR)$_2$(S$_2$CNEt$_2$)$_2$ complexes (*87*). Terminal alkynes exhibit $^1J_{CH}$ values near 210 Hz in these three-electron donor roles, similar to values reported for four-electron donor alkynes. For cationic cyclopentadienyl bisalkyne derivatives an alkyne carbon ^{13}C range of 140-180 is seen in Table VI.

Rotational barriers have been probed for a number of bisalkyne complexes (Table VII). Cationic [CpM(RC≡CR)$_2$(CO)]$^+$ complexes exhibit relatively high barriers (16-21 kcal/mol). Both standard variable-temperature NMR techniques (*94*) and two-dimensional methods (*162*) have been used to elucidate isomer interconversion schemes with two unsymmetrical alkynes in the coordination sphere. The plane of symmetry present when two symmetrical alkynes bind to a CpMX fragment is not retained in all isomers with RC≡CH ligands. The availability of distal and proximal alkyne termini locations relative to the adjacent cis ligand leads to two "cis" isomers (R and R near one another) and one "trans" isomer (Fig. 25). Rotation of only one alkyne ligand converts "cis" to "trans" and vice versa, but direct "cis" to "cis" conversion is not possible unless both alkynes rotate simultaneously.

Neutral CpM(RC≡CR)$_2$X complexes which have been studied exhibit $\Delta G^\ddagger$ values between 9 and 15 kcal/mol. Trends are difficult to rationalize as the CpMo(MeC≡CMe)$_2$X series has barriers of 9.3, 14.4, and 12.7 kcal/mol for Cl, Br, and I, respectively. The hexafluorobutyne analogs have barriers of 11.2, 11.5, and 11.1 kcal/mol, respectively (*62*).

TABLE V

NMR AND IR DATA FOR Mo(II) AND W(II) TERMINAL BISALKYNE COMPLEXES

Complex	^{1}H, ppm	^{13}C, ppm ($^1J_{CH}$, Hz)	$\nu(C{\equiv}C)$,[a] cm^{-1}	Ref.
Mo(PhC≡CH)₂(S₂CNEt₂)₂	10.5	183, 177		52,155
Mo(HC≡CH)₂(S₂CNEt₂)₂	10.5, 10.1			52
Mo(PhC≡CH)₂(S₂CNMe₂)₂	10.4	184, 176 (212)	1676	87,155
Mo(HC≡CH)₂(S₂CNC₄H₄)₂	10.3			88
Mo(n-BuC≡CH)₂(S₂CNC₄H₄)₂	9.9		1715	88
Mo(PhC≡CH)₂(S₂CNEt₂)₂	10.4		1704	87
Mo(n-BuC≡CH)₂(S₂CNMe₂)₂	9.7	188, 169 (210)	1717, 1699	87
Mo(n-BuC≡CH)₂(S₂CNEt₂)₂	9.7		1713, 1696	87
[CpMo(HC≡CPrⁱ)₂(CO)][BF₄]	10.4, 10.3, 9.4, 9.3		2040*	72
[CpMo(HC≡CMe)₂(CO)][BF₄]	8.9, 9.8		2075*	73
[CpMo(HC≡CBuᵗ)₂(CO)][BF₄]	9.0		2070*	73
[(η⁵-C₅H₇)Mo(HC≡CH)₂(CO)][BF₄]	9.5		2070*	73
[(η⁵-C₅H₇)Mo(HC≡CMe)₂(CO)][BF₄]	8.7, 9.2		2080*	73
[(η⁵-C₅H₇)Mo(HC≡CBuᵗ)₂(CO)][BF₄]	9.8		2080*	73
[CpMo(HC≡CBuᵗ)₂(NCMe)][BF₄]	11.1, 9.1			73
[CpMo(HC≡CPrⁱ)₂(NCMe)][BF₄]	11.0, 9.0			72
CpMo(CF₃C≡CH)₂Cl	10.1			72
CpW(CF₃C≡CH)(PhC≡CPh)Cl	10.2		1716	93
			1700	
			1682	93

[a] The frequency reported is assigned to $\nu(C{\equiv}C)$ here unless an asterisk indicates a CO absorption frequency.

TABLE VI

NMR AND IR DATA FOR Mo(II) AND W(II) INTERNAL BISALKYNE COMPLEXES

Complex	^{13}C, ppm	$\nu(C\equiv C)$,[a] cm^{-1}	Ref.
[W(MeC$\equiv$CMe)$_2$(CO)Br$_2$]$_2$	180, 163	2098*, 2060*	49
[W(EtC$\equiv$CEt)$_2$(CO)Br$_2$]$_2$		2078*, 2060*	49
[W(PhC$\equiv$CPh)$_2$(CO)Br$_2$]$_2$		2140*, 2004*	49
[W(PhC$\equiv$CMe)$_2$(CO)Br$_2$]$_2$		2144*, 2120*	49
W(MeC$\equiv$CMe)$_2$(CO)(CNBu^t)Br$_2$	161, 157	2092*	49
Mo(PhC$\equiv$CPh)$_2$(S$_2$CNMe$_2$)$_2$		1730	87
Mo(PhC$\equiv$CPh)$_2$(S$_2$CNEt$_2$)$_2$		1737	87
Mo(MeC$\equiv$CPh)$_2$(S$_2$CNEt$_2$)$_2$		1768	87
Mo(MeC$\equiv$CMe)$_2$(S$_2$CNC$_4$H$_4$)$_2$	180, 179		88
Mo(EtC$\equiv$CEt)$_2$(S$_2$CNC$_4$H$_4$)$_2$		1787	88
Mo(EtC$\equiv$CEt)$_2$(S$_2$CNMe$_2$)$_2$	184, 181	1795	87
Mo(EtC$\equiv$CEt)$_2$(S$_2$CNEt$_2$)$_2$		1786	87
Mo(DMAC)$_2$(S$_2$CNMe$_2$)$_2$		1819	87
Mo(DMAC)$_2$(S$_2$CNEt$_2$)$_2$		1787	87
CpMo(CF$_3$C$\equiv$CCF$_3$)$_2$CF$_3$		1775, 1760	98
CpMo(MeO$_2$CC$\equiv$CCO$_2$Me)$_2$CF$_3$		1760	98
CpMo(MeC$\equiv$CMe)(CF$_3$$\equiv$CCF$_3$)CF$_3$		1780	98
CpMo(CF$_3$C$\equiv$CCF$_3$)(PhC$\equiv$CMe)Cl		1788	93
CpMo(CF$_3$C$\equiv$CCF$_3$)(PhC$\equiv$CPh)Cl		1760	93
CpW(CF$_3$C$\equiv$CCF$_3$)(PhC$\equiv$CPh)Cl		1716	93
CpW(CF$_3$C$\equiv$CCF$_3$)$_2$Cl		1778, 1762	93
CpMo(CF$_3$C$\equiv$CCF$_3$)$_2$(SC$_6$F$_5$)		1793, 1665	95
CpW(CF$_3$C$\equiv$CCF$_3$)$_2$(SC$_6$F$_5$)		1773, 1745	95
[CpW(CO)(MeC$\equiv$CBu^n)$_2$][BF$_4$]	162, 159, 145, 142	2045*	201
[CpW(MeC$\equiv$CMe)$_2$(CH$_3$CN)][PF$_6$]	180, 160	1790	75
[CpMo(MeC$\equiv$CMe)$_2$(CH$_3$CN)][BF$_4$]	182, 162		72
[CpMo(MeC$\equiv$CMe)$_2$(CO)][BF$_4$]	165, 146	2040* 1832	72,76
[CpMo(MeC$\equiv$CMe)$_2$(CO)][PF$_6$]	164, 145	2040* 1810	75
[CpW(MeC$\equiv$CMe)$_2$(CO)][PF$_6$]	160, 142	2040* 1800	75
[CpMo(HOCH$_2$C$\equiv$CCH$_2$OH)$_2$(CO)]Cl	171, 150	2073*	94
[CpMo(PhC$\equiv$CMe)$_2$(CO)][BF$_4$]		2040*, 2060*	72
[CpMo(PhC$\equiv$CPh)$_2$(CO)][BF$_4$]		2070* 1739	73,76
[(η^5-C$_9$H$_7$)Mo(MeC$\equiv$CMe)$_2$(CO)][BF$_4$)		2075*	73
[(η^5-C$_9$H$_7$)Mo(PhC$\equiv$CPh)$_2$(CO)][BF$_4$]		2070*	73

[a] The frequency reported is assigned to $\nu(C\equiv C)$ here unless an asterisk indicates a CO absorption frequency.

TABLE VII

ALKYNE ROTATIONAL BARRIERS IN Mo(II) AND W(II) BISALKYNE COMPLEXES

Complex	$\Delta G^{\ddagger}$, kcal/mol	Ref.
$Mo(HC\equiv CH)_2(S_2CNC_4H_4)_2$	13.8	88
$Mo(MeC\equiv CMe)_2(S_2CNC_4H_4)_2$	13.7	88
$CpMo(ClCH_2C\equiv CCH_2Cl)_2Cl$	10.3	94
$CpMo(PhC\equiv CMe)_2Cl$	12.2, 13.4	94
$[CpMo(MeC\equiv CMe)_2(CO)][PF_6]$	19.4	75
$[CpW(MeC\equiv CMe)_2(CO)][PF_6]$	21.0	75
$[CpMo(MeC\equiv CMe)_2(CH_3CN)][PF_6]$	15.9	75
$[CpMo(MeC\equiv CMe)_2(CH_3CN)][BF_4]$	16.1	72
$CpMo(MeC\equiv CMe)_2SMe$	10.8	74
$CpMo(MeC\equiv CMe)_2Cl$	9.3	62
$CpMo(MeC\equiv CMe)_2Br$	14.4	62
$CpMo(MeC\equiv CMe)_2I$	12.7	62
$CpMo(CF_3C\equiv CCF_3)_2Cl$	11.2	62
$CpMo(CF_3C\equiv CCF_3)_2Br$	11.5	62
$CpMo(CF_3C\equiv CCF_3)_2I$	11.1	62
$[CpMo(CO)(HC\equiv CPh)_2][BF_4]$	16.1	162
$[CpMo(CO)(HC\equiv CCO_2Me)_2][BF_4]$	17.7	162
$[CpMo(CO)(HC\equiv CMe)_2][BF_4]$	18.1	162

Dynamic NMR studies indicate that the barrier to alkyne rotation in dithiocarbamate bisalkyne complexes is near 15 kcal/mol (87). For unsymmetrical alkyne ligands, as in $Mo(PhC\equiv CH)_2(S_2CNMe_2)_2$, several isomers are possible with like substituents either adjacent ("cis") or opposite ("trans"). Analysis of the NMR properties follows the logic presented by Faller and Murray for $CpMo(RC\equiv CR)_2Cl$ (94). The C_2 molecular symmetry dictated by the chelates can produce two different "trans" isomers with the two alkyne protons of $PhC\equiv CH$ in each isomer equivalent by C_2 rotation (Fig. 26). Only one unique "cis" isomer is possible, but the two

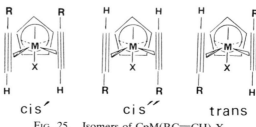

FIG. 25.   Isomers of $CpM(RC\equiv CH)_2X$.

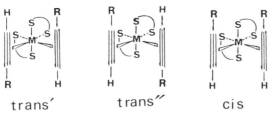

FIG. 26.   Isomers of M(RC≡CH)₂(S₂CNEt₂)₂.

acetylenic protons are NMR distinct in this isomer. Interconversion of these isomers will necessarily accompany exchange of protons by alkyne rotation, and in fact all three isomers are evident in low-temperature ^{1}H NMR spectra of Mo(PhC≡CH)₂(S₂CNMe₂)₂. The observed isomer populations place an upper limit of 0.5 kcal/mol on the ground state energy differences for the three isomers (87).

Pyrrole-N-carbodithioate bisalkyne complexes display two distinct fluxional processes. Rotation around the C—N bond of the S₂C—NC₄H₄ ligand equilibrates both halves of the pseudoaromatic NC₄H₄ ring ($\Delta G^{\ddagger}$ = 10.7 kcal/mol) (88). Alkyne rotation exchanges both ends of the alkyne ligands at somewhat higher temperatures ($\Delta G^{\ddagger}$ = 13.7 kcal/mol for MeC≡CMe, 13.8 kcal/mol for HC≡CH).

No structures of M(RC≡CR)₂L₂X₂ complexes have been published. Carbon-13 signals near 160 ppm which are assigned as alkyne carbons in W(CO)(MeC≡CMe)₂(CNBu^t)Br₂ fit nicely in the N = 3 chemical shift range (48). Dimeric [W(MeC≡CMe)₂(CO)Br₂]₂ complexes exhibit ^{13}C resonances at 180 and 163 ppm (48) which are consistent with two three-electron donor alkynes bound to each tungsten which would saturate each metal without requiring a tungsten-tungsten double bond.

The extent of alkyne $\pi_{\perp}$ donation in Mo(RC≡CR)(SBu^t)₂(CNBu^t)₂ (80) is not easily quantified as discussed in the structural and molecular orbital sections. Proton shifts for HC≡CH and PhC≡CH ligands in these complexes are near 10.4 ppm, above the N = 4 median value and approaching an N = 3 chemical shift. The ^{13}C chemical shifts range from 170 to 185 ppm, also above classic four-electron donor alkyne values, presumably reflecting competition with the two adjacent equatorial thiolates for donation to the two vacant metal $d\pi$ acceptor orbitals. Single bond $^1J_{CH}$ values of 215 and 211 Hz are typical of terminal alkynes bound to molybdenum(II) (133).

Proton and carbon chemical shift values for Cp₂Mo(RC≡CH) complexes serve as benchmarks for N = 2 alkynes in d^4 monomers: δ (^{1}H) 7.68 and 7.05 for HC≡CH and MeC≡CH; δ (^{13}C) 118 (J_{CH} = 200 Hz) and 106, 126 (J_{CH} = 198 Hz) for HC≡CH and MeC≡CH (82).

B. *Infrared Spectra*

Two potentially informative infrared probes in alkyne complexes with ancillary carbonyl ligands are (1) $\nu(C\equiv C)$ absorptions and (2) $\nu(CO)$ absorptions. The first provides direct information about the carbon-carbon multiple bond of the bound alkyne ligand, but it is usually weak and has proven particularly difficult to detect in $M(CO)(RC\equiv CR)(S_2CNEt_2)_2$ complexes (*57*). The second provides only indirect information about the metal-alkyne moiety, but its intensity and unambiguous assignment as $\nu(CO)$ have stimulated an extensive literature data bank to assist interpretation.

The carbonyl stretching frequencies for $M(CO)(RC\equiv CR)(B—B)_2$ complexes range from 1880 to 1980 cm^{-1}, well above values for related seven-coordinate monocarbonyl derivatives [1760 cm^{-1} for $W(CO)(dppe)$-$(S_2CNEt_2)_2$ (*163*), 1755 cm^{-1} for $Mo(CO)(PMePh_2)_2(S_2CNEt_2)_2$ (*164*), and 1740 cm^{-1} for $W(CO)(PMe_3)_2(S_2CNEt_2)_2$ (*165*)]. A more meaningful assessment of alkyne π acidity is provided by comparison with the average $\nu(CO)$ frequency of the parent dicarbonyl complexes, $M(CO)_2(S_2CNEt_2)_2$: 1885 cm^{-1} for M = Mo (1930, 1840) (*166*); 1865 cm^{-1} for M = W (1910, 1820) (*59*). In these six-coordinate d^4 monomers the single-faced π-acid alkyne ligand appears to be more effective at removing $d\pi$ electron density than is a second carbon monoxide ligand.

Carbonyl absorption frequencies are an unusually rich source of comparative data for $M(CO)(RC\equiv CR)L_2X_2$ derivatives since over 50 complexes have been prepared and a $\nu(CO)$ value has been reported for each (Tables II and III). A range of 1900-1980 cm^{-1} encompasses the *trans*-$L_2M(CO)$-$(RC\equiv CR)X_2$ complexes while *cis*-$L_2M(CO)(RC\equiv CR)X_2$ complexes span a higher frequency range (1960-2030 cm^{-1}). This no doubt reflects placement of L [L = CNR, $P(OMe)_3$] trans to CO in the *cis*-$L_2M(CO)$-$(RC\equiv CR)X_2$ complexes, whereas a halide trans to CO in the trans L_2 isomer allows more $d\pi$ density to drift into CO π^* orbitals. The chelating dppe derivatives presumably adopt the same structure as the cis-L_2 complexes, and indeed the $\nu(CO)$ values of 1960-2000 cm^{-1} fall at higher energies than those of the trans-L_2 cases but below the cis-L_2 cases with isonitrile and phosphite ligands.

Dicarbonyl derivatives exhibit the highest $\nu(CO)$ frequencies. Comparison of cis-$M(CO)_2$ and trans-$M(CO)_2$ frequencies is complicated by the absence of the symmetric CO stretch in IR spectra of trans isomers. The single $\nu(CO)$ which is observed for trans isomers lies above the lower energy cis absorption, reflecting π-acid competition between trans-CO ligands. It is important to recall that the diamagnetic d^4 configuration offers no inherent bonding advantage compared to the cis-$(CO)_2$ unit in contrast to the d^6 octahedral case where $d\pi$ orbital utilization favors cis π-acid ligands (see Section VI).

A weak absorption between 1620 and 1720 cm^{-1} has been assigned to the C≡C stretching mode of the bound alkyne for only a dozen M(CO)-(RC≡CR)L$_2$X$_2$ complexes (Table III). This is a drop of roughly 600 cm^{-1} from free alkyne values, consistent with substantial weakening of the triple bond on coordinaation. Assignment of C≡C stretching frequencies is also rare for Cp derivatives. Internal alkynes in the [CpMo(CO)(RC≡CR)L]$^+$ series have been reported to exhibit weak absorptions between 1640 and 1680 cm^{-1} which may be ν(C≡C) (76). For neutral CpW(CO)-(HC≡CH)R complexes C≡C stretches have been assigned to absorptions near 1700 cm^{-1} in several cases (161).

Once again it is the CO absorption in CpM(CO)(RC≡CR)X complexes that provides a marker for comparison of similar molecules and with other classes of d^4 alkyne monomers. Cationic complexes provide the highest ν(CO) frequencies, roughly 1930-1990 cm^{-1} for phosphine derivatives, [CpM(CO)(RC≡CR)(PR$_3$)]$^+$, while {CpMo(CO)(PhC≡CPh)-[P(OPh)$_3$]}$^+$ has a uniquely high CO stretch at 2011 cm^{-1} (76). That a cationic monocarbonyl complex with phosphite and diphenylacetylene ligands in the coordination sphere exhibits a CO absorption above 2000 cm^{-1} is compatible with general trends in backbonding from d^4 metals. Neutral CpMo(CO)(RC≡CR)(SR′) complexes have CO stretching frequencies between 1920 and 1980 cm^{-1} while CpW(CO)(RC≡CR)R′ complexes have CO absorptions a little lower in energy, 1910-1960 cm^{-1}. The methoxide ligand pushes the CO absorption to unusually low frequencies in CpM(CO)(RC≡CR)(OMe): 1890 cm^{-1} for Mo and 1880 cm^{-1} for W (75). Alkoxide ligands serve as π donors and are known to influence CO frequencies dramatically (167).

Comparison of carbonyl ligand frequencies for various categories of monocarbonyl alkyne complexes in the d^4 manifold of molybdenum and tungsten helps to assess ancillary ligand electron donor properties. Generally accepted trends in $d\pi$ backbonding are evident in ν(CO) values for closely related complexes: Cl > Br > I; CNR > P(OMe)$_3$ > PPh$_3$ > PEt$_3$ > py; Mo > W. Geometry is important in determining CO frequencies since cis- and trans-W(CO)(MeC≡CMe)[P(OMe$_3$)]$_2$Br$_2$ complexes differ by roughly 30 cm^{-1} in ν(CO). The data in Table III contain an extensive subset of Mo(II) carbonyl complexes with a 2-butyne ligand. The popular Mo(CO)(MeC≡CMe) moiety requires four additional ligands to complete an octahedral coordination sphere. Among cationic [CpMo(CO)-(MeC≡CMe)L]$^+$ complexes, L = PEt$_3$ pushes the CO frequency below the PPh$_3$ analog; cationic η^5-indenyl complexes exhibit CO stretching energies below their η^5-C$_5$H$_5$ counterparts. For neutral Cp complexes the lone CO frequency decreases substantially as the anionic ligand varies (CF_3 > SR > OMe). It is evident from comparison to ν(CO) for M(CO)-(RC≡CR)(S_2CNEt_2)$_2$ and M(CO)(RC≡CR)L_2X_2 that two dithiocarbamates are more electron rich than any L_2X_2 combinations reported to date.

Infrared absorptions at 1565, 1672, and 1755 cm^{-1} have been assigned to bound HC≡CH, PhC≡CH, and PhC≡CPh in Mo(RC≡CR)(SBu^t)$_2$(CNBu^t)$_2$ complexes (133). Notice that the free ligand ν(C≡C) values also differ substantially: 1974, 2105, and 2232 for HC≡CH, PhC≡CH, and PhC≡CPh, respectively (168). More useful is the comparison with Cp$_2$Mo(RC≡CR) ν(C≡C) frequencies: 1613 for HC≡CH (82), 1771 for PhC≡CPh, 1830 for MeC≡CMe (85), and 1778 cm^{-1} for Cp$_2'$Mo (CF_3C≡CCF_3) (128). Other parent acetylene complexes for which infrared assignments have been made include two-electron donor alkynes in CpW(CO)(HC≡CH)(NO) at 1723 cm^{-1} (160) and [CpMo(HC≡CH)-(η^6-C_6H_6)]$^+$ at 1689 cm^{-1} (66). Among HC≡CH ligands with N = 4, CpW(CO)(HC≡CH)[C(O)Me] (1702 cm^{-1}) and CpW(PMe$_3$)(HC≡CH)[C(O)Me] (1710 cm^{-1}) have observable infrared absorptions attributed to the C≡C stretching mode (67).

The ν(C≡C) stretching frequency has been more consistently observed in bisalkyne derivatives than in related monocarbonylalkyne derivatives. A range of 1675-1800 cm^{-1} is observed for Mo(RC≡CR)$_2$(S_2CNEt_2)$_2$ complexes [Table V (87)]. Diester alkyne ligands exhibit absorptions which can be assigned to a mode dominated by C≡C stretching in most complexes. Perturbation of the DMAC C≡C bond is reflected in frequencies tabulated as a function of formal electron donation from $\pi_\perp$. A four-electron donor DMAC (N = 4) is lowest at 1730 cm^{-1}, $N = 3\frac{1}{3}$ is next at 1760 cm^{-1}, followed by N = 3 at 1790-1820 cm^{-1} (87), and, finally, the two-electron donor DMAC ligands in W(CO)$_2$(DMAC)$_2$(dppe) absorb at 1895 cm^{-1} (30). Competition between oxo π donation and DMAC $\pi_\perp$ places MoO(DMAC)(S_2CNMe_2)$_2$ in between the three-electron and two-electron examples with ν(C≡C) observed between 1850 and 1870 cm^{-1} (140).

C. Electronic Spectra

Electronic transitions assigned to the $d\pi$ manifold produce the brilliant colors which characterize M(CO)(RC≡CR)(B—B)$_2$ complexes. A study of 20 dithiocarbamate derivatives reported transition energies ranging from 13,500 to 17,500 cm^{-1} with ϵ between 70 and 120 M^{-1} cm^{-1} for molybdenum while tungsten absorption maxima ranged from 15,800 to

18,400 cm^{-1} with ϵ between 100 and 250 M^{-1} cm^{-1} (53). Interpretation was based on a HOMO $\rightarrow$ LUMO assignment, and observed energy maxima were consistent with increasing electron donation from alkyne $\pi_\perp$ pushing the vacant d_{xy} orbital to higher energy as the M—L π antibonding counterpart of the filled $\pi_\perp$ alkyne orbital. The HOMO is independent of alkyne to first order as described in Section VI. Alkyne π acidity variations are effectively buried in the HOMO2, a three-center scheme stabilizing d_{yz}. The trend in λ_{max} with alkyne substituents suggests that this electronic transition is responsive to substituent factors as mediated by $\pi_\perp$ with σ effects relegated to a minor role.

The mixed olefin alkyne tungsten bisdithiocarbamate complexes (60) exhibit visible absorptions between 455 and 465 nm with ϵ values between 300 and 600 M^{-1} cm^{-1}. These transition energies are much higher than those observed in comparable tungsten carbonyl analogs (Table VIII). Two separate λ_{max} values were reported for $Mo(MA)(PhC_2H)(S_2CNEt_2)_2$, 600 nm ($\epsilon = 360$ M^{-1} cm^{-1}) and 810 nm ($\epsilon = 240$ M^{-1} cm^{-1}), while Mo-$(DCNE)(H\equiv CH)(S_2CNEt_2)_2$ exhibited only a single absorption at $\lambda_{max} = 550$ nm with $\epsilon = 300$ M^{-1} cm^{-1}. The lower energy of the molybdenum transitions relative to tungsten is common and has been attributed to the stronger M—L π bonding characteristic of tungsten. All of the observed visible absorptions were assigned to the d^4 $d\pi$ manifold with the one case exhibiting two bands tentatively assigned as HOMO1 $\rightarrow$ LUMO and HOMO2 $\rightarrow$ LUMO transitions.

Bisalkyne molybdenum dithiocarbamate complexes exhibit substantially higher energy electronic transitions than their monoalkyne carbonyl cousins (Table VIII). Extinction coefficients near 10^3 M^{-1} cm^{-1} are an order of magnitude larger than in $Mo(CO)(RC\equiv CR)(S_2CNEt_2)_2$ complexes. Both electrochemistry and electronic spectra are consistent with a higher lying $d\pi$ LUMO in $Mo(RC\equiv CR)_2(S_2CNEt_2)_2$ complexes, probably a result of two $\pi_\perp$ orbitals rather than just one pushing up the vacant $d\pi$ orbital. Furthermore, the HOMO is stabilized by overlap with two alkyne $\pi_\parallel^*$ orbitals rather than one CO π^* orbital. The two filled $d\pi$ orbitals in d^4 $M(RC\equiv CR)_2L_4$ need not be degenerate by symmetry as might be anticipated. To a first approximation, only at an alkyne-metal-alkyne angle of 90° would equal overlap of both $\pi_\parallel^*$ combinations with both filled $d\pi$ orbital mates produce an accidental HOMO1–HOMO2 degeneracy. The magnitude of the HOMO–LUMO gap indicated by electronic transition energies underscores the dual π acid/π base character of alkyne ligands.

Vivid green $Mo(CO)(RC\equiv CR)L_2X_2$ complexes are characterized by λ_{max} values of 14,000–17,000 cm^{-1} with $\epsilon \approx 10^2$ M^{-1} cm^{-1}; tungsten analogs absorb at higher energies (17,000–18,000 cm^{-1}) (46). Qualitative

TABLE VIII

Electronic Absorption Data for Mo(II) and W(II) Alkyne Complexes

Complex	λ_{max}, nm	E_{hv}, 10^3 cm^{-1}	ε, M^{-1} cm^{-1}	Ref.
$Mo(CO)(HC{\equiv}CPh)(PEt_3)_2Br_2$	667	15.0	260	46
$Mo(CO)(MeC{\equiv}CMe)(py)_2Cl_2$	690	14.5	130	46
$Mo(CO)(MeC{\equiv}CMe)(PEt_3)_2Cl_2$	602	16.6	40	46
$Mo(CO)(MeC{\equiv}CMe)(PEt_3)_2Br_2$	613	16.3	50	46
$Mo(CO)(PhC{\equiv}CPh)(PEt_3)_2Br_2$	658	15.2	170	46
$Mo(CO)(MeC{\equiv}CMe)(PPh_3)_2Cl_2$	629	15.9	50	46
$Mo(CO)(MeC{\equiv}CMe)(dppe)Cl_2$	621	16.1		46
$Mo(CO)(EtC{\equiv}CEt)(dppe)Cl_2$	625	16.0		46
$Mo(CO)(HC{\equiv}CBu'')(dppe)Br_2$	654	15.3		46
$Mo(CO)(EtC{\equiv}CEt)(dppe)Br_2$	633	15.8		46
$Mo(CO)(HC{\equiv}CPh)(dppe)Br_2$	680	14.7	100	46
$W(CO)(PhC{\equiv}CPh)(PEt_3)_2Cl_2$	585	17.1	290	46
$W(CO)(HC{\equiv}CPh)(PEt_3)_2Cl_2$	585	17.1		46
$W(CO)(MeC{\equiv}CMe)(PPh_3)_2Cl_2$	556	18.0		46
$Mo(CO)(DMAC)(S_2CNEt_2)_2$	720	13.9	105	53
$Mo(CO)(Me_3SiC{\equiv}CSiMe_3)(S_2CNEt_2)_2$	710	14.1	105	53
$Mo(CO)(PhC{\equiv}CPh)(S_2CNMe_2)_2$	705		140	53
$Mo(CO)(PhC{\equiv}CPh)(S_2CNEt_2)_2$	700	14.3	120	53
$Mo(CO)(HC{\equiv}CPh)(S_2CNMe_2)_2$	700		230	53
$Mo(CO)(HC{\equiv}CPh)(S_2CNEt_2)_2$	695	14.4	110	53
$Mo(CO)(HC{\equiv}CH)(S_2CNEt_2)_2$	698	14.3	110	53
$Mo(CO)(HC{\equiv}CBu'')(S_2CNMe_2)_2$	688	14.5	83	53
$Mo(CO)(HC{\equiv}CCH_2Cl)(S_2CNEt_2)_2$	685	14.6	70	53
$Mo(CO)(PhC{\equiv}CMe)(S_2CNMe)_2$	677		230	53
$Mo(CO)(EtC{\equiv}CEt)(S_2CNEt_2)_2$	660	15.2	120	53
$Mo(CO)(EtC{\equiv}CEt)(S_2CNMe_2)_2$	673		85	53
$Mo(CO)(MeC{\equiv}CMe)(S_2CNEt_2)_2$	660	15.2	110	53
$Mo(CO)(HC{\equiv}COEt)(S_2CNEt_2)_2$	645	15.5		53
$Mo(CO)(MeC{\equiv}CNEt_2)(S_2CNMe_2)_2$	580	17.2	90	53
$W(CO)(HC{\equiv}CH)(S_2CNEt_2)_2$	612	16.3	120	53
$W(CO)(DMAC)(S_2CNEt_2)_2$	630	15.9	100	53
$W(CO)(Ph_2PC{\equiv}CPPh_2)(S_2CNEt_2)_2$	630	15.9	150	53
$W(CO)(PhC{\equiv}CPh)(S_2CNEt_2)_2$	620	16.1	250	53
$W(CO)(HC{\equiv}CPh)(S_2CNEt_2)_2$	610	16.4	180	53
$W(CO)(HOCH_2C{\equiv}CH_2OH)(S_2CNEt_2)_2$	600	16.7		53
$W(CO)(EtC{\equiv}CEt)(S_2CNEt_2)_2$	583	17.2	100	53
$W(CO)(MeC{\equiv}CMe)(S_2CNEt_2)_2$	575	17.4	120	53
$W(CO)(HC{\equiv}COEt)(S_2CNEt_2)_2$	575	17.4	175	53
$W(CO)(MeC{\equiv}CNEt_2)(S_2CNEt_2)_2$	545	18.4	150	53
$[W(Bu'HNC{\equiv}CNHBu')(CNBu')_4I]I$	489	20.4	820	117
$[Mo(Bu'HNC{\equiv}CNHBu')(CNBu')_4I]I$	510	19.6	580	117

(*continued*)

TABLE VIII (*continued*)

Complex	λ_{max}, nm	$E_{h\nu}$, 10^3 cm^{-1}	ε, M^{-1} cm^{-1}	Ref.
[Mo(Bu^{t}HNC≡CNHBu^t) (CNBu^t)$_4$Cl]Cl	515	19.4	540	117
[Mo(Bu^{t}HNC≡CNHBu^t) (CNBu^t)$_4$(CN)][PF$_6$]	545	18.3	630	117
[Mo(Bu^{t}HNC≡CNHBu^t) (CNBu^t)$_5$][BPh$_4$]$_2$	564	17.7	680	117
{Mo[(C$_6$H$_{11}$)HNC≡CNH(C$_6$H$_{11}$)] [CN(C$_6$H$_{11}$)]$_4$I}I	510	19.6	600	117
[Mo(MeHNC≡CNHMe) (CNMe)$_3$(bpy)][PF$_6$]$_2$	600 529	16.7 18.9	2300	118
[Mo(EtHNC≡CNHEt)(CNEt)$_3$(bpy)][PF$_6$]$_2$	602 532	16.6 18.8	4000	118
{[HB(pz)$_3$W(CO)(MeOC≡CSMe) (PEt$_3$)}[FSO$_3$]	560	17.9		109
[CpW(CO)(MeOC≡CC$_6$H$_4$Me) (PMe$_3$)][BF$_4$]	465	21.5	1640	107
[CpW(MeOC≡CC$_6$H$_4$Me)(PMe$_3$)$_2$][BF$_4$]	552	17.8		107
Mo(HC≡CH)(SBu^t)$_2$(CNBu^t)$_2$	558	17.9	140	133

generalizations concerning color and the extent of $\pi_\perp$ donation are compatible with spectral studies to date (*48*). Monoalkyne derivatives of the type M(CO)(RC≡CR)L$_2$X$_2$ with $N = 4$ are deep purple, blue, or green. Bisalkyne products, both dimeric [WBr$_2$(CO)(RC≡CR)$_2$]$_2$ and monomeric WBr$_2$(RC≡CR)$_2$(CO)(CNBu^t), are yellow and formally have $N = 3$. The same color patterns hold for cyclopentadienyl and dithiocarbamate derivatives involving one versus two alkyne ligands.

Conversion of [CpW(CO)(PMe$_3$)(RC≡COR')]$^+$ to [CpW(PMe$_3$)$_2$-(RC≡COR')]$^+$ by phosphine replacement of CO is accompanied by a dramatic color change from yellow to violet (*107*). Extended Huckel molecular orbital calculations rationalize the red shift in λ_{max} (465 nm in the cationic CO reagent, 562 nm in the PMe$_3$ product) as largely attributable to the $d\pi$ splitting pattern. Both filled $d\pi$ levels are stabilized by the carbonyl ligand of the reagent, but in the product only the alkyne with its $\pi_\parallel^*$ orbital is a good π-acid ligand, and the HOMO of the d^4 bisphosphine complex is considerably higher in energy than in the carbonyl complex.

Electronic spectra of [M(CNR)$_4$(RNHC≡CNHR)X]$^+$ cations in methanol exhibit two major visible absorptions, with the higher energy absorption ($\lambda_{max} \approx 340$–360 nm) more intense ($\epsilon \approx 15,000$ M^{-1} cm^{-1}) than the

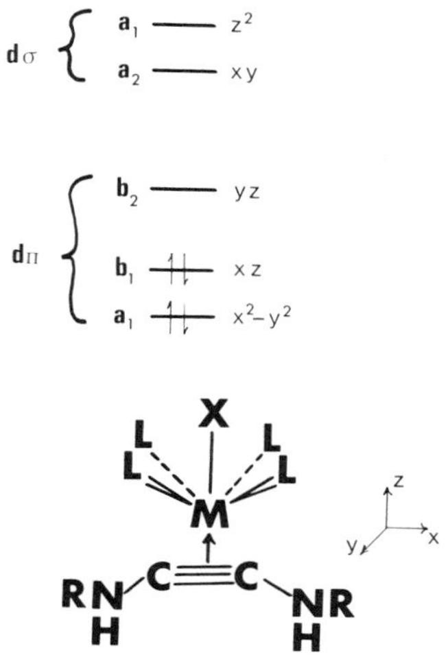

FIG. 27.   [L$_4$M(RNHC≡CNHR)X]$^+$ d orbital energies and symmetries.

lower energy one ($\lambda_{max} \approx 490–560$ nm with $\epsilon \approx 600$ M^{-1} cm^{-1}) (*117*). The absorption near 350 nm has tentatively been assigned as ligand-to-metal charge transfer (LMCT) and is not included in Table VIII.

The lowest energy absorption for each of six d^4 monomers containing a diaminoalkyne ligand has been assigned as a transition into the $d\pi$ LUMO of the complexes. In the coordinate system employed by Lippard and co-workers (Fig. 27) the b_2 symmetry of the d_{yz} LUMO restricts excitations to originating orbitals of a_1, a_2, and b_2 symmetry. The relatively low extinction coefficients favor assignment as a $d–d$ transition, and $d_{x^2-y^2}$ (a_1) to d_{yz} (b_2) is a rational assignment within the $d\pi$ manifold. Molybdenum-to-bypyridine MLCT bands near 530 and 600 nm with extinction coefficients of the order of 3000 M^{-1} cm^{-1} obscure $d–d$ transitions which may be present in [Mo(RNHC≡CNHR)(CNR)$_3$(bpy)]$^{2+}$ (*118*).

Five-coordinate Mo(RC≡CR)(SBu$'$)$_2$(CNBu$'$)$_2$ complexes exhibit strong charge transfer electronic transitions at high energy (*133*). Only the lowest energy absorption in the visible spectra of these complexes is included in Table VIII, with λ_{max} near 550 nm and $\epsilon \approx 10^2$ M^{-1} cm^{-1}. Although lucid molecular orbital descriptions of these molecules have been presented, no assignments of observed transitions have been made. The five-coordinate Mo(PhC≡CPh)(TTP) porphyrin complex has λ_{max} values of 624, 544, and 426 nm with ϵ values of 2300, 9500, and 160,000 M^{-1} cm^{-1}, respectively (*81*).

D. *Electrochemistry*

Cyclic voltammetry of a series of $M(CO)(RC{\equiv}CR)(S_2CNEt_2)_2$ complexes revealed a reversible reductive wave which was presumed to involve electron addition to the LUMO. Plots of $E_R°$ versus $E_{h\nu}$ were presented for Mo and W which suggested that indeed the LUMO energy variation, as reflected in thermodynamic potentials, accounts for most of the energy changes observed in electronic spectra (*53*).

Reversible reductions were observed for eight olefin alkyne bisdithiocarbamate derivatives of molybdenum and tungsten with potentials ranging from -0.88 to -2.02 V versus SSCE (*60*). The reduction potential was believed to correlate with the energy of the LUMO as determined by alkyne $\pi_\perp$ donation. Dialkylalkynes produced more negative reduction potentials than either $PhC{\equiv}CH$ or $HC{\equiv}CH$, and it was concluded that they were superior π donors to the metal center. Irreversible oxidations in these complexes were characterized by half-peak potentials cited as reflecting π-acid ligand properties. In this context it is noteworthy that carbon monoxide in the position cis to the alkyne produced a complex more susceptible to oxidation than MA, DCNE, or $PhC{\equiv}CH$. The TCNE complex was substantially more difficult to oxidize than any of the others. By this criterion TCNE is the strongest π-acid ligand among the set while CO is the poorest, with MA, DCNE, and $PhC{\equiv}CH$ then constituting the middle group in terms of π acidity (*60*).

Bisalkyne bisdithiocarbamate derivatives are both harder to reduce and harder to oxidize than their carbonyl analogs (*53*). Factors discussed in the Section VI and used to rationalize visible absorption spectra are also applicable here. The potentials required for oxidation and reduction reflect the strength of the alkyne–metal $d\pi$ interactions. Their properties, as well as chemical behavior, no doubt reflect the complementary nature of the complete set of σ and π metal–ligand bonds in these happy d^4 $M(RC{\equiv}CR)_2(S_2CNEt_2)_2$ complexes.

Limited electrochemical data for $Mo(CO)(RC{\equiv}CR)L_2X_2$ complexes indicate a reversible reduction at -1.18 V versus SSCE for four 2-butyne derivatives while the one phenylacetylene complex studied exhibited a reversible reduction at -1.00 V (Table IX). These results are consistent with the model developed for $Mo(CO)(RC{\equiv}CR)(S_2CNEt_2)_2$ in that the more electron-rich dialkylalkyne would be expected to push the LUMO to higher energy than $PhC{\equiv}CH$. These same complexes were characterized by an irreversible oxidation around $+0.9$ V (*46*). A preliminary report that $[CpMo[P(OMe)_3]_2(MeC{\equiv}CMe)]^+$ undergoes a reversible one-electron reduction at -1.04 V versus SSCE has been used to support the possibility of odd-electron species as reactive intermediates in this system (*74*).

TABLE IX

ELECTROCHEMICAL DATA FOR Mo(II) AND W(II) ALKYNE COMPLEXES[a]

Complex	$E_{(red)}$	$E_{(ox)}$	Ref.
Mo(CO)(HC≡CPh)(PEt$_3$)$_2$Br$_2$	−1.00	0.87*	46
Mo(CO)(MeC≡CMe)(py)$_2$Cl$_2$	−1.18	0.90	46
Mo(CO)(MeC≡CMe)(PEt$_3$)$_2$Cl$_2$	−1.18	0.87	46
Mo(CO)(MeC≡CMe)(PEt$_3$)$_2$Cl$_2$	−1.18	0.87	46
Mo(CO)(MeC≡CMe)(PPh$_3$)$_2$Cl$_2$	−1.19	0.96	46
Mo(CO)(Me$_3$SiC≡CSiMe$_3$)(S$_2$CNEt$_2$)$_2$	−1.50	0.44*	53
Mo(CO)(PhC≡CPh)(S$_2$CNEt$_2$)$_2$	−1.39	0.38*	53
Mo(CO)(HC≡CPh)(S$_2$CNEt$_2$)$_2$	−1.42	0.39*	53
Mo(CO)(HC≡CH)(S$_2$CNEt$_2$)$_2$	−1.46	0.38*	53
Mo(CO)(HC≡CCH$_2$Cl)(S$_2$CNEt$_2$)$_2$	−1.57	0.42*	53
Mo(CO)(EtC≡CEt)(S$_2$CNEt$_2$)$_2$	−1.70	0.35*	53
Mo(CO)(MeC≡CMe)(S$_2$CNEt$_2$)$_2$	−1.64	0.36*	53
W(CO)(DMAC)(S$_2$CNEt$_2$)$_2$	−1.43	0.60*	53
W(CO)(Ph$_2$PC≡CPPh$_2$)(S$_2$CNEt$_2$)$_2$	−1.59	0.35*	53
W(CO)(PhC≡CPh)(S$_2$CNEt$_2$)$_2$	−1.59	0.35*	53
W(CO)(PhC≡CH)(S$_2$CNEt$_2$)$_2$	−1.65	0.36*	53
W(CO)(MeC≡CMe)(S$_2$CNEt$_2$)$_2$	−1.83	0.31*	53
W(CO)(HC≡COEt)(S$_2$CNEt$_2$)$_2$	−1.87*	0.38*	53
W(CO)(MeC≡CNEt$_2$)$_2$(S$_2$CNEt$_2$)$_2$	−2.14*	0.24	53
Mo(DMAC)$_2$(S$_2$CNEt$_2$)$_2$	−0.88	1.07*	53
Mo(PhC≡CPh)$_2$(S$_2$CNEt$_2$)$_2$	−1.53	0.99*	53
Mo(PhC≡CMe)$_2$(S$_2$CNEt$_2$)$_2$	−1.74	0.83*	53
Mo(PhC≡CH)$_2$(S$_2$CNEt$_2$)$_2$	−1.59	0.85*	53
Mo(HC≡CBu^n)$_2$(S$_2$CNEt$_2$)$_2$	−1.92*	0.90*	53
Mo(EtC≡CEt)$_2$(S$_2$CNEt$_2$)$_2$	−2.11*	0.84*	53
[Mo(Bu^{t}HNC≡CNHBu^t)(CNBu^t)$_4$I]$^+$	−1.93*	0.59	119
[Mo(Bu^{t}HNC≡CNHBu^t)(CNBu^t)$_4$(CN)]$^+$		0.61	119
[Mo(MeHNC≡CNHMe)(CNMe)$_3$(bpy)][PF$_6$]$_2$	−1.36*	0.53	118
[Mo(EtHNC≡CNHEt)(CNEt)$_3$(bpy)][PF$_6$]$_2$	−1.37*	0.59	118
[Mo(Pr^{i}HNC≡CNHPr^i)(CNPr^i)$_3$(bpy)][PF$_6$]$_2$	−1.43*	0.63	118
[Mo(Bu^{t}HNC≡CNHBu^t)(CNBu^t)$_3$(bpy)][PF$_6$]$_2$	−1.30*	0.70	118
{MO[(C$_6$H$_{11}$)HNC≡CNH(C$_6$H$_{11}$)][CN(C$_6$H$_{11}$)]$_3$(bpy)}[PF$_6$]$_2$	−1.36*	0.65	118
[Mo(PhCH$_2$HNC≡NHCH$_2$Ph)(CNCH$_2$Ph)$_3$(bpy)][PF$_6$]$_2$	−1.29*	0.73	118
{[HB(pz)$_3$]W(CO)(MeOC≡CSMe)(PEt$_3$)}[FSO$_3$]	−1.16	1.00	109

[a] Electrochemically irreversible electron transfer redox processes are indicated with an asterisk.

Electrochemical behavior of both reagents and products of reductive isonitrile coupling reactions have been studied (119). The diaminoalkyne products undergo reversible one-electron oxidations, generally being 100–300 mV more easily oxidized than their isonitrile precursor. This suggests that the (dialkyldiamino)acetylene ligand removes less electron density than two isocyanide ligands. Alternatively, one can consider the alkyne ligand as providing more electron density to the Mo(II) center than two terminal isonitrile ligands. The electrochemical results indicate that reductive coupling is correlated with ease of oxidation of the $[Mo(CNR)_6X]^+$ reagents. This is perhaps surprising given the net reductive coupling reaction, but of course the two-electron reduction is accompanied by addition of two protons. Irreversible reduction of $[Mo(CNR)_6X]^+$ cationic reagents is sometimes observed near the solvent limit. No indication of electrochemically induced reductive coupling has been observed.

VIII

REACTIONS OF ALKYNE LIGANDS IN d^4 MONOMERS

A. Formation of η^2-Vinyl Ligands

Formation of η^2-vinyl ligands by nucleophilic attack at an alkyne carbon is a particularly noteworthy class of ligand based reactions for cyclopentadienyl alkyne monomers of Mo(II) and W(II). This well-defined class of complexes containing vinyl ligands with both carbons bound to the metal has been firmly established during the 1980s. Isolation and characterization of η^2-vinyl complexes, also called metallocyclopropenes, has important mechanistic implications since stabilization of unsaturated intermediates with vinyl ligands is clearly a viable process. Both simple vinyl groups and heteroatom-substituted vinyl ligands have been prepared; the research groups of M. Green and J. Davidson have dominated work in these areas, respectively. Green has capitalized on the positive charge of cationic monoalkyne molybdenum(II) complexes to effect addition of H^-, R^-, or Ar^- to alkynes. Davidson has utilized electron-poor hexafluorobutyne ligands to promote nucleophilic addition at carbon with phosphine, isonitrile, and thiolate reagents. We first consider formation of η^2-vinyl ligands of the type $RC\!\!=\!\!CR_2$ with R equal to hydrogen, alkyl, or aryl. A comprehensive article describing η^2-vinyl ligands accessible from $[CpMo-[P(OMe)_3]_2(RC\!\!\equiv\!\!CR)]^+$ cations was published in 1985 (169). Earlier nucleophilic addition reactions involving these reagents produced η^3-allyl

(*170*) and carbyne (*171*) ligands, and plausible mechanisms based on initial η^2-vinyl formation will be discussed after η^2-vinyl ligand chemistry is set forth. Reactions which convert alkyne complexes to compounds with multiple molybdenum–carbon bonds have been summarized recently (*10*).

Hydride addition to $[CpMoL_2(PhC\equiv CPh)]^+$ [L = P(OMe)$_3$] at $-78°C$ in THF with either Na[BH$_4$] or, preferably for solubility purposes, K[HBBu$_3^s$] as the hydride transfer reagent produces CpMoL$_2$-(η^2-CPhCHPh) (*172*). An analogous reaction occurs with Me$_2$SiC$\equiv$CH as the alkyne ligand [Eq. (50)] (*173*). These colorful reactions convert the

$$[CpL_2Mo(RC\equiv CR)]^+ + [HBBu_3^s]^- \rightarrow CpL_2Mo(\eta^2\text{-CRCHR}) \qquad (50)$$

deep purple alkyne reagents to vivid green and blue η^2-vinyl products, respectively. Both the η^2-CPhCHPh and η^2-C(SiMe$_3$)CH$_2$ complexes have been structurally characterized (Fig. 28). Other η^2-vinyl complexes in this family which have been structurally characterized include CpL$_2$Mo-(η^2-CBu^{t}CHPh), formed from $[CpL_2Mo(Bu^tC\equiv CH)]^+$ and phenyllithium, and CpL$_2$Mo(η^2-CMeCPh$_2$, generated from PhLi and $[CpL_2Mo-(MeC\equiv CPh)]^+$ [L = P(OMe)$_3$]. This set of four structures (see Table X) has been analyzed in detail (*169*).

Salient structural features of CpL$_2$Mo(η^2-CRCR$_2$) complexes include (1) the orientation of the two C$_\beta$ substituents roughly orthogonal to the MC$_\alpha$C$_\beta$ plane (this is opposite to the location of C$_\beta$ substituents of η^1-vinyl ligands which lie in the MC$_\alpha$C$_\beta$ plane), (2) short M—C$_\alpha$ distances (1.94–1.96 Å) appropriate for a molybdenum–carbon double bond (*174*),

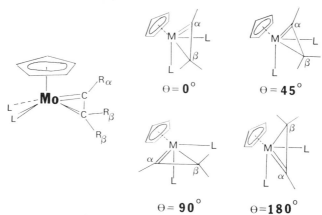

FIG. 28. CpML$_2$(η^2-CRCR$_2$) orientations defining θ, the dihedral C$_\beta$—C$_\alpha$—M—P$_1$ angle.

(3) M—C_β distances (2.25–2.30 Å) in the single bond range (174), and (4) C_α—C_β bond lengths (1.43–1.46 Å) indicative of some double bond character.

Unambiguous identification of η^2-vinyl ligands is often possible based on ^{13}C-NMR spectra; other less definitive spectroscopic probes may not reveal the presence of a metallocyclopropene. The carbenoid character of C_α is reflected in low-field chemical shifts of 230–290 ppm, whereas the four-coordinate C_β signal appears at much higher field, typically near 20–30 ppm (169). For comparison consider the chemical shifts of 151 and 128 ppm for the C_α and C_β carbons of the η^1-vinyl ligand in CpL$_3$Mo-(η^1-CHCHBu^t) [L = P(OMe)$_3$] (170).

Variable ground state η^2-vinyl orientations have been observed in solid state structures, and the rotational properties of these ligands are complex. Results of both experimental (169) and theoretical (175) studies are merging to define the qualitative rotational energy profile characterizing η^2-vinyl ligands. For CpL$_2$Mo(η^2-RCCR$_2$) [L = P(OMe)$_3$] complexes with identical substituents on C_β only the metal is chiral. If C_α and C_β do not lie in the mirror plane present for the CpL$_2$M fragment then the two P(OMe)$_3$ groups will be inequivalent. Rotation of the η^2-vinyl ligand or, more simply, a windshield wiper motion that at some point places both C_α and C_β in the mirror plane will equilibrate both phosphites. Indeed, only a single ^{31}P signal is observed for CpL$_2$Mo(η^2-R'CCR$_2$) complexes with R = Ph, R' = Me or R = H, R' = SiMe$_3$, so a fluxional process is occurring on the NMR time scale (assuming the solid state structure is indicative of the solution ground state η^2-vinyl orientation) (169). Formation of η^2-vinyl ligands with different C_β substituents creates diastereomer possibilities due to chirality at carbon as well as a chiral metal. The result is NMR-inequivalent phosphorus nuclei for a given diastereomer regardless of η^2-vinyl rotation.

Application of extended Huckel molecular orbital methods (169) to an η^2-vinyl model complex, CpMo[P(OH)$_3$]$_2$(η^2-CHCH$_2$), suggested only a small barrier to oscillation of the η^2-vinyl ligand (<5 kcal/mol), which would suffice to equilibrate the two phosphite ligands (basically either $\theta = 0°$ through 45 to 90° or $\theta = 180°$ through 225 to 270° in Fig. 28). A dihedral θ angle of 90° for C_α—C_β—Mo—P$_1$ creates an isomer identical to the $\theta = 0°$ case with P$_2$ now replacing P$_1$ as coplanar with the vinyl fragment, and it is these two equivalent isomers which are interchangeable with a low barrier. These theoretical results are compatible with the observed NMR equivalence of the two phosphites due to the windshield wiper motion ($\Delta G^\ddagger \leq 5$ kcal/mol) for equivalent C_β substituents.

A much larger barrier, about 20 kcal/mol, is calculated for complete rotation of the η^2-vinyl ligand in the model complex. In other words, a 90°

rotation away from the symmetry plane requires about four times as much energy as rotation through it. Passing from $\theta = 0$ to 270° does not interconvert equienergetic isomers, but rather reverses the location of C_α and C_β relative to the M—P axis parallel to the C—C linkage. These isomers differ in the C_α—C_β—Mo—P_1 dihedral angle by 180°, and a substantial barrier for interconversion is calculated. The isolation of two isomers of $CpL_2Mo[\eta^2\text{-}C(CH_2Ph)CHPh]$ which slowly convert to a single isomer at room temperature can be attributed to the formation of kinetic isomer which then rotates the η^2-vinyl group through the larger barrier to form the thermodynamic isomer. Alternatively, η^2-vinyl conversion to an η^1-vinyl intermediate followed by M—C_α rotation could accomplish the same net η^2-vinyl rotation. Note that Davidson has isolated discrete diastereomers of $CpClMo(CF_3C{\equiv}CCF_3)[\eta^2\text{-}CF_3CC(CF_3)(PEt_3)]$ and structurally characterized them as discussed below (176).

How do η^2-vinyl ligands fit into mechanistic schemes which result in allyl or carbyne ligands when nucleophiles are added to four-electron donor alkyne complexes? Let us first consider allyl formation. The 2-butyne ligand of $[CpL_2Mo(MeC{\equiv}CMe)]^+$ $[L = P(OMe)_3]$ adds hydride to yield (exo, anti) $CpL_2Mo(\eta^3\text{-}H_2CCHCHMe)$, and related allyl products form from other $MeC{\equiv}CR$ alkynes ($R = Et$ or Pr^i) (170). With the unsymmetrical $[CpMo(CO)(PEt_3)(MeC{\equiv}CMe)]^+$ reagent four η^3-allyl isomers are generated by hydride addition. These are believed to contain anti-$MeCHCHCH_2$ ligands with exo, endo orientations possible relative to the ring as well as proximal, distal relative to CO or PEt_3 (see Fig. 29). An X-ray structure of the analogous indenyl product has confirmed an anti, exo allyl geometry.

A reaction mechanism which accounts for allyl formation via initial nucleophilic attack at an alkyne carbon has been presented (10). Rear-

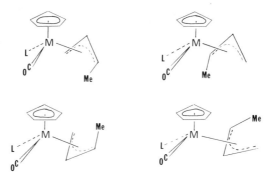

FIG. 29.   Isomers of $CpMo(CO)(PEt_3)(\eta^3\text{-}MeCHCHCH_2)$.

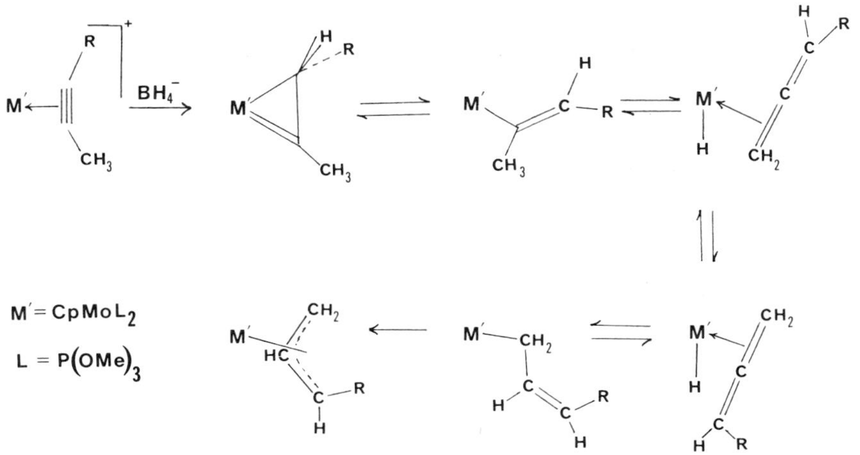

FIG. 30. Possible mechanism for allyl ligand formation from η^2-vinyl precursors with a β-H available.

rangement of an η^2-vinyl product to an unsaturated η^1-vinyl complex would allow β-hydrogen transfer to generate a metal hydride allene intermediate (Fig. 30). Rotation of the allene ligand followed by hydrogen migration to the central carbon of the allene would funnel the η^2-vinyl to the thermodynamically favored η^3-allyl complex. Deuterium labeling studies are consistent with net 1,2-hydrogen shift: NaBD$_4$ and [CpL$_2$Mo-(MeC$_2$Me)]$^+$ form η^3-anti-CH$_2$CHCDMe with deutrium exclusively syn, while the Mo(η^2-DC≡CCHMe$_2$) unit adds H$^-$ to form Mo(η^3-Me$_2$CCHCHCD) with the deuterium anti (170). The feasibility of η^1-vinyl intermediates has been probed by preparing a vinyl ligand with no β-hydrogens in order to shut down allyl formation and trapping the resultant unsaturated η^1-vinyl complex with free phosphite as CpL$_3$Mo-(η^1-CHCHBu^t) [L = P(OMe)$_3$] (170).

The absence of facile alkyl migration from metal to alkyne in related systems is important. Reaction of LiCuPh$_2$ with [CpL$_2$Mo(PhC≡CPh)]$^+$ yields the η^2-vinyl ligand η^2-CPhCPh$_2$. With LiCuMe$_2$ two products form, the analogous η^2-vinyl CpL$_2$Mo(η^2-CPhCMePh) and a molybdenum methyl alkyne product, CpL(Me)Mo(PhC≡CPh). Addition of P(OMe)$_3$ under a variety of conditions failed to convert CpLMeMo(PhC≡CPh) to the η^2-vinyl complex (169). Negative results were also obtained in efforts to convert other CpLR'Mo (RC≡CR) complexes to η^2-vinyl derivatives. Although evidence for initial attack at metal or carbon with hydride reagents is difficult to obtain, it is clear that two distinct pathways are available to alkyl reagents, and the metal–R bond does not readily transfer the alkyl to an alkyne carbon. Of course hydrogen migrations generally have lower activation energies than alkyl migrations, but these results suggest that direct formation of η^2-vinyl ligands by attack at carbon is viable, and therefore initial metal hydride formation need not be invoked as preceding η^2-vinyl formation with hydride reagents.

In addition to formation of allyl ligands from C_2 units carrying substituents with hydrogens β to the metal, metal–carbon triple bonds have resulted from nucleophilic addition to cationic alkyne complexes. An η^1-vinyl ligand can be trapped when $NaBH_4$ and $P(OMe)_3$ are added to $[CpL_2Mo(HC{\equiv}CBu^t)]^+$ at $-78°C$ in THF as mentioned above (171). The E-olefin geometry of the $Mo(\sigma{-}CH{=}CHBu^t)$ unit is compatible with cis insertion of alkyne into a Mo—H bond, but direct attack at the substituted alkyne carbon to form an η^2-vinyl which could ring open to the sterically favored trans-olefin geometry is also possible. On heating, the Mo-$(\sigma{-}CHCHBu^t)$ moiety rearranges to a $Mo{\equiv}CCH_2Bu^t$ carbyne unit with concomitant loss of phosphite (171), the first example of direct entry into carbyne chemistry from a vinyl ligand. A crossover experiment indicated extensive hydrogen–deuterium scrambling, and a C_α deprotonation–C_β portonation mechanism seems likely. Formation of carbyne ligands by protonation of metal vinylidene complexes [Eq. (51)] has now been established in this system ($177,178$) as well as several others ($179,180$). Stepwise conversion of $CpL_2Mo(Me_3SiC{\equiv}CH)]^+$ and hydride through CpL_2Mo-$(\eta^2{-}Me_3SiCCH_2)$ and on to form $CpL_2Mo{\equiv}CCH_2SiMe_3$ [L $=$ $P(OMe)_3$] has also been reported (173). The trimethylsilyl group can be removed with fluoride ion, and the $Mo{\equiv}CCH_3$ unit which results suggests that $[CpL_2Mo{=}C{=}CH_2]^-$ is being formed and protonated. Whether the metal assists in $SiMe_3$ migration is not clear.

$$L_nM{=}C{=}CR_2 + H^+ \rightarrow [L_nM{\equiv}C{-}CHR_2]^+ \tag{51}$$

The chemistry of hydride addition to the blue $[CpL_2Mo(BrC{\equiv}CPh)]^+$ complex [L $=$ $P(OMe)_3$] is unique (178). The first equivalent of $[HBBu_3^s]^-$ induces bromide migration to metal as a vinylidene ligand forms in $CpL_2{-}BrMo{=}C{=}CHPh$. A second $[HBBu_3^s]^-$ yields $CpL_2Mo{\equiv}CCH_2Ph$, possibly by S_N2' attack by H^- at C_β with concomitant displacement of bromide from the metal. Alternatively, the propensity for nucleophiles to add at C_α in vinylidene ligands could form an η^1-vinyl which then rearranges to form the carbyne product. Another route to carbyne complexes from $CpL_2{-}BrMo{=}C{=}CHPh$ is by protonation to form a seven-coordinate cationic carbyne complex, $[CpL_2BrMo{\equiv}CCH_2Ph]^+$ [cf. Eq. (51)], which can be reduced by two electrons with accompanying bromide loss to again yield $CpL_2Mo{\equiv}CCH_2Ph$.

Molecular orbital descriptions of η^2-vinyl ligands have stressed the iso-lobal relationship of η^2-CHCH$_2^-$ and four-electron donor alkyne ligands (*169,175*). The three important bonding interactions in metal–alkyne monomers with $N = 4$ are also available to η^2-vinyl ligands: σ donation (from a filled C—C π orbital), π donation (from a filled C $2p$ orbital orthogonal to the MC$_2$ plane), and π acceptance (involving a vacant C—C π^* orbital). The donor properties of $[\eta^2$-CHCH$_2]^-$ are enhanced relative to HC≡CH by the negative charge, and the localization of the π donor orbital on Cα is evident in the structural and spectral data presented above, i.e., M—C$_\alpha$ is roughly 0.3 Å shorter than M—C$_\beta$ and C$_\alpha$ resonates approximately 200 ppm below C$_\beta$ in ^{13}C NMR spectra.

The orbital analogy between $[\eta^2$-HCCH$_2]^-$ and alkynes with $N = 4$ rationalizes the formation of η^2-vinyl ligands from four-electron donor alkynes, as confirmed in the [CpL$_2$Mo(RC≡CR)]$^+$ system. For alkynes with $N = 2$ addition of a nucleophile should form η^1-vinyl ligands, and indeed Reger has effectively utilized cationic iron complexes of the type [CpL(CO)Fe(RC≡CR)]$^+$, containing an $N = 2$ alkyne ligand in the coordination sphere, to stereoselectively form a variety of η^1-vinyl ligands by addition of anionic nucleophiles (*181,182*).

What happens when nucleophiles are added to bisalkyne complexes with $N = 3$? Addition of LiMe$_2$Cu to [CpMo(CO)(MeC≡CMe)$_2$]$^+$ in THF at $-78°$C produces CpMo(CO)(MeC≡CMe)(σ-CMe=CMe$_2$) with alkyne carbons at 188 and 186 ppm in the ^{13}C-NMR spectum and η^1-vinyl carbons at 153 and 132 ppm (*183*). The formation of an η^1-vinyl here on addition of a nucleophile is reminiscent of CpFe(CO)(PPh$_3$)(σ-MeC=CMe$_2$) formation in the iron system (*182*); in both cases octahedral products form. The alkyne remaining in the Mo(II) coordination sphere shifts from $N = 3$ to $N = 4$ and provides sufficient $\pi_\perp$ electron density to stabilize the η^1-vinyl product. The 2-butyne is an electron-rich alkyne, and, in contrast to hexafluoro-2-butyne reagents discussed below, it is reluctant to function as only a two-electron donor. As a result equilibrium (52) lies to the right. Labeling experiments with LiCu(CD$_3$)$_2$ revealed CD$_3$ equally distributed in the cis and trans positions of the η^1-C(CH$_3$)=C(CH$_3$)(CD$_3$) ligand,

$$\text{CpMo(CO)(MeC≡CMe)(η^2-MeCCMe}_2) \rightleftharpoons$$

$$N = 2$$

$$\text{CpMo(CO)(MeC≡CMe)(η^1-MeCCMe}_2) \qquad (52)$$

$$N = 4$$

possibly exchanged by the intermediacy of an η^2-vinyl ligand which could ring open either way. Direct formation of the new C—C bond seems more likely than initial Mo—CH$_3$ bond formation since alkyl migrations from metal to alkyne have been difficult to achieve in related complexes (*169*).

Hydride addition to [CpMo(CO)(MeC≡CMe)₂][BF₄] under CO ulti-mately yields an alkyne insertion product, Cp(CO)₂Mo[CMeCMeCMe-(CMe=CH)O], a reaction included in Section VII,B.

A second populous class of η^2-vinyl complexes has resulted from addi-tion of phosphines, phosphites, isonitriles, and thiolates to neutral d^4 alkyne complexes. In particular, recent extensive work by the Davidson group with CpM(CF₃C≡CCF₃)₂X (97,176) and other hexafluorobutyne reagents has clarified much of this chemistry. We turn to this realm of η^2-vinyl chemistry after first considering the one class of dithiocarbamate alkyne complexes known to form η^2-vinyl ligands.

Recall that addition of PR₃, P(OR)₃, or RNC to Mo(CO)(RC≡CR)-(S₂CNEt₂)₂ leads to rapid alkyne substitution to form Mo(CO)L₂ (S₂CNEt₂)₂ products (90). On the other hand, bisalkyne dithiocarbamate complexes are reluctant to undergo either addition or substitution reac-tions with added nucleophiles, although Mo(RC≡CR)₂(S₂CNEt₂)₂ slowly undergoes sulfur atom transfer to PR₃ reagents in refluxing aromatic solvents (153). Mixed olefin–alkyne derivatives are unique in that addition of excess phosphine or phosphite to M(MA)(PhC≡CH)(S₂CNR₂)₂ (MA = maleic anhydride) leads to addition of the nucleophile to the terminal alkyne carbon (60). These reactions with neutral phosphorus nucleophiles are analogous to the formal addition of H⁻ or R⁻ to alkynes in the [(π-C₅H₅)MoL₂(RC≡CR)]⁺ system to form η^2-vinyl ligands.

Formation of the η^2-vinyl PhC=CH(PMe₃) fragment moves the termi-nal alkyne proton from the four-electron donor region (13.3 ppm) of the reagent W(MA)(PhC≡CH)(S₂CNEt₂)₂ upfield to 2.97 ppm where it appears as a doublet (29 Hz) due to coupling to phosphorus in the product (60). Carbon-13 NMR of the P(OMe)₃ adduct allowed unambiguous assignment of the vinyl β-carbon, η^2-PhC=CH[P(OMe)₃], at 11.6 ppm with $^1J_{CH} = 158$ Hz and $^1J_{CP} = 164$ Hz. The downfield chemical shift of the α-carbon (222 ppm) is compatible with considerable tungsten–carbon double bond character as represented in the metallocyclopropene reso-nance structure ii.

No reaction was observed with internal alkynes, and carbonyl alkyne analogs undergo rapid alkyne substitution at much lower temperatures as mentioned above. Only these olefin alkyne derivatives are known to pro-

            i   ii

mote nucleophilic attack at an unactivated bound alkyne to form η^2-vinyl products from a neutral complex. The one-to-one $d\pi$ orbital match with the two C_2 based ligands in the olefin–alkyne reagent is reinforced by conversion of the alkyne to the more electron-rich isolobal η^2-vinyl cousin.

Davidson has utilized electron-withdrawing alkyne substituents to promote addition of nucleophiles to neutral cyclopentadienyl alkyne derivatives. With bisalkyne reagents the conversion of one hexafluorobutyne to an η^2-vinyl ligand provides four electrons and limits the remaining alkyne to a formal two-electron donor role. The CF_3 substituents are important in both increasing the electrophilicity of the alkyne carbon and in stabilizing the $N = 2$ role for the intact alkyne retained in the product [Eq. (53)].

$$CpM(CF_3C\equiv CCF_3)_2X + L \rightarrow CpM(CF_3C\equiv CCF_3)[\eta^2\text{-}CF_3CC(CF_3)L]X \quad (53)$$

$$N = 3, N = 3 \qquad\qquad N = 2, \qquad\qquad N = 4$$

Work since 1980 has shown that some 1:1 adducts earlier assumed to be 18-electron complexes formed by ligand addition to 16-electron metal centers in alkyne complexes are actually η^2-vinyl derivatives containing η^2-$CF_3CC(CF_3)L$ units.

A particularly informative study involving the addition of monodentate ligands to $CpMo(CF_3C\equiv CCF_3)_2Cl$ was published in 1983 (176). Two isomers containing η^2-vinyl ligands resulting from PEt_3 addition were isolated and structurally characterized. The η^2-$CF_3CC(CF_3)L$ ligand orientation and the relative configuration of the C_β carbon and the metal differed in the two isomers. The kinetic isomer has a structure reminiscent of the reagent bisalkyne geometry. The two C_2 ligands, alkyne and η^2-vinyl, are nearly parallel to the cis Mo—Cl vector, and the added PEt_3 is located opposite the Cp ligand (Fig. 31). Comparison of spectroscopic data

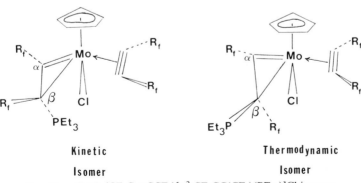

Kinetic   Thermodynamic

Isomer   Isomer

FIG. 31. $CpMo(CF_3C\equiv CCF_3)[\eta^2\text{-}CF_3CC(CF_3)(PEt_3)]Cl$ isomers.

$$\overset{\ddot{M}}{\underset{R}{\diagdown}}C{=}C\overset{R}{\underset{PR_3}{\diagup}} \quad \longleftrightarrow \quad \overset{M}{\underset{R}{\diagdown}}C{-}C\overset{R}{\underset{PR_3}{\diagup}}$$

i   ii

with other $1:1$ adducts suggested that this kinetic isomer could only be isolated with small cone angle ligands. The thermodynamic isomer of the PEt_3 addition product can be geometrically related to the kinetic isomer by retaining the $CpMo(CF_3C{\equiv}CCF_3)Cl$ entity intact, inverting configuration at C_β by switching L and CF_3, and rotating the η^2-vinyl ligand by about 30° to a less hindered position. One could invoke an intramolecular rearrangement from η^2-vinyl to η^1-vinyl and C—C bond rotation, made possible by the phosphine substituent and resonance form ii, to invert the stereochemistry at C_β (Fig. 31). Both isomers exhibit short M—C_α (1.92, 1.94 Å) and long M—C_β (2.29, 2.32 Å) bond distances (Table X) (176).

Several other simple heteroatom η^2-vinyl complexes, i.e., not chelated to the metal by additional donor atoms, have been structurally characterized. $CpW(SC_6H_4Me)(CF_3C{\equiv}CCF_3)[\eta^2\text{-}CF_3CC(CF_3)(PEt_3)]$, $CpWCl(CF_3C{\equiv}CCF_3)[\eta^2\text{-}CF_3CC(CF_3)(CNBu^t)]$ (86), and $CpW(CO)_2[\eta^2\text{-}CF_3CC(CF_3)C(O)SMe]$ (184) each attain an 18-electron count at tungsten with $N = 2$ for the alkyne adjacent to the η^2-vinyl ligand. Large structural variations are observed for the orientation of the η^2-vinyl ligand in the solid state and for the W—C_β distances while W—C_α remains close to 1.9 Å for all of these molecules (Table X). The thiolate group in CpW-$(CF_3C{\equiv}CCF_3)[\eta^3\text{-}(CF_3)CC(CF_3)(SPr^i)]$ is simultaneously bound to the tungsten and to C_β of the η^2-vinyl ligand (97).

TABLE X

STRUCTURAL PARAMETERS FOR MO AND W η^2-VINYL COMPLEXES

Complex	M—C_α, Å	M—C_β, Å	C_α—C_β, Å	Ref.
$CpMo[P(OMe)_3]_2(\eta^2\text{-}CPhCHPh)$	1.95	2.30	1.43	172
$CpMo[P(OMe)_3]_2(\eta^2\text{-}CBu^tCHPh)$	1.94	2.29	1.44	169
$CpMo[P(OMe)_3]_2(\eta^2\text{-}CMeCPh_2)$	1.96	2.25	1.46	169
$(\eta^5\text{-}C_9H_7)Mo[P(OMe)_3]_2 \ (\eta^2\text{-}CSiMe_3CH_2)$	1.96	2.26	1.44	173
$CpMoCl(CF_3C{\equiv}CCF_3)[\eta^2\text{-}CF_3CC(CF_3)(PEt_3)]$	1.92	2.29	1.43	176
$CpMoCl(CF_3C{\equiv}CCF_3)[\eta^2\text{-}CF_3CC(CF_3)(PEt_3)]$	1.91	2.32	1.42	176
$CpW(CO)_2[\eta^2\text{-}CF_3CC(CF_3)(C(O)SMe)]$	1.96	2.19	1.44	184, 190
$CpW(CF_3C{\equiv}CCF_3)(SC_6H_4Me)$ $[\eta^2\text{-}CF_3CC(CF_3)(PEt_3)]$	1.91	2.33	1.45	86
$CpW(CF_3C{\equiv}CCF_3)[\eta^3\text{-}CF_3CC(CF_3)(SPr^i)]$	1.91	2.18	1.42	97
$CpW(CF_3C{\equiv}CCF_3)Cl[\eta^2\text{-}CF_3CC(CF_3)(CNBu^t)]$	1.89	2.30	1.41	86
$CpMo(CF_3C{\equiv}CCF_3)[\eta^3\text{-}CF_3CC(CF_3)(NC_5H_4S)]$	1.91	2.12	1.39	187

Numerous η^2-vinyl complexes have been generated by the addition of neutral reagents to $CpM(CF_3C{\equiv}CCF_3)_2X$ reagents. Products formulated as $Cp(CF_3C{\equiv}CCF_3)M[\eta^2CF_3CC(CF_3)L]X$ have been characterized for Mo with $X = CF_3$, $L = CNBu^t$ (98); $X = Cl$ or SC_6F_5, $L = PEt_3$, PMe_2Ph, PPh_3, and $P(OMe)_3$; and for W with $X = Cl$, $L = CNBu^t$; $X = SC_6H_4Me$, $L = PEt_3$ (86).

The product distribution for reactions of thallium thiolates with $CpWCl(CF_3C{\equiv}CCF_3)_2$ are dependent on the organic thiolate substituent (97,185). For electron-poor thiolates ($R = C_6F_5$, Ph, C_6H_4Me in TlSR) chloride substitution at metal yields bisalkyne products [Eq. (54)]. With $R = Pr^i$ or Bu^t the SR group ends up bound to one of the reagent alkyne carbons to form C_β of the new η^2-vinyl ligand [Eq. (55)] and also binds to

$$CpW(CF_3C{\equiv}CCF_3)_2Cl + TlSR \;\rightarrow\; CpW(CF_3C{\equiv}CCF_3)_2(SR) + TlCl$$
$$R = C_6F_5,\, C_6H_5,\, C_6H_4Me \qquad (54)$$

$$CpW(CF_3C{\equiv}CCF_3)_2Cl + TlSR \;\rightarrow$$
$$R = Pr^i,\, Bu^t$$
$$CpW[\overline{\eta^2\text{-}CF_3CC(CF_3)SR}](CF_3C{\equiv}CCF_3) + TlCl \qquad (55)$$

the metal to produce a three-coordinate sulfur. For $R = Me$ the complexity of the reaction is indicated by the isolation of two isomers. One isomer exhibits four CF_3 ^{19}F signals with the other having only a single CF_3 ^{19}F-NMR signal at room temperature which represents complete exchange of the four distinct CF_3 signals observed for this second isomer at $-100°C$. The presence of two isomers with different η^2-vinyl orientations is compatible with the spectroscopic data (97), and earlier suggestions of bisalkyne and dimeric product formulations are unnecessary, although the CpW-$(SR)(CF_3C{\equiv}CCF_3)_2$ species is a likely intermediate for isomer interconversion given that it is the ground state isomer with aryl thiolates.

The isolation of these closely related thiolate complexes hints at an important role for η^2-vinyl ligands in reactions which lead to net ligand substitution at metal. The SR bridge between C_β and W may resemble a snapshot along a reaction path for alkyne insertion into a M—L bond or for transfer of L from an η^2-vinyl to metal (97). A mechanism for alkyne polymerization based on η^2-vinyl intermediates has also been constructed (186).

An example of an η^2-vinyl ligand providing a pathway to simple ligand substitution products is available in the isolation of $CpW(CO)(SC_6F_5)$-$[\eta^2\text{-}CF_3CC(CF_3)(PEt_3)]$ from the addition of PEt_3 to the carbonyl monomer (60). Heating the η^2-vinyl product promotes carbon monoxide loss to yield $CpW(SC_6F_5)(PEt_3)(CF_3C{\equiv}CCF_3)$ as the final product. Tungsten phosphorus coupling in the η^2-vinyl complex is definitive for the phosphorus carbon linkage rather than an alternative η^2-vinyl formulation with PEt_3 as a ligand and SC_6F_5 bound to C_β. To postulate that the apparently simple substitution reaction [Eq. (56)] proceeds through an isolable η^2-vinyl complex would have been heresy only a decade ago.

$$CpW(CO)(SC_6F_5)(CF_3C\equiv CCF_3) + PEt_3 \rightarrow$$

$$CpW(PEt_3)(SC_6F_5)(CF_3C\equiv CCF_3) + CO(g) \qquad (56)$$

Reaction of $CpM(CF_3C\equiv CCF_3)_2Cl$ with potentially chelating thiolates containing a heterocyclic nitrogen produces metal thiolates chelated to η^2-vinyls formed from nitrogen attack at an alkyne carbon (187,188). The N—C—S linkage of anionic pyridine-2-thiolate and related ligands replaces Cl^- with RS^- and links through nitrogen to the adjacent carbon in $Cp(CF_3C\equiv CCF_3)W[\eta^3-CF_3CC(CF_3)NC_5H_4S]$. Reaction with $S_2CNR_2^-$ presumably produces a related S—C—S link from metal to the vinyl C_β carbon, rather than a bisalkyne product (188). The use of pyridine N-oxide as a thiolate substituent produces the same product as the simple chelate with $CpMo(CF_3C\equiv CCF_3)_2Cl$, but with $CpW(CF_3C\equiv CCF_3)_2Cl$ displacement of both the cyclopentadienyl ring and the chloride is observed and bischelate bisalkyne tungsten products form (188).

Addition of hexafluorobutyne to $CpW(CO)_3SR$ (R = Me, Et, Pr^i) also yields η^2-vinyl products (189). Isolation of $Cp(CO)_2W[\eta^2-CF_3CC(CF_3)-C(O)SMe]$ reflects CO insertion in addition to η^2-vinyl formation (190). Other alkyne insertion products have been identified as well: $Cp(CO)_2\overline{W[RSCZ}=CZC(O)]$ for Z = CF_3 or CO_2Me. Addition of free ligand to η^2-vinyl complexes can promote alkyne insertion and coupling reactions. A brief survey of results in this area is presented in Section VIII,B.

Another important reaction of coordinated alkynes is deprotonation at C_β. Complete exchange of the methyl protons in $[CpMo(CO)-(MeC\equiv CMe)_2]^+$ has been effected with Et_3N as a base and with CD_3CN or $CD_3C(O)CD_3$ solvent as the D^+ source (75). The increased acidity of the alkyne methyl groups is impressive, the pK_a for free $CH_3C\equiv CCH_3$ being near 35 while $HNEt_3$ has a pK_a value near 10. Notice that deprotonation to a neutral species would form a metal allene, and one could guess that the η^2-allenyl ligand would resemble a four-electron donor alkyne just as the η^2-ketenyl and η^2-vinyl ligands do (175). Isolation of an η^2-allenyl complex in d^4 chemistry seems likely. Indeed, after the first draft of this article was submitted in early 1987, Feher, Green, and Rodrigues reported that $[CpL_2Mo(PhC\equiv CCH_2Ph)]^+$ can be deprotonated with KH to form the stable η^2-allenyl product $CpL_2Mo(\eta^2-PhC=C=CHPh)$ with L = $P(OMe)_3$ (190a).

A recapitulation of the salient points in the η^2-vinyl synthesis section is appropriate. Note that conversion of an alkyne with $N = 4$ to an η^2-vinyl ligand by nucleophilic addition to an alkyne carbon leaves the metal electron count unchanged. Only for cationic monoalkyne complexes has addition of hydride, alkyl, and aryl groups been successful. For neutral terminal alkyne complexes with an adjacent electron-poor olefin ligand addition of PR_3 or $P(OR)_3$ forms η^2-CRCH(PR_3) ligands in the bisdithiocarbamate system. For cyclopentadienyl bisalkyne reagents the addition of a nucleophile to an alkyne carbon generates adjacent vinyl and alkyne ligands which need to furnish a total of six electrons to the metal. For 2-butyne a four-electron donor alkyne and an η^1-vinyl ligand result while for hexafluoro-2-butyne a two-electron donor alkyne and an η^2-vinyl result.

B. *Formation of Alkyne Insertion Products*

Insertion and condensation reactions of alkynes have been well known for many years (*1,191*). Alkyne dimerization to form either cyclobutadiene (*192*) or metallocyclopentadiene products (*193*), trimerization to arenes (*194*), condensation of two alkynes and carbon monoxide to cyclopentadienones (*195*), and condensation of alkyne and isonitriles to cyclopentadienimines (*196*) are common reactions. This section surveys alkyne insertion products originating from monomeric Mo(II) and W(II) reagents.

Formation of five-membered oxymetallocycles from cis addition of metal acyl fragments to alkynes has been observed with numerous CpW(CO)-(RC≡CR)[C(O)R′] reagents (*197–199*). Analogous products have been formed in other metal systems, e.g., (CO)$_4$Mn(OC(R)CR=CR) was reported in 1970 (*200*). A recent full paper describes the photolytic preparation of more than 20 metallocyclic alkenylketone complexes of the type Cp(CO)$_2$M[CR=CRC(R)O] with M = Mo, W; Cp = η^5-C$_5$H$_5$, η^5-C$_5$H$_4$Me, or η^5-C$_9$H$_7$; and R = Me, Et, ^{n}Pr, or ^{n}Bu (*70*). The net conversion of CpM(CO)$_3$R reagents plus alkyne to the alkenylketone products [Eq. (57)] corresponds to 1 : 1 adduct formation, but the reaction is mechanistically complex. Formation of alkyl alkyne and acyl alkyne complexes with loss of CO precedes addition of CO to form the final product.

Two resonance structures can be drawn for these metallocycles (Fig. 32). Both structural and spectroscopic data suggest that representation **ii** is an

$$\text{CpM(CO)}_3\text{R} + \text{HC}{\equiv}\text{CH} \longrightarrow \quad \text{Cp(CO)}_2\text{M} \leftarrow \text{O} \diagdown \begin{array}{c} \text{C—R} \\ \text{C}{=}\text{C} \\ \text{H} \quad\quad \text{H} \end{array} \qquad (57)$$

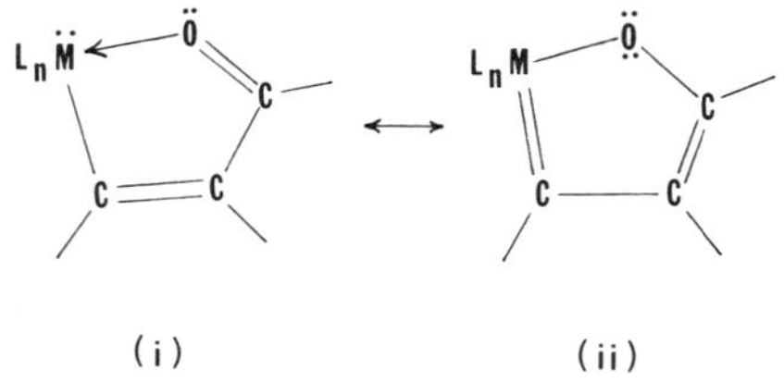

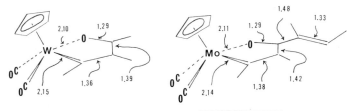

FIG. 32.   Resonance structures for $L_nM(CR=CRC(R)O)$.

important contributor to the bonding scheme (70,183). All of Alt's metallocycle complexes exhibit low-field ^{13}C-NMR signals for C_α (250 ±
15 ppm) whereas C_β is over 100 ppm further upfield (135 ± 5 ppm)
and C_γ, bound to oxygen, resonates near 200 ppm. The low-field C_α chemical shift, the $^1J_{WC}$ coupling constant of 70–80 Hz, and the short W—C_α
distance of 2.15 Å in $(\eta^5\text{-}C_5Me_5)(CO)_2W[CMe=CMeC(O)Me]$ (see
Fig. 33) all suggest double bond character between tungsten and C_α. Ring
distances also indicate extensive π delocalization throughout the five-
membered ring (C_α—C_β, 1.36 Å; C_β—C_γ, 1.39 Å; C_γ—O, 1.29 Å; O—W,
2.10 Å) (70).

An earlier structure of a vinyl-substituted alkenyl ketone metallocycle
displayed similar ring parameters and first exposed the carbenoid character
of C_α (183). The preparation of $Cp(CO)_2Mo[CMe=CMeC(O)CMe=
CHMe]$ from $[CpMo(CO)(MeC\equiv CMe)_2]^+$ and $[HBBu_3^s]^-$ at $-78°C$
under a CO atmosphere reflected trapping of a monocarbonyl species
believed to be the η^1-vinyl formed by H$^-$ addition to one of the 2-butyne
ligands. The mechanism proposed in Fig. 34 is supported by the isolation
of $Cp(CO)Mo(MeC\equiv CMe)(\sigma\text{-}MeC=CMe_2)$ which adds CO at $-78°C$ to
form the dicarbonyl alkenyl ketone complex with the vinyl substituent on
the carbonyl carbon (183).

Related   metallocycles   have   been   prepared   by   heating
$Cp(CO)_3Mo(CH_2)_nC\equiv CMe$ (n = 3, 4, 5) reagents which isomerize to

FIG. 33.   Bond distances in two $Cp(CO)_2M[CR=CRC(R)O]$ metallocycles.

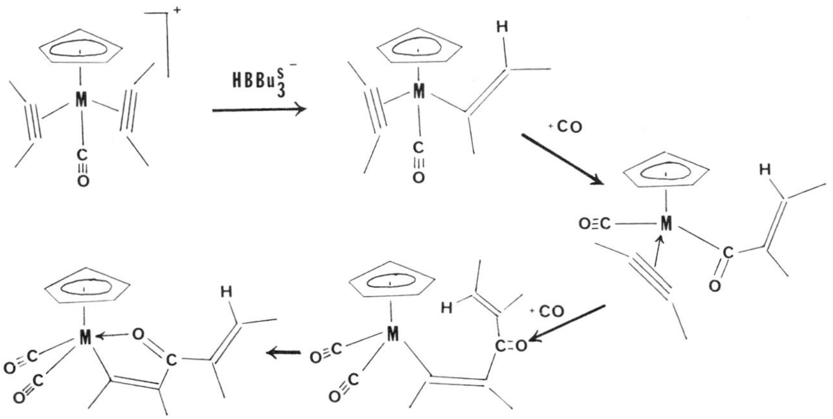
HBBu$^s_3$$^-$
+CO
+CO
M
H
O≡C
C=O

FIG. 34. Possible mechanism for alkenyl ketone metallocycle formation from $[CpMo(CO)(MeC{\equiv}CMe)_2]^+$ and $[HBR_3]^-$.

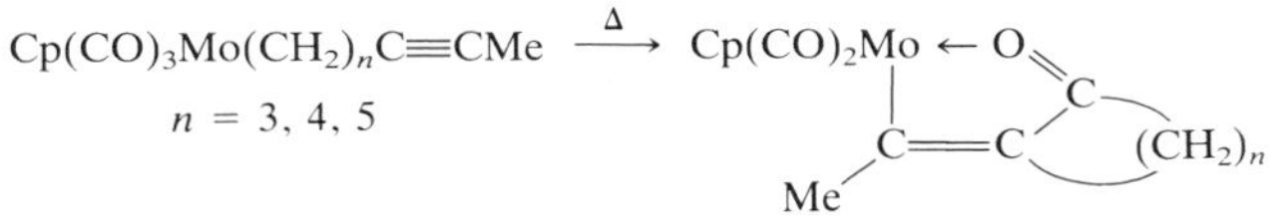
Cp(CO)₃Mo(CH₂)ₙC≡CMe
n = 3, 4, 5
Δ
Cp(CO)₂Mo ← O
C
C=C
(CH₂)ₙ
Me

(58)

dicarbonyl alkenyl ketone complexes with C_α bearing the methyl group while C_β and the carbonyl carbon, C_γ, are linked by the $(CH_2)_n$ chain [Eq. (58)] (*201*). The complex with $n = 4$ allowed isolation of free organic product on hydrogenation at 90°C to form 2-ethylcyclohexanone in 90% yield or on acidification with CF_3CO_2H to form 2-ethylidenecyclohexanone, also in 90% yield.

Reaction of $CpM(CO)_3R$ (R = Me, CH_2Ph) with 2-butyne produces an alkenyl ketone metallacycle, thermally for Mo, photochemically for W (*202*). Addition of $CNBu^t$ or PPh_3 drives CO insertion into the metal position of the metallacycle, and lactone products form [Eq. (59)]. With trifluoromethyl as the original metal alkyl group multiple alkyne and CO insertions lead to an eight-membered oxymetallacycle ring [Eq. (60)].

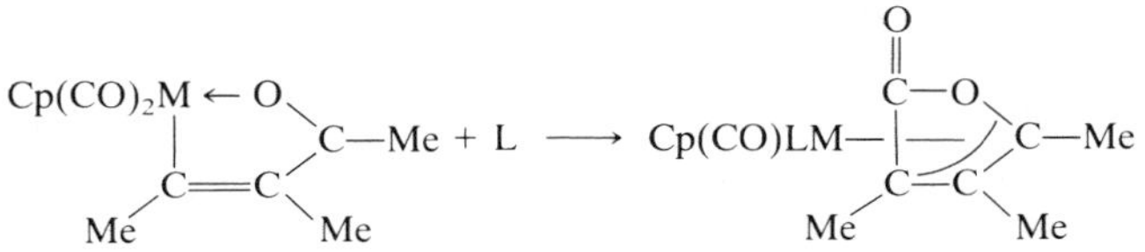
Cp(CO)₂M ← O
C—Me + L
C=C
Me
Me
Cp(CO)LM—
O
C—O
C—Me
C=C
Me
Me

(59)

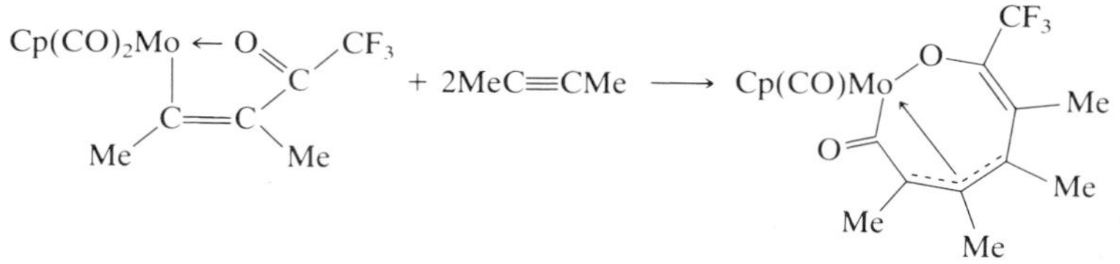

$$Cp(CO)_2Mo \leftarrow O \diagdown C \diagup CF_3 \quad + 2MeC{\equiv}CMe \quad \longrightarrow \quad \text{(ring product)} \qquad (60)$$

$$Cp(CO)_2W \leftarrow O \diagdown C-Me \quad + PMe_3 \quad \longrightarrow \quad \text{(product)} \qquad (61)$$

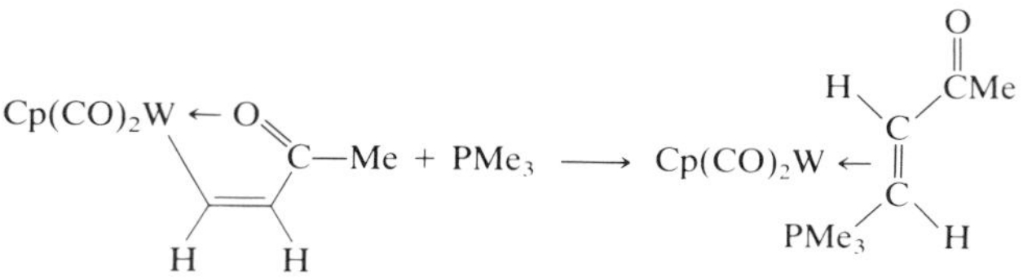

In addition to possessing physical properties compatible with carbene character at C_α, alkenyl ketone rings with a hydrogen on C_α are susceptible to nucleophilic attack by PMe$_3$ [Eq. (61)]. A full paper describing several 1:1 adducts and including the structure of Cp(CO)$_2$-W[CH(PMe$_3$)CH—C(O)Me] appeared in 1984 (203). This electrophilic behavior is reminiscent of Fischer carbene reactions (204).

Reaction of nitrosyl chloride with the $N = 4$ acetylene complex CpW-(CO)(HC{\equiv}CH)[C(O)Me] produces 20% yields of an alkenyl ketone complex, Cp(NO)(Cl)W[CH=CHCOMe], and 50% Cp(NO)(CO)W-[CHCl=CHC(O)Me] as two of the four possible diastereomers (205). A trans hydrogen arrangement is found for both olefin isomers. Although the η^2-CHCl=CHC(O)Me ligand is an olefin analog, Alt and co-workers stress the tungstenacyclopropane description of this product since the NMR parameters are reminiscent of C$_3$H$_4$ rings and addition of PMe$_3$ leads to CO substitution at metal rather than the replacement of the "olefin." As always the distinction between i and ii is a formal one, sometimes useful but more a matter of preference than substance.

$$M \leftarrow \overset{CR_2}{\underset{CR_2}{\|}} \quad \longleftrightarrow \quad M \diagup_{\diagdown}{\overset{CR_2}{\underset{CR_2}{|}}}$$

$$\quad\quad i \quad\quad\quad\quad\quad\quad\quad ii$$

Although numerous alkyne insertion products have been reported for Mo(II) and W(II), simple dimerization to form cyclobutadiene or trimerization to form an arene ligand is rare. One brief report of cyclobutadiene formation from a bisalkyne complex (206) has been followed by a full paper which suggests that an η^2-vinyl complex may be the precursor to CpM(S$_2$CNR$_2$)[η^4-C$_4$(CF$_3$)$_4$] (207).

Cyclopentadienone formation can occur with Group VI d^4 alkyne derivatives containing carbonyl ligands. Addition of CO to CpMo-$(CF_3C{\equiv}CCF_3)_2X$ reagents produces $CpX(CO)Mo[\eta^4\text{-}C_4(CF_3)_4CO]$ products [Eq. (62)] (206). When $CF_3C{\equiv}CCF_3$ is added to $CpMo(CO)_3Cl$ the

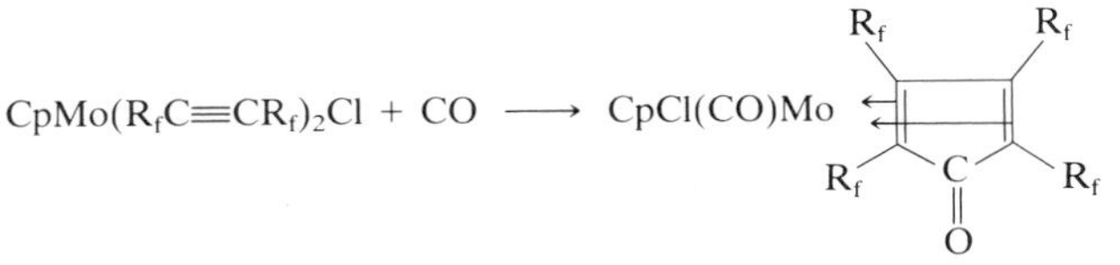

$$CpMo(R_fC{\equiv}CR_f)_2Cl + CO \longrightarrow CpCl(CO)Mo \quad (62)$$

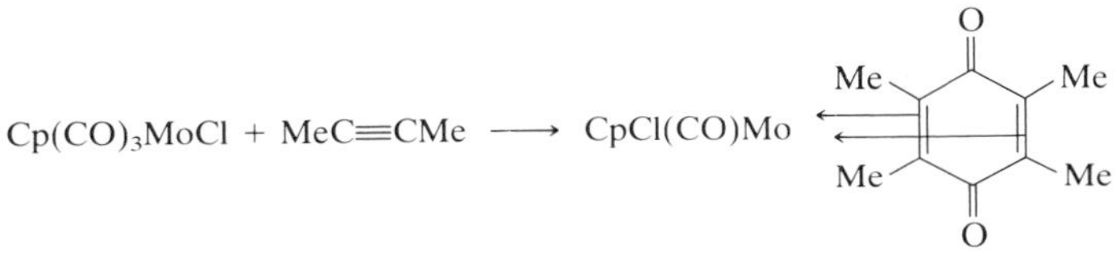

$$Cp(CO)_3MoCl + MeC{\equiv}CMe \longrightarrow CpCl(CO)Mo \quad (63)$$

same cyclopentadienone product is formed as in Eq. (62), but $MeC{\equiv}CMe$ leads to duroquinone formation as two molecules of CO condense with two 2-butyne molecules [Eq. (63)] (93). Multiple isomers of the cyclopentadienone ligand are evident when trifluoropropyne reacts with CpMo-$(CO)(CF_3C{\equiv}CH)(SCF_3)$ to form $CpMo(CO)(SCF_3)[\eta^4\text{-}C_4H_2(CF_3)_2CO]$ (63). Coupling of two cyclooctyne molecules with CO to yield cyclopentadienone complexes has been reported with $LM(CO)_2(S_2CNR_2)_2$ reagents (208).

Isonitriles also react with $CpWCl(CF_3C{\equiv}CCF_3)_2$ and CpMo-$(CF_3C{\equiv}CCF_3)_2CF_3$ to form C_5 rings (98). Condensation of the two alkynes with RNC to form η^4-cyclopentadienimines [Eq. (64)] mimics the CO

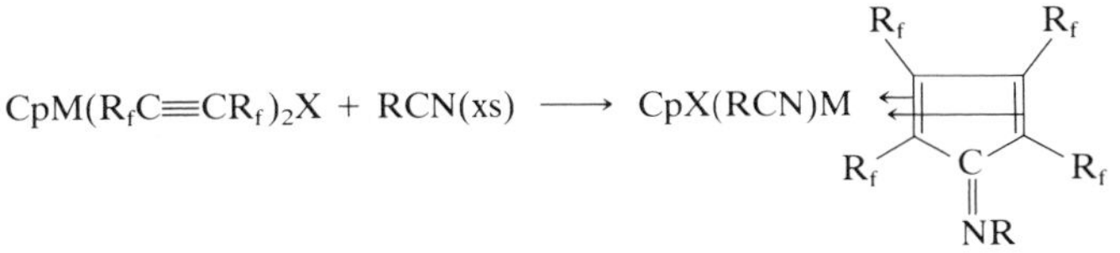

$$CpM(R_fC{\equiv}CR_f)_2X + RCN(xs) \longrightarrow CpX(RCN)M \quad (64)$$

insertion reaction. With a thiolate ligand in the coordination sphere isolation of several precursors leading to an unusual $\eta^2\text{-}C_4(CF_3)_4$ CNR ligand bound to metal through the exocyclic $C{=}N$ bond has been achieved (209). Stepwise conversion of isonitriles and $CpM(SR)(CF_3C{\equiv}CCF_3)_2$ to an η^2-vinyl, a metallocyclopentadiene, and the ultimate η^2-cyclopentadienimine product has been achieved as presented in Fig. 35. The intervention of η^2-vinyl ligands and facile CNR migration between ligand and metal make this reaction sequence noteworthy. Another example of migration of an η^2-vinyl CC_β

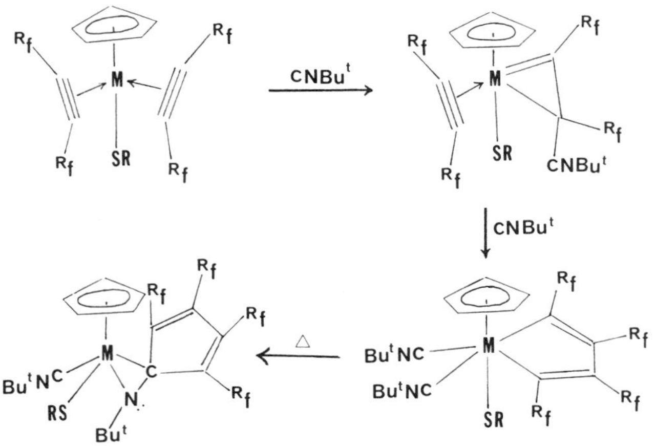

FIG. 35. Stepwise formation of an η^2-$C_4(CF_3)_4CNBu^t$ ligand from $CpMo(CF_3C\equiv CCF_3)_2(SC_6F_5)$ and $CNBu^t$.

substituent to the metal is presumbly involved in the formation of the MoC_4 ring of $CpMo(SR)(CNR)_2[\eta^2\text{-}C_4(CF_3)_4]$ from $CpMo(SR)(CF_3C\equiv CCF_3)$-$[\eta^2\text{-}CF_3CC(CF_3)(CNR)]$ and CNR (210). A metallocyclopentadiene has been trapped in the reaction of $CpMo(CF_3C\equiv CCF_3)_2Cl$ and $Co_2(CO)_8$ as $Cp(CO)_2Mo[C_4(CF_3)_4]Co(CO)_2$ forms [Eq. (65)] (211). Reaction of TlC_5H_5 and $CpMCl(CF_3C\equiv CCF_3)_2$ leads to insertion of a hexafluorobutyne link between an η^4-cyclopentadiene and the metal (92). The coordination sphere is completed with η^5-C_5H_5 and a two-electron donor $CF_3C\equiv CCF_3$ which has M—C distances of 2.14 and 2.15 Å and a $C\equiv C$ bond length of 1.28 Å.

$$CpMo(R_fC\equiv CR_f)_2Cl + Co_2(CO)_8 \longrightarrow Cp(CO)_2Mo=\!\!\!\overset{R_f}{\underset{R_f}{\diagup\!\!\!\diagdown}}\!\!\!=Co(CO)_2 \qquad (65)$$

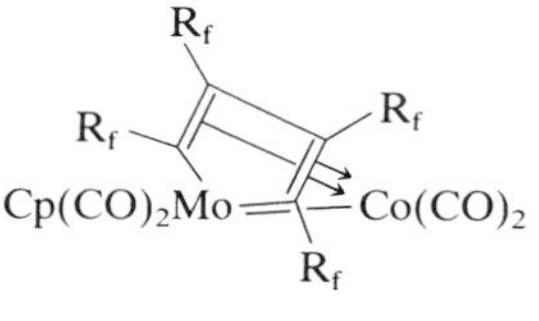

Reaction of $WBr_2(CO)(CF_3C\equiv CCF_3)_2$ with $P(OR)_3$ (R = Me, Et) (86) does not form an η^2-vinyl complex as the final product (212). Rather, loss of RBr in an Arbuzov-type reaction leads to a diethylvinylphosphonate ligand by insertion of alkyne into a W—P bond. The dimeric product which results [Eq. (66)] has one four-electron donor hexafluorobutyne bound to each tungsten with short W—C distances approaching 2.0 Å. An Arbuzov reaction also occurs with CF_3I and $CpL_2Mo\equiv CCH_2Bu^t$ [L = $P(OMe)_3$] to form a phosphonate metallocycle via phosphorus–carbyne carbon bond formation [Eq. (67)] (213).

$$W(CO)Br_2(R_fC\equiv CR_f)_2 + P(OR)_3 \longrightarrow \tag{66}$$

$$CpL_2Mo\equiv CCH_2Bu^t + CF_3I \longrightarrow CpLIMo=C-CH_2Bu^t \tag{67}$$
$$\begin{array}{c} \uparrow \quad | \\ O=P(OMe)_2 \end{array}$$

Reactions of metal thiolate complexes with activated alkynes produce several insertion products (214). A W—C(O)C(Z)=C(Z)SR metallocycle is formed from CpW(CO)$_3$(SR) and ZC≡CZ at room temperature for R = Me, Et and Z = CF$_3$, CO$_2$Me (184). Heating the Cp(CO)$_2$W—C(O)-C(CF$_3$)=C(CF$_3$)SMe product leads to isolation of an η^2-vinyl complex with CF$_3$CC(CF$_3$)C(O)SMe bound to tungsten through both carbons. Metallocycle isomerization also occurs to allow isolation of Cp(CO)$_2$-W—C(CF$_3$)=C(CF$_3$)C(O)SMe as well as the SEt analog. Further downstream are Cp(CO)$_2$W[C(CF$_3$)=C(CF$_3$)C(O)SMe] (215) and the simple substitution product, CpW(CO)(SMe) (CF$_3$C≡CCF$_3$) (190). Reaction of this product with PEt$_3$ first forms a 1 : 1 adduct, probably an η^2-vinyl complex, which loses CO to form Cp(PEt$_3$)W(SMe)(CF$_3$C≡CCF$_3$) (65).

Alkyne insertion into the Mo—H bond of CpMo(CO)$_3$H yields Cp-(CO)$_2$Mo[C(O)C(NR$_2$)=CH$_2$] when RC≡CNR$_2$ alkynes are added to the metal hydride reagent (216). Coupling of the olefin and alkyne ligands present in {Cp(MeC≡CMe)Mo[PPh$_2$(o-C$_6$H$_4$CH=CH$_2$)]}$^+$ occurs in refluxing acetonitrile and is accompanied by a 1,3-hydrogen shift to form an η^4-diene ligand. The structure of this complex has been determined (136).

Although bisalkyne bisdithiocarbamate complexes are reluctant to react with nucleophiles, protonation of M(PhC≡CPh)$_2$(S$_2$CNMe$_2$)$_2$ (M = Mo or W) yields a C$_4$ ligand, η^4-C$_4$Ph$_4$H, which serves as a five-electron donor when counted as a neutral ligand in [M(S$_2$CNMe$_2$)$_2$(η^4-C$_4$Ph$_4$H)][BF$_4$] (91). This class of butadienyl ligand had been prepared previously by hydride addition to a cationic ruthenium cyclobutadiene complex (217). A ruthenium analog with one additional ligand and so requiring only three electrons from the C$_4$R$_5$ moiety is also known (218), and substantial geometrical differences are evident in the MC$_4$ ring as a function of electron donation (see Fig. 36). More recently another example of an η^4-C$_4$R$_5$ metallocycle has been structurally characterized. The η^2-vinyl complex with sulfur bridging C$_\beta$ and W, Cp(CF$_3$C≡CCF$_3$)W[η^2-CF$_3$CC-(CF$_3$)SPr^i], inserts 2-butyne to yield Cp(CF$_3$C≡CCF$_3$)W[η^4-C(CF$_3$)C-(CF$_3$)C(Me)C(Me)SPr^i] (186). With electron-poor alkynes insertion yields hexatrienyl products with ZC≡CZ linking two hexafluorobutyne ligands.

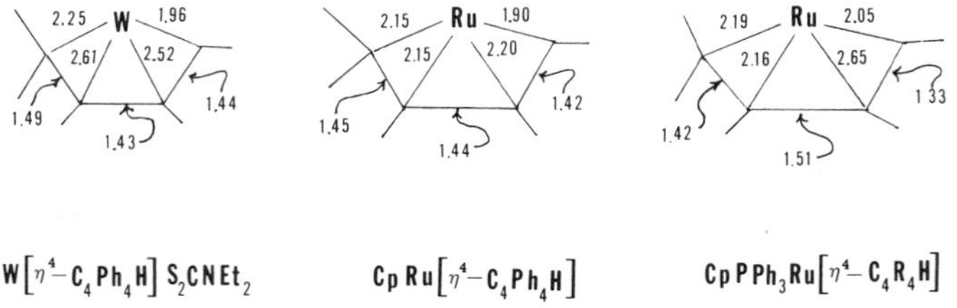

$$O W\left[\eta^{4}-C_{4}Ph_{4}H\right] S_{2}CNEt_{2} \qquad Cp Ru\left[\eta^{4}-C_{4}Ph_{4}H\right] \qquad Cp PPh_{3}Ru\left[\eta^{4}-C_{4}R_{4}H\right]$$

FIG. 36.   Ring geometries in two five-electron donor C_4R_4H and one three-electron donor C_4R_4H complexes.

A recent report by Mayr of slow polymerization of $PhC{\equiv}CH$ by $(PMe_3)_2Cl_2(PhC{\equiv}CPh)W{=}CHPh$ fulfills expectations based on the classic Chauvin mechanism for olefin metathesis (78). The presence of a carbene and a vacant coordination site are prerequisites for metallocyclobutene formation with free alkyne. Mayr has both the carbene and the alkyne initially present in the catalyst, but there is no evidence for direct involvement of the cis alkyne in the actual polymerization mechanism.

C. *Metal–Alkyne Cleavage Reactions*

Alkyne ligands have been removed from metal complexes intact, in oxidized form, and in reduced form. In spite of accomplishing alkyne–metal cleavage with different reagents in specific cases, there is no general method for systematically separating an alkyne ligand from a d^4 metal center.

Alkyne ligands are sometimes susceptible to direct ligand substitution. Conversion of bisalkyne monomers to monoalkyne complexes is particularly common for cyclopentadienyl derivatives (72) as discussed in Section II,C. Displacement of the lone alkyne from $CpML(RC{\equiv}CR)X$ complexes has rarely been reported. Similar results are characteristic of alkynes in $M(RC{\equiv}CR)_2(CO)LX_2$ (49), where one alkyne is labile, compared to $M(RC{\equiv}CR)(CO)L_2X_2$ which is resistant to alkyne substitution.

In direct contrast to the generalizations above for metal alkyne complexes reacting with nucleophiles, the situation for bisdithiocarbamate alkyne derivatives is reversed. The bisalkyne complexes $M(RC{\equiv}CR)_2$-$(S_2CNEt_2)_2$ resist alkyne substitution by phosphine and phosphite reagents, although heating solutions of PEt_3 and $Mo(RC{\equiv}CR)_2(S_2CNMe_2)_2$ produces dimeric alkyne bridged products containing no phosphorus, $Mo_2(\mu\text{-}S)(\mu\text{-}RC{\equiv}CR)(S_2CNMe_2)_3(SCNMe_2)$ (153). Recall that $M(CO)$-

$(RC{\equiv}CR)(S_2CNEt_2)_2$ monoalkyne complexes readily add phosphorus donor ligands with loss of alkyne at room temperature [Eq. (68)] (90).

$$M(CO)(RC{\equiv}CR)(S_2CNEt_2)_2 + 2\ PR_3 \rightarrow M(CO)(PR_3)_2(S_2CNEt_2)_2 + (RC{\equiv}CR) \quad (68)$$

Polar metal–carbon bonds are often susceptible to acid hydrolysis to form M—O and C—H bonds. Indeed the two-electron donor 2-butyne ligand in $Cp_2Mo(MeC{\equiv}CMe)$ yields cis-2-butene on treatment with aqueous acid (85). Similarly, $Cp_2Mo(HC{\equiv}CH)$ reacts with HCl in toluene to form ethylene (82) in analogy to the formation of ethane when HCl is added to a solution of $CpMo(H_2C{=}CH_2)$.

Acid hydrolysis of four-electron donor alkynes has rarely been reported. Formation of a trace (2%) of ethylene was reported following addition of CF_3CO_2H to $Mo(HC{\equiv}CH)(CO)(S_2CNEt_2)_2$ (52), and 20% $H_2C{=}CH_2$ was formed from $Mo(HC{\equiv}CH)(CO)(S_2PPr_2^i)_2$ and HCl in CH_2Cl_2 (54). Likewise, a small amount (3%) of stilbene forms when $Mo(PhC{\equiv}CPh)$-$(SR)_2(CNR)_3$ is acidified with CF_3CO_2H or with HCl in methylene chloride (80). These results are consistent with net donation of π-electron density from the neutral unsaturated organic molecule to the metal center and the corresponding lack of nucleophilic character for the bound alkyne. The addition of $NaBH_4$ to this same $Mo(PhC{\equiv}CPh)(SR)_2(CNR)_2$ complex produces a 98% yield of a 1:1 mixture of cis and trans stilbenes. Thus the alkyne is electrophilic in this case, perhaps surprising for a neutral low oxidation state alkyne complex (80). Indeed, the majority of reactions reported for alkyne ligands with a formal donor number of 4 involve attack at an alkyne carbon with a nucleophile (Section VIII,A). Isolation of an η^1-vinyl complex has been achieved in related Mo(IV) oxo chemistry following protonations of $Mo(O)(ZC{\equiv}CZ)(S_2CNR_2)_2$ with trifluoroacetic acid (219).

An important oxidative cleavage of a metal–alkyne moiety was mentioned earlier in Section IV,B. Aqueous peroxide reacts with the diaminoalkyne ligand, formed from isonitrile ligands by reductive coupling, to yield free oxamide (117). The yield of RNHC(O)C(O)NHR was 40% for H_2O_2 oxidation, while only 5% oxamide resulted for oxidations with Ag^+ or $Cl_2IC_6H_5$ reagents. The conversion of a four-electron donor alkyne to an oxamide is reminiscent of oxidations of carbene ligands to generate ketones (220). This chemistry suggests that the dicarbene resonance form is useful for anticipating some reactions of d^4 alkyne complexes. A note of caution is in order for representing alkyne complexes where $\pi_\perp$ donation is important with resonance form iii. The number of electrons *explicitly depicted* in the metal–alkyne bonding scheme increases from 2 to 4 to 8 as one considers resonance forms, i, ii, and iii, respectively. In this context a d^4

$$
\begin{array}{ccc}
\mathrm{M} \longleftarrow \!\!\! \Vert \begin{array}{l} \mathrm{CR} \\ \mathrm{CR} \end{array} &
\mathrm{M} \!\! < \!\! \begin{array}{l} \mathrm{CR} \\ \Vert \\ \mathrm{CR} \end{array} &
\mathrm{M} \!\! \lessgtr \!\! \begin{array}{l} \mathrm{CR} \\ \mathrm{CR} \end{array} \\
\mathbf{i} & \mathbf{ii} & \mathbf{iii}
\end{array}
$$

metal–alkyne can formally be represented by all three, but a d^2 metal center would not have sufficient valence electrons available to fill the bonding orbitals associated with resonance form **iii**.

IX

CONCLUSION

The chemistry which has developed for molybdenum(II) and tungsten(II) alkyne monomers encompasses syntheses, structures, spectra, molecular orbital descriptions, and reactions. The sheer volume of literature reports germane to this seemingly narrow topic is surprising. The Mo(II) and W(II) complexes addressed here are not unique in terms of alkyne $\pi_\perp$ donation. Related alkyne chemistry is appearing for d^4 metals other than molybdenum and tungsten, as well as for d^2 complexes in general.

Alkyne adducts of $MoCl_4$ were reported in 1973 by Greco et al. (*221*). An array of monomeric (*222*) and dimeric (*223*) d^2 W(IV) complexes related to $[W(ClC\equiv CCl)Cl_5]^-$ has been reported by Dehnicke. Schrock has prepared Mo(IV) and W(IV) alkyne derivatives with halide, alkoxide (*224*), and thiolate (*225*) ligands present. Other d^2 monomers with alkyne ligands have terminal oxo (*226,227*), nitrene (*228*), and sulfido (*61*) ligands.

Chromium alkyne chemistry also reflects the importance of $\pi_\perp$ donation, but the stoichiometries differ from those of heavier Group VI monomers. Isolation of the unusual four-coordinate $Cr(CO)_2(Me_3SiC\equiv CSiM_3)_2$ monomer (*229*) has been followed by reaction with diphenylacetylene to yield $Cr(CO)_2(PhC\equiv CPh)(\eta^4\text{-}C_4Ph_4)$ with $N = 4$ which can be reduced sequentially to a dianion with $N = 2$ (*230*).

Nearby elements also display chemistry reflecting alkyne $\pi_\perp$ donation to a vacant metal $d\pi$ orbital. In Group V, Lippard's coupled carbonyl product $Ta(Me_3SiOC\equiv COSiMe_3)(dmpe)_2Cl$ is a d^4 Ta(I) alkyne monomer (*116*), similar in electron count to $CpV(CO)_2(RC\equiv CR)$ complexes (*231*). The preparative route and physical properties of a series of $Ta(CO)_2\text{-}(RC\equiv CR)(I)L_2$ d^4 monomers are compatible with a four-electron donor description for the alkyne ligands (*231a*). The d^2 configuration has also produced Group V alkyne adducts such as $[TaCl_4(PhC\equiv CPh)(py)]^-$ (*232*) and $TaCl_3L_2(PhC\equiv CPh)$ (*233*). $CpMX_2(RC\equiv CR)$ complexes have been prepared (*234*) as have related benzyne adducts (*235*). Rhenium alkyne complexes reflect π donation in both high $[Re(V), d^2]$ (*236*) and low (d^4) (*237*) oxidation states. The $ReO(RC\equiv CR)_2X$ system (*238*) is proving to be a useful vehicle to exciting rhenium oxo complexes (*239*). A d^8 iron complex with a four-electron donor alkyne was reported recently by Boncella, Green, and O'Hare, $[MeSi(CH_2PMe_2)_3]Fe(PhC\equiv CPh)$ (*240*), and no doubt $[P(OMe)_3]_3Fe(PhC\equiv CPh)$ also has $N = 4$ although definitive spectral or structural data are not available for the four-coordinate phosphite derivative (*241*).

ACKNOWLEDGMENTS

Preparation of this article was aided by a grant from the U.S. Department of Energy (85ER13430). The efforts of Ms. Evelyn Kidd in cheerful and efficient preparation of the manuscript are acknowledged.

REFERENCES

1. M. A. Bennett, *Chem. Rev.* **62**, 611 (1962).
2. F. L. Bowden and A. B. P. Lever, *Organomet. Chem. Rev.* **3**, 227 (1968).
3. R. G. Guy and B. L. Shaw, *Adv. Inorg. Chem. Radiochem.* **4**, 77 (1962).
4. F. R. Hartley, *Chem. Rev.* **69**, 799 (1969).
5. F. R. Hartley, *Angew. Chem. Int. Ed. Engl.* **11**, 596 (1972).
6. L. D. Pettit and D. S. Barnes, *Fortschr. Chem. Forsch.* **28**, 85 (1972).
7. A. I. Gusev and Yu. T. Struchkov, *J. Sruct. Chem.* **11**, 340 (1970).
8. E. O. Greaves, C. J. L. Lock, and P. M. Maitlis, *Can. J. Chem.* **46**, 3879 (1968).
9. S. Otsuka and A. Nakamura, *Adv. Organomet. Chem.* **14**, 245 (1976).
10. M. Green, *J. Organomet. Chem.* **300**, 93 (1986).
11. T. Masuda and T. Higashimura, *Accts. Chem. Res.* **17**, 51 (1984).
12. C. I. Simionescu and V. Percec, *Prog. Polym. Sci.* **8**, 133 (1982).
13. T. J. Katz, S. M. Haker, R. D. Kendrick, and C. S. Yannoni, *J. Am. Chem. Soc.* **107**, 2182 (1985).
14. R. R. Schrock, *Accts. Chem. Res.* **19**, 342 (1986).
15. M. I. Bruce and A. G. Swincer, *Adv. Organomet. Chem.* **22**, 59 (1983).
16. M. H. Chisholm and H. C. Clark, *Accts. Chem. Res.* **6**, 202 (1973).
17. K. R. Birdwhistell, S. J. N. Burgmayer, and J. L. Templeton, *J. Am Chem. Soc.* **105**, 7789 (1983).
18. J. Wolf, H. Werner, O. Serhadli, and M. L. Ziegler, *Angew. Chem. Int. Ed. Engl.* **22**, 414 (1983).
19. M. H. Chisholm, K. Folting, D. W. Hoffman, and J. C. Huffman, *J. Am. Chem. Soc.* **106**, 6794 (1984).
20. M. H. Chisholm, D. M. Hoffman, and J. C. Huffman, *Chem. Soc. Rev.* **14**, 69 (1985).
21. K. P. C. Vollhardt, *J. Org. Chem.* **59**, 1574 (1984).
22. L. S. Liebeskind and C. F. Jewell, Jr., *J. Organomet. Chem.* **285**, 305 (1985).
23. K. H. Dotz, *Angew. Chem. Int. Ed. Engl.* **23**, 587 (1984).

24. W. D. Wulff, S. R. Gilbertson, and J. P. Springer, *J. Am. Chem. Soc.* **108**, 520 (1986).
25. M. Green, J. A. K. Howard, A. P. James, A. N. M. Jelfs, C. M. Nunn, and F. G. A. Stone, *J. Chem. Soc., Chem. Commun.*, 1623 (1984).
26. F. G. A. Stone, *Angew. Chem. Int. Ed. Engl.* **23**, 89 (1984).
27. D. P. Tate and J. M. Augl, *J. Am. Chem. Soc.* **85**, 2174 (1963).
28. D. P. Tate, J. M. Augl, W. M. Ritchey, B. L. Ross, and J. G. Grasselli, *J. Am. Chem. Soc.* **86**, 3261 (1964).
29. K. H. Theopold, S. J. Holmes, R. R. Schrock, *Angew. Chem. Int. Ed. Engl.* **22**, 1010 (1983).
30. K. R. Birdwhistell, T. L. Tonker, and J. L. Templeton, *J. Am Chem. Soc.* **109**, 1401 (1987).
31. S. J. Landon, P. M. Shulman, and G. L. Geoffroy, *J. Am. Chem. Soc.* **107**, 6739 (1985).
32. R. B. King, *Inorg. Chem.* **7**, 1044 (1968).
33. R. M. Laine, R. E. Moriarty, and R. Bau, *J. Am. Chem. Soc.* **94**, 1402 (1972).
34. R. Tsumura and N. Hagihara, *Bull. Chem. Soc. Jpn.* **38**, 1901 (1965).
35. A. N. Nesmeyanov, A. J. Gusev, A. A.Pasynskii, K. N. Anisimov, N. E. Kolobova, and Yu. T. Struchkov, *J. Chem. Soc., Chem. Commun.*, 1365 (1968).
36. A. N. Nesmeyanov, A. J. Gusev, A. A. Pasynskii, K. N. Anisimov, N. E. Kolobova, and Yu. T. Struchkov, *J. Chem. Soc., Chem. Commun.*, 739 (1969).
37. J. A. Connor and G. A. Hudson, *J. Organomet. Chem.* **160**, 159 (1978).
38. J. A. Connor and G. A. Hudson, *J. Organomet. Chem.* **185**, 385 (1980).
39. K. J. Odell, E. M. Hyde, B. L. Hyde, B. L. Shaw, and I. Shepherd, *J. Organomet. Chem.* **168**, 103 (1979).
40. K. W. Chiu, D. Lyons, G. Wilkinson, M. Thornton-Pett, and M. B. Hursthouse, *Polyhedron* **2**, 803 (1983).
41. R. Bowerbank, M. Green, H. P. Kirsch, A. Mortreux, L. E. Smart, and F. G. A. Stone, *J. Chem. Soc., Chem. Commun.* 245 (1977).
42. G. A. Carriedo, J. A. K. Howard, D. B. Lewis, G. E. Lewis, and F. G. A. Stone, *J. Chem. Soc., Dalton Trans.*, 905 (1985).
43. J. M. Maher, J. R. Fox, B. M. Foxman, and N. J. Cooper, *J. Am. Chem. Soc.* **106**, 2347 (1984).
44. P. Umland and H. Vahrenkamp, *Chem. Ber.* **115**, 3580 (1982).
45. P. B. Winston, S. J. N. Burgmayer, and J. L. Templeton, *Organometallics* **2**, 167 (1983).
46. P. B. Winston, S. J. N. Burgmayer, T. L. Tonker, and J. L. Templeton, *Organometallics* **5**, 1707 (1986).
47. M. A. Bennett and I. W. Boyd, *J. Organomet. Chem.* **290**, 165 (1985).
48. J. L. Davidson and G. Vasapollo, *Polyhedron* **2**, 305 (1983).
49. J. L. Davidson and G. Vasapollo, *J. Chem. Soc., Dalton Trans.*, 2239 (1985).
50. P. K. Baker and E. M. Keys, *Inorg. Chim. Acta* **116**, L49 (1986).
51. P. K. Baker and E. M. Keys, *Polyhedron* **5**, 1233 (1986).
52. J. W. McDonald, W. E. Newton, C. T. C. Creedy, and J. L. Corbin, *J. Organomet. Chem.* **92**, C25 (1975).
53. J. L. Templeton, R. S. Herrick, and J. R. Morrow, *Organometallics* **3**, 535 (1984).
54. J. W. McDonald, J. L. Corbin, and W. E. Newton *J. Am. Chem. Soc.* **97**, 1970 (1975).
55. W. E. Newton, J. L. Corbin, and J. W. McDonald, *Inorg. Synth.* **18**, 53 (1978).
56. C. Y. Chou and E. A. Maatta, *Inorg. Chem.* **23**, 2912 (1984).
57. L. Ricard, R. Weiss, W. E. Newton, G. J.-J. Chen, and J. W. McDonald, *J. Am. Chem. Soc.* **100**, 1318 (1978).
58. B. C. Ward and J. L. Templeton, *J. Am. Chem. Soc.* **102**, 1532 (1980).

59. J. A. Broomhead and C. G. Young, *Aust. J. Chem.* **35,** 277 (1982).
60. J. R. Morrow, T. L. Tonker, and J. L. Templeton, *J. Am. Chem. Soc.* **107,** 6956 (1985).
61. D. C. Brower, T. L. Tonker, J. R. Morrow, D. S. Rivers, and J. L. Templeton, *Organometallics* **5,** 1093 (1986).
62. J. L. Davidson and D. W. A. Sharp, *J. Chem. Soc., Dalton Trans.*, 2531 (1975).
63. P. S. Braterman, J. L. Davidson, and D. W. A. Sharp, *J. Chem. Soc., Dalton Trans.*, 241 (1976).
64. J. E. Guerchais, J. L. LeQuere, F. Y. Petillon, L. Manojlovic-Muir, K. W. Muir, and D. W. A. Sharp, *J. Chem. Soc., Dalton Trans.*, 283 (1982).
65. J. L. Davidson, *J. Chem. Soc., Dalton Trans.*, 2423 (1986).
66. M. L. H. Green, J. Knight, and J. A. Segal, *J. Chem. Soc., Dalton Trans.*, 2189 (1977).
67. H. G. Alt, *J. Organomet. Chem.* **127,** 349 (1977).
68. H. G. Alt and J. A. Schwarzle, *J. Organomet. Chem.* **155,** C65 (1978).
69. H. G. Alt, *J. Organomet. Chem.* **288,** 149 (1985).
70. H. G. Alt, H. E. Engelhardt, U. Thewalt, and J. Riede, *J. Organomet. Chem.* **288,** 165 (1985).
71. F. R. Kreissl, G. Reber, and G. Muller, *Angew. Chem. Int. Ed. Engl.* **25,** 643 (1986).
72. S. R. Allen, P. K. Baker, S. G. Barnes, M. Green, L. Trollope, L. M. Muir, and K. W. Muir, *J. Chem. Soc., Dalton Trans.*, 873 (1981).
73. M. Bottrill and M. Green, *J. Chem. Soc., Dalton Trans.*, 2365 (1977).
74. S. R. Allen, T. H. Glauert, M. Green, K. A. Mead, N. C. Norman, A. G. Orpen, C. J. Schaverien, and P. Woodward, *J. Chem. Soc., Dalton Trans.*, 2747 (1984).
75. P. L. Watson and R. G. Bergman, *J. Am. Chem. Soc.* **102,** 2698 (1980).
76. K. Sunkel, U. Nagel, and W. Beck, *J. Organomet. Chem.* **222,** 251 (1981).
77. H. G. Alt, *J. Organomet. Chem.* **256,** C12 (1983).
78. A. Mayr, K. S. Lee, M. A. Kjelsberg, and D. Van Engen, *J. Am. Chem. Soc.* **108,** 6079 (1986).
79. H. C. Foley, L. M. Strubinger, T. S. Targos, and G. L. Geoffroy, *J. Am. Chem. Soc.* **105,** 3064 (1983).
80. M. Kamata, T. Yoshida, S. Otsuka, K. Hirotsu, T. Higuchi, M. Kido, K. Tatsumi, and R. Hoffmann, *Organometallics* **1,** 227 (1982).
81. A. DeCian, J. Colin, M. Schappacher, L. Ricard, and R. Weiss, *J. Am. Chem. Soc.* **103,** 1850 (1981).
82. J. L. Thomas, *Inorg. Chem.* **17,** 1507 (1978).
83. A. Nakamura and S. Otsuka, *J. Am Chem. Soc.* **94,** 1886 (1972).
84. K. L. Tang Wong, J. L. Thomas, and H. H. Brintzinger, *J. Am. Chem. Soc.* **96,** 3694 (1974).
85. J. L. Thomas, *J. Am. Chem. Soc.* **95,** 1838 (1973).
86. J. L. Davidson, G. Vasapollo, L. Manojlovic-Muir, and K. W. Muir, *J. Chem. Soc., Chem. Commun.*, 1025 (1982).
87. R. S. Herrick and J. L. Templeton, *Organometallics* **1,** 842 (1982).
88. R. S. Herrick, S. J. Burgmayer, and J. L. Templeton, *Inorg. Chem.* **22,** 3275 (1983).
89. R. S. Herrick and J. L. Templeton, *Inorg. Chem.* **25,** 1270 (1986).
90. R. S. Herrick, D. M. Leazer, and J. L. Templeton, *Organometallics* **2,** 834 (1983).
91. J. R. Morrow, T. L. Tonker, and J. L. Templeton, *J. Am. Chem. Soc.* **107,** 5004 (1985).
92. J. L. Davidson, M. Green, D. W. A. Sharp, F. G. A. Stone, and A. J. Welch, *J. Chem. Soc., Chem. Commun.*, 706 (1974).
93. J. L. Davidson, M. Green, F. G. A. Stone, and A. J. Welch, *J. Chem. Soc., Dalton Trans.*, 738 (1976).
94. J. W. Faller and H. H. Murray, *J. Organomet. Chem.* **172,** 171 (1979).

95. J. L. Davidson, M. Green, F. G. A. Stone, and A. J. Welch, *J. Chem. Soc., Dalton Trans.*, 287 (1977).

96. J. L. Davidson, *J. Organomet. Chem.* **186**, C19 (1980).

97. L. Carlton, J. L. Davidson, J. C. Miller, and K. W. Muir, *J. Chem. Soc., Chem. Commun.*, 11 (1984).

98. J. L. Davidson, M. Green, J. Z. Nyathi, F. G. A. Stone, and A. J. Welch, *J. Chem. Soc., Dalton Trans.*, 2246 (1977).

99. W. Beck and K. Schloter, *Z. Naturforsch.* **33B**, 1214 (1978).

100. E. O. Fischer and P. Friedrich, *Angew. Chem. Int. Ed. Engl.* **18**, 327 (1979).

101. M. R. Churchill, H. J. Wasserman, S. J. Holmes, and R. R. Schrock, *Organometallics* **1**, 766 (1982).

102. F. R. Kreissl, W. Uedelhoven, and K. Eber, *Angew. Chem. Int. Ed. Engl.* **17**, 859 (1978).

103. F. R. Kreissl, A. Frank, U. Schubert, T. L. Lindner, and G. Huttner, *Angew. Chem. Int. Ed. Engl.* **15**, 632 (1976).

104. F. R. Kriessl, P. Friedrich, and G. Huttner, *Angew. Chem. Int. Ed. Engl.* **16**, 102 (1977).

105. F. R. Kreissl, W. J. Sieber, and M. Wolfgruber, *Z. Naturforsch.* **38B**, 1419 (1983).

106. F. R. Kreissl, W. J. Sieber, and M. Wolfgruber, *Angew. Chem. Int. Ed. Engl.* **22**, 493 (1983).

107. F. R. Kreissl, W. J. Sieber, P. Hofmann, J. Riede, and M. Wolfgruber, *Organometallics* **4**, 788 (1985).

108. J. C. Jeffery, J. C. V. Laurie, I. Moore, and F. G. A. Stone, *J. Organomet. Chem.* **258**, C37 (1983).

109. H. P. Kim, S. Kim, R. A. Jacobson, and R. J. Angelici, *Organometallics* **5**, 2481 (1986).

110. J. B. Sheridan, G. L. Geoffroy, and A. L. Rheingold, *Organometallics* **5**, 1514 (1986).

111. A. Mayr, G. A. McDermott, A. M. Dorries, A. K. Holder, W. C. Fultz, and A. L. Rheingold, *J. Am. Chem. Soc.* **108**, 310 (1986).

112. E. O. Fischer, A. C. Filippou, H. G. Alt, and K. Ackermann, *J. Organomet. Chem.* **254**, C21 (1983).

113. A. Mayr, G. A. McDermott, A. M. Dorries, and D. Van Engen, *Organometallics* **6**, 1503 (1987).

114. K. R. Birdwhistell, T. L. Tonker, and J. L. Templeton, *J. Am. Chem. Soc.* **107**, 4474 (1985).

115. C. T. Lam, P. W. R. Corfield, and S. J. Lippard, *J. Am. Chem. Soc.* **99**, 617 (1977).

116. P. A. Bianconi, I. D. Williams, M. P. Engeler, and S. J. Lippard, *J. Am. Chem. Soc.* **108**, 311 (1986); P. A. Bianconi, R. N. Vrtis, C. P. Rao, I. D. Williams, M. P. Engeler, and S. J. Lippard, *Organometallics* **6**, 1968 (1987).

117. C. M. Giandomenico, C. T. Lam, and S. J. Lippard, *J. Am. Chem. Soc.* **104**, 1263 (1982).

118. S. Warner and S. J. Lippard, *Organometallics* **5**, 1716 (1986).

119. C. Caravana, C. M. Giandomenico, and S. J. Lippard, *Inorg. Chem.* **21**, 1860 (1982).

120. R. Hoffmann, C. N. Wilker, S. J. Lippard, J. L. Templeton, and D. C. Brower, *J. Am. Chem. Soc.* **105**, 146 (1983).

121. J. Chatt, A. J. L. Pombeiro, R. L. Richards, G. H. D. Royston, K. W. Muir, and R. Walker, *J. Chem. Soc., Chem. Commun.* **708** (1975).

121a. R. N. Vrtis, C. P. Rao, S. Warner and S. J. Lippard, *J. Am. Chem. Soc.* **110**, 2669 (1988).

122. G. A. McDermott and A. Mayr, *J. Am. Chem. Soc.* **109**, 580 (1987).

123. R. Hoffmann, C. N. Wilker, and O. Eisenstein, *J. Am. Chem. Soc.* **104**, 632 (1982).
124. C. N. Wilker, R. Hoffmann, and O. Eisenstein, *Nouv. J. Chim.* **7**, 535 (1983).
125. D. C. Brower, J. L. Templeton, and D. M. P. Mingos, *J. Am. Chem. Soc.* **109**, 5203 (1987).
126. B. Capelle, M. Dartiguenave, Y. Dartiguenave, and A. L. Beauchamp, *J. Am. Chem. Soc.* **105**, 4662 (1983).
127. B. Capelle, A. L. Beauchamp, M. Dartiguenave, and Y. Dartiguenave, *J. Chem. Soc., Chem. Commun.*, 566 (1982).
128. J. L. Petersen and J. W. Egan, Jr., *Inorg. Chem.* **20**, 2883 (1981).
129. P. W. R. Corfield, L. M. Baltusis, and S. J. Lippard, *Inorg. Chem.* **20**, 922 (1981).
130. M. R. Chruchill and H. J. Wasserman, *Inorg. Chem.* **22**, 41 (1983).
131. W. P. Griffith, *Coord. Chem. Rev.* **5**, 459 (1970).
132. W. P. Griffith, *Coord. Chem. Rev.* **8**, 369 (1972).
133. M. Kamata, K. Hirotsu, T. Higuchi, M. Kido, K. Tatsumi, T. Yoshida, and S. Otsuka, *Inorg. Chem.* **22**, 2416 (1983).
134. J. R. Morrow, T. L. Tonker, and J. L. Templeton, *Organometallics* **4**, 745 (1985).
135. J. A. K. Howard, R. F. D. Stansfield, and P. Woodward, *J. Chem. Soc., Dalton Trans.*, 246 (1976).
136. S. R. Allen, M. Green, G. Moran, A. G. Orpen, and G. E. Taylor, *J. Chem. Soc., Dalton Trans.*, 441 (1984).
137. K. A. Mead, H. Morgan, and P. Woodward, *J. Chem Soc., Dalton Trans.*, 271 (1983).
138. N. E. Kolobova, V. V. Skripkin, T. V. Rozantseva, Yu. T. Struchkov, G. G. Aleksandrov, and V. G. Andrianov, *J. Organomet. Chem.* **218**, 351 (1981).
139. K. Tatsumi and R. Hoffmann, *Inorg. Chem.* **19**, 2656 (1980).
140. W. E. Newton, J. W. McDonald, J. L. Corbin, L. Ricard, and R. Weiss, *Inorg. Chem.* **19**, 1997 (1980).
141. K. Tatsumi, R. Hoffmann, and J. L. Templeton, *Inorg. Chem.* **21**, 466 (1982).
142. M. J. S. Dewar, *Bull. Soc. Chim. Fr.* **18**, C71 (1951).
143. J. Chatt and L. H. Duncanson, *J. Chem Soc.*, 2939 (1953).
144. T. A. Albright, R. Hoffmann, J. C. Thibeault, and D. L. Thorn, *J. Am. Chem. Soc.* **101**, 3801 (1979).
145. B. E. R. Schilling, R. Hoffmann, and D. L. Lichtenberger, *J. Am. Chem. Soc.* **101**, 585 (1979).
146. B. E. R. Schilling, R. Hoffmann, and J. W. Faller, *J. Am. Chem. Soc.* **101**, 592 (1979).
147. J. L. Templeton, P. B. Winston, and B. C. Ward, *J. Am. Chem. Soc.* **103**, 7713 (1981).
148. E. A. Robinson, *J. Chem. Soc., Dalton Trans.*, 2373 (1981).
149. P. Kubacek and R. Hoffmann, *J. Am. Chem. Soc.* **103**, 4320 (1981).
150. J. L. Templeton and B. C. Ward, *J. Am. Chem. Soc.* **102**, 6568 (1980).
151. R. Colton and C. J. Rix, *Aust. J. Chem.* **21**, 1155 (1968).
152. J. K. Burdett, *Inorg. Chem.* **20**, 2067 (1981).
153. R. S. Herrick, S. J. N. Burgmayer, and J. L. Templeton, *J. Am. Chem. Soc.* **105**, 2599 (1983).
154. J. L. Templeton, B. C. Ward, G. J.-J. Chen, J. W. McDonald, and W. E. Newton, *Inorg. Chem.* **20**, 1248 (1981).
155. J. L. Templeton and B. C. Ward, *J. Am. Chem. Soc.* **102**, 3288 (1980).
156. M. H. Chisholm, J. C. Huffman, and J. A. Heppert, *J. Am. Chem. Soc.* **107**, 5116 (1985).
157. J. C. Jeffery, I. Moore, H. Razay, and F. G. A. Stone, *J. Chem. Soc., Dalton Trans.*, 1581 (1984).

158. M. Green, J. A. K. Howard, A. P. James, A. N. M. Jelfs, C. M. Nunn, and F. G. A. Stone, *J. Chem. Soc., Chem. Commun.,* 1778 (1985).
159. H. G Alt and H. I. Hayen, *Angew Chem. Suppl.*, 1364 (1983).
161. H. G. Alt, *Z. Naturforsch.* **32B,** 1139 (1977).
162. A. M. Kook, P. N. Nicklas, J. P. Selegue, and S. L. Smith. *Organometallics* **3,** 499 (1984).
163. B. C. Ward and J. L. Templeton, *J. Am. Chem. Soc.* **103,** 3743 (1981).
164. B. A. L. Crichton, J. R. Dilworth, C. J. Pickett, and J. Chatt, *J. Chem. Soc., Dalton Trans.,* 892 (1981).
165. E. Carmona, K. Doppert, J. M. Marin, M. L. Poveda, L. Sanchez, and R. Sanchez-Delgado, *Inorg. Chem.* **23,** 530 (1984).
166. R. Colton and G. R. Scollary, *Aust. J. Chem.* **21,** 1427 (1968).
167. M. H. Chisholm, J. C. Huffman, and R. L. Kelly, *J. Am. Chem. Soc.* **101,** 7615 (1979).
168. T. B. Grindley, K. F. Johnson, A. R. Katritzky, H. J. Keogh, C. Thirkette, and R. D. Topsom, *J. Chem. Soc., Perkin Trans.* 2, 282 (1974).
169. S. R. Allen, R. G. Beevor, M. Green, N. C. Norman, A. G. Orpen, and I. D. Williams, *J. Chem. Soc., Dalton Trans.,* 435 (1985).
170. S. R. Allen, P. K. Baker, S. G. Barnes, M. Bottrill, M. Green, A. G. Orpen, I. D. Williams, and A. J. Welch, *J. Chem. Soc., Dalton Trans.,* 927 (1983).
171. M. Bottrill and M. Green, *J. Am. Chem. Soc.* **99,** 5795 (1977).
172. M. Green, N. C. Norman, and A. G. Orpen, *J. Am. Chem. Soc.* **103,** 1267 (1981).
173. S. R. Allen, M. Green, A. G. Orpen, and I. D. Williams, *J. Chem. Soc., Chem. Commun.,* 826 (1982).
174. M. R. Churchill and W. J. Youngs, *Inorg. Chem.* **18,** 2454 (1979).
175. D.C. Brower, K. R. Birdwhistell, and J. L. Templeton, *Organometallics* **5,** 94 (1986).
176. J. L. Davidson, W. F. Wilson, L. Manojlovic-Muir, and K. Muir, *J. Organomet. Chem.* **254,** C6 (1983).
177. D. S. Gill and M. Green, *J. Chem. Soc., Chem. Commun.,* 1037 (1981).
178. R. G. Beevor, M. Green, A. G. Orpen, and I. D. Williams, *J. Chem. Soc., Chem. Commun.,* 673 (1983).
179. A. Mayr, K. C. Schaefer, and E. Y. Huang, *J. Am. Chem. Soc.* **106,** 1517 (1984).
180. K. R. Birdwhistell, S. J. N. Burgmayer, and J. L. Templeton, *J. Am. Chem. Soc.* **105,** 7789 (1983).
181. D. L. Reger and P. J. McElligott, *J. Am. Chem. Soc.* **102,** 5923 (1980).
182. D. L. Reger, K. A. Belmore, E. Mintz, and P. J. McElligott, *Organometallics* **3,** 134 (1984).
183. S. R. Allen, M. Green, N. C. Norman, K. E. Paddick, and A. G. Orpen, *J. Chem. Soc., Dalton Trans.,* 1625 (1983).
184. J. L. Davidson, M. Shiralian, L. Manojlovic-Muir, and K. W. Muir, *J. Chem. Soc., Chem. Commun.,* 30 (1979).
185. L. Carlton and J. L. Davidson, *J. Chem. Soc., Dalton Trans.,* 895 (1987).
186. L. Carlton, J. L. Davidson, P. Ewing, L. Manojlovic-Muir, and K. W. Muir, *J. Chem. Soc., Chem. Commun.,* 1474 (1985).
187. J. L. Davidson, I. E. P. Murray, P. N. Preston, M. V. Russo, L. Manojlovic-Muir, and K. W. Muir, *J. Chem. Soc., Chem. Commun.,* 1059 (1981).
188. J. L. Davidson, I. E.P. Murray, P. N. Preston, and M. V. Russo, *J. Chem. Soc., Dalton Trans.,* 1783 (1983).
189. J. L. Davidson, *J. Chem. Soc., Chem. Commun.,* 597 (1979).
190. J. L. Davidson, M. Shiralian, L. Manojlovic-Muir, and K. W. Muir, *J. Chem. Soc., Dalton Trans.,* 2167 (1984).

190a. F. J. Feher, M. Green, and R. A. Rodrigues, *J. Chem. Soc., Chem. Commun.*, 1206 (1987).
191. R. F. Heck, "Organotransition Metal Chemistry." Academic Press, New York, 1974.
192. A. Efraty, *Chem. Rev.* **77,** 691 (1977).
193. H. Yamazaki and N. Hagihara, *J. Organomet. Chem.* **21,** 431 (1970).
194. P. M. Maitlis, *Accts. Chem. Res.* **9,** 93 (1976).
195. R. S. Dickson and P. J. Fraser, *Adv. Organomet. Chem.* **12,** 323 (1974).
196. J. L. Davidson, M. Green, J. A. K. Howard, S. A. Mann, J. Z. Nyathi, F. G. A. Stone, and P. Woodward, *J. Chem. Soc., Chem. Commun.*, 803 (1975).
197. H. G. Alt, *Angew. Chem. Int. Ed. Engl.* **15,** 759 (1976).
198. H. G. Alt, M. E. Eichner, and B. M. Jansen, *Angew, Chem. Int. Ed. Engl.* **21,** 861 (1982).
199. H. G. Alt, M. E. Eichner, and B. M. Jansen, *Angew. Chem. Suppl.*, 1826 (1982).
200. B. L. Booth and R. G. Hargreaves, *J. Chem. Soc. A.*, 308 (1970).
201. P. L. Watson and R. G. Bergman, *J. Am. Chem. Soc.* **101,** 2055 (1979).
202. J. L. Davidson, M. Green, J. Z. Nyathi, C. Scott, F. G. A. Stone, A. J. Welch, and P. Woodward, *J. Chem. Soc., Chem. Commun.*, 714 (1976).
203. H. G. Alt and U. Thewalt, *J. Organomet. Chem.* **268,** 235 (1984).
204. E. O. Fischer, *Adv. Organomet. Chem.* **14,** 1 (1976).
205. H. G. Alt, H. I. Hayen, H. P. Klein, and U. Thewalt, *Angew. Chem. Int. Ed. Engl.* **23,** 809 (1984).
206. J. L. Davidson, *J. Chem. Soc., Chem. Commun.*, 113 (1980).
207. J. L. Davidson, *J. Chem. Soc., Dalton Trans.*, 2715 (1987).
208. M. A. Bennett, I. W. Boyd, G. B. Robertson, and W. A. Wickramasinghe, *J. Organomet. Chem.* **290,** 181 (1985).
209. J. L. Davidson, W. F. Wilson, and K. W. Muir, *J. Chem. Soc., Chem. Commun.*, 460 (1985).
210. J. L. Davidson and W. F. Wilson, *J. Chem. Soc., Dalton Trans.*, 27 (1988).
211. J. L. Davidson, L. Manojlovic-Muir, K. W. Muir, and A. N. Keith, *J. Chem. Soc., Chem. Commun.*, 749 (1980).
212. J. L. Davidson, G. Vasapollo, J. C. Millar, and K. W. Muir, *J. Chem. Soc., Dalton Trans.*, 2165 (1987).
213. P. K. Baker, G. K. Barker, M. Green, and A. J. Welch, *J. Am. Chem. Soc.* **102,** 7812 (1980).
214. F. Y. Petillon, F. Le Floch-Perennou, J. E. Guerchais, D. W.A. Sharp, L. Manojlovic-Muir, and K. W. Muir, *J. Organomet. Chem.* **202,** 23 (1980).
215. L. J. Manojlovic-Muir and K. W. Muir, *J. Organomet. Chem.* **168,** 403 (1979).
216. H. Brix and W. Beck, *J. Organomet. Chem.* **234,** 151 (1982).
217. M. Crocker, M. Green, A. G. Orpen, H. P. Neumann, and C. J. Schaverien, *J. Chem. Soc., Chem. Commun.*, 1351 (1984).
218. T. Blackmore, M. I. Bruce, and F. G. A. Stone, *J. Chem. Soc., Dalton Trans.*, 106 (1974).
219. G. J.-J. Chen, J. W. McDonald, and W. E. Newton, *Organometallics* **4,** 422 (1985).
220. F. J. Brown, *Prog. Inorg. Chem.* **27,** 1 (1980).
221. A. Greco, F. Pirinoli, and G. Dall'asta, *J. Organomet. Chem.* **60,** 115 (1973).
222. M. Kersting, K. Dehnicke, and D. Fenske, *J. Organomet, Chem.* **309,** 125 (1986).
223. K. Stahl, F. Weller, and K. Dehnicke, *Z. Anorg. Allg. Chem.* **533,** 73 (1986).
224. K. H. Theopold, S. J. Holmes, and R. R. Schrock, *Angew. Chem. Suppl.*, 1409 (1983).
225. E. C. Walborsky, D. E. Wigley, E. Roland, J. C. Dewan, and R. R. Schrock, *Inorg. Chem.* **26,** 1615 (1987).

226. N. G. Bokiy, Yu. V. Gatilov, Yu. T. Struchkov, and N. A. Ustynyuk, *J. Organomet. Chem.* **54**, 213 (1973).
227. E. A. Maatta, R. A. D. Wentworth, W. E. Newton, J. W. McDonald, and G. D. Watt, *J. Am. Chem. Soc.* **100**, 1320 (1978).
228. D. D. Devore, E. A. Maatta, and F. Takusagawa, *Inorg. Chim. Acta* **112**, 87 (1986).
229. K. H. Dotz and J. Muhlemeier, *Angew. Chem. Suppl.*, 2023 (1982).
230. D. J. Wink, J. R. Fox, and N. J. Cooper, *J. Am. Chem. Soc.* **107**, 5012 (1985).
231. D. F. Foust and M. D. Rausch, *J. Organomet. Chem.* **239**, 321 (1982).
231a. M. J. McGeary, A. S. Gamble and J. L. Templeton, *Organmetallics* **7**, 271 (1988).
232. F. A. Cotton and W. T. Hall, *Inorg. Chem.* **19**, 2354 (1980).
233. F. A. Cotton and W. T. Hall, *J. Am. Chem. Soc.* **101**, 5094 (1979).
234. M. D. Curtis and J. Real, *Organometallics* **4**, 940 (1985).
235. S. J. McLain, R. R. Schrock, P. R. Sharp, M. R. Churchill, and W. J. Youngs, *J. Am. Chem. Soc.* **101**, 263 (1979).
236. E. Hey, F. Weller, and K. Dehnicke, *Z. Anorg. Allg. Chem.* **514**, 25 (1984).
237. D. B. Pourreau, R. R. Whittle, and G. L. Geoffroy, *J. Organomet. Chem.* **273**, 333 (1984).
238. J. M. Mayer, D. L. Thorn, and T. H. Tulip, *J. Am. Chem. Soc.* **107**, 7454 (1985).
239. J. M. Mayer, T. H. Tulip, J. C. Calabrese, and E. Valencia, *J. Am. Chem. Soc.* **109**, 157 (1987).
240. J. M. Boncella, M. L. H. Green, and D. O'Hare, *J. Chem. Soc., Chem. Commun.*, 618 (1986).
241. T. V. Harris, J. .W. Rathke, and E. L. Muetterties, *J. Am. Chem. Soc.* **100**, 6966 (1978).

ADVANCES IN ORGANOMETALLIC CHEMISTRY, VOL. 29

Unsaturated Dimetal Cyclopentadienyl Carbonyl Complexes

MARK J. WINTER

Department of Chemistry
University of Sheffield
Sheffield S3 7HF, England

I

INTRODUCTION

The reactions and nature of multiple bonds between two metal atoms form a lively area of research and debate *(1,2)*. Many examples of triple bonds between metals are known, including the extensively studied molybdenum or tungsten alkoxide, $M_2(OR)_6$, and amido, $M_2(NR_2)_6$, complexes *(3)*. Classic examples of quadruple bonds are epitomized by $[Re_2Cl_8]^{2-}$ *(4)* and molybdenum acetate, $Mo_2(OCOMe)_4$ *(5)*.

Another highly important class of unsaturated dimetal compounds is based on the general formula $Cp_2M_2(CO)_x$ ($Cp = C_5H_5$) and forms the subject of this article. In this class, perhaps the complexes $Cp_2M_2(CO)_4$ **1–3** attracted most early attention. Of these, **2** is most studied. Although not the first discovered, it is undoubtedly the easiest to make in large quantities, and this fact is not the least important reason for the attention it has received.

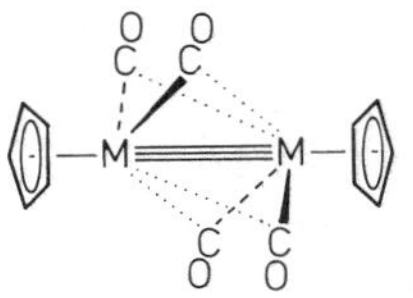

1, M = Cr
2, M = Mo
3, M = W

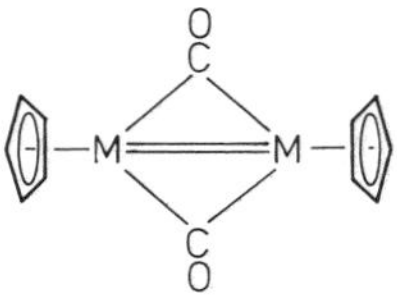

4, M = Co
5, M = Rh
6, M = Ir

The advent of convenient high yield sources of Cp^*H (C_5Me_5H) *(6)* opened up new vistas of cyclopentadienyl metal chemistry research. In this article, complexes X^* with Cp* ligands are distinguished from the corresponding Cp complexes X by an asterisk. The differing electronic and steric requirements of the Cp and Cp* ligands can have quite dramatic effects on reactivity at the metal atom to which they are bound. Further, the ability to isolate compounds is sometimes affected. This helps explains why the permethyl derivatives $Cp^*_2M_2(\mu\text{-}CO)_2$ **4*** and **5*** are well known while the cyclopentadienyl analogs **4** and **5** are scarcely reported. Some In [indenyl, $(\eta^5\text{-}C_9H_7)$] and Tp [hydrido(trispyrazolyl)borate, $HB(C_3H_3N_2)_3$] derivatives are known, but they are much less common; they are mentioned where appropriate.

One useful physical effect of the Cp* group is the greater solubility of compounds generally displayed as compared to their Cp analogs. However, the differences in chemistry often demonstrated by the Cp* compounds may or may not be considered desirable. Enhanced solubility relative to the Cp compounds is also often achieved by using the Cp' (C_5H_4Me) (complexes denoted as **1'**, etc.) ligand. While the solubility of the Cp' species is often higher than that of the Cp compounds, the chemistry is generally unaffected by the single methyl substituent.

II

BOND ORDER FORMALISMS IN Cp$_2$M$_2$(CO)$_n$ COMPLEXES

The $CpM(CO)_2$ (M = Group 6 metal, d^6) unit is a 15-electron fragment. If the 18-electron rule is to be followed, then two such fragments must be linked by a triple bond.

On this basis, formally at least, the series **1–3** contain metal–metal triple bonds. As we shall see, the nature of the metal–metal bonding is quite different from that of carbon–carbon bonding in ethyne. Similarly, it is useful to regard the metal–metal bond order in the series **4–6** as double. However, the nature of the M=M bond is quite different from the C=C bond in ethene. The carbonyls are semi-bridging or bridging in **1–6** and are extensively involved in the metal–metal interactions. This has resulted in some differences of opinion as to whether one should really regard the metal–metal bonds as multiple.

While the unsaturation displayed by **1–6** is clearly different in nature from that in ethyne and ethene, writing three or two lines, as appropriate, to connect the metals is nevertheless a good formalism. They are a useful pointer to the *nature of their chemical reactivity* and are not intended as an indication of the *nature of the bonding*.

III

STOICHIOMETRIES AND OCCURRENCE

Let us delay, for the present, consideration of the nature of the unsaturation in these complexes $Cp_2M_2(CO)_x$. Connection of metals by lines indicating covalent two-electron bonds allows us to construct a series (Table I) of simple structures with formula $Cp_2M_2(CO)_x$. The 18-electron rule is satisfied for each metal by writing in single, double, or triple metal–metal bonds. Arbitrarily, the number of bridging carbonyls is kept to a minimum. We are concerned here with the nature and reactivity of unsaturated dimetal $Cp_2M_2(CO)_x$ compounds, and so a discussion of the complexes with single metal–metal bonds, while interesting, is not included.

Viewed simplistically, the stoichiometry $Cp_2M_2(CO)_5$ could reasonably apply to complexes $Cp_2M_2(CO)_4(\mu\text{-}CO)$ (M≡M, M = V, Nb, and Ta) and $Cp_2M_2(CO)_4(\mu\text{-}CO)$ (M=M, M = Cr, Mo, and W). The only reports of this stoichiometry for the Group 6 elements concern $Cp_2W_2(CO)_5$ (7) and $Cp_2Mo_2(CO)_5$ (8), but these fascinating structures probably do not contain metal–metal double bonds. This leaves the Group 5 elements. The niobium and tantalum species are unknown, but $Cp_2V_2(CO)_5$ (7) (9) and $Cp^*_2V_2(CO)_5$ (7*) (10) are known. The actual structure, often represented by 7a or 7b, is rather different from the simplistic molecule $Cp_2V_2(CO)_4$-$(\mu\text{-}CO)$ containing a V≡V bond.

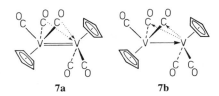

7a   7b

With the same simplistic ideas, the metal–metal triple bonded molecules $Cp_2M_2(CO)_4$ 1–3 and metal–metal double bonded $Cp_2M_2(CO)_4$ (M = Mn, Tc, and Re) are predicted. No molecules of this formula are known for the Group 7 metals. The Group 6 molecules 1–3 (11–13) are very well known. The Cp*derivatives 1*–3* (14–16) are also well known for the Group 6 elements, and their chemistry shows interesting parallels and differences with the Cp analogs. The formulas $Cp_2M_2(CO)_3$ (M≡M, M = Mn, Tc, Re; M=M, M = Fe, Ru, Os) are to be expected. Of the two series 8–10 and 11–13, only the Cp* complexes 8* (17), 10* (18), and 11* (19) have been isolated. Each has three bridging carbonyls rather

TABLE I

STOICHIOMETRIES AND SIMPLISTIC STRUCTURES OF COMPLEXES WITH FORMULA
$L_2M_2(CO)_x$ (L = Cp, Cp', OR Cp*)

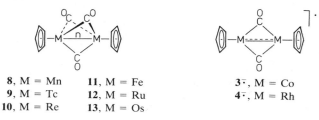

n	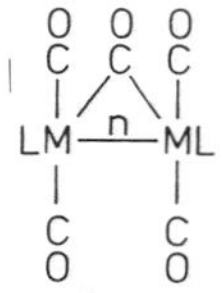	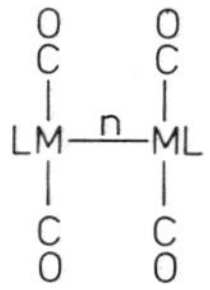	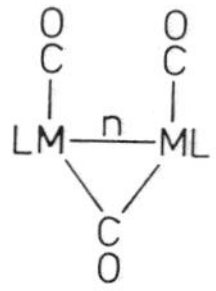	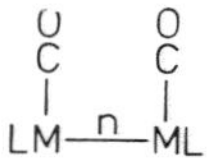
3	V, Nb, Ta	Cr, Mo, W	Mn, Tc, Re	Fe, Ru, Os
2	Cr, Mo, W	Mn, Tc, Re	Fe, Ru, Os	Co, Rh, Ir
1	Mn, Tc, Re	Fe, Ru, Os	Co, Rh, Ir	Ni, Pd, Pt

than the single one shown in Table I. The parent **11** (*20*) has not been isolated but is recognized as a transient intermediate in photolyses of $Cp_2Fe_2(CO)_4$.

8, M = Mn   **11**, M = Fe   **3⁻**, M = Co
9, M = Tc   **12**, M = Ru   **4⁻**, M = Rh
10, M = Re   **13**, M = Os

The metal–metal triple bonded complexes $Cp_2M_2(CO)_2$ (M = Fe, Ru, and Os) and the metal–metal double bonded compounds $Cp_2M_2(CO)_2$ (M = Co, Rh, and Ir) are anticipated for the dicarbonyl stoichiometry. The Group 8 series is unknown. An extensive chemistry exists for the Group 9 series although mainly for their Cp* analogs **4*** and **5*** (*21,22*). The iridium species **6*** has only been mentioned briefly (*23*). The parent cobalt complex **4** is known (*24*) and isolable, but little of its chemistry is known. The rhodium derivative **5** was reported in 1986 (*25*). It is thermally unstable at temperatures over 200 K. The Group 9 stoichiometries are further represented by radical monoanions $[Cp_2M_2(\mu\text{-}CO_2]^{\overline{\cdot}}$. The cobalt derivatives **3⁻** (*26,27*) and **3*⁻** (*28*) are particularly well known. The rhodium analog **4*⁻** is a relatively recent discovery (*29*). These have a formal metal–metal bond order of 1.5.

None of the monocarbonyl complexes $Cp_2M_2(CO)$ (M=M, M = Co, Rh, and Ir) or (M=M, M = Ni, Pd, and Pt) appear to be known, but the two complexes $L_2Pt_2(CO)$ (L = Cp and Cp*) were very recently characterized by fast time-resolved IR spectroscopy (*29a*).

IV

STRUCTURES

A. Pentacarbonyls

The X-ray crystal structure of $Cp_2V_2(CO)_5$ (**7**) (Fig. 1) has been determined twice, one of these analyses at $-150°C$ (*30–33*). It is quite clearly different from that of $Cp_2Re_2(CO)_5$ (**14**) (*34*) with which it might have been expected to be isostructural. The IR spectrum of $Cp_2V_2(CO)_5$ (*31*) is also

14

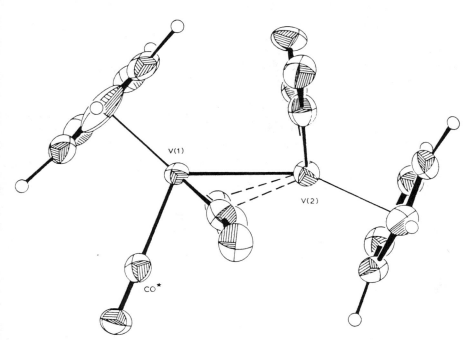

FIG. 1.   X-Ray crystal structure of $Cp_2V_2(CO)_5$ (**7**), taken with permission from Ref. *33*, copyright Elsevier Sequoia S.A.

quite clearly different from that of $Cp_2W_2(CO)_5$ (**7**). If it had been iso-structural with **14**, a formal V≡V triple bond could have been written as in Table I.

There is some confusion as to a suitable representation for the actual structure of **7** shown in Fig. 1. The metal–metal bond is 2.462 Å long and seems too short for a V—V single bond. Certainly, the reactivity of **7** is consistent with multiple bonding. What is at question is its nature and the role of the semi-bridging carbonyls.

Its structure is represented as **7a** in more recent papers (*33,35*), but **7b** (*31,36*) and other representations have also been used. The two semi-bridging carbonyls are 2.42 Å (average) from the second V atom (*32*). While this distance is clearly too long for a full bond there must be some orbital overlap with the second metal. The two semi-bridging carbonyls, taken together, are suggested to function as a net two-electron donor (*32*) to the second, formally electron deficient, vanadium. This would allow the 18-electron rule to be satisfied for both metals provided there is a V=V bond. Structure **7b** requires that the two semi-bridging carbonyls are *electron acceptors* (*31,36*). Unfortunately, it is not easy to say whether a particular semi-bridging carbonyl is functioning as an electron donor or acceptor to a second metal. For instance, the linear semi-bridge Mo—Mo—CO bond angles in $Cp_2Mo_2(CO)_4$ (see below), for which calculations suggest the carbonyls are electron acceptors, are not strikingly different from the V—V—CO bond angles of 65° (average) in **7**.

Given that 1 mol of $Cp_2V_2(CO)_5$ reacts with 3 mol of PEt_2Ph to cleave the VV interaction (see below), one might be forgiven for suggesting that from a chemical point of view the VV bond order is effectively three! From a chemical point of view structure **7c** is useful. There are parallels between such a structure and the crystallographically characterized **15** which has a metal–metal single dative bond and semi-bridging electron acceptor carbonyl ligands (*37*).

7c   15

B. *Tetracarbonyls*

The X-ray crystal structures of $Cp_2M_2(CO)_4$ (M = Cr and Mo) (Fig. 2) and $Cp^*_2M_2(CO)_4$ (M = Cr and Mo) are all known. There are significant

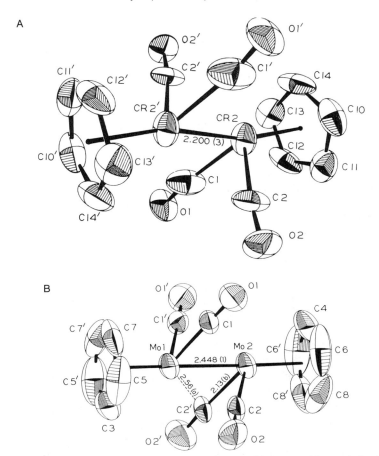

FIG. 2.   X-Ray crystal structures of (A) $Cp_2Cr_2(CO)_4$ (**1**) (taken with permission from Ref. *38*, copyright Elsevier Sequoia S.A.) and (B) $Cp_2Mo_2(CO)_4$ (**2**) (taken with permission from Ref. *41*, copyright American Chemical Society).

differences between the four structures (*38–44*) (Table II). The Cp_2-$Mo_2(CO)_4$ molecule is the odd one out. The CpMoMoCp unit in **2** is virtually linear (*41*) whereas the ring ligands bend away from the MM axis in all the other cases (*38–40,42–44*) including the closely related In_2Mo_2-$(CO)_4$ molecule (*44*).

There are even quite striking differences in the geometries of **2** (*41*) and **2*** (*42*). The four carbonyls bend back over the MoMo bond in **2** in an equivalent fashion with CO—Mo—Mo bond angles averaging 67°. In **2***,

TABLE II

PERTINENT X-RAY CRYSTAL DATA FOR $Cp_2M_2(CO)_4$ COMPLEXES[a]

Complex	MM distance (Å)	Cp—M—M'(°)	M—C—O (°)	M'—M—CO (°)	Refs.
$Cp_2Cr_2(CO)_4$ (1)	2.200(3)	159	171, 173	75, 75	38
	2.230(3)	165	171, 168	72, 72	
$Cp*_2Cr_2(CO)_4$ (1*)	2.276(2)	159	171, 175	73, 79	39,40
$Cp_2Mo_2(CO)_4$ (2)	2.4477(12)	Nearly 180	176	67	41
$Cp*_2Mo_2(CO)_4$ (2*)	2.488(3)	168	154, 169	57, 70	42
$Tp_2Mo_2(CO)_4$	2.507(1)	153	170, 174	70, 83	43
$In_2Mo_2(CO)_4$	2.500(1)		171, 174	71, 75	44

[a] Values are given for the two independent molecules that crystallize within the unit cell for $Cp_2Cr_2(CO)_4$. For all molecules except $Cp_2Mo_2(CO)_4$, where all four carbonyls can be grouped together, the four carbonyls are best grouped into two pairs. Quoted angles are averages for each pair. Some of the carbonyl angles are unreliable owing to disorder problems.

and the other molecules, the degree of bridging is markedly different for the two pairs of carbonyls. The structures of the two chromium species 1 (38) and 1* (39,40) are far more closely related to each other but show distinct differences from the two molybdenum structures. In no cases are any of the carbonyls symmetrically bridging as in $Cp_2Fe_2(CO)_4$.

The origin of the "lean back" displayed by the carbonyls is interesting. In the sense that all the M'—M—CO angles in Table I are less than 90°, they can be said to "semi-bridge" (36). There is some debate as to whether the origin of the nonlinearity of the M—C—O angles is steric or electronic. Both factors are likely to have influence. Certainly in the $Cp*_2Cr_2$-$(CO)_4$ case, for instance, the direction of the bend is that expected for nonbonded interactions between the carbonyls and the Cp* ring of the second metal (39,40). The bending has also been rationalized, however, by postulation of a donor interaction from the carbonyl orbitals into orbitals of the second metal (38,41). A $(\sigma + \pi)$ interaction would result in the carbonyls acting as formal four-electron donors. Set against this, extended Huckel (45) and Fenske–Hall (46) molecular orbital calculations suggest that the carbonyls are not donors to the second metal at all. Rather, the metals donate electrons back into the π^* orbitals of the semi-bridging carbonyls. The Mo≡Mo bond can then be thought of as a little longer than expected due to this slight interaction with the CO π^* orbitals. This concept is more in keeping with the original proposals concerning the electronic nature of semi-bridging carbonyls (36). This is not to say that a carbonyl cannot act as a four-electron $(\sigma + \pi)$ donor. Well-documented examples of such include $Mn_2(CO)_4(\sigma,\pi\text{-}CO)(Ph_2CH_2CH_2PPh_2)$ (47–49)

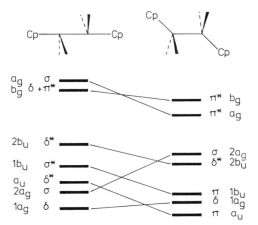

FIG. 3. Schematic representation of the frontier orbitals of $Cp_2Mo_2(CO)_4$ (2) corresponding to the observed geometry (left) and an idealized octahedral geometry (right) (45).

and $CpMo(CO)_2(\sigma, \pi\text{-}CO)NbCp_2$ (50). The $\sigma + \pi\text{-}CO$ in these cases however is manifest by ν_{CO} stretches at 1665 and 1560 cm^{-1}, respectively, whereas those of $Cp_2Mo_2(CO)_4$ are rather higher (ν_{CO} 1900 and 1850 cm^{-1}) (12).

The extended Hückel calculations (45) start from an unbridged geometry where the OC—Mo—CO and Mo—Mo—C bond angles are both 90° while the Mo—Mo—Cp bond angle is set to 125.3°. This leads to a bond description including characteristic "five below two" frontier orbitals shown in Fig. 3. The five highest occupied molecular orbitals are π, δ, π, δ^*, and σ and allow a correlation with the triple bond assigned on the basis of the 18-electron rule.

In moving from the idealized structure to the real structure, a slight change in OC—Mo—CO angle is necessary but a large change is required in Mo—Mo—C angle. The latter necessitates quite a large change in the MoMoCp angle toward linear. Although the geometric change is quite large, there is only a 2 kcal mol^{-1} difference in energy between the idealized and real structures. The overall energy surface is very "soft," accounting for the considerable variation in Cp—M—M' angle seen in Table I. The calculations show that in moving to the real geometry, the metals lose electron density to the carbonyls. The metal orbitals act as donors to carbonyl π^* acceptor orbitals. This is illustrated for just one carbonyl. The carbonyl participates in a three-center–two-electron bond in an interaction with the filled symmetric combination of metal π orbitals.

The effect of positioning the carbonyls over the MoMo axis is to increase backbonding at the expense of metal–metal π bonding interactions. The π a_u bond becomes rather δ^* in character while the $1b_u$ π orbital becomes σ^* in character. The resultant bonding metal to carbonyl interactions are shown in Fig. 4. The top overlap approaches δ^* in nature and that beneath σ^*. It is still not clear why the carbonyls are linear, but perhaps any deviation from linearity would diminish the metal donor to CO π^* interaction (46).

This understanding that the MoMo bond order in **2** and similar species should be taken as rather less than three does not detract from the usefulness of writing a triple bond. Compound **2** and relatives display many chemical characteristics of multiple bonding. Their nature is to react as unsaturated molecules.

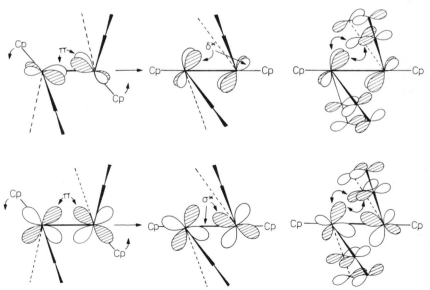

FIG. 4.   The change from π bonding to δ^* and σ^* on going from a bent CpMoMoCp axis to linear in $Cp_2Mo_2(CO)_4$ (**2**) (45).

C. Tricarbonyls

The X-ray crystal structures of $Cp^*_2Mn_2(\mu\text{-}CO)_3$ (8^*) (51) [Mn≡Mn = 2.170(1) Å], $Cp^*_2Re_2(\mu\text{-}CO)_3$ (10^*) (18) [Re≡Re = 2.411(1) Å], and $Cp^*_2Fe_2(\mu\text{-}CO)_3$ (11^*) (19) [Fe=Fe = 2.265(1) Å] (Fig. 5) are clearly closely related to each other. The $M_2(CO)_3$ cores are highly symmetrical with approximate D_{3h} symmetry. The Group 7 molecules are 30-electron systems with electronic similarities (52) to $Fe_2(CO)_9$ which also has three bridging carbonyls (53). The Fe=Fe distance at 2.265(1) Å in 8^*

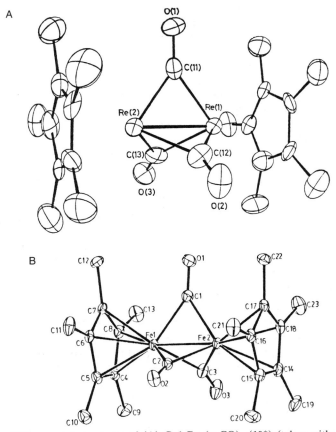

FIG. 5.   X-Ray crystal structures of (A) $Cp^*_2Re_2(\mu\text{-}CO)_3$ (10^*) (taken with permission from Ref. 18, copyright Royal Society of Chemistry) and (B) $Cp^*_2Fe_2(\mu\text{-}CO)_3$ (11^*) (taken with permission from Ref. 19, copyright American Chemical Society).

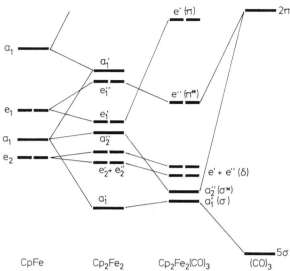

FIG. 6.   Qualitative molecular orbital diagram for the $Cp_2M_2(\mu\text{-}CO)_3$ molecules **8–13** based on calculations for $Cp_2Fe_2(\mu\text{-}CO)_3$ (**8**) (*19*). The δ level is the HOMO for the Group 7 molecules whereas there are two unpaired electrons in the π^* level for the Group 8 species.

is clearly shorter than that normally associated with the Fe—Fe single bond often written for $Fe_2(CO)_9$ (FeFe 2.52 Å) (*52,53*). Based on the 18-electron rule, one can write a triple metal–metal bond for the Group 7 species and a double bond for the iron compound.

As for the Group 6 molecules, these formalisms are useful but not particularly realistic. The energy level diagram of the 30-electron compounds $Cp^*_2M_2(\mu\text{-}CO)_3$ (M = Mn and Re) (Fig. 6) is similar to that of the 34-electron $Fe_2(CO)_9$ (*54*). The three bridging groups are heavily implicated in the bonding and help to render a simple assignment of metal–metal bonding orbitals unrealistic. In molecular orbital schemes, specific metal–metal bonds are not readily assigned (*54*). If four extra electrons were to be placed in the 30-electron compounds, they would have to be placed in an e'' level that is antibonding with respect to metal–metal bonding. The fact that those four electrons are not there helps to account for the short metal–metal distance.

The 32-electron iron compound **11*** is paramagnetic (*19*). This is ascribed to the ground state having a triplet structure with the HOMO π^* (e'', overlap of metal d_{xz} and d_{yz} orbitals) relative to FeFe bonding (*19*), and only half occupied by two electrons.

D. Dicarbonyls

The $Cp_2M_2(CO)_2$ structure is characterized by two bridging carbonyls. The structures of $Cp^*_2Co(\mu\text{-}CO)_2$ (4*) (21,28,55) and $Cp^*_2Rh_2(\mu\text{-}CO)_2$ (5*) (56) and the heterobimetallic $Cp^*_2CoRh(CO)_2$ are all known (Table III, Fig. 7). The last structure is very similar to that of $Cp^*_2Rh_2(CO)_2$ (5*), but the Co=Rh bond is shorter than the arithmetic mean of the corresponding homodinuclear complexes (57).

Some of these molecules undergo one-electron reduction to species of general formula $[L_2M_2(CO)_2]^{\overline{\cdot}}$ (L = Cp or Cp*). Complexes $4^{\overline{\cdot}}$ and $4^{*\overline{\cdot}}$ are crystallographically characterized. Their formal metal–metal bond order is 1.5 (see below). Also included in Table III are data for Cp_2Co_2-(CO)(NO) (isoelectronic to $4^{\overline{\cdot}}$), $Cp_2Fe_2(NO)_2$ (Fe=Fe bond, isoelectronic to 4), $Cp_2Co_2(NO)_2$ with a single Co—Co bond, and a rhodium dianion $[Cp^*_2Rh_2(CO)_2]^{2-}$ (Rh—Rh). A number of points arise. Provided we consider similar metals, the metal–metal bond lengths show less than dramatic changes on going from bond orders 2 to 1.5 to 1. The reasons for such small changes in these distances are as yet unexplained. Extended Huckel calculations predict larger changes than those observed (61,62).

The nature of the metal–metal interaction is fascinating. In a Cp_2M_2 fragment, the metal d_{xz} and d_{yz} orbitals form two π and two π^* bonds. The two π bonds are the HOMOs (Fig. 8) in a d^8–d^8 Cp_2M_2 system. This results in a bond order for Cp_2M_2 of two between the metals, but made up from two π bonds rather than a σ and π bond. There are metal–metal σ

TABLE III

KEY CRYSTALLOGRAPHIC DATA FOR SOME $Cp_2M_2(\mu\text{-}CO)_2$ (M = Co AND Rh) COMPLEXES AND RELATED NITROSYL DERIVATIVES[a]

Complex	M=M (Å)	M—E—M[a] (°)	M—E—O (°)	E—M—E' (°)	Refs.
$Cp^*_2Co_2(CO)_2$	2.338	78	141	102	21,28,55
	2.327		140 (av)		
$Cp^*_2Rh_2(CO)_2$	2.564				56
$[Cp_2Co_2(CO)_2]^{\overline{\cdot}}$	2.364	81	139	99	27,28,57
$[Cp^*_2Co_2(CO)_2]^{\overline{\cdot}}$	2.372	81	139	98	28,55
$[Cp^*_2Rh_2(CO)_2]^{2-}$	2.606				29
	2.613				
$Cp^*_2CoRh(CO)_2$	2.404				56
$Cp_2Co_2(CO)(NO)$	2.370	81	139	99	58,59
$Cp_2Co_2(NO)_2$	2.372	81	139	99	58
$Cp_2Fe_2(NO)_2$	2.326	82	139	98	60

[a] E = C or N as appropriate.

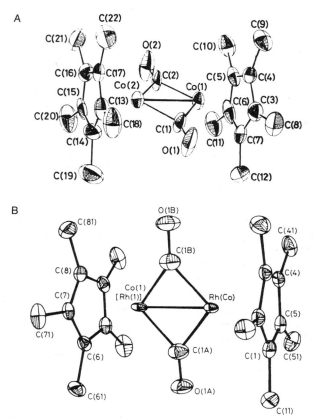

Fig. 7. X-Ray crystal structures of (A) $Cp^*_2Co_2(\mu\text{-}CO)_2$ (4*) (taken with permission from Ref. 55, copyright Royal Society of Chemistry) and (B) $Cp^*_2CoRh(\mu\text{-}CO)_2$ (taken with permission from Ref. 57, copyright Royal Society of Chemistry).

and σ^* orbitals. Both are occupied and lower in energy than the bonding π orbitals: the σ metal interactions are not great. If the σ metal–metal interaction were greater, then the σ^* orbital would be destabilized over the π^* orbitals. For a d^8–d^8 system, occupancy would then be $\sigma^2\delta^4\delta^{*4}\pi^4\pi^{*2}$. One of the metal–metal bonds would be σ as perhaps generally envisaged when writing a line between two metals. At this stage, the nature of the double bond is similar to that of O_2.

Crucial interactions occur when two carbonyls are placed in the xz plane. The π_x orbital is destabilized over the π^* orbitals through an interaction with carbonyl σ orbitals while the π_x^* and π_y orbitals are stabilized by backbonding. The effect is to give a predicted occupancy of $(\pi^*_x)^2(\pi_y)^2$ for

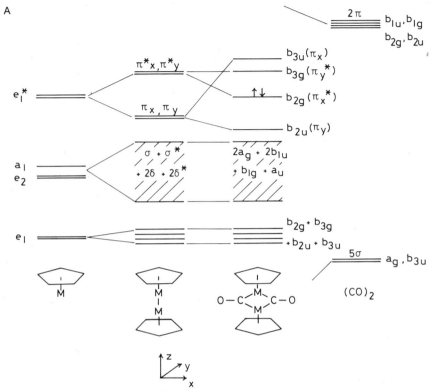

FIG. 8.   Molecular orbital scheme for a $Cp_2M_2(\mu\text{-}CO)_2$ system. (A) An energy level scheme derived by successive perturbations while (B) the diagrams represent the bridging orbitals. Taken with permission from Ref. 62, copyright Royal Society of Chemistry.

the two HOMOs. Photoelectron spectra of $Cp^*_2M_2(CO)_2$ (M = Co and Rh) lend support to these predictions (62). The point is that the metal–metal interaction is now seen as made up from two highly delocalized orbitals of π type symmetry involving the bridging groups. Effectively, there are no direct metal–metal bonds. The two lines written between the two metals for these dicarbonyl dimers are therefore not to be confused with the two lines written between the carbon atoms in ethene.

The monoanions $[Cp_2M_2(CO)_2]^{\bar{}}$ have an extra electron compared to the parent neutral molecules $Cp_2M_2(CO)_2$. Various other directly analogous substituted cyclopentadienyl anions including $[(C_5H_4R)_2\text{-}Co_2(\mu\text{-}CO)_2]^{\bar{}}$ (R = CO_2Me, $SiPh_2Me$, $SiMe_3$, and CH_2CO_2Me) are known (63). The extra electron resides in the out of plane π^*_{yz} made up

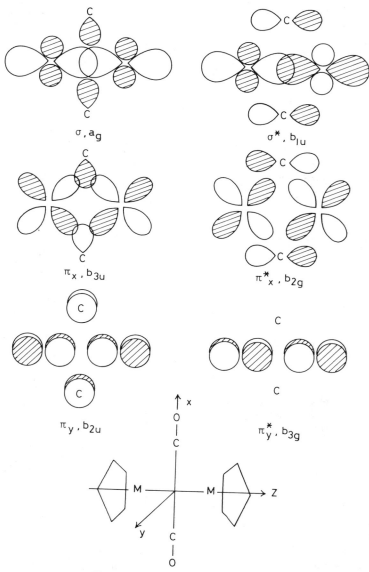

B

σ , a_g

σ^* , b_{1u}

π_x , b_{3u}

π^*_x , b_{2g}

π_y , b_{2u}

π^*_y , b_{3g}

FIG. 8B. See legend on p. 115.

from the out of phase combination of d_{yz} orbitals (Fig. 8). This reduces the formal metal–metal bond order to 1.5 although there is little change in the metal–metal distances. It is noticeable that the monoanion $4^{\overline{\cdot}}$ is less inclined to decomposition than the neutral species **4**. This is attributable to greater d orbital backbonding to the carbonyls (*64,65*).

V

OTHER SYSTEMS RELATED TO Cp$_2$M$_2$(CO)$_n$

We are concerned in this article with cyclopentadienyl systems. It should be recognized, however, that a d^n fragment CpM is electronically related to a d^{n-1} metal–arene fragment $M(\eta\text{-}C_6R_6)$ as well as a d^{n+1} metal–cyclobutadiene $M(\eta\text{-}C_4R_4)$ fragment. As examples, based on the existence of Cp*$_2$Mn$_2(\mu\text{-}CO)_3$, one might expect the formally equivalent molecules Cr$_2(\mu\text{-}CO)_3(\mu\text{-}C_6R_6)_2$ and Fe$_2(\mu\text{-}CO)_3(\eta\text{-}C_4R_4)$ to exist. In fact, examples of both exist and include **16** (M = Cr, R = H) (*66*) and **17** [M = Fe, R_4 = Ph$_2$Bu$'_2$ (*67*), Ph$_4$ (*67*), H$_4$ (*68*), or Me$_4$ (*69*)]. All but **17** (R_4 = H$_4$) which is written as $(\eta\text{-}C_4H_4)_2Fe_2(CO)_2(\mu\text{-}CO)$ have a symmetrical $(\mu\text{-}CO)_3$ system bridging a metal–metal triple bond. Clearly, a number of series of such molecules can be formulated, but they do not all yet exist. Another arene complex is $(\eta\text{-}C_6H_6)_2V_2(CO)_4$ (*70*) which has a V≡V bond (2.15 Å) and is isoelectronic with Cp$_2$Cr$_2$(CO)$_4$. There is certainly potential for the synthesis of some interesting molecules based on these relationships.

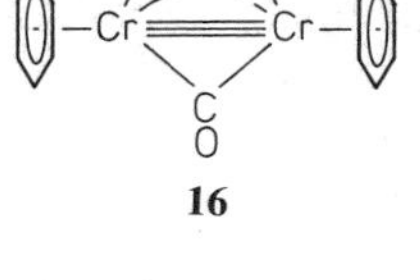

16 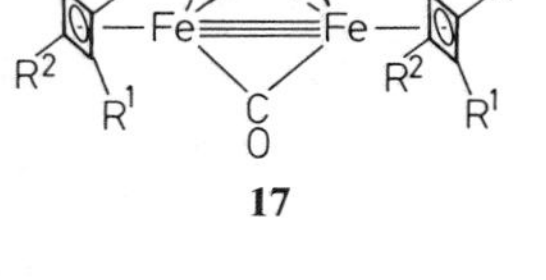   **17**

Mixed systems exist. Treatment of the 20-electron compound Fe-(C$_6$Me$_6$)$_2$ with Cp*M(CO)$_2$ (M = Co, Rh, or Ir) leads to **18–20** which are clearly analogous to the **4–6** series (*71*). More exotic molecules include $(\eta\text{-}C_4Me_4)Fe(\mu\text{-}CO)_3MnCp^*$ (**21**) (*69*), $(\eta\text{-}C_4Me_4)Co(\mu\text{-}CO)_3MCp$ (**22** and **23**) (*72,73*), and CpMoMn(CO)$_7$ (**24**) (*74*). Based on the 18-electron formalism, Cp*$_2$Os$_2(\mu\text{-}H)_2(\mu\text{-}CO)_3$ (*75*) has an osmium–osmium triple bond and is formally equivalent to Cp*$_2$Re$_2(\mu\text{-}CO)_3$.

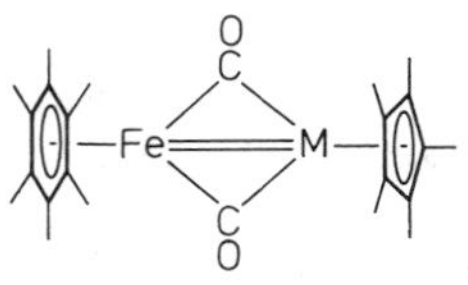

18, M = Co
19, M = Rh
20, M= Ir

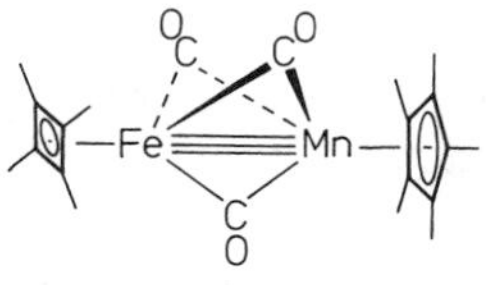

21

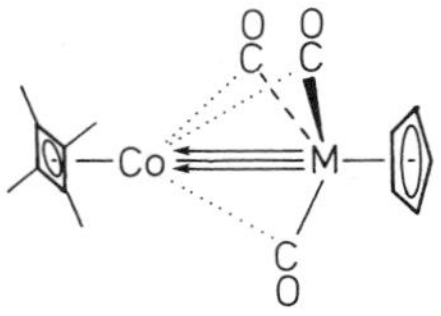

22, M = Mo
23, M = W

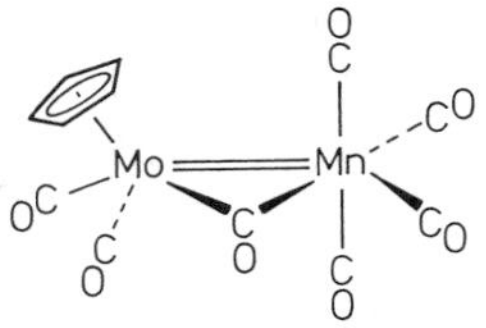

24

VI

SYNTHESES AND SPECTROSCOPIC PROPERTIES

A. *Pentacarbonyls*

The first reported synthesis (*9,76*) (78%) of $Cp_2V_2(CO)_5$ (**7**) involved protonation of the dianion $[CpV(CO)_3]^{2-}$ made by sodium amalgam reduction of $CpV(CO)_4$. No vanadium hydride is isolated in this reaction. A slightly better yield (89%) results from the photolysis of $CpV(CO)_4$ in THF using a falling-film photoreactor (*77*). The product **7** is produced via dimerization of an intermediate solvent complex $CpV(CO)_3(THF)$ followed by a further carbonyl loss (*35*). The corresponding Cp* complex **7*** is also available by an exactly analogous photolysis of $Cp^*V(CO)_4$ or by protonation of $[Cp^*V(CO)_3]^{2-}$ (*10*).

The two semi-bridging carbonyls are manifest by two low frequency stretches, ν_{CO}(hexane) 1869 and 1828 cm^{-1} (*9*). A single Cp signal in the ^1H NMR spectrum shows the molecule is fluxional (*9*).

B. *Tetracarbonyls*

Early syntheses of **2*** involved high temperature reactions of $Mo(CO)_6$ with acetylpentamethylcyclopentadiene (*14,78,79*) or Cp*H (*15,16*) itself. These methods are now generally superseded. All species **1–3**, **1'–3'**, and

1*–3* are best made by pyrolysis of the hexacarbonyls $L_2M_2(CO)_6$ (M = Cr, Mo, and W; L = Cp, Cp', and Cp*) while purging the solution with an inert gas to sweep away evolved CO. Toluene is generally adequate for chromium (11) and molybdenum (80), but xylene is sometimes preferred for tungsten (13). More recently, diglyme is suggested to offer advantages (81). The molybdenum and tungsten $L_2M_2(CO)_6$ species are straightforward to make while the chromium species is a little trickier since it is rather sensitive. The best route involves addition of acidified Fe-$(III)_2(SO_4)_3$ to the anion $[LM(CO)_3]^-$ (L = Cp, Cp', or Cp*) (82). While the chromium derivative $Cp_2Cr_2(CO)_4$ is obtained in high yield by heating $Cp_2Cr_2(CO)_6$ in toluene (11), the method suffers a little because of the inconvenience in actually isolating $Cp_2Cr_2(CO)_6$. A useful one-pot method [70% from $Cr(CO)_6$] for the synthesis of $Cp_2Cr_2(CO)_4$ involves heating a mixture in toluene of the readily available $Cr(NCMe)_3(CO)_3$ and CpH (83). On a small scale 1–3 can also be made by photolysis of the corresponding hexacarbonyl (13).

The indenyl complex $In_2Mo_2(CO)_4$ is also made by thermolysis of the corresponding hexacarbonyl (44,84), while the hydridotrispyrazolylborato complex, $Tp_2Mo_2(CO)_4$, is best made by the one-electron oxidation of $[TpMo(CO)_3]^-$ with $[Cp_2Fe]^+$ (43,85). The heterobimetallic complex Cp_2-$MoW(CO)_4$ can be made in respectable yield by thermolysis of a mixture of $Cp_2Mo_2(CO)_6$ and $Cp_2W_2(CO)_6$ (81,86).

Complexes 2 and 3 show single carbonyl and cyclopentadienyl resonances in their ^1H-NMR spectra (81). The mixed metal species Cp_2-$MoW(CO)_4$ shows two Cp resonances but a single carbonyl signal indicates carbonyl scrambling (81). The ^{95}Mo-NMR spectra of 2 in CH_2Cl_2 at ambient temperature (δ 182) and 2* (δ 133) show very deshielded signals relative to their respective hexacarbonyls $[Cp_2Mo_2(CO)_6, (\delta -1856); Cp^*_2Mo_2(CO)_6, (\delta -1701)]$ (87).

C. Tricarbonyls

Photolysis of $Cp^*Mn(CO)_3$ in THF leads to the solvent complex Cp^*Mn-$(CO)_2(THF)$. Removal of solvent at $-20°C$ followed by warming to room temperature while maintaining reduced pressure results in dimerization of solvent complex, decarbonylation, and solvent loss to form the air-sensitive 8* (17,51,88). While not isolated, the related Cp complex 8 has been observed in the gas phase. It is seen in the electron-impact mass spectrum of the THF complex $CpMn(CO)_2(THF)$, which shows a molecular ion and cracking pattern assignable to 8 rather than the THF complex itself (51). The rhenium complex 9* is formed on photolysis of Cp^*Re-$(CO)_3$ in THF (18) and in the carbonylation (15–20 atm, THF or toluene) of $Cp^*_2Re_2(O)_2(\mu\text{-}O)_2$ (89,90).

The only isolated Group 8 metal species in this series is **11*** (*19*). The dimer $Cp^*_2Fe_2(CO)_4$, which exists as purely the trans isomer, undergoes photolysis (355 nm) in an alkane while purging the solution with argon to form isolable quantities of air- and water-sensitive **11***. Photolysis of $Cp^*FeH(CO)_2$ is actually a better route since the hydride is more soluble. Photolysis of the related $Cp_2Fe_2(CO)_4$ in PVC films at 12–77 K or alkane glasses at 12 K also proceeds by CO loss, in this case to form **11**. This material has not been isolated (*20*). Fast time resolved IR spectroscopy in cyclohexane at room temperature shows that photolysis of $Cp_2Fe_2(CO)_4$ gives the two primary photoproducts $CpFe(CO)_2$ and **11**, the latter with a much longer lifetime ($t_{1/2}$ 0.6 second in the absence of CO under Ar) (*91,92*). The parent dimer $Cp_2Fe_2(CO)_4$ exists as a mixture of cis and trans isomers. Only the trans isomer undergoes reaction to form **11** (*19,20*). Similar results are obtained for the corresponding indenyl derivatives (*93*).

The bridging carbonyls in the two series **8–10** and **11–13** correlate with low frequency IR bands [**9***, ν_{CO} (hexane) 1748 cm^{-1} (*18*); **8***, ν_{CO} (hexane) 1785 cm^{-1} (*17*); and **11**, 1812 cm^{-1} (*20,91,92*). Isotopic labeling studies with ^{13}CO-enriched material show a spectrum best fitted (*20*) to the local D_{3h} structure indicated in the X-ray crystal structure of **11*** (Fig. 5).

D. *Dicarbonyls*

Of the three cyclopentadienyl complexes $Cp_2M_2(CO)_2$ (M = Co, Rh, and Ir), only the cobalt species is at all well known. Photolysis of toluene solutions of $CpCo(CO)_2$ at −78°C while purging the solution with N_2 gives the unsaturated 16-electron species $CpCo(CO)$ (*24*). This dimerizes to form a 6% isolated yield of a complex $Cp_2Co_2(CO)_2$ that is sufficiently stable to sublime at 40°C under high vacuum. The predominant product of the solution photolysis is $Cp_2Co_2(\mu\text{-}CO)(CO)_2$ (Co—Co) formed when $CpCo(CO)$ reacts with unreacted $CpCo(CO)_2$. This molecule is very reactive and thermally decarbonylates to form a cluster $Cp_3Co_3(\mu\text{-}CO)_2$ (μ_3-CO). Subsequent research demonstrated that photolysis of $CpCo(CO)_2$ in Ar, CH_4, or N_2 matrices at low dilution gives $Cp_2Co_2(CO)_2$ (*94*). At high dilution, monocobalt species result. Photolysis of Cp_2Co_2-$(\mu\text{-}CO)(CO)_2$ in a methylcyclohexane glass gives **4** cleanly (*25*). The tricarbonyl is photosensitive to the point that photolysis of $CpCo(CO)_2$ in a similar glass leads to prompt formation of **4** without any observation of $Cp_2Co_2(\mu\text{-}CO)(CO)_2$. Complex **4** is a common product in the reactions of **4**$^{\overline{\cdot}}$ with alkyl halides, for instance a 50% yield of **4** is obtained in the reaction of **4**$^{\overline{\cdot}}$ with α,α'-dibromo-o-xylene (*95*).

Sodium amalgam reduction of $CpCo(CO)_2$ affords the radical anion Na—**4**$^{\overline{\cdot}}$ with a CoCo bond order of 1.5 (*26–28*). An unwelcome by-product is $Na[Co(CO)_4]^-$. One electron oxidation of this $4^{\cdot-}$ in THF by $FeCl_3$ gives **4** in high yield but contaminated by Cp_2Fe (27,57).

The Cp* complex **4*** is far easier to make. All the methods rely on $Cp*Co(CO)_2$ as starting material. Oxidation of Na—$4^{\cdot-}$ by $FeCl_3$ in THF gives **4** in high yield (28). The usefulness of this route is diminished by the rather moderate yield of the anion (50–60%) obtained by sodium reduction of $Cp*Co(CO)_2$. Thermolysis of $Cp*Co(CO)_2$ in toluene also gives **4*** (62%, 24 hours) (21). Photolysis of $Cp*Co(CO)_2$ in THF is also useful, giving a 93% yield in 4 days provided that the reaction solution is continually purged with N_2 to prevent facile recombination of evolved CO with **4*** as it forms (21). The same product together with ethene is formed in 80% yield in the reaction between $Cp*Co(CO)_2$ and $Cp*CO(\eta-C_2H_4)_2$ (96). The major product of the reaction between $Cp*Co(CO)_2$ and $SiMe_3N_3$ is **4*** (97).

The rhodium Cp complex **5** is virtually unknown. It is made by the photolysis of $CpRh(CO)_2$ at 93 K in concentrations higher than 5 mM (25). The Cp* species **5***, however, is very well known. Sublimation of $Cp*Rh(CO)_2$ at low pressures ($\sim10^{-3}$ torr) gives unchanged material, but at higher pressures (10–20 torr) **5*** is formed (22). Perhaps the best synthesis is by treatment of $Cp*Rh(CO)_2$ with $Me_3NO\cdot2H_2O$ in acetone which gives 88–95% yields after a 3-hour reaction (98,99). The iridium compound **6*** is virtually unknown, but addition of CCl_4 to $Cp*Ir(H)$-(Ph)CO gives the expected $Cp*Ir(Cl)(Ph)CO$ as the predominant product together with a 9% yield of **6*** (23).

Heterobimetallic complexes are best made by a derivation of one of the methods indicated above. For instance, treatment of $Cp*M^1(\eta-C_2H_4)_2$ (M^1 = Co or Rh) with $Cp*M^2(CO)_2$ (M^2 = Rh or Ir) leads to $Cp*_2M^1M^2$-$(CO)_2$ in high yield [M^1 = Co, M^2 = Rh, 80% (56); M^1 = Co, M^2 = Ir, 80% (69); M^1 = Rh, M^2 = Ir, 60% (100)]. Finally, the rhodium radical anion $5^{\cdot-}$ is available through the reduction of **5** with Na/K alloy or sodium amalgam (29). Prolonged reaction leads to a further reduction forming the diamagnetic $[Cp_2Rh_2(\mu-CO)_2]^{2-}$.

VII

REACTIONS

A. Pentacarbonyls

The chemistry of **7** is relatively unexplored as compared to the Group 6 and 9 molecules. It is dominated by reactions with phosphines and alkynes. Pyrolysis of $Cp_2V_2(CO)_5$ in THF in the absence of phosphines gives predominantly a tetravanadium cluster, $Cp_4V_4(CO)_4$, with all four carbonyls terminal, together with traces of $Cp_3V_3(CO)_9$ and some $CpV(CO)_4$ (*101*). Complex 7 reacts with PPh_3 by carbonyl substitution to form Cp_2V_2-$(CO)_4(PPh_3)$ as major product and by cleavage of the VV interaction to form traces of $CpV(CO)_3(PPh_3)$ and $CpV(CO)_2(PPh_3)_2$ (*9,32,76*). Analogous results occur with arsines (*76*). Under photochemical conditions with PPh_3 as incoming ligand, only the substitution product $Cp_2V_2(CO)_4(PPh_3)$ is observed (*32*). In the absence of a ligand, photolysis in THF at $-78°C$ allows the observations of $Cp_2V_2(CO)_4(THF)$ (*35*). The X-ray crystal structure of $Cp_2V_2(CO)_4(PPh_3)$ (*32*) shows a close relationship to that of the pentacarbonyl 7 and suggests that the CO* ligand in 7 (Fig. 1) is that labilized by photolysis. This point of view is reinforced by the long V—CO* bond of 1.972 Å (*30–32*) relative to the other terminal average V—CO lengths of 1.916 Å There is a thermal equilibrium between 7 and $Cp_2V_2(CO)_4$ + CO as shown by the results of labeling studies involving ^{13}CO. It is the tetracarbonyl (solvent stabilized in donor solvents) that reacts with donor ligands to form V_2 species.

Stronger nucleophiles, such as the more basic phospines (PBu^n_3 or PEt_2Ph), react with cleavage of the V=V bond to form mixtures of CpV-$(CO)_2(L)_2$ and $CpV(CO)_3(L)$ (L + PEt_2Ph or PBu^n_3) (*9,35,76*). Cleavage of the V=V bond also occurs on treatment with NO when the product is the dinitrosyl $CpV(CO)(NO)_2$ (*76*).

The complex $Cp_2V_2(CO)_5$ does not react with 1,4-cyclohexadiene, ethene, 2,5-norbornadiene, or allene at ambient temperature, but does react with 1,3-cyclohexadiene with V=V cleavage to form $CpV(CO)_2(\eta^4$-$C_6H_8)$ (*33*). However, it does react with a number of alkynes (*33*). In most cases, photolytic reactions proceed by V=V cleavage to form $CpV(CO)_2$-(R^1CCR^2) ($R^1 = R^2 = $ H, Me, and Ph; $R^1 = $ Ph, $R^2 = $ Me), but cyclobutadiene complexes $CpV(CO)_2(C_4Ph_2R_2)$ (R = Ph or Me) are also formed in the cases of PhCCPh and PhCCMe when using hexane, but not THF, as solvent. It is notable that the corresponding photolysis reactions of $CpV(CO)_4$ with alkynes lead to bisalkyne complexes $CpV(CO)_2$ $(RCCR)_2$ but no cyclobutadiene species at all (*102*). The inference is that cyclobutadiene formation must occur at a divanadium intermediate rather than a monovanadium species since otherwise cyclobutadiene complexes would form in the $CpV(CO)_4$ reactions. Low temperature photolysis of 7 in THF in the presence of PhCCMe gives an intermediate with a low frequency IR carbonyl absorption (ν_{CO} 1842 cm^{-1}) which may be Cp_2V_2-$(CO)_4(\eta$-PhCCMe) with a structure analogous to the dimolybdenum complexes $Cp_2Mo_2(CO)_4(\eta$-RCCR) (see below). Stereoselective catalytic thermal hydrogenation of PhCCPh to *cis*-HPhC=CHPh occurs in the presence of 7 as a catalyst precursor.

B. *Tetracarbonyls*

The chemistry of $Cp_2Mo_2(CO)_4$, in particular, is extensive and the subject of recent reviews (*103,104*).

1. *Addition of Two-Electron Donor Ligands*

a. Carbonylation.   All the complexes **1–3** carbonylate to generate the hexacarbonyls $L_2M_2(CO)_6$ from which they came. Not all, however, do so readily. Although $Cp_2Cr_2(CO)_6$ readily decarbonylates on warming in toluene (*11*), it is rather less inclined to carbonylate, requiring 100 atm at 25°C. However, irradiation of **1** or **1*** under an atmosphere of ^{13}CO does result in carbonyl exchange with ^{13}CO forming labeled **1** or **1***, but not in any labeled hexacarbonyls (*105*). The other Cp systems both carbonylate easily. The ease of carbonylation of the Cp* species decreases in the order Mo > W > Cr (Table IV). It is apparent that the Cp* species carbonylate less easily than their Cp analogs, perhaps due, in part, to steric factors. Carbonylation of mixtures of $Cp_2Mo_2(CO)_4$ and $Cp'_2Mo_2(CO)_4$ gives no crossover products (*106*), confirming that carbonylation does not require any metal–metal bond cleavage. This is relevant to the mechanism of photodecarbonylation of $Cp_2Mo_2(CO)_6$ to form **2**. The lack of crossover product suggests that carbonylation proceeds by direct addition of two molecules of CO without any Mo—Mo bond scission.

In fact, the Mo≡Mo bond shows a very marked tendency to abstract CO from any CO source. This can have unexpected results. One might have expected **2** to function as an "inorganic alkyne" in its reaction with $Cp_2Ti(CO)_2$. Instead, the reaction of $Cp_2Ti(CO)_2$ with **2** apparently

TABLE IV

CONDITIONS UNDER WHICH CARBONYLATION OCCURS[a]

Complex	Pressure	Temperature
$Cp_2Cr_2(CO)_4$	100 atm	25°C
$Cp^*_2Cr_2(CO)_4$	300 atm	25°C
$Cp_2Mo_2(CO)_4$	1 atm	Ambient
$Cp^*_2Mo_2(CO)_4$	1 atm	Ambient
$Cp_2W_2(CO)_4$	1 atm	Ambient
$Cp^*_2W_2(CO)_4$	50 atm	82°C

[a] Taken from data presented in Refs. *13, 80,* and *106*.

gives $Cp_2Mo_2(CO)_6$ and "Cp_2Ti." These react in THF to form para-magnetic **25** containing an isocarbonyl interaction (*107*). In noncoordinat-ing toluene, the result is $[Cp_2TiOCMo(CO)_2Cp]_2$ (**26**) containing the 12-membered Ti—O—C—Mo—C—O—Ti—O—C—Mo—C—O ring (*108*). Addition of THF ruptures this ring, producing **25**.

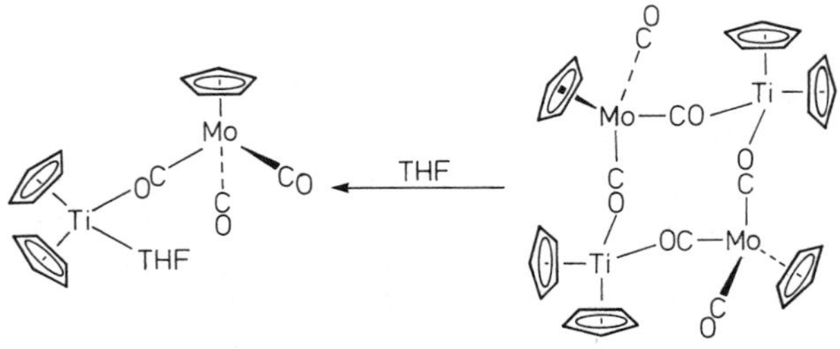

25   **26**

b. Phosphines and Isonitriles. Phosphines readily add across the Mo≡Mo bond of **2** to form $Cp_2Mo_2(CO)_4(PR_3)_2$ [R_3 = Ph_3, $(OMe)_3$, H_3, H_2Me, or HMe_2] (**27**) (*80,109*). Isonitriles react in similar fashion to form $Cp_2Mo_2(CO)_4(CNR)_2$ (**27**) (R = Me, Bu^t) but via isolable $Cp_2Mo_2(CO)_4(\sigma + \pi\text{-CNR})$ (see below) (*110*). The PH_3, PH_2Me, and $PHMe_2$ complexes (*109*) eliminate 1 equiv of phosphine (233 K for the PH_3 species) to form hydridophosphido compounds **28**. Phosphine, PH_3, reacts in an exactly analogous fashion with **3** to form the tungsten analog

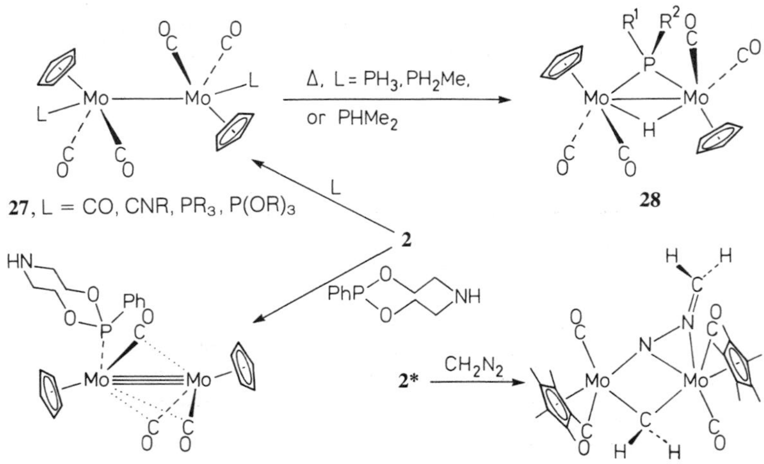

27, L = CO, CNR, PR_3, $P(OR)_3$

2

28

2*

29   **30**

$Cp_2W_2(\mu\text{-}H)(\mu\text{-}PH_2)(CO)_4$, the PH_3 elimination now proceeding at ambient temperature.

The reaction of **2** with $PPh(\overline{OCH_2CH_2)_2NH}$ is rather different. Here the predominant product is the *substitution* product **29** although quantities of $Cp_2Mo_2(CO)_4[PPh(\overline{OCH_2CH_2)_2NH}]_2$ are also formed (*111,112*). X-Ray analysis of **29** reveals a structure rather closer in nature to that of **1** than **2** in the sense that the CpMoMoCp axis is bent. Complex **29** reacts with $P(OMe)_3$ by substitution of the $PPh(\overline{OCH_2CH_2)_2NH}$ ligand to form $Cp_2\text{-}Mo_2(CO)_3[P(OMe)_3]$, whose solid state structure is very similar to that of **29**. Complex **2*** reacts with $PPh(\overline{OCH_2CH_2)_2NH}$ to form **29***, and it is interesting to note that **2*** reacts with $P(OMe)_3$ to form the analogous $Cp^*_2Mo(CO)_3[P(OMe)_3]$ rather than **27*** [L = $P(OMe)_3$] (*113*).

Diazoalkanes are good sources of alkylidene fragments $:CR_2$ which are two-electron donors to metals. However, the reactions of diazoalkanes CR_2N_2 with **2**, for instance, do not give any products $Cp_2Mo_2(CO)_4(CR_2)_x$ ($x = 1$ or 2). Reaction of **2** with CH_2N_2 itself gives polymethylene but no isolable metal complex (*114*). The Cp* species (*115*) reacts with an excess of diazomethane by addition of both CH_2 and CH_2N_2 across the triple bond to form a molecule (**30**) in which the MoMo bond is cleaved completely (a total six-electron donor capacity is required) and the two $CpMo(CO)_2$ units are held together by CH_2 and CH_2N_2 units. Since these reactions are not characterized as simple two-electron donor additions, they are considered in more detail below.

2. Reactions with Diazoalkanes

Diazoalkanes are important sources of reactive $:CR_2$ fragments in alkylidene chemistry. Bridging alkylidenes are extensively reviewed (*116–118*). A number of diazoalkanes add across the metal–metal triple bond associated with **2**. The reactions are, however, highly complex (as shown by the formation of **30** above), with reaction products heavily dependent on the R substituents of R_2CN_2 as well as the substituents on the cyclopentadienyl ring. While many interesting products arise in these reactions, the consequence is a minimum ability to predict reaction products.

A one-to-one adduct (**31**) is formed in near quantitative yield in the reaction of Me_2CN_2 with **2** (*114*). The analogous **31*** is formed on treatment of **2*** with Me_2CN_2 and is crystallographically characterized (*119*). The ^1H- and ^{13}C-NMR spectra of **31** are consistent with the presence of two isomers in solution (*114*).

The reactions of diaryldiazomethanes are noteworthy. They react with **2** or **2'** to form fluxional one-to-one adducts **32** or **32'** in which the N_2

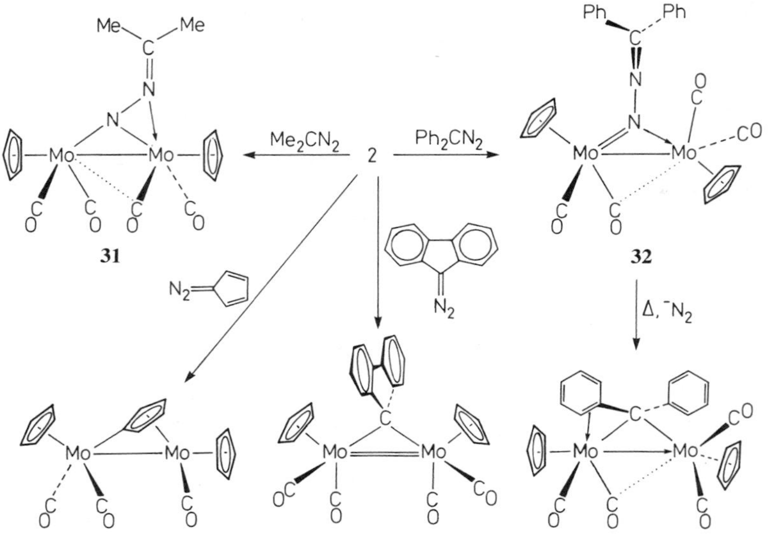

fragment is retained (*114,120,121*). The corresponding Cp* derivative **32***
is available in a similar fashion but is nonfluxional (*114*). These molecules
undergo thermal or photochemical N_2 loss to give bridging diarylmethane
adducts **33**. The interest of this structure (*120,121*) is that one of the aryl
groups is coordinated to a metal, making the :CAr_2 group a four-electron
donor. Electron counting requires a single MoMo bond in **33** rather than a
double bond as might have been expected if the :CAr_2 group were to
function as a two-electron donor.

The diazo compound 9-diazofluorene behaves differently, and the
Mo=Mo complex **34** (Fig. 9) is formed directly without observation of any
nitrogen-containing adduct (*114,121,122*). The physical constraint of the
two aryl groups being linked means that the :CAr_2 function must remain a
two-electron donor. Despite the metal–metal unsaturation of **34**, it does
not react with CO, and this is ascribed to a high lying LUMO which is
therefore unable to bind CO (*122*). Neither is a simple bridging alkylidene
formed in the reaction of **2** with $C_4H_4CH_2$. In this case the product is **35**,
formed without observation of any intermediate (*123*).

The reaction of $CH(CO_2Et)N_2$ with **2** is clearly highly complicated. The
most interesting product is a crystallographically characterized Mo_3 cluster
(**36**) (*124*) in which the three molybdenum atoms are bound to nitrogen
in a **T** coordination geometry. The source of the oxygen ligand is un-

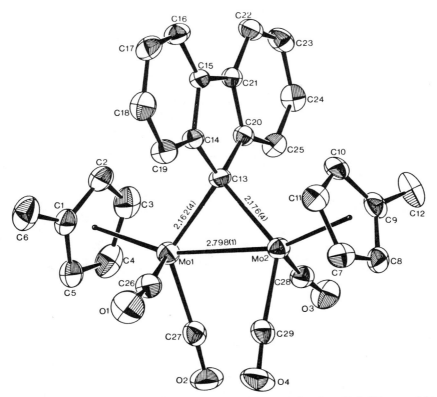

FIG. 9.   X-Ray crystal structure of **34′**, taken with permission from Ref. *121*, copyright American Chemical Society.

clear. On the other hand, reaction of $CH(CO_2Et)N_2$ with **2*** gives the crystallographically characterized **37** (*114*) in which the bonding mode of the ligand has some similarities to that of the Me_2CN_2 adduct **31**. The reaction of **2** with $C(CO_2Et)_2N_2$ gives yet a different product; here, cleavage of the Mo≡Mo bond gives **38** in 45% yield (*114*).

Cleavage of the Mo≡Mo bond in **2*** is also effected by addition of α-ketodiazoalkanes. Products **39*** (*125*) are clearly very related to structures **38**. It is likely that these reactions proceed by intermediates related to **31, 32,** and **37**. Compound **2** reacts with $C(Ph)(COPh)N_2$ to give an exactly related product (**39**) (*126*). Addition of aryl azides $RC_6H_4N_3$ (R = H, Me, Bu^t) with **2** or RN_3 (R = Me, Et, C_3H_5) with **2*** gives nitrene (M=NAr) complexes (**40** or **40***) (*127,128*). The nitrene group may be formed by elimination of N_2 from a bridging azide structure (*128*).

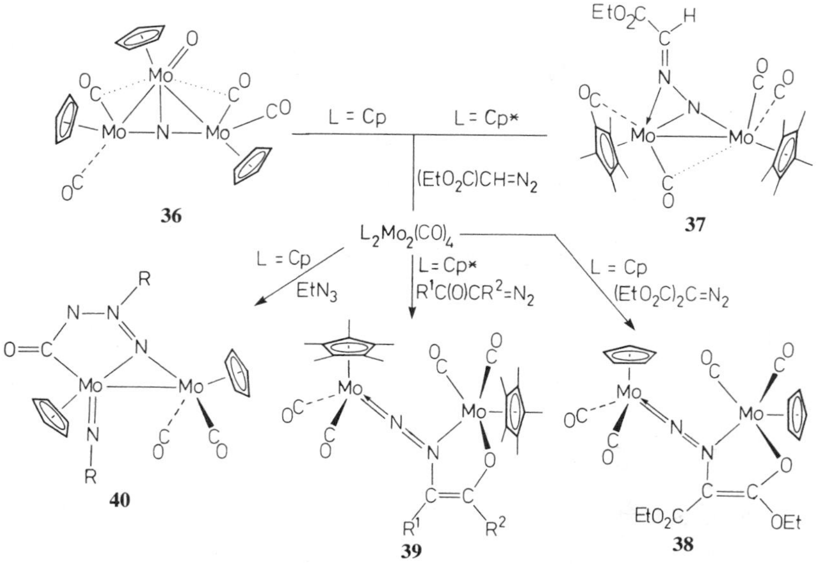

3. Addition of Four-Electron Donor Ligands

a. Alkynes. Addition of alkynes across the MoMo bond of **2** is a characteristic general reaction that results in molecules of type **41** containing pseudotetrahedral Mo_2C_2 "dimetallatetrahedrane" cores (*13,80,129–136*). The tungsten species **3**, although less extensively studied, behaves similarly (*13,135,137*). Most of the resulting molecules possess a very characteristic structure (Fig. 10) with a semi-bridging carbonyl. The semi-bridging carbonyl is reflected in a low frequency band in the 1830–1860 cm^{-1} range.

Very sterically demanding alkynes such as $Me_3SiCCSiMe_3$ react to form complexes such as **41b** ($R^1 = R^2 = SiMe_3$) that have a subtly different geometry illustrated in Fig. 10 (*133*). There are no bridging carbonyls, and

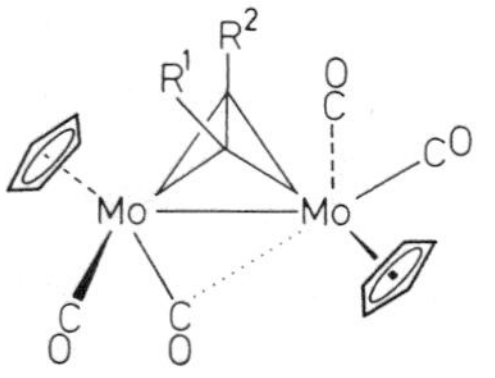

41a 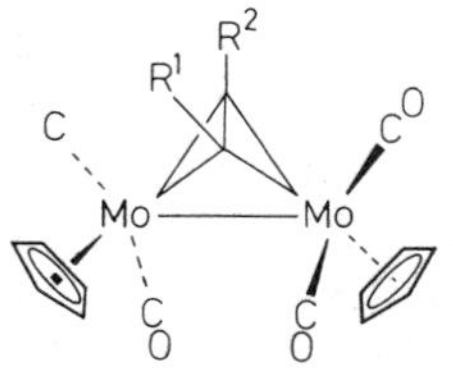   **41b**

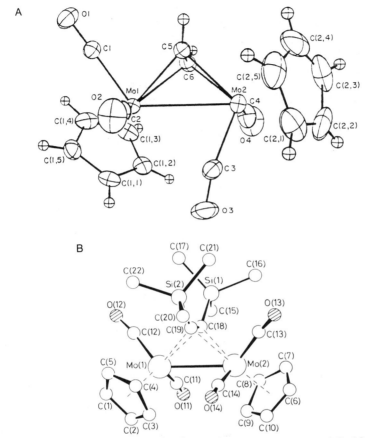

FIG. 10. X-Ray crystal structures of the two distinct structural types of Cp$_2$Mo$_2$(CO)$_4$-(μ-R^{1}CCR^2) molecules (**41a** and **41b**). (A) **41a** (R^1 = R^2 = H), taken with permission from Ref. *131*, copyright American Chemical Society. (B) **41b** (R^1 = R^2 = SiMe$_3$), taken with permission from Ref. *133*, copyright Royal Society of Chemistry.

the molecule possesses C_2 symmetry unlike the far more common asymmetric molecules. Complexes **41** are fluxional, and it is likely that the fluxional process which averages all four carbonyls and the cyclopentadienyl groups involves interconversion of the two isomeric forms **41a** and **41b** (129–131,133,135).

Alkynes do not generally add across the CrCr bond in Cp$_2$Cr$_2$(CO)$_4$ to form isolable chromium analogs of **41**, although the PhCCPh adduct is formed on photolysis of **1** in the presence of PhCCPh (*138*).

The M_2C_2 unit is very common, and the alkyne ligand can be modified (*133*). Measurements of the enthalpies of reaction of **2** with alkynes indicate there is a strong interaction between the alkyne and the Mo_2 core with measured values [-33.9($R^1 = R^2 = H$) and -31.6 kcal mol^{-1} ($R^1 = Ph$, $R^2 = H$)] nearly equal to that of 2 mol of CO (*136*). There is quite a driving force toward the M_2C_2 unit to the extent that polyalkenes such as cot and cod undergo hydrogen shifts and eliminations while adding across the $Mo \equiv Mo$ bond to give products **42** and **43** based on the Mo_2C_2 unit (*139–141*).

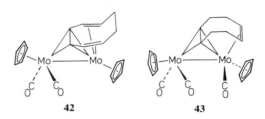

        42   **43**

While the chromium complex **1** does not react with alkynes to form simple 1 : 1 adducts, it does react with alkynes. Products **44** (Fig. 11) are formed in high yield and consist of a $Cr\equiv Cr$core bridged by a linked two-alkyne ligand (*138,142,143*). The structure of the Cr_2C_4 unit is clearly highly delocalized, but the NMR and X-ray data suggest that the bisalkylidene structure written is important relative to a chromacyclopentadiene structure.

Alkyne linkage is also displayed by the molybdenum systems. Complex **45** ($R^1 = R^2 = Ph$) (*138*) is obtained in a thermal reaction of **41a** ($R^1 = R^2 = Ph$), while **45** ($R^1 = R^2 = Et$) can be made from **41a** ($R^1 = R^2 = Et$) both thermally and photochemically (*144*). Chain extension continues to complexes **46** (Fig. 11) containing four linked alkynes in reactions involving the highly active $MeO_2CC\equiv CCO_2Me$ (*138,142,146,147*).

The reactions of **2*** with alkynes are quite different from those of **2**. The photochemical reaction of HCCH with **2*** gives the bridging vinylidene complex **47*** (*145*). The vinylidene is a four-electron ligand and is bound in a characteristic "side on" fashion displayed by a number of other ligand types (see below). The alkylidene nature of the bridging carbon is shown by a high frequency signal at δ 337.3. Thermal isomerization of **47*** to **41*** proceeds in toluene, and **47*** also reacts with CH_2N_2 to form a bridging allene complex **48***. The allene complex **48*** can also be made by direct

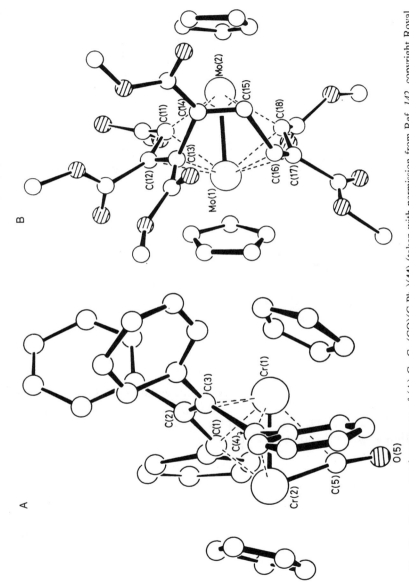

Fig. 11. X-Ray crystal structures of (A) $Cp_2Cr_2(CO)(C_4Ph_4)$ (**44**) (taken with permission from Ref. *142*, copyright Royal Society of Chemistry) and (B) $Cp_2Mo_2[(RCCR)(HCCH)(RCCR)_2]$ (**46**) (R = CO_2Me, taken with permission from Ref. *147*, copyright Royal Society of Chemistry).

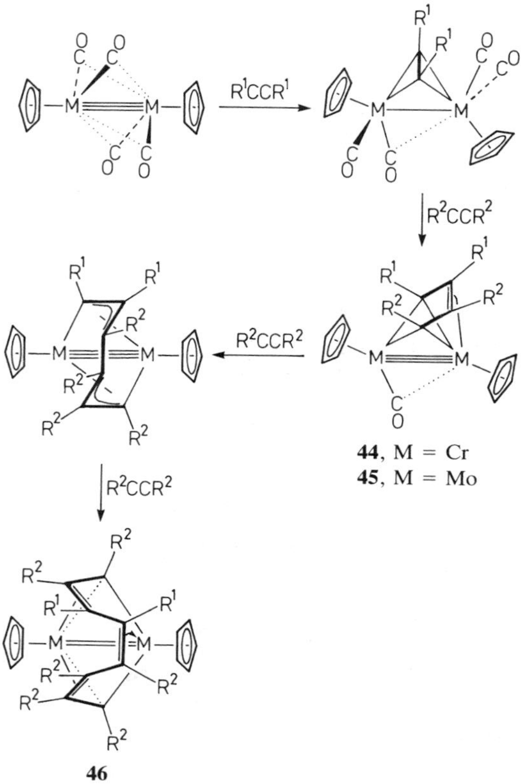

44, M = Cr
45, M = Mo

46

treatment of **2*** with allene. Complex **48** is available in similar fashion (*148*). The X-ray crystal structure of **48** (Fig. 12) (*149*) shows an approximately C_2 structure and suggests that the allene ligand is best regarded as two orthogonal C=C bonds each interacting with one metal.

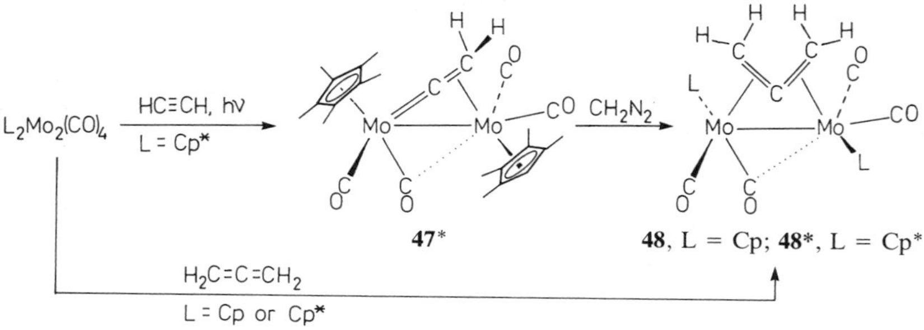

47*

48, L = Cp; **48***, L = Cp*

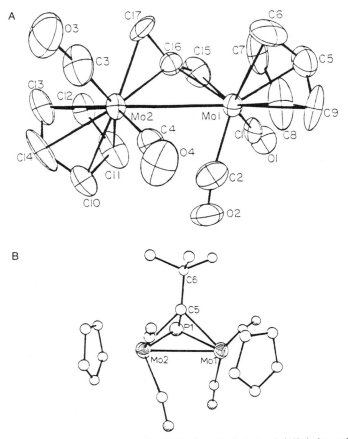

FIG. 12.   X-Ray crystal structures of (A) $Cp_2Mo_2(CO)_4$(allene) (48) (taken with permission from Ref. *149*, copyright American Chemical Society) and (B) $Cp_2Mo_2(CO)_4(\mu-\eta^2-Bu^tCCP)$ (49) (taken with permission from Ref. *151*, copyright Verlag Chemie GmbH).

b. Heteroalkynes.   Stable phosphaalkynes are relatively recent discoveries, but it is clear that in some ways they can mimic alkynes. The heteroalkyne $Bu^tC\equiv P$ adds across the $Mo\equiv Mo$ bond in **2** to form Cp_2-$Mo_2(CO)_4(\mu-\eta^2-Bu^tCCP)$ (49) (*150–152*). The PC bond in **49** is still quite short (1.719 Å), about the length of a PC double bond. The extra interest of **49** is that it has a phosphorus atom that in principle can still function as a two-electron ligand. This is illustrated by compound **50** which is simply made by the addition of $W(CO)_5(THF)$ to **49** (*150,152*).

Rather remarkably, reaction of P_4 with **2** results in formation of Cp_2-$Mo_2(CO)_4(\mu-\eta^2-P_2)$ (51) in which the Mo_2P_2 group closely resembles the

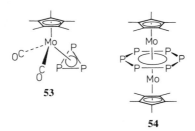

Mo$_2$C$_2$ unit in **41** (*153*). In this molecule *both* phosphorus atoms display an ability to bind metals. The simplest of these involves Cr(CO)$_5$ units at each phosphorus but the P$_2$ unit will also bridge the two rhenium atoms of Re$_2$(CO)$_6$(μ-Br)$_2$ as in compound **52** (*154*).

Phosphorus also reacts with **2***, but under more extreme conditions (*155*). The expected **51*** is formed together with a number of by-products including **53*** and **54***. The hexaphosphabenezene compound **54*** (Fig. 13) contains a planar P$_6$ unit sandwiched "triple decker" fashion between two Cp*Mo units. The Cp* ring appears as a septet at δ 0.47 in the ^{1}H-NMR spectrum because of coupling to the six phosphorus atoms and a P$_6$ ring–current effect.

The related μ-As$_2$ species Cp*$_2$Mo$_2$(CO)$_4$(μ–η^2-As$_2$) is formed in a directly analogous fashion from As$_4$ and **2*** (*156*). Both products are also formed in the reaction of **2*** with As$_4$S$_4$ (*157*). The Cp complex Cp$_2$-Mo$_2$(CO)$_4$(μ–η^2-As$_2$) is known but made by the reaction of Cp$_2$Mo$_2$(CO)$_6$ with As$_4$ (*158*).

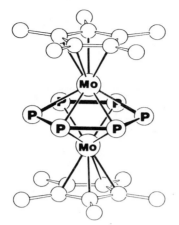

FIG. 13.   X-Ray crystal structure of Cp*$_2$Mo$_2$(P$_6$) (**54***), taken with permission from Ref. *155*, copyright VCH Verlagsgesellschaft mbH.

c. Isonitriles and Other $\sigma + \pi$ Ligands.   Compound **47** resulting from isomerization of HCCH displays a $\sigma + \pi$ coordination mode characteristic of a number of other ligands. Addition of isonitriles CNR to **2** or **2*** gives a series of complexes **55** or **55*** (*110,159–161*). (Fig. 14, Table V) in which the isonitrile functions as a four-electron donor to the dimetal system. This

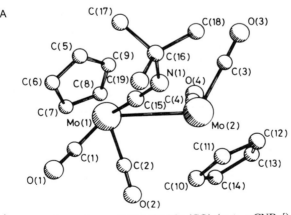

FIG. 14.   X-Ray crystal structures of (A) Cp$_2$Mo$_2$(CO)$_4$($\sigma + \pi$-CNBu^t) (**55**, R = Bu^t) (taken with permission from Ref. *110*, copyright Royal Society of Chemistry) and (B) [Cp$_2$Mo$_2$(CO)$_4$($\sigma + \pi$-CN)]$^-$ (**56**) (taken with permission from Ref. *166*, copyright American Chemical Society).

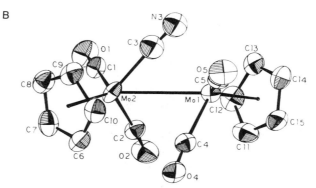

Fɪɢ. 14B.   See legend on p. 135.

is formally achieved by donation of two electrons from carbon to one metal and donation of two electrons from the CN π bond to the second metal. The resulting CN bond is rather longer than found in corresponding terminal isonitriles, reflecting loss of CN bonding electron density. The drop in formal CN bond order is reflected in the low observed ν_{CN} values in the range 1650–1740 cm^{-1}. These compounds are also formed as by-products in the reactions of Me=N=C=S with **2** * (*162*) or with PhN=C=NPh (*163–165*).

Cyanide also adds across the Mo≡Mo bond in analogous fashion to form the anion **56** (Fig. 14) (*132,166*). This molecule is fluxional, with the

TABLE V

EXAMPLES OF $\sigma + \pi$ COMPLEXES SYNTHESIZED FROM **2**, **2***, AND **3**[a]

XYZ	M	L	M¹–M²	M¹–X	M²–X	M²–Y	X–Y	M¹–X–Y	X–Y–Z	Refs.
CNBuᶠ	Mo	Cp	3.215	1.953	2.327	2.205	1.212	168	136	110,159
CNBuᶠ	W	Cp								110
CNMe	Mo	Cp								110,159
CNMe	W	Cp								110
CNPh	Mo	Cp	3.212	1.942	2.247	2.207	1.244	168	135	159
CNPh	Mo	Cp	3.238	1.932	2.248	2.185	1.234	170	137	163–165
CNtol	Mo	Cp								163
CNMe	Mo	Cp*	3.240	1.93	2.21	2.14	1.17	158	138	162
CNCF₃	Mo	Cp*								160,161
CN⁻	Mo	Cp	3.139	1.94	2.42	2.46	1.07	167	—	166,132
=C=CPhR	Mo	Cp	3.120	1.909	2.179	2.443	1.380	168		167
CCPh⁻	Mo	Cp								167
CCPh⁻	Mo	In								167
CCMe⁻	Mo	Cp								168
=C=C=CMe₂	Mo	Cp	3.145	1.912	2.209	2.240	1.336	167	144	168
NCNMe₂	Mo	Cp	3.056	2.056	2.149	2.103	1.335	135	135	169
N₂CMe₂	Mo	Cp*	3.050	2.120	2.134	2.126	1.369	113	155	119
=C=CH₂	Mo	Cp*	3.082							145

[a] Pertinent crystallographic data for $L_2M_2(CO)_4(\sigma + \pi\text{-XYZ})$ are included where the X-ray crystal structures have been determined. M^2 is the metal receiving XY π-electron density, X bridges both M^1 and M^2, Z is the substituent. R = $(CH_2)_4OMe$. Bond lengths are given in angstroms, angles in degrees.

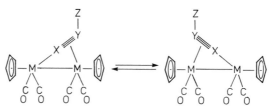

FIG. 15.   Oscillation displayed by a number of Cp$_2$Mo$_2$(CO)$_4$(σ + π-XYZ) complexes (55–59).

cyanide ligand oscillating across the MoMo bond (Fig. 15). Similar behavior is displayed by some of the isonitrile species but at higher temperatures (*110*). Acetylides CCR$^-$ [R = Ph (*167*) or Me (*168*)] form similarly
fluxional adducts (**57**). These undergo protonation or alkylation at the
β-carbon atom to form σ + π vinylidene derivatives (**58**). Other molecules
that add across the Mo≡Mo triple bond include NCNMe$_2$ (*169*), where
the product is **59**, as well diazoalkanes such as N$_2$CMe$_2$, where the product
is **31** (*114,170*). Ring opening of cyclopropene occurs to give a μ–σ:η^3
complex (**60**) (*171,172*).

    d. Four-Electron Thioketone Ligands and Related Species.   A number
of thioketones (*173*), thioesters (*174*), thiolactones (*174*), and dithioesters
(*175,176*) add across the Mo≡Mo or W≡W bonds of **2**, **2'**, or **3** to

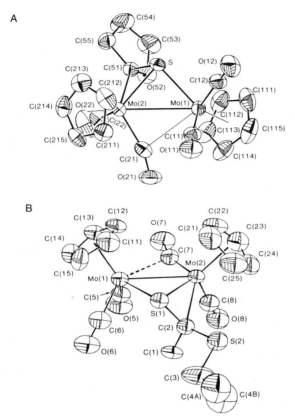

F IG. 16. X-Ray crystal structures of (A) Cp$_2$Mo$_2$(CO)$_4$(SC$_4$H$_6$O) (**61**) (R^{1}R^2 = —CH$_2$CH$_2$CH$_2$O—, taken with permission from Ref. *174*, copyright American Chemical Society) and (B) Cp$_2$Mo$_2$(CO)$_4$[MeC(S)SEt] (**61**) (R^1 = Me, R^2 = SEt, taken with permission from Ref. *177*, copyright American Chemical Society).

give complexes of structural type **61**. The common feature of these molecules is a characteristic M$_2$CS pseudotetrahedrane core (Fig. 16). The thiocamphor complex is also crystallographically characterized (*173*). The bonding can be regarded as donation of two electrons from the C=S bond to one metal and of a sulfur lone pair to the second. These molecules are also characterized by a semi-bridging carbonyl with an appropriately low carbonyl stretching frequency. Aldehydes and ketones do not add across the MoMo bond of **2** directly although an acetaldehyde analog of **61** is known (*177*).

Somewhat unexpectedly, the reaction conditions for the dithioester reactions are critical (*175*). The expected product **61** is formed at ambient temperature, but the same reagents at reflux in toluene produce **62**. Thermolysis of **61** (R^2 = SR) does not lead to formation of **62**.

Related complexes (**63**) are formed in the reaction of [P(C_6H_4OMe-p)(=S)(μ-S)]$_2$ (so-called Lawesson's reagent) with **2** (*178, 179*). The analogous tungsten species is made similarly. Here the bridging ligand is (C_6H_4OMe-p)PS, but the M_2SP core is structurally very similar to the M_2CS core of **61**. Different results are obtained in the reaction of **2** with Ph_3P=S. The resultant product (**64**) (*180*) has two Ph_3P=S ligands coordinated to one metal, a process accompanied by the migration of one CO ligand to the second metal. With CS_2 moderate yields of the η^2-CS_2 complex **65** are obtained (*181*). The CS_2 ligand is a two-electron donor and fails to mimic the coordination mode of allene in **48**.

4. *Electrophilic Addition of X$_2$ and HX*

It is fair to say the chemistry of **1–3** and their derivatives is dominated by reactions with nucleophiles. The molybdenum species $Cp_2Mo_2(CO)_4$ does, however, also react with halogens and hydrogen halides (*182, 183*).

Iodine reacts with **2** at low temperature to form a red–brown compound (**66a**) that isomerizes on warming to the crystallographically characterized violet species $Cp_2Mo_2(\mu$-I)$_2$(CO)$_4$ (**66b**) (*182*) (Fig. 17). The red–brown compound is not crystallographically characterized because of its low solubility. Solubility, quite often a problem with Cp_2Mo_2 compounds, can be alleviated by going to Cp′ systems. Generally, the chemistry is unaffected by this change. In this case, however, this useful generalization fails since the Cp′ reaction shows no such red–brown primary product. Perhaps the

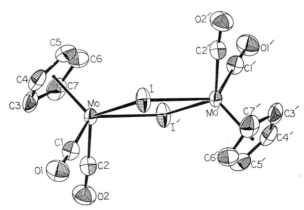

Fig. 17. X-Ray crystal structure of $Cp_2Mo_2(CO)_4(\mu$-I)$_2$ (**66b**), taken with permission from Ref. *182*, copyright American Chemical Society.

red–brown Cp species is a kinetic product trapped by precipitation at the low temperatures necessarily used. The corresponding chloride is best made by use of $PhICl_2$, a mild Cl_2 phase transfer agent, but the structure of the chloride (67) is subtly different, with a single Cl bridge and a mixed valence Mo(I)Mo(III) structure.

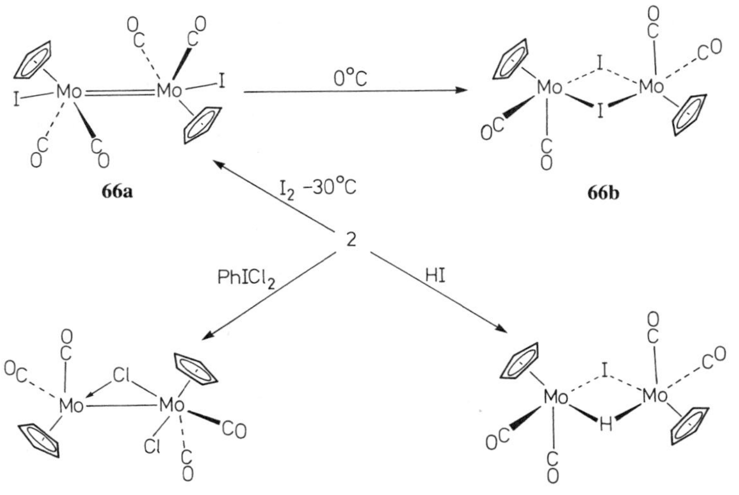

66a   0°C   **66b**

I_2 −30°C

2

$PhICl_2$ HI

67   **68**

A single equivalent of HI adds across the Mo≡Mo bond of **2** to form **68** while excess gives **66b**. This is reversed by the hydride source $LiAlH(OBu^t)_3$, sequential reactions leading back to **2**. Related behavior is observed for HCl, but the required reaction conditions are very specific.

5. Reactions of Elemental Sulfur and Selenium

The reactions of **1–3** and their Cp* analogs with sulfur are very complicated. Complex **2** reacts with a single equivalent of sulfur in acetone to form a cation (**69**) with approximate C_3 symmetry whose counterion is $[CpMo(CO)_3]^-$ (*184*).

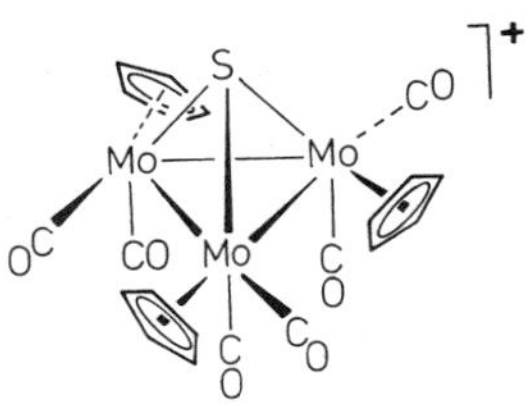

69 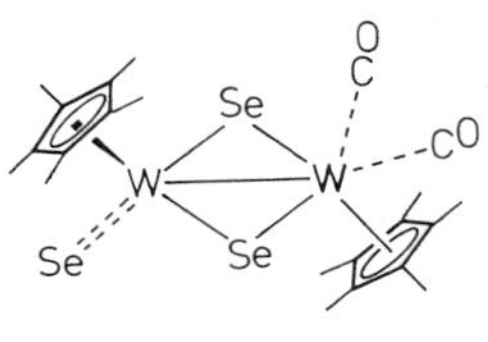   **71***

Multiple products with formula $Cp^*_2Mo_2S_4$ arise in the reaction of 2^* with sulfur (185). One of these has structure 70^*. A similarly complex reaction occurs for 3^*, while the product of reacting 3^* with selenium is 71^* (186). On the other hand, a pentasulfide (72^*) is formed (187) when 1^* is treated with an excess of sulfur. Its structure is interesting: the five sulfur atoms adopt a planar configuration. Eighteen-electron configurations are attained in 72^* provided a $Cr=Cr$ bond is accepted.

If the pentasulfide 72^* is photolyzed in the presence of 1, 1^*, or 2, the products are cubane structures (188). Similar products are obtained from 70^* and any of 1, 1^*, or 2 (188) or from the reaction of $2'$ and Cp_2-$Mo_2(SC_3H_6S)_2$, in which case propene is evolved (189).

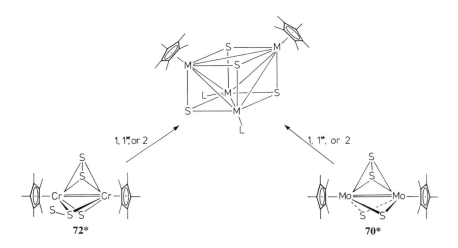

6. Reactions of $ArNO_2$, NO_2^-, and NO

Complete cleavage of the $Cr\equiv Cr$ bond to form mixtures of CpCr-$(CO)_2(NO)$ and $CpCr(NO)_2(NO_2)$ is observed when 1 is treated with NO gas (190). Any of NO_2^-, NO_3^-, and NO_2 react with 2 or 2^* under photochemical conditions by complete cleavage of the MM bond to form $CpMo(CO)_2(NO)$ or its Cp^* analog together with a number of oxide species (191). These results suggest that the $Mo\equiv Mo$ bond can effect one-electron reduction of NO_2^- to NO as well as three-electron reduction of NO_3^- to NO.

The reaction of $PhNO_2$ with 2 or 3 is complex but remarkably facile (3 hours at ambient temperature) (192). Quite respectable yields (40–46%) of $[CpM(=O)]_2(\mu\text{-}O)(\mu\text{-}NPh)$ are obtained in addition to some Cp_2-$M_2(CO)_6$. Nitrosoarenes ArNO give the same products (193), suggesting that the reaction with $PhNO_2$ proceeds by initial decarbonylation of the NO_2 group to form ArNO.

7. Cluster Formation

Complex **2** is a potent cluster precursor. The isolobal relationship (*194*) between diarylalkynes and the alkylidynes $CpW(CO)_2(\equiv CR)$ is now quite well known (*195*). Since **2** and **3** readily add alkynes, one might reason that alkylidynes $CpW(CO)_2(\equiv CR)$ will also add across the $M\equiv M$ bonds of **2** or **3**. In fact, this happens, and in quantitative yield for **2**, when the product is **73**. The yield of **74** in the tungsten reaction is less impressive, but it is obtained in higher yield in the reaction of $Cp_2W_2(CO)_4$-$(\mu-\eta^2\text{-}RCCR)$ with $CpW(CO)_2(\equiv CR)$. The behavior of the chromium system is quite different, the product being the μ-alkyne species Cp_2-$W_2(CO)_4(\mu-\eta^2\text{-}RCCR)$. Remarkably, **1** is catalytic for this dimerization (*196*).

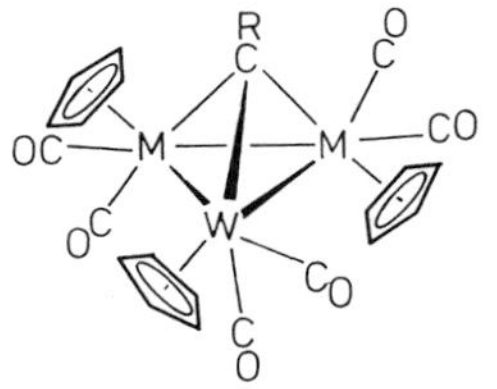

73, M = Mo
74, M = W

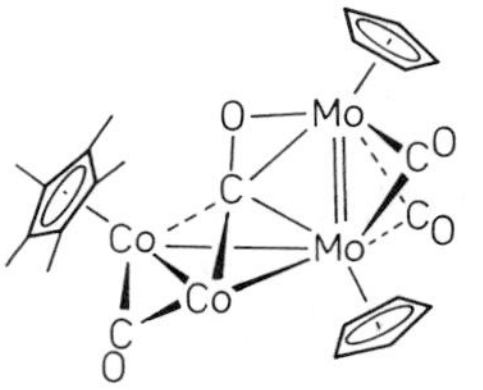

75

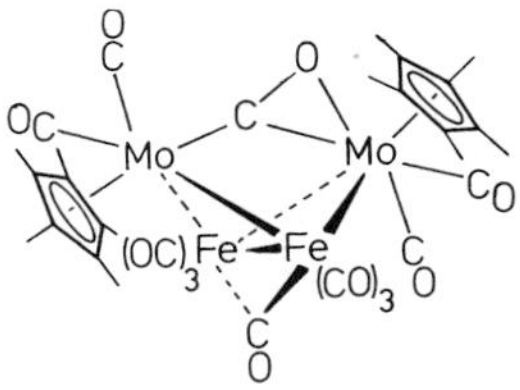

76

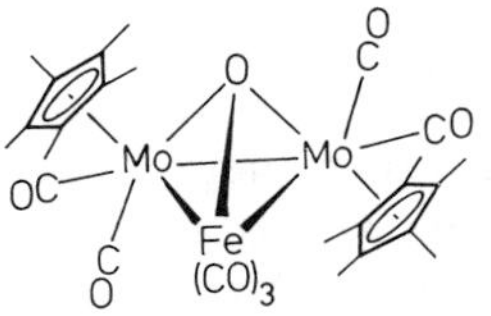

77

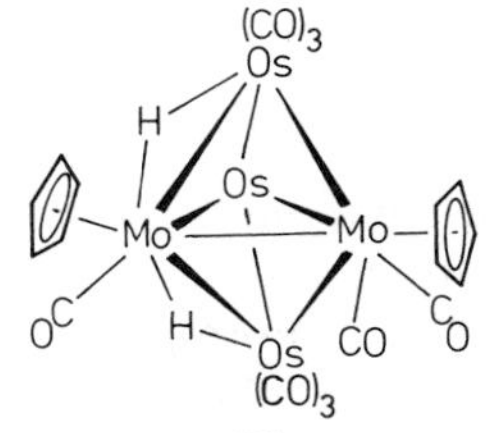

78

While metal clusters frequently form in reactions of **2** it is noticeable that **2** never dimerizes to form $Cp_4Mo_4(CO)_8$. The complex **75** (*197*) is formed in the reaction of **2** with $Cp^*Co(C_2H_4)_2$ and is remarkable for the η^2-(μ_4-CO) CO bonding mode and for retaining a Mo≡Mo bond. A related species (**76***) is formed in the reaction of **2*** with $Fe_2(CO)_9$ (*198*). If there is oxygen present, a different cluster, the 48-electron **77***, is formed (*199*). This neatly demonstrates the importance of the reaction atmosphere. Another example is reflected in compound **78**, one of the species obtained in the reaction between **2** and $Os_3(\mu$-$H)_2(CO)_{10}/H_2$ (*200*).

Heterometallic clusters are not always formed. The reaction between **2** and $RuCo_2(CO)_{11}$ leads only to $Ru_3(CO)_{12}$, $CO_4(CO)_{12}$, $CpMo(CO)_3$—$Co(CO)_4$, and $Cp_2Mo_2(CO)_6$ (*201*).

C. *Tricarbonyls*

The rhenium compound **10*** is rather unreactive. Carbonylation (100°C, 200 atm) gives $Cp^*Re(CO)_3$, while treatment with I_2 affords $Cp^*Re(CO)_2$-$(I)_2$. It does not react with ethyne, diazomethane, or phosphines under conditions where other dimetal complexes with multiple MM bonds do so (*18*). The manganese complex **8***, also somewhat unreactive, reacts with phosphites by Mn≡Mn cleavage. In the case of $P(OEt)_3$ the two products $Cp^*Mn(CO)_2[P(OEt)_3]$ and $CpMn(CO)[P(OEt)_3]_2$ form in a 1:1 ratio (*51*).

Both complexes **11** and **11*** show a marked tendency to bind two-electron donor ligands. Carbonylation causes reversion to $Cp_2Fe_2(CO)_4$ or $Cp^*_2Fe_2(CO)_4$ (*19,20,91–93*). Other donor ligands (MeCN, PPh_3) also react with **10*** (*19*). The Cp derivative reacts with MeCN, PBu^n_3, and PPh_3 but not $P(O$—$tolyl)_3$ or THF to form $Cp_2Fe_2(\mu$-$CO)_3(L)$ species (*202*). The activation energies are apparently a consequence of the spin-forbidden nature of the addition of a ligand to the presumably triplet structure $Cp_2Fe_2(\mu$-$CO)_3$.

D. *Dicarbonyls*

1. *Reactions with Diazoalkanes*

It was pointed out earlier that diazo compounds R^1R^2C=N=N often function as effective sources of carbene fragments :CR^1R^2 (*115–118*). The reactions of diazoalkanes with **4*** and **5*** are more predictable than those of the corresponding reactions of **2**. The general consequence of addition of $R^1R^2CN_2$ to **4*** or **5*** is loss of N_2 and addition of the carbene fragment into the metal–metal double bond. This generally results in a bridging alkylidene, or dimetallapropane, structure. This can be compared to cyclopropanation reactions by diazo compounds.

In the cobalt case, the primary products 79* generally still possess two bridging carbonyl groups (203–206), but when the carbene fragment is CHCF₃ the two carbonyls adopt an unbridged structure 80*. The crystallographically characterized parent methylene complex exists as a mixture of carbonyl bridged and unbridged forms 79* and 80* (205) that interconvert on the NMR time scale. The CH_2, CHMe, and CPh_2 complexes undergo thermal decarbonylation to form the Co=Co species 81*. The parent methylene complex is crystallographically characterized (207), and its structure (Fig. 18) is very similar to that of 4*. Addition of $CHRN_2$ to 81* (R = H or Me) results in a further alkylidene addition to form double alkylidene complexes 82*.

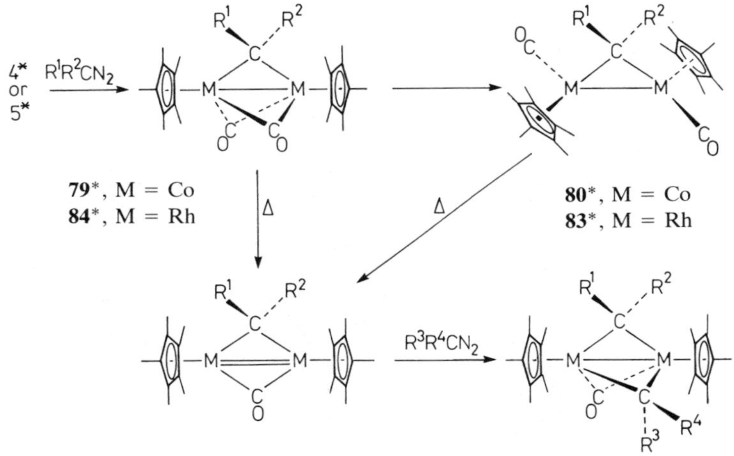

79*, M = Co
84*, M = Rh

80*, M = Co
83*, M = Rh

81*, M = Co
85*, M = Rh

82*, M = Co
86*, M = Rh

The corresponding rhodium compounds undergo related additions. Many diazoalkanes, including sterically demanding species, add across the Rh=Rh bond in very high yield to form a series of compounds (83*) (207–213). Nitrogen elimination prior to product formation is very rapid; no intermediates are apparently detectable. However, dimetallacyclic Rh—N=N—CR^1R^2—Rh (210) intermediates are put forward as plausible intermediates, and bridged species (84*) are probably formed first. When the alkylidene substituents form a cycle (for instance, —C—CCl=CCl—CCl=CCl—), this rearrangement does not always

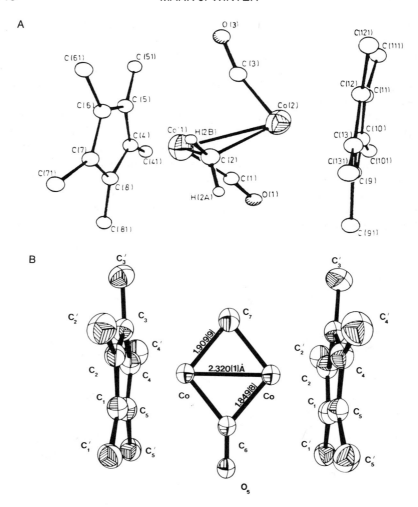

FIG. 18.   X-Ray crystal structures of (A) Cp*$_2$Co$_2$(CO)$_2$(μ-CH$_2$) (**80***) (taken with permission from Ref. *205*, copyright Elsevier Sequoia S.A.) and (B) Cp*$_2$Co$_2$(μ-CO)(μ-CH$_2$) (**81***) (taken with permission from Ref. *207*, copyright American Chemical Society).

proceed (*209,211*). Decarbonylation is an option for these compounds, however, in which case the Rh=Rh complexes **85*** arise. The rhodium complexes **85*** are also reactive toward diazoalkanes. Thus, addition of CH$_2$N$_2$ to **85*** results in the double alkylidene **86*** (R^1 = R^2 = Ph, R^3 = R^4 = H).

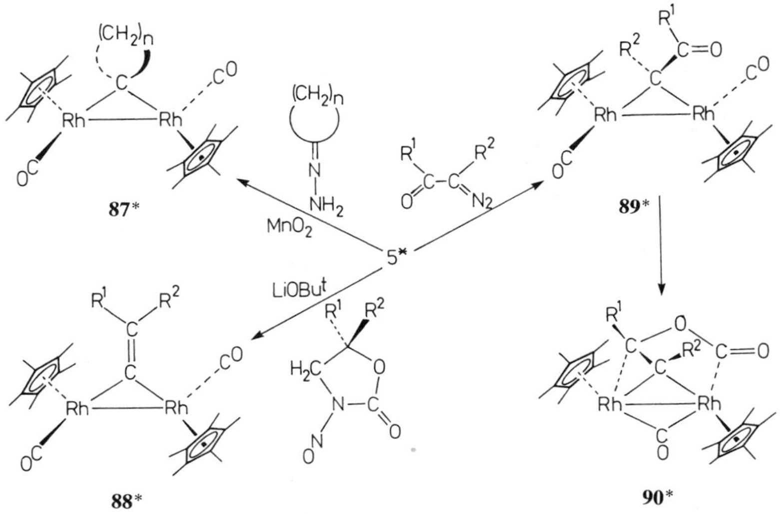

A number of modifications to the diazoalkane approach are possible. For instance, a number of cyclic hydrazones $CH_2(CH_2)_{n-1}C{=}NNH_2$ give intermediate diazo species $CH_2(CH_2)_{n-1}CN_2$ on dehydration with MnO_2 that react with 5^* to form products of type 87^*, effectively dimetallaspiranes (*214–217*). The cobalt complexes are similarly accessible, but, as expected, all the cobalt species possess two bridging carbonyls. Thermolysis of the six-carbon ring complex affords cyclohexene.

Diazoalkenes also add across the Rh=Rh bond of 5^*. Diazoalkenes are produced *in situ* from the corresponding cyclic N-nitrosourethanes by the addition of $LiBu^t$ (*218,219*). The effect is addition of $C{=}CR^1R^2$ fragments across the Rh=Rh bond, so forming vinylidene compounds 88^*. Diazoketones $R^1C(=O){-}CR^2{=}N{=}N$ also react with 5^*; the primary products are the expected products $Cp^*_2Rh_2(\mu\text{-}CO)[\mu\text{-}CR^2(COR^1)]$ (89^*), but these are isolable only in certain cases (*220–222*). Rearrangement by intramolecular cycloaddition reactions generally proceed rather easily for these molecules, resulting in compounds 90^*. These arise through a nucleophilic attack of the bridging function keto groups on a rhodium carbonyl group.

2. Alkynes and Heteroalkynes

A number of alkynes add across the Co=Co bonds of 4^* (*223–226*). In such cases the addition is followed by an insertion reaction to give dimetal complexes 91^* bridged by a $-C(=O){-}CR{=}CR-$ unit. Addition of ethyne itself is reversible and **91*** readily eliminates HCCH (*224*). Alkynes such as MeCCMe, CF_3CCCF_3, and $C_6F_5CCC_6F_5$ react with **4*** with Co≡Co bond cleavage to give ultimately cyclopentadienone derivatives such as $Cp^*Co(\eta^4\text{-}C_4R_4O)$ (R = CF_3, Me, C_6F_5) or arene complexes $Cp^*Co(\eta^4\text{-}C_6R_6)$ (R = CF_3). These reactions are likely to proceed through type **91*** intermediates (*226*).

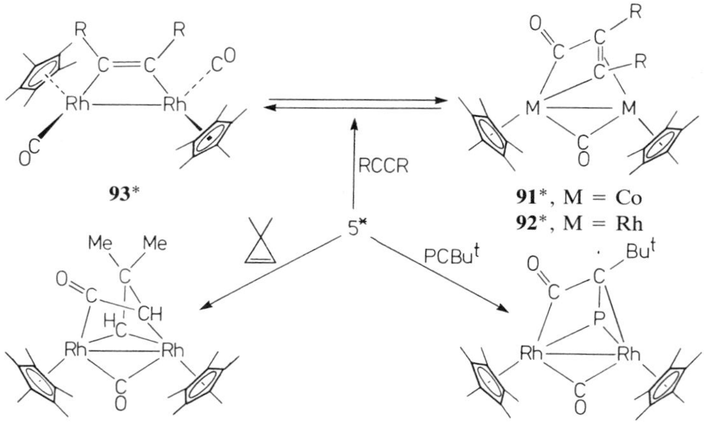

The corresponding reactions of **5*** are generally a little different. Addition of ethyne gives complex **92*** (R = H), but this is in equilibrium in solution with an isomeric ketenylmethylene species (*225*). Regiospecific addition is observed for PhCCH and other asymmetric alkynes (*226*). A number of the metallaenone structures exist in equilibrium with the dimetallacyclobutene structures $Cp^*_2Rh_2(CO)_2(\mu-\eta^1:\eta^1\text{-}R^1CCR^2)$ (**93***). These isomers are distinguishable by their IR and NMR spectroscopic properties (*226,227*). In all cases, except the $PhCCC_6F_5$ complex, the solid state structure is entirely as the metallaenone form. The $PhCCC_6F_5$ species exists as a mixture of the two forms. The CF_3CCCF_3 complex **92*** is crystallographically characterized (*226*) as is the Cp analog of **93*** (R = CF_3) (*228*). Other products also form in low yield in these reactions, including monorhodium cyclopentadienones and η^5-arenes as well as a number of dirhodium structures.

The heteroalkyne :PCBu^t behaves in a remarkably similar fashion. Addition across the Rh≡Rh bond in **5*** affords the rigid metallaenone **94*** (*151*). While dimethylcyclopropene is not an alkyne, the reaction of **5*** with dimethylcyclopropene is related. The product is **95*** and is derived by cleavage of the C≡C of the three-carbon ring. The analogous dicobalt and CoRh heterobimetallic species are formed similarly (*228,229*).

3. Other Ligands

Examples of other interesting additions (225) across the Rh=Rh bond include complexes of SO$_2$ (96*) (225,230,231), selenium (97*) (225,232), tellurium (225), sulfur (225,233), and AuCl [from AuCl(CO)] (225,230). The latter reaction results in a Rh$_2$Au triangle and is just one of a large number of heteropolymetallic cluster complexes built up from 4* and 5* (see below).

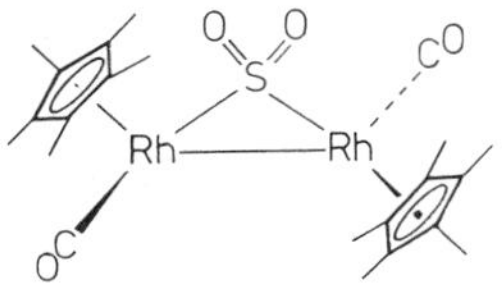

96* 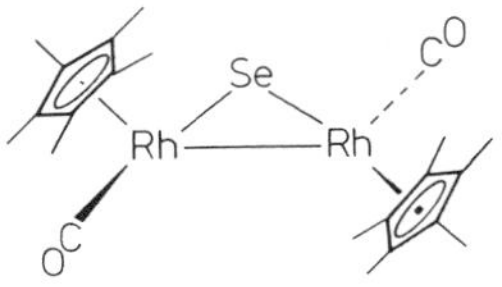   97* 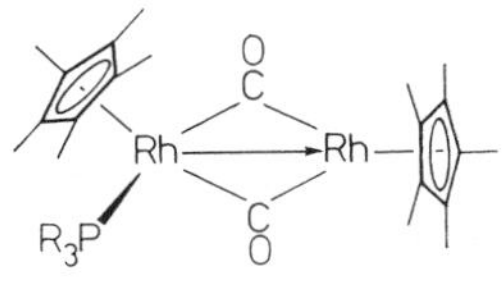   98*

Phosphines (PMe$_3$, PMe$_2$H) and phosphites [P(OMe)$_3$] do not cleave the Rh=Rh bond in 5*; instead, a single ligand adds to form Cp*$_2$Rh$_2$-(μ-CO)$_2$(PR$_3$) 98* in which the ligand apparently binds at a single rhodium, requiring that a donor Rh—Rh bond be written if the 18-electron rule is to stand (234). The metal–metal bonds are not cleaved by excess ligand to form Cp*Rh(CO)(PR$_3$), although such species are accessible through reactions of CpRh(CO)$_2$. Chelating phosphines dppe and dmpe do, however, cleave the Co=Co bond (235). The products are Cp*Co-(dppe) or Cp*Co(dmpe). The Cp*, Co, and the two phosphorus atoms are trigonally arranged. Photoelectron spectra suggest that the product metal atoms are very electron-rich centers.

4. Cluster Forming Reactions

One of the more elegant and useful concepts to arise recently in inorganic chemistry is that of the isolobal analogy (194). This is beyond the scope of this article, but in essence parallels are drawn between the nature of the frontier orbitals of particular metal and organic fragments. These isolobal relationships allow many seemingly complex inorganic structures to be compared to much simpler organic species and sometimes permit a rationalization of apparently unrelated inorganic reactions.

An isolobal relationship exists between 5 and ethene. The logic goes that it should therefore be possible to construct molecules containing the Cp$_2$-Rh$_2$(CO)$_2$ unit that are formally analogous to ethene complexes. In practice, 5* is used because of the essential nonavailability of the Cp species. Many compounds have been prepared by using this approach (195,236).

As examples, addition of Pt(η-C$_2$H$_4$)(PPh$_3$)$_2$ to **5*** gives the trimetallic species PtL[η^2-Cp*$_2$Rh$_2$(CO)$_2$] (**99***) (*237*) which is formally equivalent to the well-known PtL(η-C$_2$H$_4$) complexes. In similar fashion, addition of 2 equiv of **5*** to Pt(η-C$_2$H$_4$)$_3$ gives the PtRh$_4$ species **100*** (*238,239*). This complex is analogous to the D_{2d} species Pt(η-PhCCPh)$_2$. In closely related reactions, a number of metal fragments formally able to bind alkenes also react with **4*** or **5*** to form clusters that are formally equivalent to simple metal alkene complexes. These include derivatives of M(CO)$_5$ (**101***) [from Mo(CO)$_5$(NCMe), Cr(CO)$_5$(THF), or W(CO)$_5$(THF)] (*240,241*), Cp'Mn(CO)$_2$ (**102***) (*242*), and (arene)Cr(CO)$_2$ (*240,242*). Some of these compounds are fluxional. For example, the Cp$_2$CoRh(CO)$_2$ unit in the CoRH analog of **101*** rotates about the Mo—CoRH axis in analogous fashion to alkenes (*241*). Silver salts also react with **5*** to form (μ-AgX)[Cp*$_2$Rh$_2$(CO)$_2$] or (μ_4-Ag[Cp*$_2$Rh$_2$(CO)$_2$]$_2$)$^+$ species where the Rh=Rh unit effectively again functions as an alkene (*243*).

The Cp*Cu fragment is isolobal to methylene. One would therefore expect sources of Cp*Cu to add the Cp*Cu group across the Rh=Rh bond of **5*** in much the same way as diazoalkanes. This is illustrated by complex **103***, which is the product of reaction between Cp*Cu(THF) and **5*** (*244,245*).

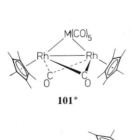

101*

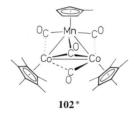

102*

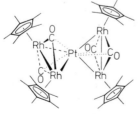

100*

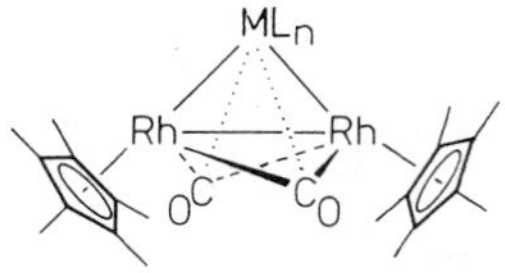

99*, ML$_n$ = Pt(CO)(PPh$_3$)
103*, ML$_n$ = CuCp*

5. Reactions of the Radical Anion $[Cp_2Co_2(\mu\text{-}CO)_2]^{\cdot-}(4^{\cdot-})$

In common with many other transition metal anions, the radical anion $[Cp_2Co_2(CO)_2]^{\cdot-}$ ($4^{\cdot-}$) reacts with alkyl halides. These reactions are particularly significant because of subsequent reactions of the products (246). The reaction of 2 equiv of MeI with $4^{\cdot-}$ gives the highly thermally unstable bismethyl derivative **104** (247) together with I^- and free iodine. This species decomposes at room temperature in THF or benzene to form acetone in greater than 85% yield (246–249). Acetone is formed more rapidly under a CO atmosphere. Crossover experiments involving **104** and d_6-**104** or **104** and the related ethyl species $Cp_2Co_2Et_2(\mu\text{-}CO)_2$ show that acetone formation is intermolecular. There appear to be at least two mechanisms operating, depending on the presence of CO: one is via an intermediate postulated as $CpCoMe_2(CO)$ and the other via dicobalt species *cis*- and *trans*-**105**.

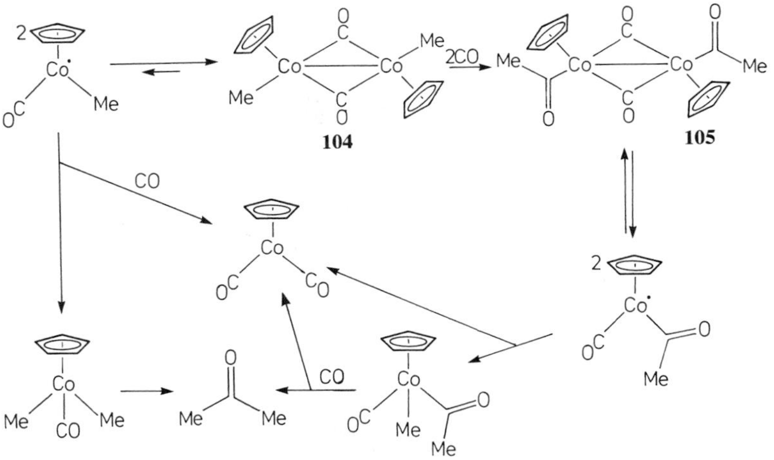

The radical anion $4^{\cdot-}$ also reacts with a single equivalent of dihaloalkanes. Thus, reaction with CH_2I_2 gives the μ-methylene complex **80** ($R^1 = R^2 = H$) (250–252) together with some **4**. This cis-to-trans ratio for **80** is 95:5. The situation is more complex with other alkylidenes derived from $CR^1R^2I_2$ (e.g., $R^1 = R^2 = Me$, Et; $R^1 = H$, $R^2 = Et$, Bu^n; $R^1 = Me$, $R^2 = Pr^n$). The trans isomer still predominates, but a third isomer **79** with bridging carbonyls is also present. When R^1 and R^2 are both alkyl groups, small amounts of Co=Co complexes **81** are also formed. These are derived from **79** and **80** by irradiation or warming.

Heating the CMe_2 complex in benzene at 80°C gives propene. Curiously, a small amount of the CHEt complex also forms under these conditions. Mechanistic studies show that hydrogen migration rather than carbon migrations are responsible (along with the requisite metal migrations). These results can be linked to an alkene isomerization pathway.

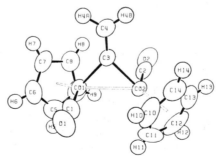

An extension of this synthetic methodology leads to μ-vinylidene complexes. Reaction of $H_2C{=}CBr_2$ with $\mathbf{4^{\cdot}}$ leads to complex **106** (Fig. 19) (253,254), containing only terminal carbonyls, together with **4**. The vinylic hydrogens are nearly coplanar with the Co—Co bond in the solid state. Successive reactions of **106** with acid and hydride give the alkylidene **107**.

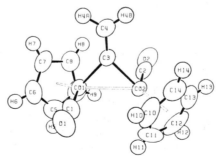

FIG. 19.   X-Ray crystal structure of $Cp_2Co_2(CO)_2(\mu\text{-}C{=}CH_2)$ (**106**), taken with permission from Ref. *253*, copyright American Chemical Society.

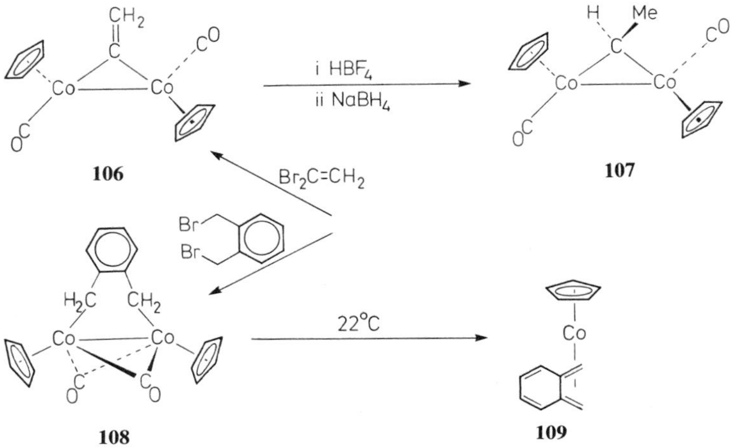

106   **107**

108   **109**

Direct conversion of the vinylidene to the μ-alkylidene is achieved under a hydrogen atmosphere, but the maximum buildup of the alkylidene never exceeds 43% since it, in turn, reacts with H_2 to form ethane.

A further extension leads to a dimetallacyclohexene complex **108**. This is achieved through the reaction of **4**$^{\overline{}}$ with $o\text{-}C_6H_4(CH_2Br)_2$ (225,256). This molecule is fluxional, the fluxional process apparently involving the interconversion of two (presumably boat) conformations of the metallacycle. Complex **108** undergoes a first-order process generating the mononuclear o-xylylene compound **109** together with $CpCo(CO)_2$. Carbonylation of **108** releases **4** and $CpCo(CO)_2$ together with two o-xylylene dimers.

Reaction of **4**$^{\overline{}}$ with $I(CH_2)_3I$ results in the trimethylene bridged complex **110**, effectively a dimetallacyclopentane (257,258). In contrast to the dimethyl complex **104**, **110** is surprisingly thermally stable. It decomposes only slowly in benzene at 100°C in a sealed tube. The organic products of thermal degradation are primarily (73%) propene, but cyclopropane is a substantial by-product (17%). The degradation is intramolecular. On the other hand, reaction of **110** with I_2 leads to greater quantities of cyclopropane (71%) and much lesser quantities of propene (9%). Mechanistic studies using 2,3-diiodopentane suggest that formation of **110** proceeds initially by a electron transfer process to form **4**, I^-, and a $[CH_2]_3I$ radical. The two pathways shown lead to **110**.

Very recently a number of phosphorus derivatives have become available from **4**$^{\overline{}}$. Reaction with $(Me_3Si)_2CHPCl_2$ gives **111** which has a bridging $PHCH(SiMe_3)_2$ group (259). The formation of **111** is mechanistically obscure, but undoubtedly the source of the $Co(CO)_3$ unit is $[Co(CO)_4]^-$, which is always present in solutions of **4⁻̇**. On the other hand, reaction of **4⁻̇** with ArPCl$_2$ (Ar = 2,4,6-C$_6$H$_2$Bu$^t_3$) does not have these particular problems. The product is the "open" phosphinidene **112** in which no metal–metal bond remains (*260,261*).

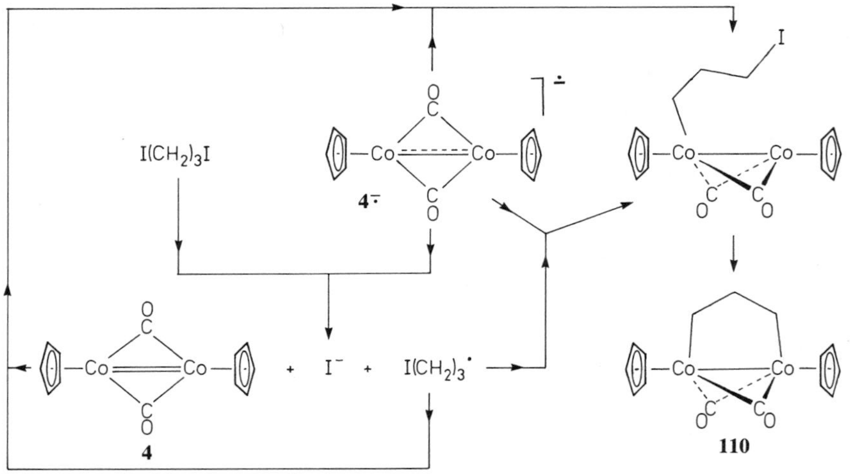

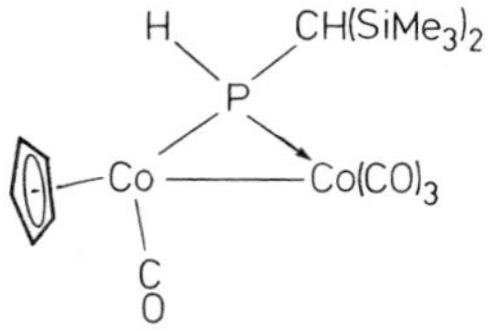

ACKNOWLEDGMENTS

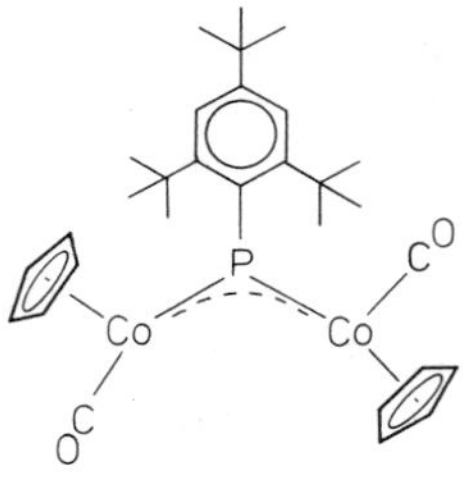

I am grateful to all researchers who supplied reprints of articles and to Professor L. F. Dahl for kindly providing results prior to publication.

REFERENCES

1. F. A. Cotton and R. A. Walton, "Multiple Bonds Between Metal Atoms." Wiley, New York, 1982.
2. M. H. Chisholm, Ed., "Reactivity of Metal–Metal Bonds." ACS Symposium Series 155, American Chemical Society, Washington, D.C., 1981.
3. M. H. Chisholm and D. W. Reichert, *J. Am. Chem. Soc.* **96**, 1249 (1974).
4. F. A. Cotton and C. B. Harris, *Inorg. Chem.* **4**, 330 (1965).

5. D. Lawton and R. Mason, *J. Am. Chem, Soc.* **87**, 921 (1965).
6. F. X. Kohl and P. Jutzi *in* "Organometallic Syntheses" (R. B. King and J. J. Eisch, Eds), Vol. 3, p. 489. Elsevier Sequoia, Amsterdam, 1987.
7. R. H. Hooker, K. A. Mahmoud, and A. J. Rest, *J. Organomet. Chem.* **254**, C25 (1983).
8. J. L. Hughey, C. R. Bock, and T. J. Meyer, *J. Am. Chem. Soc.* **97**, 4440 (1975).
9. E. O. Fischer and R. J. J. Schneider, *Angew. Chem. Int. Ed. Engl.* **6**, 569 (1967) [Ger.: *Angew. Chem.* **79**, 537 (1967)].
10. W. A. Herrmann and W. Kalcher, *Chem. Ber.* **115**, 3886 (1982).
11. P. Hackett, P. S. O'Neill, and A. R. Manning, *J. Chem. Soc., Dalton Trans.*, 1625 (1974).
12. R. C. Job and M. D. Curtis, *Inorg. Chem.* **12**, 2510 (1973).
13. D. S. Ginley, C. R. Bock, and M. S. Wrighton, *Inorg. Chim. Acta* **23**, 85 (1977).
14. R. B. King and A. Efraty, *J. Am. Chem. Soc.* **93**, 4950 (1971).
15. R. B. King and M. B. Bisnette, *J. Organomet. Chem.* **8**, 287 (1967).
16. R. B. King, M. Z. Iqbal, and A. D. King, Jr., *J. Organomet. Chem.* **171**, 53 (1979).
17. W. A. Herrmann, R. Serrano, and J. Weichmann, *J. Organomet. Chem.* **246**, C57 (1983).
18. J. K. Hoyano and W. A. G. Graham, *J. Chem. Soc., Chem. Commun.*, 27 (1982).
19. J. P. Blaha, B. E. Bursten, J. C. Dewan, R. B. Frankel, C. L. Randolph, B. A. Wilson, and M. S. Wrighton, *J. Am. Chem. Soc.* **107**, 4561 (1985).
20. R. H. Hooker, K. A. Mahmoud, and A. J. Rest, *J. Chem. Soc., Chem. Commun.*, 1022 (1983).
21. W. I. Bailey, Jr., D. M. Collins, F. A. Cotton, J. C. Baldwin, and W. C. Kaska, *J. Organomet. Chem.* **165**, 373 (1979).
22. A. Nutton and P. M. Maitlis, *J. Organomet. Chem.* **166**, C21 (1979).
23. P. A. Chetcuti and M. F. Hawthorne, *J. Am. Chem. Soc.* **109**, 942 (1987).
24. W.-S. Lee and H. H. Brintzinger, *J. Organomet. Chem.* **127**, 87 (1977).
25. F. R. Anderson and M. S. Wrighton, *Inorg. Chem.* **25**, 112 (1986).
26. N. E. Schore, C. S. Ilenda, and R. G. Bergman, *J. Am. Chem. Soc.* **98**, 255 (1976).
27. N. E. Schore, C. S. Ilenda, and R. G. Bergman, *J. Am. Chem. Soc.* **98**, 256 (1976).
28. L. M. Cirjak, R. E. Ginsburg, and L. F. Dahl, *Inorg. Chem.* **21**, 940 (1982).
29. M. J. Krause and R. G. Bergman, *J. Am. Chem. Soc.* **107**, 2972 (1985).
29a. A. J. Dixon, S. Firth, A. Haynes, M. Poliakoff, J. J. Turner, and N. M. Boag, *J. Chem. Soc., Dalton Trans.*, 1501 (1988).
30. F. A. Cotton, B. A. Frenz, and L. Kruczynski, *J. Am. Chem. Soc.* **95**, 951 (1973).
31. F. A. Cotton, B. A. Frenz, and L. Kruczynski, *J. Organomet. Chem.* **160**, 93 (1978).
32. J. C. Huffmann, L. N. Lewis, and K. G. Caulton, *Inorg. Chem.* **19**, 2755 (1980).
33. L. N. Lewis and K. G. Caulton, *J. Organomet. Chem.* **252**, 57 (1983).
34. A. S. Foust, J. K. Hoyano, and W. A. G. Graham, *J. Organomet. Chem* **32**, C65 (1971).
35. L .N. Lewis and K. G. Caulton, *Inorg. Chem.* **19**, 1840 (1980).
36. F. A. Cotton, *Prog Inorg. Chem.* **21**, 1 (1976).
37. M. L. Aldridge, M. Green, J. A. K. Howard, G. N. Pain, S. J. Porter, F. G. A. Stone, and P. Woodward, *J. Chem. Soc., Dalton Trans.*, 1333 (1982).
38. M. D. Curtis and W. M. Butler, *J. Organomet. Chem.* **155**, 131 (1978).
39. J. Potenza, P. Giordano, D. Mastropaolo, A. Efraty, and R. B. King, *J. Chem. Soc., Chem. Commun.*, 1333 (1972).
40. J. Potenza, P. Giordano, D. Mastropaolo, and A. Efraty, *Inorg. Chem.* **13**, 2540 (1974).
41. R. J. Klingler, W. M. Butler, and M. D. Curtis, *J. Am. Chem. Soc.* **100**, 5034 (1978).

42. J.-S. Huang and L. F. Dahl, *J. Organomet. Chem.* **243,** 57 (1983).
43. M. D. Curtis, K-B. Shiu, W. M. Butler, and J. C. Huffman, *J. Am. Chem. Soc.* **108,** 3335 (1986).
44. M. A. Greaney, J. S. Merola, and T. R. Halbert, *Organometallics* **4,** 2057 (1985).
45. E. D. Jemmis, A. R. Pinhas, and R. Hoffmann, *J. Am. Chem. Soc.* **102,** 2576 (1980).
46. B. J. Morris-Sherwood, C. B. Powell, and M. B. Hall, *J. Am. Chem. Soc.* **106,** 5079 (1984).
47. C. J. Commons and B. F. Hoskins, *Aust. J. Chem.* **28,** 1663 (1975).
48. R. Colton and C. J. Commons, *Aust. J. Chem.* **28,** 1673 (1975).
49. R. Colton, C. J. Commons, and B. F. Hoskins, *J. Chem. Soc., Chem. Commun.*, 363 (1975).
50. A. A. Pasynskii, Yu. V. Skripkin, I. L. Eremenko, V. T. Kalinnikov, G. G. Aleksandrov, V. G. Andrainov, and Yu. T. Struchov, *J. Organomet. Chem.* **165,** 49 (1979).
51. I. Bernal, J. D. Korp, W. A. Herrmann, and R. Serrano, *Chem. Ber.* **117,** 434 (1984).
52. R. H. Summerville and R. Hoffmann, *J. Am. Chem. Soc.* **101,** 3821 (1979).
53. F. A. Cotton and J. M. Troup, *J. Chem. Soc., Dalton Trans.*, 800 (1980).
54. J. W. Lauher, M. Elian, R. H. Summerville, and R. Hoffmann, *J. Am. Chem. Soc.* **98,** 3219 (1976).
55. R. E. Ginsburg, L. M. Cirjak, and L. F. Dahl, *J. Chem. Soc., Chem. Commun.*, 468 (1979).
56. M. Green, D. R. Hankey, J. A. K. Howard, P. Louca, and F. G. A. Stone, *J. Chem. Soc., Chem. Commun.*, 757 (1983).
57. N. E. Schore, C. S. Ilenda, and R. G. Bergman, *J. Am. Chem. Soc.* **99,** 1781 (1977).
58. I. Bernal, J. D. Korp, G. M. Reisner, and W. A. Herrmann, *J. Organomet. Chem.* **139,** 321 (1977).
59. W. A. Herrmann and I. Bernal, *Angew. Chem. Int. Ed. Engl.* **16,** 172 (1977) [Ger.: *Angew. Chem.* **89,** 186 (1977)].
60. J. M. Calderon, S. Fontana, E. Frauendorfer, V. W. Day, and S. D. Iske, *J. Organomet. Chem.* **64,** C16 (1974).
61. A. R. Pinhas and R. Hoffmann, *Inorg. Chem.* **18,** 654 (1979).
62. N. Dudeney, J. C. Green, O. N. Kirchner, and F. St. J. Smallwood, *J. Chem. Soc., Dalton Trans.*, 1883 (1984).
63. N. E. Schore, *J. Organomet. Chem.* **173,** 301 (1979).
64. K. A. Schugart and R. F. Fenske, *J. Am. Chem. Soc.* **108,** 5094 (1986).
65. K. A. Schugart and R. F. Fenske, *J. Am Chem. Soc.* **108,** 5100 (1986).
66. L. Knoll, K. Reiss, J. Schafer, and P. Klufers, *J. Organomet. Chem.* **193,** C40 (1980).
67. S.-I. Murahashi, T. Mizoguchi, T. Hosokawa, I. Moritani, Y. Kai, M. Kohara, N. Yasuoka, and N. Kasai, *J. Chem. Soc., Chem. Commun.*, 563 (1974).
68. I. Fischler, K. Hildenbrand, and E. K. von Gustorf, *Angew. Chem. Int. Ed. Engl.* **14,** 54 (1975) [Ger: *Angew. Chem.* **87,** 35 (1975)].
69. W. A. Herrmann, C. E. Barnes, R. Serrano, and B. Koumbouris, *J. Organomet. Chem.* **256,** C30 (1983).
70. J. D. Atwood, T. S. Janik, J. L. Atwood, and R. D. Rogers, *Synth. React. Inorg. Met.-Org. Chem.* **10,** 397 (1980).
71. R. Hörlein and W. A. Herrmann, *J. Organomet. Chem.* **303,** C38 (1986).
72. P. Härter, H. Pfisterer, and M. L. Ziegler, *Angew. Chem. Int. Ed. Engl.* **25,** 839 (1986) [Ger: *Angew. Chem.* **98,** 812 (1986)].
73. P. Härter and W. A. Herrmann, *in* "Organometallic Syntheses" (R. B. King and J. J. Eisch Eds.), Vol. 3, p. 309. Elsevier Sequoia, Amsterdam, 1987.
74. K. R. Pope and M. S. Wrighton, *Inorg. Chem.* **26,** 2321 (1987).

75. J. K. Hoyano and W. A. G. Graham, *J. Am. Chem. Soc.* **104**, 3722 (1982).
76. E.O. Fischer and R. J. J. Schneider, *Chem. Ber.* **103**, 3684 (1970).
77. W. A. Herrmann and J. Plank, *Chem. Ber.* **112**, 392 (1979).
78. R. B. King, *Coord. Chem. Rev.* **20**, 155 (1976).
79. R. B. King and A. Efraty, *J. Am. Chem. Soc.* **94**, 3773 (1972).
80. R. J. Klingler, W. Butler, and M. D. Curtis, *J. Am. Chem. Soc.* **97**, 3535 (1975).
81. M. D. Curtis, N. A. Fotinos, L. Messerle, and A. P. Sattelberger, *Inorg. Chem.* **22**, 1559 (1983).
82. R. Birdwhistell, P. Hackett, and A. R. Manning, *J. Organomet. Chem.* **157**, 239 (1978).
83. S. A. R. Knox, R. F. D. Stansfield, F. G. A. Stone, M. J. Winter, and P. Woodward, *J. Chem. Soc., Dalton Trans.*, 173 (1982).
84. R. B. King and M. B. Bisnette, *Inorg. Chem.* **4**, 475 (1965).
85. K.-B. Shiu, M. D. Curtis, and J. C. Huffman, *Organometallics* **2**, 936 (1983).
86. M. H. Chisholm, M. W. Extine, R. L. Kelly, W. C. Mills, C. A. Murillo, L. A. Rankel, and W. W. Reichert, *Inorg. Chem.* **17**, 1673 (1978).
87. C. G. Young, M. Minelli, J. H. Enemark, G. Miessler, N. Janietz, H. Kauermann, and J. Wachter, *Polyhedron* **5**, 407 (1986).
88. W. A. Herrmann and R. Serrano, *in* "Organometallic Syntheses" (R. B. King and J. J. Eisch Eds.), Vol. 3, p. 43. Elsevier Sequoia, Amsterdam, 1987.
89. W. A. Herrmann, R. Serrano, U. Küsthardt, M. L. Ziegler, E. Guggolz, and T. Zahn, *Angew. Chem. Int. Ed. Engl.* **23**, 515 (1984) [Ger.: *Angew. Chem.* **96**, 498 (1984)].
90. W. A. Herrmann, R. Serrano, and H. Bock, *Angew. Chem. Int. Ed. Engl.* **23**, 383 (1984) [Ger.: *Angew. Chem.* **96**, 364 (1984)].
91. A. F. Hepp, J. P. Blaha, C. Lewis, and M. S. Wrighton, *Organometallics* **3**, 174 (1984).
92. B. D. Moore, M. B. Simpson, M. Poliakoff, and J. J. Turner, *J. Chem. Soc., Chem. Commun.*, 972 (1984).
93. Y.-M. Wuu, C. Zou, and M. S. Wrighton, *J. Am. Chem. Soc.* **109**, 5861 (1987).
94. O. Crichton, A. J. Rest, and D. J. Taylor, *J. Chem. Soc., Dalton Trans.*, 167 (1980).
95. W. H. Hersh, F. J. Hollander, and R. G. Bergman, *J. Am. Chem. Soc.* **105**, 5834 (1983).
96. R. G. Beevor, S. A. Frith, and J. L. Spencer, *J. Organomet. Chem.* **221**, C25 (1981).
97. R. L. Bedard, A. D. Rae, and L. F. Dahl, *J. Am. Chem. Soc.* **108**, 5924 (1986).
98. W. A. Herrmann, J. Plank, C. Bauer, M. L. Ziegler, E. Guggolz, and R. Alt, *Z. Anorg. Allg. Chem.* **487**, 85 (1982).
99. W. A. Herrmann, J. Plank, and C. Bauer, *in* "Organometallic Syntheses" (R. B. King and J. J. Eisch Eds.), Vol. 3, p. 48. Elsevier Sequoia, 1987.
100. A. C. Bray, M. Green, D. R. Hankey, J. A. K. Howard, O. Johnson, and F. G. A. Stone, *J. Organomet. Chem.* **281**, C12 (1985).
101. W. A. Herrmann, J Plank, and B. Reiter, *J. Organomet. Chem.* **164**, C25 (1979).
102. S. Otsuka nad A. Nakamura, *Adv. Organomet. Chem.* **14**, 245 (1976).
103. M. D. Curtis, L. Messerle, N. A. Fotinos, and R. F. Gerlach, *in* "Reactivity of Metal–Metal Bonds" (M. H. Chisholm, Ed.), p. 221. ACS Symposium Series 155, American Chemical Society, Washington, D.C., 1981.
104. M. D. Curtis, *Polyhedron* **6**, 759 (1987).
105. J. L. Robbins and M. S. Wrighton, *Inorg. Chem.* **20**, 1133 (1981).
106. N. N. Turaki and J. M. Huggins, *Organometallics* **4**, 1766 (1985).
107. J. S. Merola, R. A. Gentile, G. B. Ansell, M. A. Modrick, and S. Zentz, *Organometallics* **1**, 1731 (1982).
108. J. S. Merola, K. S. Campo, R. A. Gentile, M. A. Modrick, and S. Zentz, *Organometallics* **3**, 334 (1984).

109. E. A. V. Ebsworth, A. P. McIntosh, and M. Schröder, *J. Organomet. Chem.* **312,** C41 (1986).
110. H. Adams, N. A. Bailey, C. Bannister, M. A. Faers, P. Fedorko, V. A. Osborn, and M. J. Winter, *J. Chem. Soc., Dalton Trans.,* 341 (1987).
111. J. Wachter, A. Mitschler, and J. G. Riess, *J. Am. Chem. Soc.* **103,** 2121 (1981).
112. J. Wachter, J. G. Riess, and A. Mitschler, *Organometallics* **3,** 714 (1984).
113. J. G. Riess, U. Klement, and J. Wachter, *J. Organomet. Chem.* **280,** 215 (1985).
114. M. D. Curtis, L. Messerle, J. J. D'Errico, W. M. Butler, and M. S. Hay, *Organometallics* **5,** 2283 (1986).
115. W. A. Herrmann and L. K. Bell, *J. Organomet. Chem.* **239,** C4 (1982).
116. W. A. Herrmann, *Adv. Organomet. Chem.* **20,** 159 (1982).
117. W. A. Herrmann, *Pure Appl. Chem.* **54,** 66 (1982).
118. W. A. Herrman, *J. Organomet. Chem.* **250,** 319 (1983).
119. L. K. Bell, W. A. Herrmann, G. W. Kriechbaum, H. Pfisterer, and M. L. Ziegler, *J. Organomet. Chem.* **240,** 381 (1982).
120. L. Messerle and M. D. Curtis, *J. Am. Chem. Soc.* **102,** 7789 (1980).
121. M.D. Curtis, L. Messerle, J. J. D'Errico, H. E. Solis, I. D. Barcelo, and W. M. Butler, *J. Am. Chem. Soc.* **109,** 3603 (1987).
122. J. J. D'Errico and M. D. Curtis, *J. Am. Chem. Soc.* **105,** 4479 (1983).
123. W. A. Herrmann, G. W. Kriechbaum, C. Bauer, E. Guggolz, and M. L. Ziegler, *Angew. Chem. Int. Ed. Engl.* **20,** 815 (1981) [Ger.: *Angew. Chem.* **93,** 836 (1981)].
124. N. D. Feasey, S. A. R. Knox, and A. G. Orpen, *J. Chem. Soc., Chem. Commun.,* 75 (1982).
125. L. K. Bell, W. A. Herrmann, M. L. Ziegler, and H. Pfisterer, *Organometallics* **1,** 1673 (1982).
126. W. A. Herrmann, G. W. Kriechbaum, M. L. Ziegler, and H. Pfisterer, *Angew. Chem. Int. Ed. Engl.* **21,** 707 (1982) [Ger.: *Angew. Chem.* **94,** 713 (1981)].
127. W. A. Herrmann, G. W. Krichbaum, R. Dammel, H. Bock, M. L. Ziegler, and H. Pfisterer, *J. Organomet. Chem.* **254,** 219 (1983).
128. M. D. Curtis, J. J. D'Errico, and W. M. Butler, *Organometallics* **6,** 2151 (1987).
129. W. I. Bailey, Jr., F. A. Cotton, J. D. Jamerson, and J. R. Kolb, *J. Organomet. Chem.* **121,** C23 (1976).
130. W. I. Bailey, Jr., D. M. Collins, and F. A. Cotton, *J. Organomet. Chem.* **135,** C53 (1977).
131. W. I. Bailey, Jr., M. H. Chisholm, F. A. Cotton, and L. A. Rankel, *J. Am. Chem. Soc.* **100,** 5764 (1978).
132. M. D. Curtis and R. J. Klingler, *J. Organomet. Chem.* **161,** 23 (1978).
133. J. A. Beck, S. A. R. Knox, R. F. D. Stansfield, F. G. A. Stone, M. J. Winter, and P. Woodward, *J. Chem. Soc., Dalton Trans.,* 195 (1982).
134. R. F. Gerlach, D. N. Duffy, and M. D. Curtis, *Organometallics* **2,** 1172 (1983).
135. P. Bougeard, S. Peng, M. Mlekuz, and M. J. McGlinchey, *J. Organomet. Chem.* **296,** 383 (1985).
136. S. P. Nolan, R. L. de la Vega, and C. D. Hoff, *Inorg. Chem.* **25,** 4446 (1986).
137. D. S. Ginley, C. R. Bock, M. S. Wrighton, B. Fischer, D. L. Tipton, and R. Bau, *J. Organomet. Chem.* **157,** 41 (1978).
138. S. A. R. Knox, R. F. D. Stansfield, F. G. A. Stone, M. J. Winter, and P. Woodward, *J. Chem. Soc., Dalton Trans.,* 173 (1982).
139. R. Goddard, S. A. R. Knox, F. G. A. Stone, M. J. Winter, and P. Woodward, *J. Chem. Soc., Chem. Commun.,* 559 (1976).

140. R. Goddard, S. A. R. Knox, R. F. D. Stansfield, F. G. A. Stone, M. J. Winter, and P. Woodward, *J. Chem. Soc., Dalton Trans.,* 147 (1982).
141. M. Griffiths, S. A. R. Knox, R. F. D. Stansfield, F. G. A. Stone, and M. J. Winter, *J. Chem. Soc., Dalton Trans.,* 159 (1982).
142. S. A. R. Knox, R. F. D. Stansfield, F. G. A. Stone, M. J. Winter, and P. Woodward, *J. Chem. Soc., Chem. Commun.,* 221 (1978).
143. J. S. Bradley, *J. Organomet. Chem.* **150,** C1 (1978).
144. S. Salter and E. L. Muetterties, *Inorg. Chem.* **20,** 946 (1981).
145. N. M. Doherty, C. Eischenbroich, H.-J. Kneuper, and S. A. R. Knox, *J. Chem. Soc., Chem. Commun.,* 170 (1985).
146. M. Green, N. C. Norman, and A. G. Orpen, *J. Am Chem. Soc.* **103,** 1269 (1981).
147. A. M. Boileau, A. G. Orpen, R. F. D. Stansfield, and P. Woodward, *J. Chem. Soc., Dalton Trans.,* 187 (1982).
148. M. H. Chisholm, L. A. Rankel, W. I. Bailey, Jr., F. A. Cotton, and C. A. Murillo, *J. Am. Chem. Soc.* **99,** 1261 (1977).
149. W. I. Bailey, Jr., M. H. Chisholm, F. A. Cotton, C. A. Murillo, and L. A. Rankel, *J. Am Chem. Soc.* **100,** 802 (1978).
150. J. C. T. R. Burckett-St. Laurent, P. B. Hitchcock, H. W. Kroto, M. F. Meidine, and J. F. Nixon *J. Organomet. Chem.* **238,** C82 (1982).
151. G. Becker, W. A. Herrmann, W. Kalcher, G. W. Kriechbaum, C. Pahl, C. T. Wagner, and M. L. Ziegler, *Angew. Chem. Int. Ed. Engl.* **22,** 413 (1983) [Ger.: **95,** 417 (1983)].
152. O. J. Scherer, H. Sitzmann, and G. Wolmershäuser, *J. Organomet. Chem.* **268,** C9 (1984).
153. P. B. Hitchcock, M. F. Meidine, and J. F. Nixon, *J. Organomet. Chem.* **333,** 337 (1987).
154. O. J. Scherer, H. Sitzmann, and G. Wolmershäuser, *Angew. Chem. Int. Ed. Engl.* **23,** 968 (1984) [Ger.: *Angew. Chem.* **96,** 979 (1984)].
155. O. J. Scherer, H. Sitzmann, and G. Wolmershäuser, *Angew. Chem. Int. Ed. Engl.* **24,** 351 (1985) [Ger.: *Angew. Chem.* **97,** 358 (1985)].
156. O. J. Scherer, H. Sitzmann, and G. Wolmershäuser, *J. Organomet. Chem.* **309,** 77 (1986).
157. I. Bernal, H. Brunner, W. Meier, H. Pfisterer, J. Wachter, and M. L. Ziegler, *Angew. Chem. Int. Ed. Engl.* **23,** 438 (1984) [Ger.: *Angew. Chem.* **96,** 428 (1984)].
158. P. J. Sullivan and A. L. Rheingold, *Organometallics* **1,** 1547 (1982).
159. R. D. Adams, D. A. Katahira, and L.-W. Yang, *Organometallics* **1,** 231 (1982).
160. D. Lentz, I. Brüdgam, and H. Hartl, *J. Organomet. Chem.* **299,** C38 (1986).
161. D. Lentz, I. Brüdgam, and H. Hartl, *Angew. Chem. Int. Ed. Engl.* **23,** 525 (1984) [Ger.: *Angew. Chem.* **96,** 511 (1984)].
162. H. Brunner, H. Buchner, J. Wachter, I. Bernal, and W. H. Ries, *J. Organomet. Chem.* **244,** 247 (1983).
163. H. Brunner, B. Hoffmann, and J. Wachter, *J. Organomet. Chem.* **252,** C35 (1983).
164. I. Bernal, H. Brunner, and J. Wachter, *J. Organomet. Chem.* **277,** 395 (1984).
165. I. Bernal, M. Draux, H. Brunner, B. Hoffmann, and J. Wachter, *Organometallics* **5,** 655 (1986).
166. M. D. Curtis, K. R. Han, and W. M. Butler, *Inorg. Chem.* **19,** 2096 (1980).
167. R. J. Mercer, M. Green, and A. G. Orpen, *J. Chem. Soc. Chem. Commun.,* 567 (1986).
168. S. F. T. Froom, M. Green, R. J. Mercer, K. R. Nagle, A. G. Orpen, and S. Schwiegk, *J. Chem. Soc., Chem. Commun.,* 1666 (1986).

169. M. H. Chisholm, F. A. Cotton, M. W. Extine, and L. A. Rankel, *J. Am. Chem. Soc.* **100,** 807 (1978).
170. J. J. D'Errico, L. Messerle, and M. D. Curtis, *Inorg. Chem.* **22,** 849 (1983).
171. G. K. Barker, W. E. Carroll, M. Green, and A. J. Welch, *J. Chem. Soc., Chem. Commun.,* 1071 (1980).
172. W. E. Carroll, M. Green, A. G. Orpen, C. J. Schaverien, I. D. Williams, and A. J. Welch, *J. Chem. Soc., Dalton Trans.,* 1021 (1986).
173. H. Alper, N. D. Silavwe, G. I. Birnbaum, and F. R. Ahmed, *J. Am. Chem. Soc.* **101,** 6582 (1979).
174. H. Alper, F. W. B. Einstein, R. Nagal, J.-F. Petrignani, and A. C. Wills, *Organometallics* **2,** 1291 (1983).
175. H. Alper, F. W. B. Einstein, F. W. Hartstock, and A. C. Willis, *J. Am. Chem. Soc.* **107,** 173 (1985).
176. H. Alper, F. W. B. Einstein, F. W. Hartstock, and A. C. Willis, *Organometallics* **5,** 9 (1986).
177. H. Adams, N. A. Bailey, J. T. Gauntlett, and M. J. Winter, *J. Chem. Soc., Chem. Commun.,* 1360 (1984).
178. H. Alper, F. W. B. Einstein, J.-F. Petrignani, and A. C. Willis, *Organometallics* **2,** 1422 (1983).
179. S. Scheibye, S.-O. Lawesson, and C. Romming, *Acta Chem. Scand., Ser. B* **B35,** 239 (1981).
180. H. Alper and J. Hartgerink, *J. Organomet. Chem.* **190,** C25 (1980).
181. H. Brunner, W. Meier, and J. Wachter, *J. Organomet. Chem.* **210,** C23 (1981).
182. M. D. Curtis, N. A. Fotinos, K. R. Han, and W. M. Butler, *J. Am. Chem. Soc.* **105,** 2686 (1983).
183. R. B. King, A. Efraty, and W. M. Douglas, *J. Organomet. Chem.* **60,** 125 (1973).
184. M. D. Curtis and W. M. Butler, *J. Chem. Soc., Chem. Commun.,* 998 (1980).
185. H. Brunner, W. Meier, J. Wachter, E. Guggolz, T. Zahn, and M. L. Ziegler, *Organometallics* **1,** 1107 (1982).
186. H. Brunner, J. Wachter, and H. Wintergerst, *J. Organomet. Chem.* **235,** 77 (1982).
187. H. Brunner, J. Wachter, E. Guggolz, and M. L. Ziegler, *J. Am. Chem. Soc.* **104,** 1765 (1982).
188. H. Brunner, H. Kauermann, and J. Wachter, *J. Organomet. Chem.* **265,** 189 (1984).
189. P. D. Williams and M. D. Curtis, *Inorg. Chem.* **25,** 4562 (1986).
190. T. J. Greenhough, B. W. S. Kolthammer, P. Legdzins, and J. Trotter, *Inorg. Chem.* **18,** 3543 (1979).
191. K. Isobe, S. Kimura, and Y. Nakamura, *J. Chem. Soc., Chem. Commun.,* 378 (1985).
192. J.-F. Petrignani and H. Alper, *Inorg. Chim. Acta* **77,** L243 (1983).
193. H. Alper, J-F. Petrignani, F. W. B. Einstein, and A. C. Willis, *J. Am. Chem. Soc.* **105,** 1701 (1983).
194. R. Hoffmann, *Angew. Chem. Int. Ed. Engl.* **21,** 711 (1982) [Ger.: *Angew. Chem.* **94,** 725 (1982)].
195. F. G. A. Stone, *Angew. Chem. Int. Ed. Engl.* **23,** 89 (1984) [Ger.: *Angew. Chem.* **96,** 85 (1984)].
196. M. Green, S. J. Porter, and F. G. A. Stone, *J. Chem. Soc., Dalton Trans.,* 513 (1983).
197. P. Brun, G. M. Dawkins, M. Green, A. D. Miles, A. G. Orpen, and F. G. A. Stone, *J. Chem. Soc., Chem. Commun.,* 926 (1982).
198. C. P. Gibson and L. F. Dahl, *Organometallics* **7,** in press (1988).
199. C. P. Gibson, J.-S. Huang, and L. F. Dahl, *Organometallics* **5,** 1676 (1986).
200. L.-Y. Hsu, W.-L. Hsu, D.-Y. Jan, and S. G. Shore, *Organometallics* **5,** 1041 (1986).

201. E. Roland, W. Bernhardt, and H. Vahrenkamp, *Chem. Ber.* **119**, 2566 (1986).
202. A. J. Dixon, M. A. Healy, M. Poliakoff, and J. J. Turner, *J. Chem. Soc., Chem. Commun.*, 994 (1986).
203. W. A. Herrmann, J. M. Huggins, B. Reiter, and C. Bauer, *J. Organomet. Chem.* **214**, C19 (1981).
204. W. A. Herrmann, J. M. Huggins, C. Bauer, H. Pfisterer, and M. L. Ziegler, *J. Organomet. Chem.* **226**, C59 (1982).
205. W. A. Herrmann, C. Bauer, J. M. Huggins, H. Pfisterer, and M. L. Ziegler, *J. Organomet. Chem.* **258**, 81 (1983).
206. W. A. Herrmann, J. M. Huggins, C. Bauer, M. L. Ziegler, and H. Pfisterer, *J. Organomet. Chem.* **262**, 253 (1984).
207. T. R. Halbert, M. E. Leonowicz, and D. J. Maydonovitch, *J. Am. Chem. Soc.* **102**, 5101 (1980).
208. A. D. Clauss, P. A. Dimas, and J. R. Shapley, *J. Organomet. Chem.* **201**, C31 (1980).
209. C. Bauer and W. A. Herrmann, *J. Organomet. Chem.* **209**, C13 (1981).
210. W. A. Herrmann, C. Bauer, J. Plank, W. Kalcher, D. Speth, and M. L. Ziegler, *Angew. Chem. Int. Ed. Engl.* **20**, 193 (1981) [Ger.: *Angew. Chem.* **93**, 212 (1981)].
211. M. Green, R. M. Mills, G. N. Pain, F. G. A. Stone, and P. Woodward, *J. Chem. Soc., Dalton Trans.*, 1309 (1982).
212. W. A. Herrmann, C. Bauer, G. Kriechbaum, H. Kunkely, M. L. Ziegler, D. Speth, and E. Guggolz, *Chem. Ber.* **115**, 878 (1982).
213. W. A. Herrman and C. Bauer, *in* "Organometallic Syntheses" (R. B. King and J. J. Eisch, Eds.), Vol. 3, p. 231. Elsevier Sequoia, Amsterdam, 1987.
214. W. A. Herrmann, C. Bauer, and K. K. Mayer *J. Organomet. Chem.* **236**, C18 (1982).
215. W. A. Herrmann, C. Weber, M. L. Ziegler, and C. Pahl, *Chem. Ber.* **117**, 875 (1984).
216. W. A. Herrmann, E. Herdtweck, and C. Weber, *Angew. Chem. Int. Ed. Engl.* **25**, 563 (1986) [Ger.: *Angew. Chem.* **98**, 557 (1986).
217. W. A. Herrmann, C. Bauer, and C. Weber, "Organometallic Syntheses" (R. B. King and J. J. Eisch, Eds.), Vol. 3, p. 234. Elsevier Sequoia, Amsterdam, 1987.
218. W. A. Herrmann and C. Weber, *J. Organometal. Chem.* **282**, C31 (1985).
219. W. A. Herrmann, C. Weber, M. L. Ziegler, and O. Serhaldi, *J. Organometal. Chem.* **297**, 245 (1985).
220. C. Bauer, E. Guggolz, W. A. Herrmann, G. Kriechbaum, and M. L. Ziegler, *Angew. Chem., Int. Ed. Engl.* **21**, 212 (1982) [Ger.: *Angew. Chem.* **94**, 209 (1982)].
221. W. A. Herrmann, G. W. Kriechbaum, C. Bauer, B. Koumbouris, H. Pfisterer, E. Guggolz, and M. L. Ziegler, *J. Organomet. Chem.* **262**, 89 (1984).
222. W. A. Herrmann, C. Bauer, M. L. Ziegler, and H. Pfisterer, *J. Organomet. Chem.* **243**, C54 (1983).
223. R. S. Dickson, G. S. Evans, and G. D. Fallon, *J. Organomet. Chem.* **236**, C49 (1982).
224. W. A. Herrmann, C. Bauer, and J. Weichmann, *J. Organomet. Chem.* **243**, C21 (1983).
225. W. A. Herrmann, C. Bauer, and A. Schäfer, *J. Organomet. Chem.* **256**, 147 (1983).
226. R. S. Dickson, G. S. Evans, and G. D. Gallon, *Aust. J. Chem.* **38**, 273 (1985).
227. R. S. Dickson, G. D. Fallon, S. M. Jenkins, and R. J. Nesbit, *Organometallics* **6**, 1240 (1987).
228. C. J. Schaverien, M. Green, A. G. Orpen, and I. D. Williams, *J. Chem. Soc., Chem. Commun.*, 912 (1982).
229. M. Green, A. G. Orpen, C. Schaverien, and I. D. Williams, *J. Chem. Soc., Dalton Trans.*, 2483 (1985).
230. W. A. Herrmann, C. Bauer, and J. Weichmann, *Chem. Ber.* **117**, 1271 (1984).
231. W. A. Herrmann and C. Bauer, *in* "Organometallic Syntheses" (R. B. King and J. J. Eisch Eds.), Vol. 3, p. 291. Elsevier Sequoia, Amsterdam, 1987.

232. W. A. Herrmann and J.Weichmann *in* "Organometallic Syntheses" (R. B. King and J. J. Eisch, Eds.), Vol. 3, p. 287. Elsevier Sequoia, Amsterdam, 1987.

233. H. Brunner, N. Janietz, W. Meier, G. Sergeson, J. Wachter, T. Zahn, and M. L. Ziegler, *Angew. Chem. Int. Ed. Engl.* **24**, 1060 (1985) [Ger.: *Angew. Chem.* **97**, 1056 (1985)].

234. H. Werner and B. Klingert, *J. Organomet. Chem.* **233**, 365 (1982).

235. N. Dudeney, J. C. Green, P. Grebenik, and O. N. Kirchner, *J. Organomet. Chem.* **252**, 221 (1983).

236. T. V. Ashworth, M. J. Chetcuti, L. J. Farrugia, J. A. K. Howard, J. C. Jeffrey, R. Mills, G. N. Pain, F. G. A. Stone, and P. Woodward, *in* "Reactivity of Metal–Metal Bonds" (M. H. Chisholm, Ed.), p. 299. ACS Symposium Series 155, American Chemical Society Washington, D.C., 1981.

237. N. M. Boag, M. Green, R. M. Mills, G. N. Pain, F. G. A. Stone, and P. Woodward, *J. Chem. Soc., Chem. Commun.*, 1171 (1980).

238. M. Green, J. A. K. Howard, R. M. Mills, G. N. Pain, F. G. A. Stone, and P. Woodward, *J. Chem. Soc., Chem. Commun.*, 869 (1981).

239. M. Green, J. A. K. Howard, G. N. Pain, and F. G. A. Stone, *J. Chem. Soc., Dalton Trans.*, 1327 (1982).

240. R. D. Barr, M. Green, K. Marsden, F. G. A. Stone, and P. Woodward, *J. Chem. Soc., Dalton Trans.*, 507 (1983).

241. R. D. Barr, M. Green, J. A. K. Howard, T. B. Marder, A. G. Orpen, and F. G. A. Stone, *J. Chem. Soc., Dalton Trans.*, 2757 (1984).

242. L. M. Cirjak, J.-S. Huang, Z.-H. Zhu, and L. F. Dahl, *J. Am Chem. Soc.* **102**, 6623 (1980).

243. W. A. Herrmann and W. Kalcher, *Chem. Ber.* **118**, 3861 (1985).

244. G. A. Carriedo, J. A. K. Howard, and F. G. A. Stone, *J. Organomet. Chem.* **250**, C28 (1983).

245. G. A. Carriedo, J. A. K. Howard, and F. G. A. Stone, *J. Chem. Soc., Dalton Trans.*, 1555 (1984).

246. R. G. Bergman, *Acc. Chem. Res.* **13**, 113 (1980).

247. N. E. Schore, C. S. Ilenda, and R. G. Bergman, *J. Am. Chem. Soc.* **98**, 7436 (1976).

248. M. A. White and R. G. Bergman, *J. Chem. Soc., Commun.*, 1056 (1979).

249. N. E. Schore, C. S. Ilenda, M. A. White, H. E. Bryndza, M. G. Matturro, and R. G. Bergman, *J. Am. Chem. Soc.* **106**, 7451 (1984).

250. K. H. Theopold and R. G. Bergman, *J. Am Chem. Soc.* **103**, 2489 (1981).

251. K. H. Theopold and R. G. Bergman, *Organometallics* **1**, 219 (1982).

252. K. H. Theopold and R. G. Bergman, *J. Am. Chem. Soc.* **105**, 464 (1983).

253. E. N. Jacobsen and R. G. Bergman, *Organometallics* **3**, 329 (1984).

254. E. N. Jacobsen and R. G. Bergman, *J. Am. Chem. Soc.* **107**, 2023 (1985).

255. K. H. Theopold, W. H. Hersh, and R. G. Bergman, *Isr. J. Chem.* **22**, 27 (1982).

256. W. H. Hersh and R. G. Bergman, *J. Am. Chem. Soc.* **103**, 6992 (1981).

257. K. H. Theopold and R. G. Bergman, *J. Am. Chem. Soc.* **102**, 5694 (1980).

258. G. K. Yang and R. G. Bergman, *J. Am. Chem. Soc.* **105**, 6045 (1983).

259. A. M. Arif, A. H. Cowley, M. Pakulski, M. B. Hursthouse, and A. Karauloz, *Organometallics* **4**, 2227 (1985).

260. A. M. Arif, A. H. Cowley, N. C. Norman, A. G. Orpen, and M. Pakulski, *J. Chem. Soc., Chem. Commun.*, 1267 (1985).

261. A. H. Cowley, *Phosphorus Sulphur* **26**, 31 (1986).

Organometallic Chemistry of Arene Ruthenium and Osmium Complexes

HUBERT LE BOZEC, DANIEL TOUCHARD, and PIERRE H. DIXNEUF

Laboratoire de Chimie
Unité Associée au Centre National de la Recherche Scientifique (CNRS) 415
Université de Rennes I
F-35042 Rennes Cédex, France

I

INTRODUCTION

Arene ruthenium and osmium complexes play an increasingly important role in organometallic chemistry. They appear to be good starting materials for access to reactive arene metal hydrides or 16-electron metal(0) intermediates that have been used recently for carbon–hydrogen bond activation. Various methods of access to cyclopentadienyl, borane, and carborane arene ruthenium and osmium complexes have been reported.

η^6-Arene ruthenium and osmium offer specific properties for the reactivity of arene ligand. The activation toward nucleophiles or electrophiles is controlled mainly by the oxidation state of the metal (II or 0). Recently, from classic organometallic arene ruthenium and osmium chemistry has grown an area making significant contributions to the chemistry of cyclophanes. These compounds are potential precursors of organometallic polymers which show interesting electrical properties and conductivity.

The possibility of coordination of a two-electron ligand, in addition to arene, to the ruthenium or osmium atom provides a route to mixed metal or cluster compounds. Cocondensation of arene with ruthenium or osmium vapors has recently allowed access to new types of arene metal complexes and clusters. In addition, arene ruthenium and osmium appear to be useful and specific catalyst precursors, apart from classic hydrogenation, for carbon–hydrogen bond activation and activation of alkynes; such compounds may become valuable reagents for organic syntheses.

Arene ruthenium (*1*) and osmium (*2*) complexes have been exhaustively reviewed in *Comprehensive Organometallic Chemistry*, and since the early 1980s the above-mentioned new developments in their chemistry have appeared. Thus, after briefly mentioning the early methods of preparation and the most significant initial chemistry, we focus our attention on the organometallic chemistry and new aspects of arene ruthenium and osmium complexes.

II

η^6-ARENE RUTHENIUM(II) AND OSMIUM(II) COMPLEXES

A. Preparation of η^6-Arene Ruthenium(II) Derivatives

1. General Methods from RuCl$_3$

The first general route to (η^6-arene)ruthenium(II) complexes, employing Ru(arene)$_2^{2+}$ cations, has been developed by E. O. Fischer and co-workers. The cations were prepared by displacement of halides from RuCl$_3$ with Lewis acids, under reducing conditions, in the presence of arenes (3–5).

$$RuCl_3 + 2\ arene \xrightarrow[\text{2. H}_2\text{O/X}^-]{\text{1. AlCl}_3/\text{Al}} Ru(\eta^6\text{-arene})_2^{2+}(X^-)_2$$

One of the most general and useful methods of access to (η^6-arene)ruthenium(II) complexes is based on the dehydrogenation of cyclohexadiene derivatives by ethanolic solutions of RuCl$_3 \cdot x$H$_2$O (6). This method has been developed for the synthesis of binuclear complexes **1** by Zelonka and Baird (7) for benzene (**1a**) and by Bennett and co-workers (8–10) for *para*-cymene (**1b**), a useful starting material (10), and substituted benzene derivatives, such as xylene, mesitylene (9), or 1,3,5-triphenylbenzene (11) [Eq. (1)]. The chloride ligands of complexes **1** can

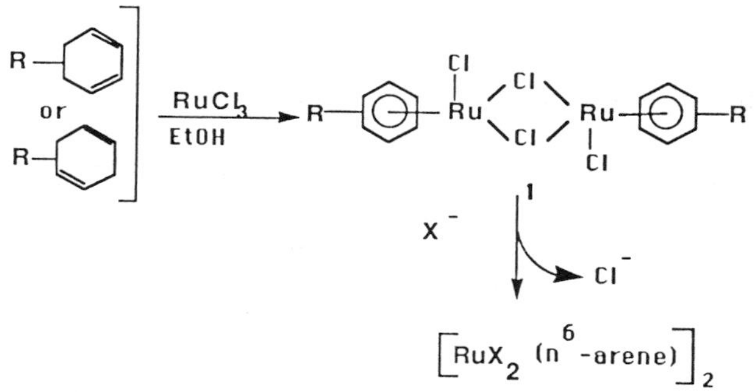

$$(1)$$

arene: C$_6$H$_6$ (**1a**), *p*-Me-C$_6$H$_4$-*i*Pr (**1b**), C$_6$Me$_6$ (**1c**) be exchanged with $X^- = I^-$, Br^-, SCN^- by treatment with the corresponding salts (7,9). Complexes **1** may also be obtained by addition of HCl to $(\eta^6$-arene$)(\eta^4$-diene$)Ru(0)$ complexes, in two or three steps from $RuCl_3$ (12).

2. General Methods from [RuCl₂(arene)]₂

a. Arene Ligand Exchange. $[RuCl_2(C_6Me_6)]_2$ (**1c**) is a good precursor for η^6-hexamethylbenzene ruthenium derivatives. It is readily obtained (80%) by displacement of the labile *p*-cymene ligand of **1b**, by fusing **1b** with hexamethylbenzene (10,13). Analogously, complex **1b** undergoes arene exchange leading to other complexes **1** on heating with neat arene (1,2,4,5-$Me_4C_6H_2$) (14), 1,2,3,4-$Me_4C_6H_2$(94%), 1,3,5-$Et_3C_6H_3$ (84%), or

$$[RuCl_2(arene)]_2 \xleftarrow[\substack{p\text{-cymene}}]{\substack{\text{arene}}} [RuCl_2(MeC_6H_4\text{-}i\text{-}Pr)]_2 \xrightarrow[\substack{p\text{-cymene}}]{\substack{C_6Me_6}} [RuCl_2(C_6Me_6)]_2$$

$$\mathbf{1} \qquad\qquad\qquad \mathbf{1b} \qquad\qquad\qquad \mathbf{1c}$$

1,3,5-i-$Pr_3C_6H_3$(35%)(15). Photochemical or thermal exchange of the *p*-cymene ligand of $RuCl_2(PBu_3)(MeC_6H_4\text{-}i\text{-}Pr)$ with benzene or hexamethylbenzene has also been shown to take place, but in moderate yields (9).

b. Addition of Ligands to [RuCl₂(arene)]₂. Addition of a variety of two-electron ligands L, such as phosphine, phosphite, arsine, stilbine, pyridine (7,9), isonitrile (16), carbon monoxide (17), or dimethyl sulfoxide (11), leads to cleavage of the halo bridges of complexes **1** and affords mononuclear complexes of type **3** in good yield. The X-ray structures of two complexes **3**, $RuCl_2(PMePh_2)(C_6H_6)$ and $RuCl_2(PMePh_2)(MeC_6H_4$-i-Pr), have been determined and show a slight tilting from planarity of the arene ligand (8). Bidentate phosphines or arsines (L⌒L) also cleave the halo bridges to produce binuclear complexes of type **4** (7,18). In polar solvents, complexes **1** can add two basic ligands L per ruthenium atom to afford cationic derivatives of general structure **5** $[(L)_2 = (PR_3)_2$, (P-$(OR)_3)_2$ (19,20), diphosphines, diarsines (18), bipyridines (21), or pyrazoles (22)] [Eq. (2)].

$$[RuX_2(\eta^6\text{-arene})]_2 \xrightarrow{2\,L} 2\,RuX_2(L)(\eta^6\text{-arene})$$

$$\mathbf{1} \qquad\qquad\qquad\qquad \mathbf{3}$$

$$4\,L \downarrow \qquad\qquad \searrow L⌒L$$

$$2\,RuX(L)_2(arene)^+X^- \qquad (arene)RuCl_2L⌒LCl_2Ru(arene) \qquad (2)$$

$$\mathbf{5} \qquad\qquad\qquad\qquad \mathbf{4}$$

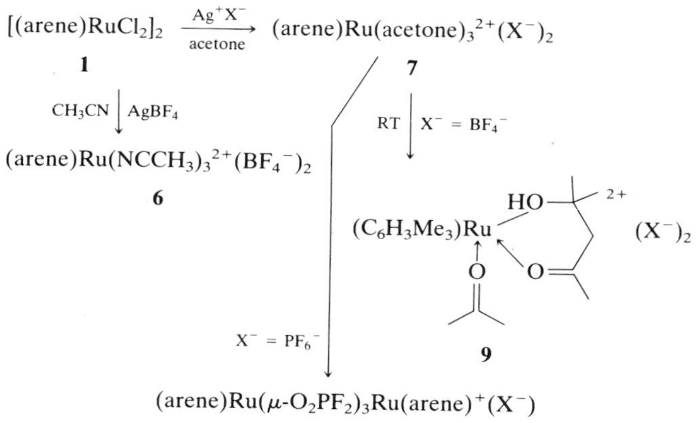

$$[(arene)RuCl_2]_2 \xrightarrow[\text{acetone}]{Ag^+X^-} (arene)Ru(acetone)_3^{2+}(X^-)_2$$

1   **7**

$$CH_3CN \downarrow AgBF_4$$

$$(arene)Ru(NCCH_3)_3^{2+}(BF_4^-)_2$$

6

RT $\bigg|$ X$^-$ = BF$_4^-$

$$(C_6H_3Me_3)Ru \underset{O}{\overset{HO-\underset{}{\diagup}}{\diagdown}} \quad (X^-)_2$$

9

X$^-$ = PF$_6^-$

$$(arene)Ru(\mu\text{-}O_2PF_2)_3Ru(arene)^+(X^-) \tag{3}$$

8

Treatment of complexes **1** with silver salts, in the presence of coordinating solvents, leads to cations **6** (CH$_3$CN) (*9*) or **7** (acetone) (*23*) [Eq. (3)]. Complex **7** evolves above 0°C according to the nature of X$^-$ (PF$_6^-$ or BF$_4^-$) to give derivative **8**, via the intermediate cation **9** for which an X-ray structure has been determined (*23*). Treatment of complexes of type **3** with phosphines L^1 and NH$_4$PF$_6$ in methanol allows the isolation of complexes **10** containing mixed ligands L and L^1 bonded to ruthenium(II) (*17,19,20*) [Eq. (4)]. Reaction of **11** with silver salts in acetone leads to the acetone complex **12**. The acetone ligand is displaced by ethylene to afford the olefin ruthenium(II) derivative **13** (*24*) [Eq. (5)].

$$RuCl_2(L)(\eta^6\text{-arene}) \xrightarrow[\text{NH}_4\text{PF}_6/\text{MeOH}]{L^1} RuCl(L)(L^1)(\eta^6\text{-arene})^+PF_6^- \tag{4}$$

3   **10**

arene = C$_6$H$_6$,   L = L^1 = PMe$_3$ or P(OMe)$_3$
        MeC$_6$H$_4$-*i*-Pr,   L = PMe$_2$Ph, L^1 = PMe$_3$
        C$_6$Me$_6$ L = P(OMe)$_3$, L^1 = PMe$_3$
                     L = CO, L^1 = PMe$_3$

$$RuCl_2(PMe_3)(C_6H_6) \xrightarrow[\text{acetone}]{AgPF_6} RuCl(acetone)(PMe_3)(C_6H_6)^+PF_6^-$$

11   **12**

$$\bigg| CH_2{=}CH_2$$

$$RuCl(CH_2{=}CH_2)(PMe_3)(C_6H_6)^+PF_6^- \tag{5}$$

13

Complexes **1a** and **1b** react in methanol or THF with hydrazine or 1,1-dimethylhydrazine to give cations Ru(H$_2$NNH$_2$)$_3$(arene)$^{2+}$ (**14**) or Ru-(H$_2$NNMe$_2$)$_3$(arene)$^{2+}$. Hydrazine is displaced from **14** by pyridine or

4-methylpyridine to afford **15**. Ethylenediamine with **1a** and **1b** in methanol leads to the cations $Ru(H_2NCH_2CH_2NH_2)(Cl)(arene)^+$ (25).

$$Ru(H_2NNH_2)_3(arene)(BPh_4)_2 \xrightarrow[\text{acetone}]{3\ L} Ru(L)_3(arene)(BPh_4)_2$$

$$\qquad\quad \textbf{14} \qquad\qquad\qquad\qquad\qquad\qquad \textbf{15}$$

arene = benzene, *p*-cymene, L = pyridine, 4-methylpyridine

Reaction of $[RuCl_2(C_6H_6)]_2$ (**1a**), in acetonitrile, with pyrazole, 3,5-
dimethylpyrazole, or potassium tris(3,5-dimethylpyrazolyl)borate, K[HB-
$(Me_2pz)_3$], proceeds by substitution of all chlorides and affords complex
16 (Scheme 1), resulting from the coupling of acetonitrile with pyrazole
(22). The reaction of **1a** with tetrakis(1-pyrazolyl)borate ion in boiling
acetonitrile, however, led to the trisubstituted derivative **17**, the structure
of which was established by an X-ray diffraction study (26).

  c. *Preparation of Ru(arene¹)(arene²)²⁺ Complexes.*   A general route to
$Ru(arene^1)(arene^2)^{2+}$ (**18**) has been developed, starting directly with the
easily accessible complexes **1a–1c**. Complexes **1a–1c** are first reacted with
$AgBF_4$ in acetone at room temperature, and then intermediates of type **7**
are heated with an excess of arene² in the presence of an acid, i.e., HBF_4,

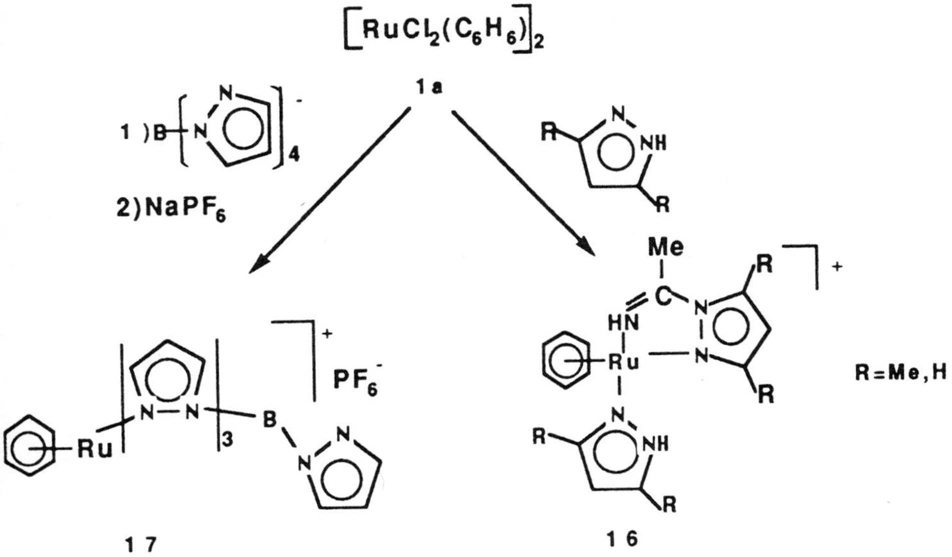

SCHEME 1

$$[RuCl_2(arene^1)]_2 \xrightarrow[AgBF_4]{acetone} \xrightarrow[heating]{arene^2/HX} Ru(arene^1)(arene^2)^{2+}(X^-)_2$$

1   **18**

arene^1 = C$_6$H$_6$ (**1a**), MeC$_6$H$_4$-*i*-Pr (**1b**), C$_6$Me$_6$ (**1c**)
arene^2 = C$_6$H$_6$, C$_6$H$_3$Me$_3$, C$_6$Me$_6$, C$_{10}$H$_8$, PhCOCH$_3$, Ph—Ph, PhNMe$_2$, PhCO$_2$Me, PhOH, PhCF$_3$

HPF$_6$, or CF$_3$CO$_2$H. The yield increases for arene^1 in the order C$_6$Me$_6$ >- MeC$_6$H$_4$-*i*-Pr > C$_6$H$_6$ (*27*). The role of the acid is probably removal of the Me$_2$CHOHCH$_2$COCH$_3$ or O$_2$PF$_2$ groups in the intermediate **8** or **9** [Eq. (3)]. Deprotonation of Ru(η^6-C$_6$H$_3$Me$_3$)(η^6-C$_6$H$_5$OH)(BF$_4$)$_2$ leads to the η^6-phenoxo complex Ru(η^6-C$_6$H$_3$Me$_3$)(η^6-C$_6$H$_5$O)(BF$_4$) (*27*). Ru(η^6-*p*-MeC$_6$H$_4$-*i*-Pr)(η^5-C$_4$Me$_4$S)(PF$_6$)$_2$ has also been obtained from **1b** by reaction with a silver salt in the presence of tetramethyl-thiophene (*28*).

(η^6-Octamethylnaphthalene)(η^6-arene)Ru^{2+} complexes of type **18a** [Eq. (6)] were prepared using chloride scavengers and three alternative

1   **18a**

reagents: AgPF$_6$ and trifluoroacetic acid (arene = C$_6$Me$_6$, C$_6$H$_3$Et$_3$, C$_6$H$_3$-*i*-Pr$_3$), AlCl$_3$ in dichloromethane (arene = C$_6$H$_6$, MeC$_6$H$_4$-*i*-Pr, C$_6$H$_3$Me$_3$, C$_6$H$_2$Me$_4$), or a molten mixture of octamethylnaphthalene and AlCl$_3$ in which [RuCl$_2$(C$_6$Me$_6$)]$_2$ **1c** is dissolved (*15*). A more convenient method of preparing complexes **18** was developed, without using silver salts but employing reflux of **1** with arenes in trifluoroacetic acid. This reaction is supposed to occur via initial formation of the trichloro-bridged cation **19**, which is formed in the absence of arene. By heating **19** with mesitylene in trifluoroacetic acid, complex **18** is obtained in quantitative yields (*29*) [Eq. (7)].

$$\textbf{1a} \xrightarrow{CF_3CO_2H} [Ru_2(\eta^6\text{-}C_6H_6)_2(\mu\text{-}Cl)_3]^+ \xrightarrow[CF_3CO_2H]{arene} Ru(\eta^6\text{-}C_6H_6)(arene)^{2+}$$

19   **18**

RuCl$_2$(PPh$_3$)C$_6$Me$_6$ (**20**) reacts with AgPF$_6$ in acetone to give intermediate **21**. Treatment of **21** with CH$_2$Cl$_2$ affords two products in low yield, the triply chloro-bridged cation **19** and the bisarene ruthenium dication **18**, the X-ray structural analysis of which reveals the η^6-bonding mode of one phenyl ring of PPh$_3$ (*30*).

$$RuCl_2(PPh_3)(\eta^6\text{-}C_6Me_6) \xrightarrow[\text{acetone}]{AgPF_6} Ru(\eta^6\text{-}C_6Me_6)Cl(PPh_3)(acetone)^+PF_6^-$$

$$\mathbf{20} \qquad\qquad\qquad\qquad\qquad\qquad \mathbf{21}$$

$$\Big\downarrow {\scriptstyle CH_2Cl_2}$$

$$(\eta^6\text{-}C_6Me_6)Ru(\mu\text{-}Cl)_3Ru(\eta^6\text{-}C_6Me_6)^+ + Ru(\eta^6\text{-}C_6Me_6)(\eta^6\text{-}C_6H_5PPh_2)^{2+}$$

$$\mathbf{19} \qquad\qquad\qquad\qquad\qquad \mathbf{18}$$

d. Formation of (Arene)Ru(μ-X)₃Ru(arene)⁺ Cations. Reaction of **1a** and **1b** with hot water, following treatment with ammonium hexafluorophosphate, affords the triply chloro-bridged ruthenium binuclear complex of type **19** in 41–68% yield (9). The same complexes **19** can be

$$[RuCl_2(\eta^6\text{-arene})]_2 \xrightarrow[\text{2. NH}_4PF_6]{\text{1. H}_2O/\text{heating}} (\eta^6\text{-arene})Ru{\overset{\displaystyle Cl}{\underset{\displaystyle Cl}{-Cl-}}}Ru(\eta^6\text{-arene})^+PF_6^-$$

$$\mathbf{1a\text{--}1c} \qquad\qquad\qquad\qquad\qquad\qquad \mathbf{19}$$

more readily prepared, in more than 90% yield, by shaking **1** with an excess of NH_4PF_6 in methanol (30–32). Complexes **19** (arene = benzene, mesitylene) have also been obtained in good yield by reacting **1** with HPF_6 or CF_3CO_2H [Eq. (7)] (29). However, if **1a** is shaken with HCl in ethanol in the presence of an excess of cesium chloride, the monomeric anion $[Ru(\eta^6\text{-}C_6H_6)Cl_3]^-$, analogous to complexes of type **3**, is obtained quantitatively (31,32). Another convenient synthesis of **19** (arene = benzene, mesitylene) has been achieved by Stephenson and co-workers by protonation with HBF_4 of an equimolar mixture of neutral and cationic pyridine arene ruthenium monomers **22** and **23** in methanol (33,34).

$$Ru(\eta^6\text{-arene})(py)Cl_2 + Ru(\eta^6\text{-arene})(py)_2Cl^+ \xrightarrow[\text{MeOH}]{HBF_4} (arene)Ru{\overset{\displaystyle Cl}{\underset{\displaystyle Cl}{-Cl-}}}Ru(arene)^+$$

$$\mathbf{22} \qquad\qquad\qquad\qquad \mathbf{23} \qquad\qquad\qquad\qquad\qquad \mathbf{19}$$

The reaction of an aqueous solution of **1a** with an excess of NaOH or Na_2CO_3 gives two cationic products **24** and **25**. Complex **25** is the minor product, but it can be completely and irreversibly formed by recrystallization of **24** in acetone (35). The structure of **25**, initially proposed as the binuclear tris-hydroxo-bridged ruthenium complex **25A** (Scheme 2), has

$$\mathbf{1a} \xrightarrow[\text{H}_2O]{NaOH \text{ or } Na_2CO_3} \xrightarrow{NaBPh_4} [(C_6H_6(OH)Ru(\mu\text{-}OH)_2Ru(H_2O)(C_6H_6)]^+ + \mathbf{25}$$

$$\mathbf{24}$$

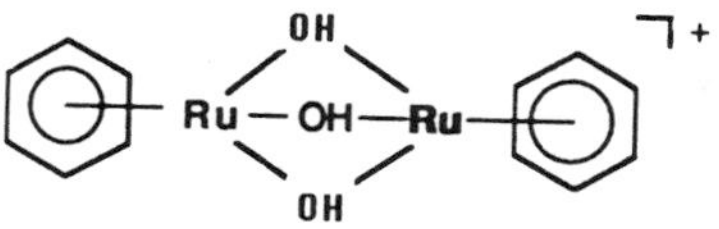

25A 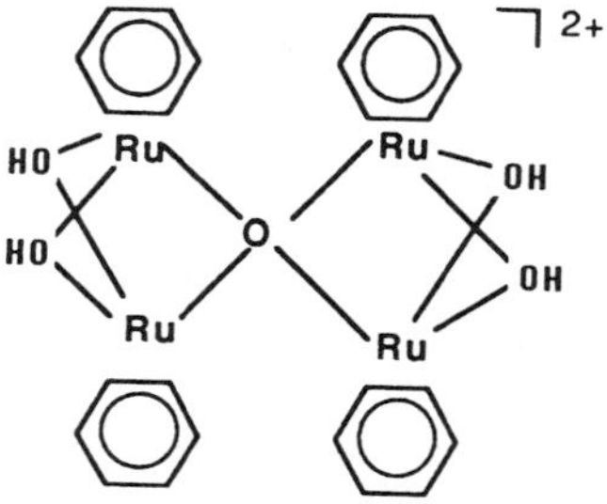   25B

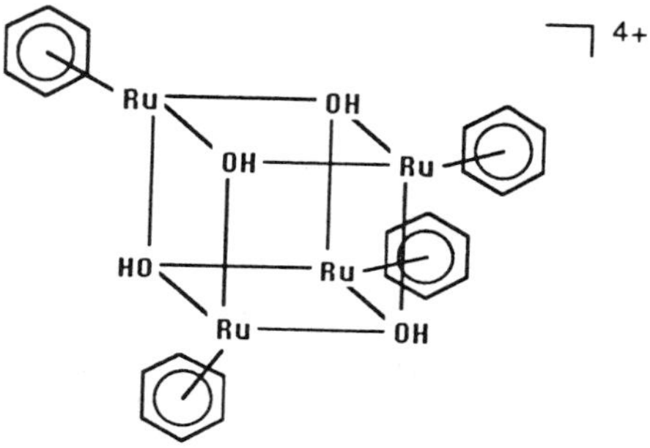

26

SCHEME 2

been reformulated, aided by an X-ray structural analysis, as the tetranuclear complex **25B** containing a μ_4-bridging O^{2-} ion (*36*). In contrast,
reaction of other complexes of type **1** (arene = mesitylene, hexamethylbenzene, *p*-cymene) with an excess of NaOH or Na_2CO_3 gives only **25A**
containing conventional triple hydroxo bridges, as demonstrated by X-ray
analysis (arene = mesitylene) (*35,36*). When **1a** is treated with aqueous
Na_2CO_3 in 1 : 2 molar ratio, a tetrameric cation (**26**) was formed in 25%
yield (*37*).

Reaction of **1** with NaOR/ROH and NaBPh$_4$ gives triply alkoxo-bridged
complexes of type **27**. These compounds can also be prepared from **25** in
methanol or ethanol at reflux. Attempts to synthesize longer alkoxide

$$\mathbf{1} \xrightarrow[\text{NaBPh}_4]{\text{NaOR/ROH}} (\text{arene})\text{Ru} \overset{\text{OR}}{\underset{\text{OR}}{-\text{OR}-}} \text{Ru}(\text{arene})^+ \xleftarrow[\text{reflux}]{\text{ROH}} \mathbf{25}$$

27

arene = benzene, mesitylene, hexamethylbenzene, R = Me, Et, Ph

chains such as ^{i}PrO or ^{n}BuO have been unsuccessful (35). Complex **27** (arene = benzene, p-cymene) has been obtained by reaction of **1** with NaNH$_2$ in CH$_3$CN followed by treatment with an alcoholic solution of NaBPh$_4$(R = Me, Et). The structure of the benzene complex (R = Me) is based on an X-ray structural analysis (38). Triply thioato-bridged complexes [Ru$_2$(arene)$_2$((SEt)$_3$](BPh$_4$) have been synthesized by an analogous way from **1** (arene = benzene, p-cymene) and Pb(SEt)$_2$ in CH$_3$CN (38).

3. Derivatives of [RuX$_2$(arene)]$_2$ and RuX$_2$(L)(arene) Complexes

Binuclear [RuX$_2$(arene)]$_2$ (**1**) and mononuclear RuX$_2$(L)(arene) (**3**) derivatives have been shown to be useful precursors for access to alkyl-or hydrido(arene)ruthenium complexes. The latter are key compounds for the formation of arene ruthenium(0) intermediates capable of C—H bond activation leading to new hydrido and cyclometallated ruthenium arene derivatives. Arene ruthenium carboxylates appear to be useful derivatives of alkyl–ruthenium as precursors of hydrido–ruthenium complexes; their access is examined first.

a. Arene Ruthenium Carboxylates. Bis(acetato)- or bis(trifluoroacetato)ruthenium arene complexes **28** are prepared by treatment of **1** with a large excess of silver carboxylate for a variety of arenes [C$_6$H$_6$, MeC$_6$H$_4$-i-Pr, C$_6$H$_3$Me$_3$, C$_6$H$_2$Me$_4$ (39), C$_6$Me$_6$ (39,40)]. In complexes of type **28** one carboxylate group is unidentate and the other is bidentate, although they appear equivalent in NMR (39,40). Complex **29** similarly leads to the carboxylates **30** and **31** (41), whereas complex **32**, analogous to **30**, results from the addition of PPh$_3$ to the dicarboxylate **28** (39). The bistrifluoroacetate complex (**28**, R = CF$_3$) is also obtained by treatment of **28** (R = Me) with trifluoroacetic acid (39).

$$Ru(O_2CMe)_2(PPh_3)(C_6Me_6)$$

32

$$\uparrow PPh_3$$

$$[RuCl_2(arene)]_2 \xrightarrow[R\ =\ CH_3,\ CF_3]{AgO_2CR} Ru(O_2CR)_2(arene)\cdot H_2O$$

1   **28**

$$RuCl_2(PMe_3)(\eta^6\text{-}C_6Me_6) \xrightarrow{AgO_2CMe} Ru(O_2CR)_2(PMe_3)(\eta^6\text{-}C_6Me_6)$$

29   **30**

$$\downarrow KPF_6$$

$$Ru(O_2CMe)(PMe_3)(\eta^6\text{-}C_6Me_6)^+PF_6^-$$

31

Trifluoroacetato ruthenium (II) complexes **34** and **35** have been obtained
by H. Werner and co-workers, by selective cleavage of the $Ru-CH_3$ bonds
of dialkylruthenium(II) species **33**. One trifluoroacetato ligand of **35** can be
displaced by phosphine or carbon monoxide to give complexes **36** or **37**
(*42*). Complex **33** ($PR_3 = PPh_3$) reacts with HBF_4 in the presence of
propionic anhydride to afford the bidentate carboxylato ruthenium cation
38 (*43*).

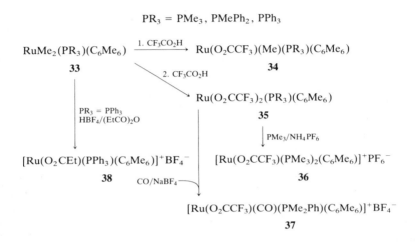

Alternatively, complexes **39** and **40** are prepared from **1** using three
methods, depending on the nature of the arene or the halide ligands: by
treatment with a mixture $RCO_2H/(RCO)_2O$, by addition of 2 equiv of
silver carboxylate, or by reaction of sodium acetate in acetone (*39*). The
trifluoroacetato chloro ruthenium complex **40** is a good starting material
for the preparation of cations **41**, by displacement of the trifluoroacetato
ligand by 2 equiv of ligand (L = $PEtPh_2$, pyridine, pyrazine, 1,3-dithiane,
4,4′-bipyridine, or $L_2 = 2,2′$-bipyridine). The X-ray structure of one of
these (**41**, L = pyrazine, $X^- = PF_6^-$) has been determined (*39*).

$$[RuX_2(arene)]_2 \xrightarrow{R = CH_3, CF_3} RuX(\eta^2-O_2CCR)(arene)$$

1 (X = Cl, Br)   **39**

$$\downarrow CF_3CO_2H$$

$$RuCl(L)_2(arene)^+X^- \xleftarrow{2L} RuCl(O_2CCF_3)(arene)$$

41   **40**

arene = C_6H_6, MeC_6H_4-*i*-Pr, $C_6H_3Me_3$, $C_6H_2Me_4$, C_6Me_6

b. *Arene Ruthenium Alkyl Derivatives.*   Reaction of precursor **1a** with HgMe$_2$ or HgPh$_2$ in acetonitrile, followed by addition of PPh$_3$, leads to monoalkylation or arylation and the formation of complexes **42** (63%) or **43** (51%). Reaction of **1a** with tetraallyltin produces the η^3-allyl complex **44** (72%) (*44*). Other η^3-allyl ruthenium complexes of type **44** have been prepared from **1a** with allylmercury chlorides (allyl, methallyl, crotyl, 1- and 2-phenylallyl, and 1-acetyl-2-methylallyl) (*45*).

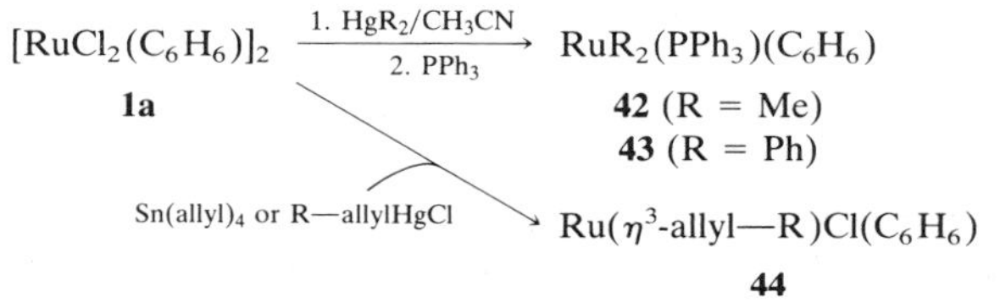

$$[RuCl_2(C_6H_6)]_2 \xrightarrow[\text{2. PPh}_3]{\text{1. HgR}_2/\text{CH}_3\text{CN}} RuR_2(PPh_3)(C_6H_6)$$

1a

42 (R = Me)
43 (R = Ph)

Sn(allyl)$_4$ or R—allylHgCl $\longrightarrow$ Ru(η^3-allyl—R)Cl(C$_6$H$_6$)

44

Benzene complexes RuX$_2$(L)(C$_6$H$_6$) (**3**) on reaction with MeLi or MeMgI afford only small amounts of RuMe$_2$(L)C$_6$H$_6$ and Ru(CH$_3$)-X(L)C$_6$H$_6$ (X = Cl, I, L = PMe$_2$Ph, PPh$_3$) (*9*). Similar complexes of structure **3** but containing hexamethylbenzene, however, when treated with methyllithium give better yields of dimethylruthenium(II) complexes **33** [PR$_3$ = PMe$_3$ (79%), PMePh$_2$ (57%), PPh$_3$ (63%) (*42*). Reaction of 1 equiv of MeLi affords **45** in 46% yield (*46*). Reaction of cation **46** with methyllithium gives not reduction products but chloride substitution and cation **47** (*19*).

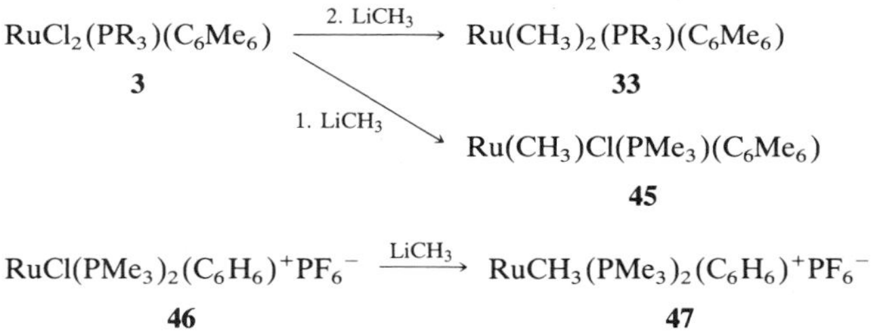

$$RuCl_2(PR_3)(C_6Me_6) \xrightarrow{\text{2. LiCH}_3} Ru(CH_3)_2(PR_3)(C_6Me_6)$$

3

33

1. LiCH$_3$ $\longrightarrow$ Ru(CH$_3$)Cl(PMe$_3$)(C$_6$Me$_6$)

45

$$RuCl(PMe_3)_2(C_6H_6)^+PF_6^- \xrightarrow{\text{LiCH}_3} RuCH_3(PMe_3)_2(C_6H_6)^+PF_6^-$$

46   **47**

Cleavage of Ru—CH$_3$ bonds by acids has been used by Werner *et al.* for the selective introduction of small molecules at the ruthenium center. Addition of HBF$_4$ to complex **33**, in the presence of carbon monoxide or ethylene, allows the coordination of these molecules in complexes **48, 49,** and **50** (*43*). Complexes of type **33** give carboxylate complexes **34, 35,** and **38** on treatment with carboxylic acids (*42,43*). A bis(alkyl)ruthenium(II) complex (**52**) was also obtained by addition of PMe$_3$ to the ethylene complex **51** (*24,26*).

$$Ru(CH_3)_2(PR_3)(C_6Me_6) \xrightarrow[\text{L/THF}]{\text{HBF}_4} RuCH_3(L)(PR_3)(C_6Me_6)^+BF_4^-$$

33

	L	PR$_3$
48	CO	PPh$_3$
49	C$_2$H$_4$	PPh$_3$
50	C$_2$H$_4$	PMe$_3$

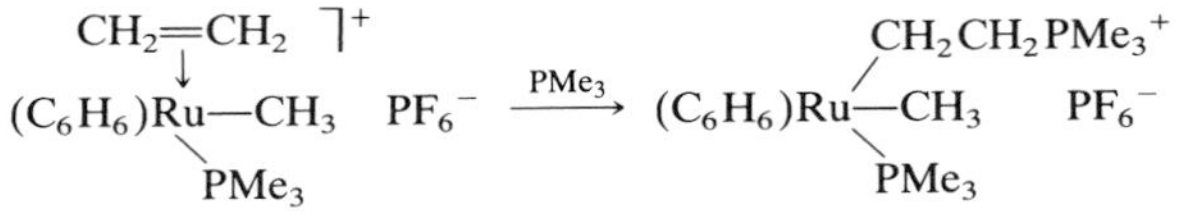

51　　　　　　　　　　　**52**

An interesting aspect of the bis(methyl)ruthenium complexes **33** is their tendency to allow hydride abstraction with (Ph$_3$C)PF$_6$. They lead to ethylene complexes **55** and **56** (*46,47*).The X-ray structure of **55** shows bond distances of 1.411(13) Å for C=C and 1.50 Å for Ru—H (*46,47*). The reaction probably proceeds via intermediates **53** and **54**, although it was not established whether the transformation **33** → **53** involves a mono-electronic process as with WMe$_2$(C$_5$H$_5$)$_2$ (*48*). A similar reaction from the deuterated derivative **57** gives exclusively complex **58**, configurationally

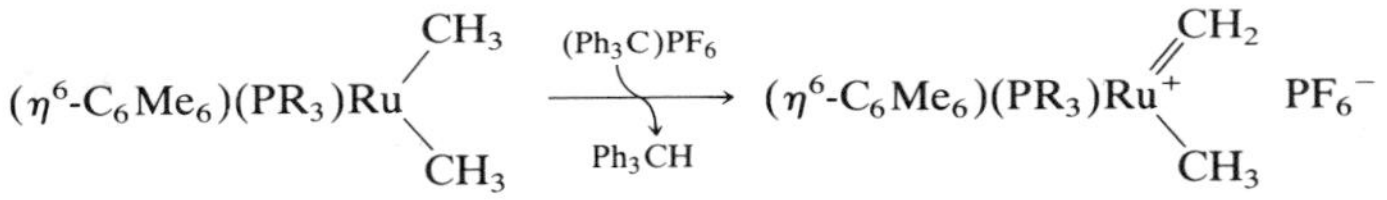

33 (PR$_3$ = PMe$_2$Ph, PPh$_3$)　　　　　　　　**53**

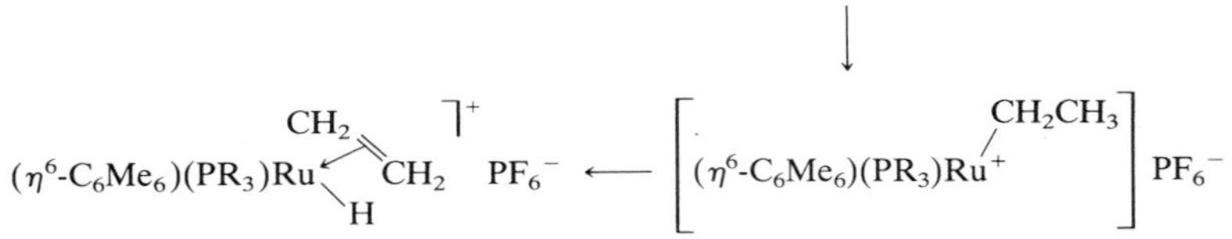

55 (PR$_3$ = PPh$_3$)　　　　　　　　　　**54**
56 (PR$_3$ = PMe$_2$Ph)

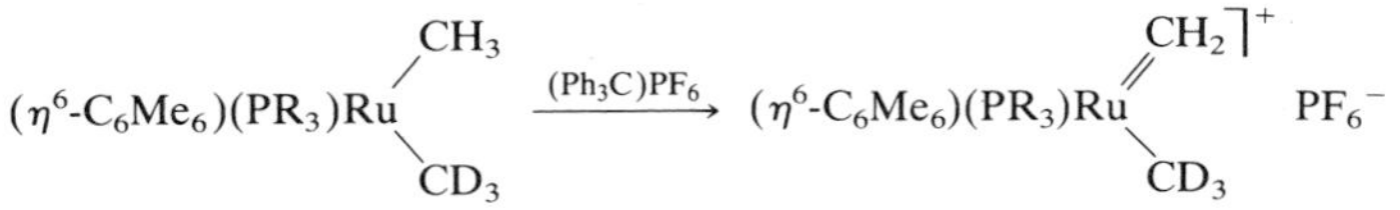

57

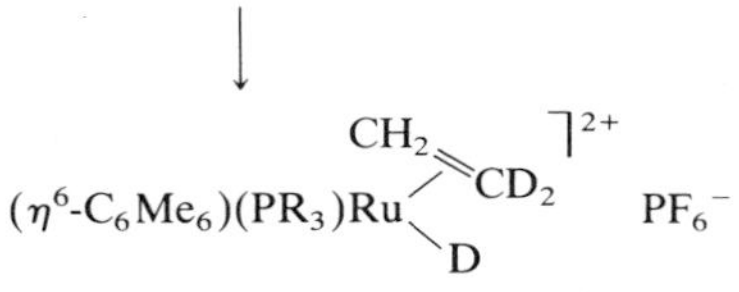

58

stable (-80 to $35°C$), indicating that there is no equilibrium between **58** and the possible intermediate $RuCD_2CH_2D^+$ of type **54** (*46*).

It is noteworthy that reaction of cation **45** with $(Ph_3C)PF_6$ gives methyl rather than hydride abstraction and affords, in the presence of PMe_2Ph, cation **59** (*46*).

$$RuCl(CH_3)(PMe_3)(\eta^6\text{-}C_6Me_6) \xrightarrow{(Ph_3C)PF_6}$$
45

$$RuCl(PMe_2Ph)(PMe_3)(\eta^6\text{-}C_6Me_6)PF_6^- + Ph_3CCH_3$$
59

Another general route to σ-alkylruthenium(II) complexes has been developed by H. Werner and co-workers by alkylation of (η^6-arene)-ruthenium(0) complexes. Complexes of type **60**, containing identical or different phosphine or phosphite ligands (*19,49*), PMe_3 and carbonyl ligands (*24,50*), or PMe_3 and ethylene groups (*17, 49*) react with methyl iodide to give methylruthenium(II) complexes (**61**). Ethyl iodide with the

$$(\eta^6\text{-arene})Ru \underset{L^2}{\overset{L^1}{\diagdown}} \xrightarrow[NH_4^+PF_6^-]{CH_3I} \left[(\eta^6\text{-arene})Ru \underset{L^2}{\overset{L^1}{\diagdown}} CH_3 \right]^+ PF_6^-$$

60   **61**

arene = C_6H_6, $MeC_6H_4\text{-}i\text{-}Pr$, C_6Me_6
$L^1 = L^2 = P(OMe)_3$, PMe_3, PMe_2Ph; $L^1 = PMe_3$, $L^2 = CO$, C_2H_4

trimethylphosphine benzene derivative **60a** gives an analogous alkyl complex **62a**, whereas with the hexamethylbenzene complex **60c** a mixture of **62a** and $[Ru(I)(PMe_3)_2(C_6Me_6)]^+PF_6^-$ is obtained. Hydride abstraction from the ethyl complex **62c** with $Ph_3C^+PF_6^-$ leads to the dicationic ethylene ruthenium(II) derivative **63** (*49*). Diazadiene ruthenium(0) complexes **64** are also readily alkylated by reaction with CH_3I or C_2H_5I (Scheme 3). The benzene ligand of complex **65a** is easily displaced in acetonitrile, and the ethyl derivative **65b**, via hydride abstraction, affords the dicationic ethylene complex **66** analogous to **63** (*51*).

$$Ru(PMe_3)_2(\eta^6\text{-}C_6R_6) \xrightarrow[NH_4^+PF_6^-]{C_2H_5I} Ru(C_2H_5)(PMe_3)_2(C_6R_6)^+PF_6^-$$

60a (R = H)   **62a** (R = H)
60c (R = CH₃)   **62c** (R = CH₃)

$$\downarrow Ph_3C^+PF_6^-$$

$$Ru(CH_2{=}CH_2)(PMe_3)_2(C_6Me_6)^{2+}(PF_6^-)_2$$

63

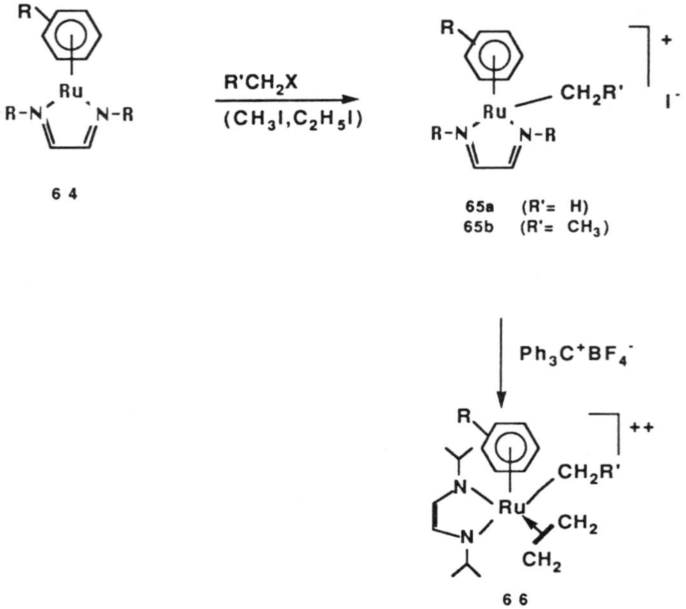

6 4

R'CH$_2$X
(CH$_3$I, C$_2$H$_5$I)

65a (R'= H)
65b (R'= CH$_3$)

Ph$_3$C$^+$BF$_4^-$

6 6

Scheme 3

c. *Hydrido(arene)ruthenium Complexes.* Hydrido(arene)ruthenium complexes are useful derivatives for access to (arene)ruthenium(0) intermediates and C—H bond activation. Most have been made either by selective reduction of arene ruthenium(II) complexes or by protonation of arene ruthenium(0) complexes, giving monohydrido or bishydrido ruthenium derivatives.

i. *Monohydrido(arene)ruthenium from ruthenium(II) complexes.* RuH-(Cl)(PPh$_3$)(C$_6$Me$_6$) (**68**), a hydrogenation catalyst, is prepared from **67** either by reaction with borohydride in THF (45%) or by heating in 2-propanol with aqueous Na$_2$CO$_3$ (98%). The mesitylene complex **69** is

$$\text{RuCl}_2(\text{PPh}_3)(\text{arene}) \xrightarrow[\text{aq Na}_2\text{CO}_3]{\text{2-propanol}} \text{RuH(Cl)(PPh}_3)(\text{arene})$$

67

68 (arene = C$_6$Me$_6$)
69 (arene = C$_6$H$_3$Me$_3$)

obtained similarly (*52a*). Reaction of **1c** with 2-propanol and Na$_2$CO$_3$ gave [(η^6-C$_6$Me$_6$)Ru(μ-H)$_2$(μ-Cl)Ru(η^6-C$_6$Me$_6$)]Cl (*52b*), but these results appear to be irreproducible (*14, 53*).

Complex **68** is obtained directly from **1c** by heating with anhydrous 2-propanol and Na$_2$CO$_3$, with a 3-fold excess of ligand (L = PPh$_3$) (*54*).

This direct route from **1c** to the analogous complexes **70** appears general for a variety of ligands [L = AsPh$_3$, SbPh$_3$, P(Ph-*p*-F)$_3$, P(Ph-*p*-Me)$_3$, P(Ph-*p*-OMe)$_3$, PPh$_2$-*t*-Bu, PPh$_2$-i-Pr, P-i-Pr$_3$, and PCy$_3$ (35–70%] but not for PPh(*t*-Bu)$_2$. The corresponding complex RuCl$_2$(PCy$_3$)(C$_6$Me$_6$) could not be obtained from **1c** and PCy$_3$ (*54*).

$$[RuCl_2(\eta^6\text{-}C_6Me_6)]_2 \xrightarrow[\text{2-propanol/Na}_2\text{CO}_3]{3L} RuH(Cl)(L)(C_6Me_6)$$

$$\textbf{1c} \qquad\qquad\qquad \textbf{70} \quad (L = PR_3)$$

Complexes **70** have also been made from **1c** [L = PMe(*t*-Bu)$_2$, PEt(*t*-Bu)$_2$, PPr(*t*-Bu)$_2$] or from RuCl$_2$(L)(C$_6$Me$_6$) (**3c**) (L = PMe$_3$, PMe$_2$Ph, PEtPh$_2$), using the same methods (*55*). Other derivatives of type **70**, containing ligand L^2 = PMePh$_2$, PPrPh$_2$, P(OPh)$_3$, P(OMe)$_3$, PEt$_3$ or PPh$_2$(C$_6$H$_4$-*o*-Me), have been obtained by displacement of the ligand L^1 of similar complexes **70** (L^1 = AsPh$_3$,SbPh$_3$) (*55*). Reduction of RuCl$_2$-(PHCy$_2$)(C$_6$Me$_6$) with 1 equiv of Red-Al leads to the formation of RuHCl(PHCy$_2$)(C$_6$Me$_6$), a species of type **70** (*56*).

$$RuH(Cl)(L^1)(C_6Me_6) \xrightarrow[L^1]{L^2} RuH(Cl)(L^2)(C_6Me_6)$$

$$\textbf{70} \quad (L^1 = AsPh_3, SbPh_3) \qquad\qquad \textbf{70} \quad (L^2 = PR_3)$$

A general route to bridging monohydridoruthenium(II) complexes **71** was found by reaction of **1** with hydrogen (3–4 atm), in dichloromethane containing NEt$_3$, at room temperature for 2–3 days, under conditions similar to those of heterolytic hydrogen cleavage (*14*). The reaction is general for a variety of arenes [benzene, durene, mesitylene, *p*-cymene and hexamethylbenzene (40–60%)]. When treated with NaPF$_6$, complexes **71** lead to cationic triply bridged derivatives **72**. The (μ-hydrido) ^{1}H-NMR resonance of compounds **71** (−7 to −11.6 ppm) is shifted to higher field by

$$[RuX_2(\text{arene})]_2 + H_2 \xrightarrow[\text{H}^+\text{NEt}_3, \text{Cl}^-]{\text{NEt}_3} (\text{arene})Ru\overset{\displaystyle H\quad Cl}{\underset{\displaystyle Cl\quad Cl}{\diagdown\diagup}}Ru(\text{arene})$$

$$\textbf{1} \qquad\qquad\qquad\qquad\qquad\qquad\qquad \textbf{71}$$

$$\text{NaPF}_6 \downarrow$$

$$(\text{arene})Ru\overset{\displaystyle H}{\underset{\displaystyle Cl}{\diagdown\diagup}}Cl\overset{}{\text{—}}Ru(\text{arene})^+PF_6{}^-$$

$$\textbf{72}$$

approximately 3 ppm in cations **72**. Only complex **71** (arene $= C_6Me_6$) is satisfactorily obtained from **1c** by reaction with 2-propanol and Na_2CO_3 (70%) (*14*).

The bisacetato ruthenium complex **28**, on heating in 2-propanol, leads to the bridged hydrido dinuclear complexes **73** and **74**. The bistrifluoroacetato complex **28** also leads to complex **73**. The η^2-acetato complex **39** was transformed in hot 2-propanol to another bridged hydrido derivative (**75**, arene = durene, mesitylene, *p*-cymene, hexamethylbenzene; 60–70%). The introduction of alkyl substituents on the benzene ring is reflected by a shift of the μ-^{1}H resonance toward high field (*14,53*).

$$[(C_6Me_6)Ru(\mu\text{-}H)(\mu\text{-}O_2CR)_2Ru(C_6Me_6)] \cdot H(O_2CR) \cdot H_2O$$
73

$Ru(O_2CR)_2(C_6Me_6)$ — 2-propanol (60–80°C)
28

$(R = CH_3, CF_3)$ | $NH_4{}^+PF_6{}^-$

$$[(C_6Me_6)Ru(\mu\text{-}H)(\mu\text{-}O_2CR)_2Ru(C_6Me_6)]^+PF_6{}^-$$
74

$Ru(\eta^2\text{-}O_2CMe)Cl(arene) \longrightarrow [(arene)Ru(\mu\text{-}H)(\mu\text{-}Cl)(\mu\text{-}O_2CMe)Ru(arene)]X$
39   **75** $(X^- = Cl^-, PF_6{}^-)$

Two types of (η^6-arene)Ru-H complexes are obtained by coordination of arene to a ruthenium moiety. The tetraphenylborate complex **77** is produced by reaction of the salt **76** with PPh_3 in polar solvents (*57*). $RuH_2(PPh_3)_4$ reacts with a variety of arenes (C_6H_5Y) used as solvents (Y = Me, Et, *i*-Pr, $COCH_3$, OMe, Cl, F) in the presence of the protonating reagent $CH_2(O_2SCF_3)_2$ to give complexes **78**. The X-ray structure of the toluene complex **78** has been established (*58*) [Eq. (8)].

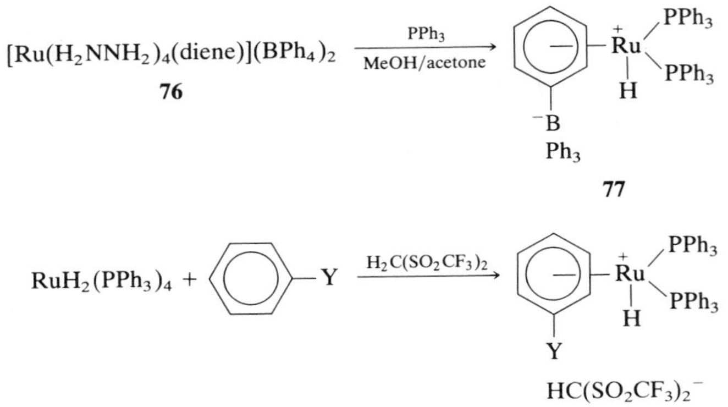

$[Ru(H_2NNH_2)_4(diene)](BPh_4)_2$ $\xrightarrow[\text{MeOH/acetone}]{PPh_3}$
76

77

(8)

$RuH_2(PPh_3)_4$ + ⬡—Y $\xrightarrow{H_2C(SO_2CF_3)_2}$

$HC(SO_2CF_3)_2{}^-$
78

ii. Polyhydrido(arene)ruthenium from ruthenium(II) complexes. Werner and co-workers have shown that complex **35** reacts with Red-Al [NaAlH$_2$-(OCH$_2$CH$_2$OMe)$_2$] in THF to give the dihydrido ruthenium derivatives **79**

$$Ru(OCOCF_3)_2(PR_3)(C_6Me_6) \xrightarrow[\text{THF}]{\text{Red-Al}} RuH_2(PR_3)C_6Me_6$$

35   **79**

$$\downarrow \text{HBF}_4$$

[RuH$_3$(PR$_3$)C$_6$Me$_6$]BF$_4$

80

[PR$_3$ = PMe$_3$ (31%), PMePh$_2$ (30%), PPh$_3$ (30%) (*42,59*), and PHCy$_2$ (90%) (*56*)]. Reaction of complexes **79** (PR$_3$ = PPh$_3$) with an excess of MeI yields RuI$_2$(PPh$_3$)C$_6$Me$_6$, but with HBF$_4$ the trihydride cation **80** is obtained [PMe$_3$ (64%), PPh$_3$ (43%)]. The analogous compound **82** is obtained from **1a** via intermediate **81** (*42,59*). An interesting aspect of the

$$[RuCl_2(C_6H_6)]_2 \xrightarrow[\text{2. Red-Al}]{\text{1. P-}i\text{-Pr}_3} RuH_2(P\text{-}i\text{-Pr}_3)C_6H_6$$

1a   **81**

$$CF_3CO_2H \downarrow PF_6^-$$

[RuH$_3$(P-i-Pr$_3$)C$_6$H$_6$]PF$_6$

82

$$RuH_2(PMe_3)C_6Me_6 \xrightarrow[\text{NH}_4\text{PF}_6]{\text{Me}_2\text{C}=\text{O}} (C_6Me_6)(PMe_3)Ru^+ \underset{NH_2CH(CH_3)_2}{\overset{H}{<}} \quad PF_6^-$$

79   **83**

reactivity of derivative **79** (PR$_3$ = PMe$_3$) is its reaction with acetone and NH$_4$PF$_6$. The amine complex **83** (59%), which has been characterized by X-ray analysis, is obtained. The isopropylamine ligand arises from reductive amination of acetone via the trihydride intermediate of type **80** (*60*).

Other dihydrido complexes of structure **79**, but containing bulky phosphines, have been obtained by Bennett and Latten by borohydride reduction of the chloride derivatives (*55*).

$$RuCl_2(L)C_6Me_6 \xrightarrow{\text{NaBH}_4} RuH_2(L)C_6Me_6 \xleftarrow{\text{NaBH}_4} RuHCl(L)C_6Me_6$$

L = PMe$_2$Ph, PMe$_2$(*t*-Bu)   **79**   L = PEt(*t*-Bu)$_2$

iii. Hydridoruthenium(II) derivatives from arene ruthenium(0) complexes. A variety of hydridoruthenium(II) compounds have been obtained by H. Werner and co-workers by protonation of (η^6-arene)(L)$_2$Ru(0) complexes. Complexes **60** containing identical or different phosphines are protonated

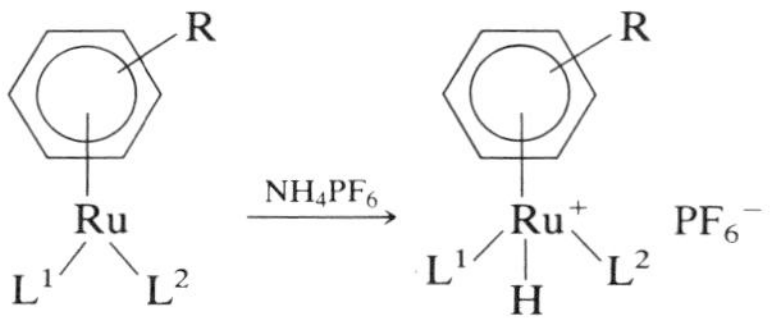

$$\text{(9)}$$

C$_6$H$_6$, MeC$_6$H$_4$-i-Pr, C$_6$Me$_6$, L^1 = L^2 = PMe$_3$
C$_6$H$_6$, L^1 = L^2 = PMePh$_2$, PMe$_2$Ph, PPh$_3$
C$_6$H$_6$, L^1 = PMe$_3$, L^2 = PPh$_3$, PMe$_2$Ph, P(OMe)$_3$
C$_6$Me$_6$, L^1 = PMe$_3$, L^2 = CO

by NH$_4$PF$_6$, and cations **84** are obtained in good yield [Eq. (9)]. However, the hydride complex **86**, containing less basic ligands P(OMe$_3$), could be obtained by protonation of **85** only with a stronger acid such as trifluoroacetic acid (*17,19,49*).

$$\text{Ru[P(OMe)}_3]_2(\text{C}_6\text{H}_6) \xrightarrow[\text{PF}_6^-]{\text{CF}_3\text{CO}_2\text{H}} \text{Ru(H)[P(OMe)}_3]_2(\text{C}_6\text{H}_6)^+\text{PF}_6^-$$

$$\quad\quad\quad\;\; \mathbf{85} \quad\quad\quad\quad\quad\quad\quad\quad\quad\quad\quad\quad \mathbf{86}$$

The ethylene ruthenium(0) complex **87** on protonation with CF$_3$CO$_2$H affords the ethylene hydrido derivative **88**. NMR analysis has shown complex **88** to be in equilibrium with the [Ru(C$_2$H$_5$)(PMe$_3$)(C$_6$H$_6$)]$^+$ intermediate, which in the presence of PMe$_3$ gives the ethyl derivative **89** (*24,50*). This reaction contrasts with the addition of PMe$_3$ to the corresponding methyl or ethylene cation (**51** → **52**).

$$\text{Ru(C}_2\text{H}_4)(\text{PMe}_3)\text{C}_6\text{H}_6 \xrightarrow[\text{2. NH}_4\text{PF}_6]{\text{1. CF}_3\text{CO}_2\text{H}} \text{Ru(H)(C}_2\text{H}_4)(\text{PMe}_3)(\text{C}_6\text{H}_6)^+\text{PF}_6^-$$

$$\quad\quad\;\; \mathbf{87} \quad\quad\quad\quad\quad\quad\quad\quad\quad\quad\quad\quad\quad \mathbf{88}$$

$$\Big\downarrow \text{PMe}_3$$

$$\text{Ru(CH}_2\text{CH}_3)(\text{PMe}_3)_2(\text{C}_6\text{H}_6)^+\text{PF}_6^-$$

$$\mathbf{89}$$

iv. C—H bond activation and cyclometallated arene ruthenium complexes. Photolysis of dihydridoruthenium complexes **90** and **91** in benzene and toluene leads to **92** and **93**, resulting from H$_2$ elimination and aromatic C—H bond activation (Scheme 4). Complex **93** is also obtained by reduction of **95** with borohydride. In contrast, photolysis of **90** in cyclohexane leads to the cyclometallated product **96**, arising from intramolecular C—H bond activation. Complex **96** in benzene or benzene-d_6 gives **92** and **97**, respectively (*61*).

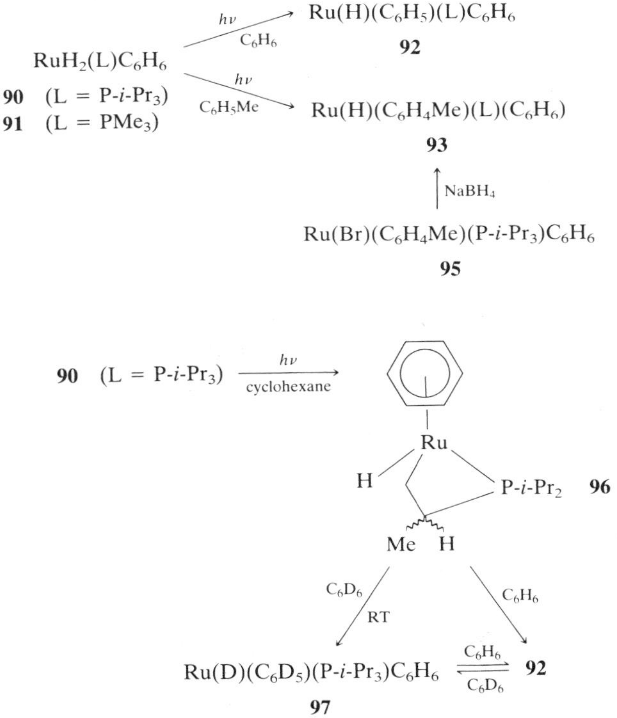

$$RuH_2(L)C_6H_6 \quad \xrightarrow[C_6H_6]{h\nu} \quad Ru(H)(C_6H_5)(L)C_6H_6$$

92

90 (L = P-*i*-Pr₃)

91 (L = PMe₃)

$$\xrightarrow[C_6H_5Me]{h\nu} \quad Ru(H)(C_6H_4Me)(L)(C_6H_6)$$

93

$\uparrow$ NaBH₄

$$Ru(Br)(C_6H_4Me)(P\text{-}i\text{-}Pr_3)C_6H_6$$

95

90 (L = P-*i*-Pr₃) $\xrightarrow[\text{cyclohexane}]{h\nu}$

96

$$\swarrow^{C_6D_6}_{RT} \qquad \searrow^{C_6H_6}$$

$$Ru(D)(C_6D_5)(P\text{-}i\text{-}Pr_3)C_6H_6 \quad \underset{C_6D_6}{\overset{C_6H_6}{\rightleftarrows}} \quad \textbf{92}$$

97

SCHEME 4

Complexes **98** [L = PPh₃, P(Ph-*p*-F)₃, P(Ph-*p*-Me)₃] react with methyl-lithium to give, after methanolysis, the orthometallated complexes **99** (Scheme 5). Complex **98** (L = PPh₃) also leads to **99** by reaction with phenyllithium or Red-Al (*54*). The formation of **99** suggests that the initial reduction of **98** leads to a 16-electron ruthenium(0) intermediate followed by C—H bond activation as for the transformations of **90** and **91**. Treatment of complex **98** (L = P-*i*-Pr₃) with methyllithium produces the cyclometallated diastereoisomers **100**. Complexes **101** and **102** are obtained by treatment of **98** (L = PPh₂-*t*-Bu) with methyllithium at −78°C and at +70°C, respectively. Complex **101** isomerizes to **102** by a first-order process ($k \sim 0.2$ hour⁻¹ in C₆D₆ at 50°C) when L is PPh₂-*i*-Pr; **98** leads to **103** which isomerizes to the orthometallated complex **104** (*54*).

Under the conditions leading to RuH(Cl)(L)(arene) complexes, the derivative **105** gives the cyclometallated compound **106** (*55*) [Eq (10)]. The bidentate complex **108** is formed from the evolution of the methyl derivative **49** in solution (Scheme 6). The reaction is consistent with ethylene

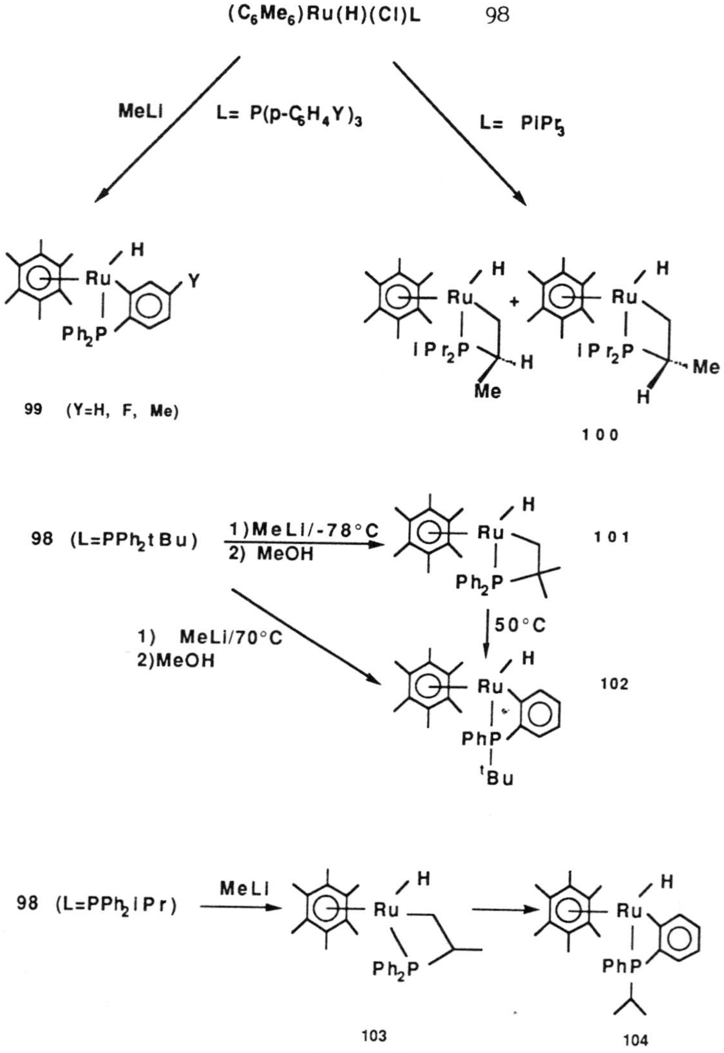

$(C_6Me_6)Ru(H)(Cl)L$ 98

MeLi   L= P(p-C$_6$H$_4$Y)$_3$ L= PIPr$_3$

99   (Y=H, F, Me)

100

98 (L=PPh$_2$tBu) $\xrightarrow{\begin{array}{c}1)MeLi/-78°C\\2)\ MeOH\end{array}}$ 101

1) MeLi/70°C
2)MeOH   50°C

102

98 (L=PPh$_2$iPr) $\xrightarrow{MeLi}$ →

103   104

SCHEME 5

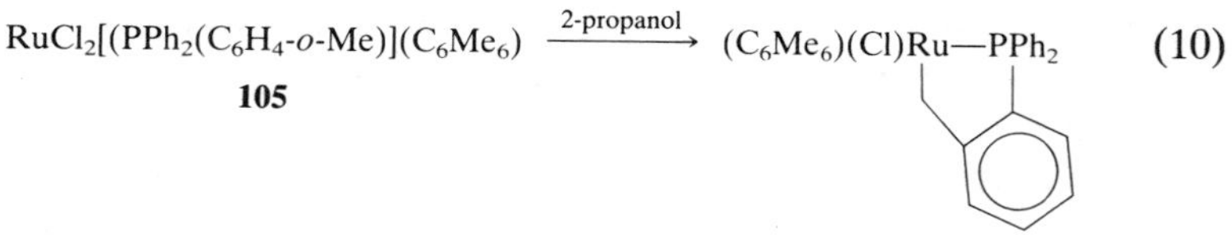

$$RuCl_2[(PPh_2(C_6H_4\text{-}o\text{-}Me)](C_6Me_6) \xrightarrow{\text{2-propanol}} (C_6Me_6)(Cl)Ru\text{—}PPh_2 \quad (10)$$

105

106

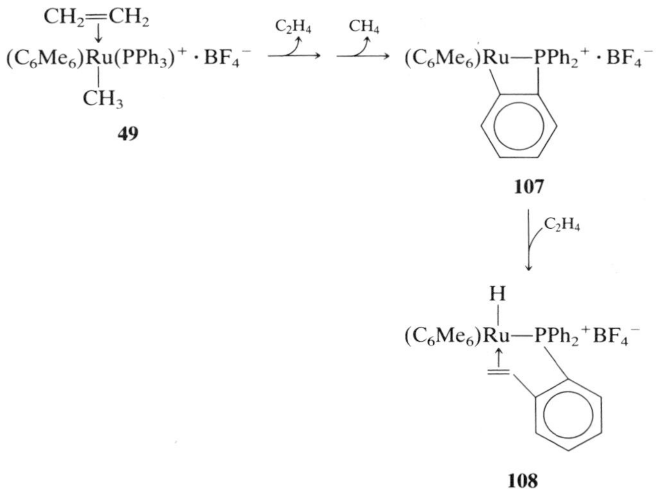

$$CH_2{=}CH_2$$
$$(C_6Me_6)\underset{\underset{\displaystyle CH_3}{|}}{\overset{\overset{\displaystyle \downarrow}{|}}{Ru}}(PPh_3)^+ \cdot BF_4^- \quad \xrightarrow{\;\;C_2H_4\;\;} \quad \xrightarrow{\;\;CH_4\;\;} \quad (C_6Me_6)Ru{-}PPh_2^+ \cdot BF_4^-$$

49

107

$\Big\downarrow C_2H_4$

$$(C_6Me_6)\overset{\overset{\displaystyle H}{|}}{\underset{\underset{\displaystyle \|}{\uparrow}}{Ru}}{-}PPh_2{}^+BF_4^-$$

108

SCHEME 6

insertion into the Ru—aryl bond of intermediate **107** followed by β-elimination (43). Finally, a complex of type **92**, Ru(H)(C$_6$H$_5$)(PMe$_3$)-(C$_6$H$_6$), has been obtained by cocondensation of ruthenium atom vapor with benzene and PMe$_3$ (62) (Section VII,A).

d. Arene Ruthenium Carbene Derivatives. Although many (η^5-C$_5$R$_5$)-(R$_3$P)$_2$Ru—carbene complexes have been reported (63) the isolectronic (η^6-C$_6$R$_6$)(R$_3$P)(X)Ru—carbene derivatives have been obtained only recently (64). Methylidene ruthenium arene species **53** are intermediates for the formation of Ru(H)(η^2-CH$_2$=CH$_2$)$^+$ cations **55** and **56** by hydride abstraction from the corresponding Ru(CH$_3$)$_2$ complexes **33** (46,47).

Complex **3c**, a catalytic precursor for addition reactions to alkynes (65), reacts at room temperature with a variety of terminal alkynes in alcohols to produce stable alkoxyl alkyl carbene ruthenium(II) derivatives **109** in good yields (Scheme 7). Reaction of **3c** (L = PMe$_3$), with trimethylsilyacetylene in methanol gives the carbene ruthenium complex **110**, by protonolysis of the C—Si bond, whereas with 4-hydroxy-1-butyne in methanol the cyclic carbene complex **111** is obtained (65,66).

The formation of complexes **109** has been shown to proceed via a vinylidene ruthenium intermediate (**112**), which has been indirectly isolated by protonation of an acetylide-ruthenium complex (**112**). Arene ruthenium vinylidene complexes **113** appear to be much more reactive than their isoelectronic (C$_5$H$_5$)(R$_3$P)$_2$Ru=C=CHR$^+$ complexes (63,66).

$$(C_6Me_6)(PMe_3)ClRu-Cl \xrightarrow{\text{LiC} \equiv \text{C}-\text{Ph}} (C_6Me_6)(PMe_3)ClRu-C \equiv C-Ph$$

3c   **112**

$$\downarrow \text{HBF}_4, \text{Et}_2\text{O}$$

$(C_6Me_6)(PMe_3)ClRu^+ = C \begin{smallmatrix} \diagup OMe \\ \diagdown CH_2Ph \end{smallmatrix} \quad BF_4^- \quad \xleftarrow{\text{MeOH}} \quad (C_6Me_6)(PMe_3)ClRu = C = C \begin{smallmatrix} \diagup H \\ \diagdown Ph \end{smallmatrix} \quad BF_4^-$

109   **113**

e. Stereochemical Aspects of Arene Ruthenium Complexes. The above reactions involving the complexes $[RuX_2(\text{arene})]_2$ have been used to study three stereochemical aspects of (arene)ruthenium(II) derivatives. Complex **1a** in methanol reacts with amino acid anions, potassium glycinate or alanate, to give chiral complexes **114** and **115** (Scheme 8). When R is a methyl group two diastereoisomers related to the pair *RR/SS* and *RS/SR* of enantiomers can be observed by ^{1}H and ^{13}C NMR, indicating that epimerization at the chiral ruthenium center is slow. Complex **115** represents only the *SS* enantiomer (*67*).

The first example of an optically active arene ruthenium complex was obtained by Brunner and Gastinger (*68*). The reaction of **1a** with HgMe$_2$ (*44*) and then (*R*)-(+)Ph$_2$PNHCH(Me)Ph afforded a pair of diastereoisomers, **116** and **117**, which were separated by chromatography (*68,69*) (Scheme 9). Complexes **116** and **117** (X = Cl) react with SnCl$_2$ to give complexes **118** and **119**, but the stereoselectivity depends on reaction

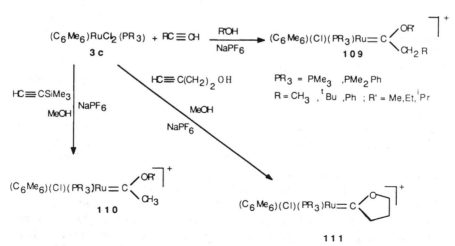

SCHEME 7

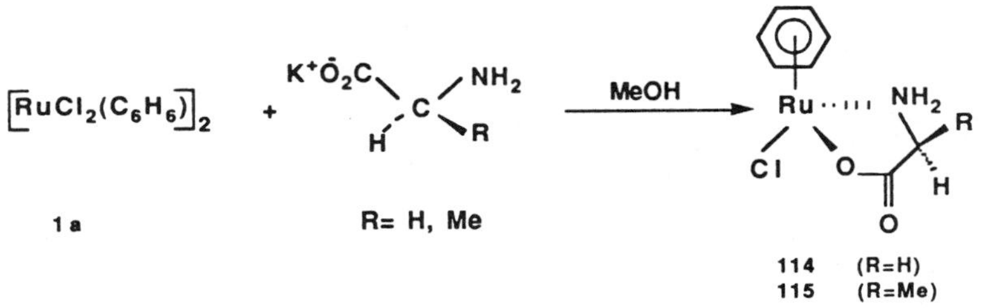

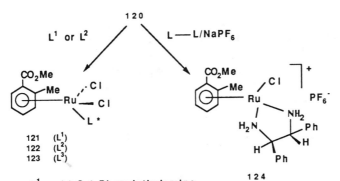

$$\left[RuCl_2(C_6H_6)\right]_2 + \quad K^+\bar{O}_2C\overset{NH_2}{\underset{H}{\overset{|}{C}}}R \quad \xrightarrow{MeOH}$$

1a   R= H, Me

114   (R=H)
115   (R=Me)

SCHEME 8

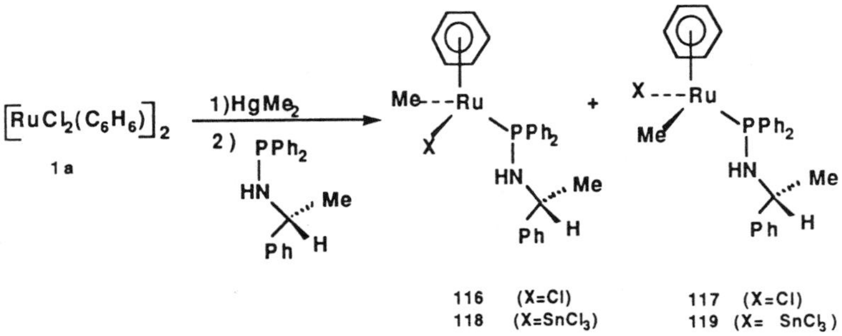

$$\left[RuCl_2(C_6H_6)\right]_2 \quad \xrightarrow[\text{2) PPh}_2]{\text{1)HgMe}_2}$$

1a

116   (X=Cl)
118   (X=SnCl₃)

117   (X=Cl)
119   (X= SnCl₃)

SCHEME 9

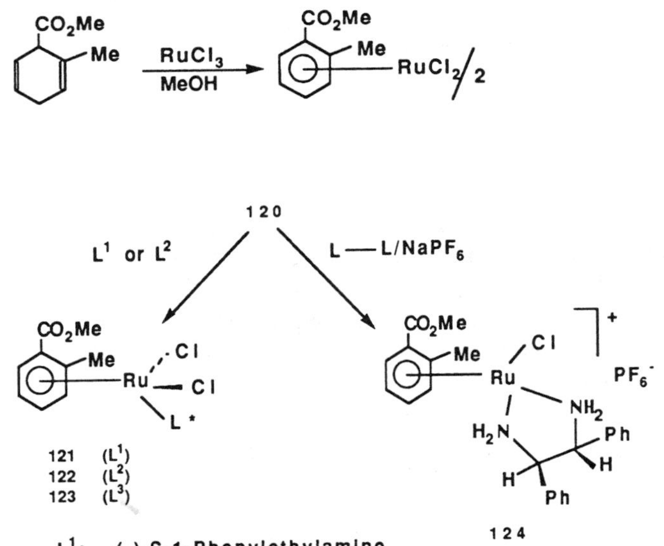

120

L¹ or L²   L—L/NaPF₆

121   (L¹)
122   (L²)
123   (L³)

124

L¹: (-)-S-1-Phenylethylamine
L²: (+)-dehydroabiethylamine
L³: (+)-Neomenthyldiphenylphosphine
L-L   :(-)-(R)(R)-1,2-diphenyl-1,2-diaminoethane

SCHEME 10

conditions (*69*). The X-ray structure of one of the latter diastereoisomers which rotates (−) at 436 nm has established the *R* configuration at the ruthenium center, e.g., **119** (*70*).

(Arene)ruthenium(II) complexes **120–124**, having a planar chirality, have been obtained recently (*71,72*). Complexes **121–123** are formed by addition of optically active amines (L^1 and L^2) or phosphine (L^3) to the dinuclear complex **120** (Scheme 10). They exist as a mixture of two configurationally stable diastereoisomers. The absolute configuration of one of the diastereoisomers of **123** has been determined, and complex **124** containing a chelating optically active diamine has also been isolated (*71,72*).

4. Arene Cyclopentadienyl Ruthenium(II) Complexes

a. Introduction of a Cyclopentadienyl Ligand.   Zelonka and Baird have described an easy method of preparing the benzene ruthenium complex **125** from **1a** and thallium cyclopentadienide (*44*). This reaction has been extended by Stephenson to other complexes of type **1** and provides a general route to several arene cyclopentadienyl ruthenium complexes **125** (*73*).

$$[Ru(arene)Cl_2]_2 + TlC_5H_5 \xrightarrow[CH_3CN]{Y^-} Ru(arene)(C_5H_5)^+Y^-$$

$$\textbf{1} \qquad\qquad\qquad\qquad \textbf{125}$$

$$arene = C_6H_6,\ p\text{-}MeC_6H_4CHMe_2,\ C_6H_5OMe,\ C_6Me_6$$

$$Y^- = BPh_4^-,\ PF_6^-$$

b. Introduction of an Arene Ligand.   Hexamethylbenzene and mesitylene ruthenium cyclopentadienyl complexes **125** have been prepared in moderate yield (30–50%) and under drastic conditions by ligand exchange of ruthenocene **126** (*74*). A rather similar method has been used by Nesmeyanov to prepare several complexes of type **125** (arene = C_6H_6, $C_6H_5CH_3$, C_6H_5—C_6H_5, C_6H_5Cl) in 5–7% yield (*75*).

$$Ru(\eta^5\text{-}C_5H_5)_2 + arene \xrightarrow[\text{decalin, 190°C}]{AlCl_3/Al/H_2O} \xrightarrow{NH_4PF_6} Ru(arene)(C_5H_5)^+PF_6^-$$

$$\textbf{126} \qquad\qquad\qquad\qquad\qquad\qquad \textbf{125}$$

$$arene = C_6Me_6,\ C_6H_3Me_3$$

Photolysis of complex **125** (arene = benzene) in acetonitrile gives a quantitative yield of cyclopentadienyl tris(acetonitrile) ruthenium complex

127 (*76,77*). Compound **127** is a useful starting material in preparing a variety of arene cyclopentadienyl ruthenium complexes of type **125** by

$$125 \xrightarrow[h\nu]{CH_3CN} Ru(C_5H_5)(CH_3CN)_3{}^+PF_6{}^- \xrightarrow{arene} Ru(arene)(C_5H_5)^+PF_6{}^-$$

$$\textbf{127}\textbf{125}$$

$$C_6H_6 \Delta;\ 1,2\text{-}C_2H_4Cl_2$$

arene = p-$C_6H_4Cl_2$; $[2_2]$ (1,4)cyclophane; 4 or 5-chloroindole

thermal substitution of CH_3CN ligands (*76,78*). Cyclopentadienyl cyclooctadiene chlororuthenium(II) complex **128** (X = Cl) has been shown to be a highly reactive precursor of **125**-type compounds by thermal displacement of the cycloocta-1,5-diene and chloro ligands with aromatics (*79*) [Eq. (11)].

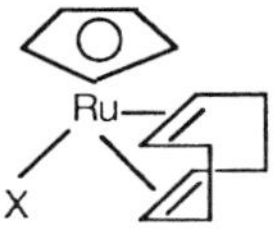

$$\text{arene} = C_6H_6,\ C_6Me_6,\ \text{indene},\ C_6H_5CH_2Cl,\ N(C_6H_5)_3$$

Complex **125** in which one phenyl group of $BPh_4{}^-$ is π bonded to ruthenium, has been synthesized in low yield (20%) by reaction of compound **129** with sodium tetraphenylborate in boiling methanol (*80,81*) [Eq. (12)]. The zwitterionic structure has been confirmed by X-ray structural analysis (*82*).

c. Transformation of a Ligand to an Arene Ligand. The labile complex **128** (X = Br) undergoes facile oxidative addition with 3-bromocyclohexene to give the ruthenium(IV) complex **130**, which on warming in ethanol yields complex **125** in high yield (*83*) (Scheme 11). This elegant transformation **130** → **125** occurs by spontaneous dehydrohalogenation

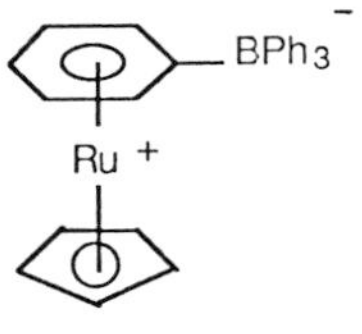

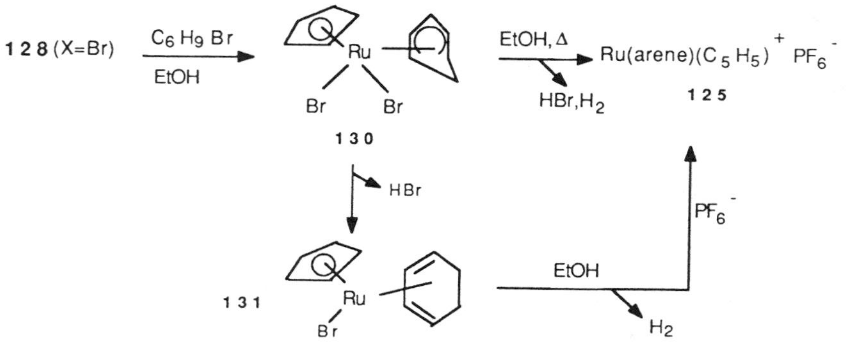

SCHEME 11

and dehydrogenation. A possible intermediate is the cyclohexadiene complex **131**, which also spontaneously dehydrogenates to **125** on warming in ethanol (*84*).

Cyclopentadienyl dicarbonyl ruthenium dimer **132** reacts with silver tetrafluoroborate and diphenylacetylene to afford the cyclobutadiene ruthenium complex **133** (Scheme 12). Irradiation of **133** in dichloromethane in the presence of several alkynes leads to the arene cyclopentadienyl ruthenium complexes **125** in high yield. This reaction appears to be a general route to sterically crowded ruthenium arene cations (*85*).

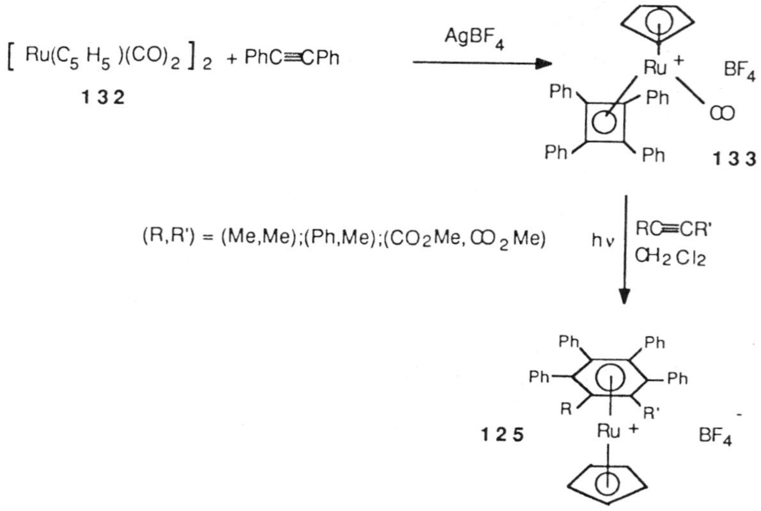

SCHEME 12

A new route to $Ru(\eta^5\text{-}C_5H_5)(\eta^6\text{-}C_6H_5\text{—}R)$ (**125**) has been found by reaction of the diene ligand of **134** (R = H, Me) with acetylene and propyne [Eq. (13)]. The toluene ligand is formed by addition of either acetylene to the isoprene ligand or propyne to the butadiene ligand. The reaction corresponds to a $[\pi 4s + \pi 2s]$ cycloaddition followed by dehydrogenation (86)

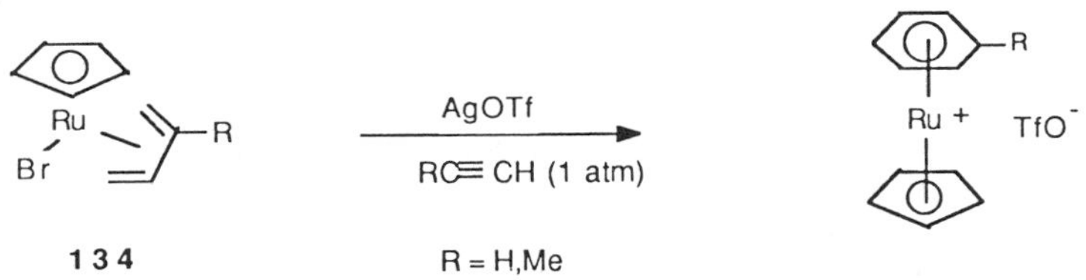

134   R = H,Me

125

d. Transformation of a Ligand to a Cyclopentadienyl Ligand. The intramolecular condensation of an η^3-allyl ligand and an alkyne has recently been reported (Scheme 13). This reaction appears to be quite general and gives the corresponding substituted cyclopentadienyl complexes **125** in good yield from **135** (87).

e. Introduction of an Arene and a Cyclopentadienyl Ligand. Arene pentamethylcyclopentadienyl ruthenium complexes of type **125** have been prepared in good yield and in one step by reacting the hexaaquo ruthenium(II) complex **136** with pentamethylcyclopentadiene in the presence of

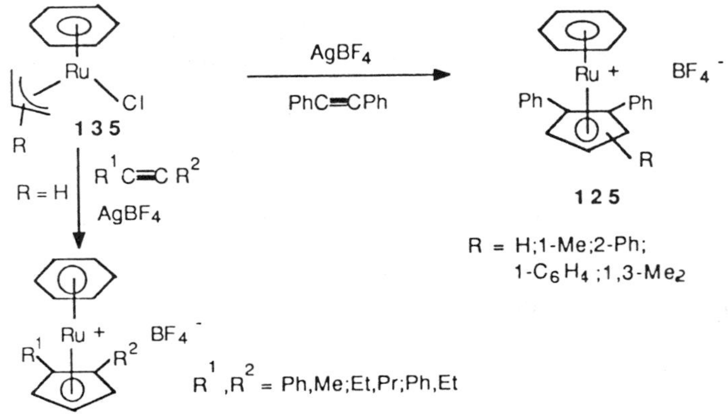

Scheme 13

arene. This procedure has also been used to prepare a series of arene cyclooctadienyl and arene 2,4-dimethylpentadienyl ruthenium(II) complexes (*88*). This method, however, requires the previous synthesis of derivatives **136** from the toxic precursor ruthenium tetraoxide.

$$Ru(H_2O)_6{}^{2+}(X^-)_2 + C_5Me_5H \xrightarrow[arene]{EtOH, \Delta} Ru(arene)(C_5Me_5)^+X^-$$
$$\textbf{136} \qquad\qquad\qquad\qquad\qquad \textbf{125}$$

$$X^- = TfO^-, Tos^-, arene = C_6H_6, C_6Me_6, C_6H_3Me_3, tosylate$$

The best method of preparing pentamethylcyclopentadienyl arene ruthenium complexes is that of Kaganovich *et al.* (*89*). Three complexes of type **125** have been prepared in good yield directly from the commercial reagent ruthenium trichloride, pentamethylcyclopentadiene, and arenes (*89*).

$$RuCl_3, 3\ H_2O \xrightarrow[reflux]{EtOH} \xrightarrow[arene]{C_5Me_5H} \xrightarrow{NH_4PF_6} Ru(arene)(C_5Me_5)^+PF_6{}^-$$
$$\textbf{125}$$

$$arene = C_6H_6, C_6H_3Me_3, C_6H_5Me$$

These two procedures seem to be specific for introducing the pentamethylcyclopentadienyl ligand only.

5. *Ruthenium Borane and Carborane Complexes*

$[RuCl_2(\eta^6\text{-arene})]_2$ complexes have been used as organometallic synthons for access to metallacarboranes and boranes. A few examples of their formation are indicated here.

a. Synthesis of Ruthenacarboranes. Complex **1** (arene = benzene) reacts with $B_9C_2H_{11}{}^{2-}$ dianion to form the air-stable complex **137** in 32%

$$[Ru(C_6H_6)Cl_2]_2 + Tl[3,1,2\text{-}TlC_2B_9H_{11}] \xrightarrow{THF} 3,1,2\text{-}(C_6H_6)RuC_2B_9H_{11}$$
$$\textbf{1} \qquad\qquad\qquad\qquad\qquad\qquad \textbf{137}$$

yield (*90*). Reaction of ruthenacarborane **137** with ethanolic KOH in refluxing ethylene glycol results in polyhedral contraction to form complexes **138** and **139** in very low yield. Complex **139** has been characterized by X-ray structural analysis (*91*).

$$\textbf{137} + KOH/EtOH \xrightarrow[reflux]{\substack{ethylene \\ glycol}} 1,2,4\text{-}(C_6H_6)RuC_2B_8H_{10} + 2,5,6\text{-}(C_6H_6)RuC_2B_7H_{11}$$
$$\textbf{138} \qquad\qquad\qquad\qquad \textbf{139}$$

b. Synthesis of Ruthenaboranes. A large number of polyhedral ruthenaboranes have been synthesized in 1 to more than 80% yield, by reaction of **1** (arene = hexamethylbenzene) with polyhedral borane mono-anions ($B_3H_8^-$, $B_4H_9^-$, $B_6H_{11}^-$, $B_9H_{12}^-$, $B_9H_{14}^-$, $B_{10}H_{13}^-$) or dianions ($B_{10}H_{10}^{2-}$, $B_{10}H_{14}^{2-}$). Compounds formed include arachno-, nido-, and closo-type clusters (*92–96*). Several complexes have been structurally characterized by X-ray diffraction studies (*93–96*).

Reaction between the *arachno*-[$B_3H_8^-$] anion and complex **1** gives the arachno four-vertex complex **140** in 66% yield (*92,97*). Treatment of **140** with *closo*-[$B_{10}H_{10}$]$^{2-}$ in refluxing ethanol affords as the major product (32%) the unexpected triruthenium decaboron double-cluster compound **141**, which has been analyzed by single-crystal X-ray diffraction (*97*).

$$1 + Tl[B_3H_8] \xrightarrow{CH_2Cl_2} arachno\text{-}[2\text{-}(C_6Me_6)\text{-}2\text{-}Cl\text{-}2\text{-}RuB_3H_8]$$

140

$$\downarrow$$

$$1\text{-}[(C_6Me_6)_2RuH_4]\text{-}isocloso\text{-}1\text{-}RuB_{10}H_8\text{-}2,3\text{-}(OEt)_2 + \ldots$$

141

B. η^6-Arene Osmium(II) Complexes

1. General Methods from $OsCl_6^{2-}$, OsO_4, and $OsCl_3 (H_2O)_3$

Cationic bis benzene osmium(II) complex **142** has been obtained by a route similar to that of Ru(arene)$_2^{2+}$ complexes starting with Na_2OsCl_6 (*98,99*). Polymeric neutral benzene osmium(II) complex **143** has been

$$Na_2OsCl_6 + C_6H_6 \xrightarrow[\text{2. NH}_4\text{PF}_6]{\text{1. Al/AlCl}_3,\ 150°C} [(C_6H_6)_2Os](PF_6)_2$$

142

prepared by reaction of osmium tetraoxide with concentrated HCl, followed by heating with 1,3-cyclohexadiene in ethanol (*100,101*). The same reaction using 1,3,5-trimethyl-1,4-cyclohexadiene gives a compound whose composition varies from the expected binuclear complex **144** to the polymeric compound $Os_4Cl_9(C_6H_3Me_3)_3$ (**145**) (*13,102*). *p*-Cymene osmium complex [$OsCl_2$(Me-C_6H_4-*i*-Pr)$_2$] (**144b**) has been prepared in more than 80% yield by dehydrogenation of α-phellandrene with Na_2OsCl_6 (*34,103*) or $OsCl_3(H_2O)_3$ (*95*) in refluxing ethanol. Complex [$OsCl_2(C_6H_6)$]$_2$ (**144a**) has also been obtained by this method but in lower yield (33%) (*34*).

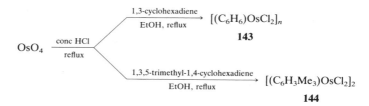

2. *General Methods from [OsCl₂(arene)]η*

a. Arene Ligand Exchange. *p*-Cymene ligands in complex **144b** can be displaced by hexamethylbenzene at 180°C to afford **144c** in 30% yield, but the displacement is not as easy as with the ruthenium species (*95*).

$$[(p\text{-cymene})OsCl_2]_2 \xrightarrow[\;180°C\;]{C_6Me_6} [(C_6Me_6)OsCl_2]_2$$
$$\textbf{144b} \qquad\qquad\qquad \textbf{144c}$$

b. Ligand Substitution or Addition. Complex **143** undergoes halogen exchange to give the iodo derivative $[(C_6H_6)OsI_2]_2$ (*100*). Halogeno bridges are easily cleaved at room temperature by two-electron ligands to afford the neutral mononuclear complexes **146** in high yield (*103,104–108*).

$$[(arene)OsX_2]_2 \xrightarrow{\;L\;} (arene)OsX_2L$$
$$\textbf{143} \qquad\qquad\qquad \textbf{146}$$

arene = benzene, *p*-cymene, X = Cl, I

L = CO, CH₃CN, CNR, DMSO, PMe₃, PMe₂H, PMe'Bu₂, PPh₃, P'Pr₃, P(OPh)₃

$(C_6H_6)OsCl_2(CH_3CN)$ has also been prepared directly from osmium tetraoxide using the same procedure as for the formation of **143** and by extraction of the polymeric intermediate in acetonitrile (*13*). Complexes **143** (X = I) or **144** (arene = *p*-cymene, X = Cl) react with monodentate ligands such as phosphite (*106*) and pyridine (*34*) and with potentially binucleating ligands such as pyrazine and 1,3-dithiane (*39*) in refluxing methanol or ethanol to give the cationic complexes of type **147**.

$$[(arene)OsX_2]_2 + \text{excess } L \xrightarrow[\text{2. NH}_4\text{PF}_6]{\text{1. MeOH, reflux}} (arene)OsCl(L)_2{}^+PF_6{}^-$$
$$\textbf{147}$$

143 arene = benzene, X = I   L = P(OMe)₃, PMe₃

144 arene = *p*-cymene, X = Cl   L = pyrazine, pyridine, 1,3-S₂C₄H₈

c. Formation of (Arene)OsX₃Os(arene)⁺. Triply chloro-, hydroxo-, and alkoxo-bridged osmium cations have been prepared as the ruthenium

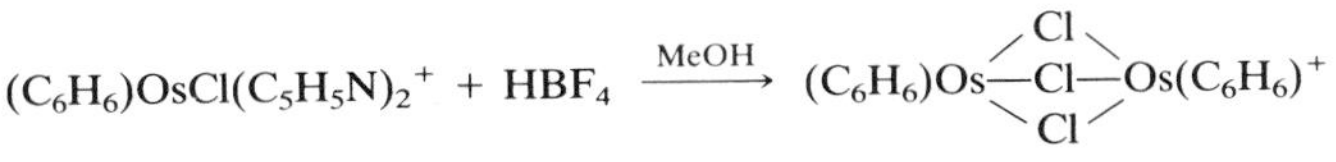

$$(C_6H_6)OsCl(C_5H_5N)_2^+ + HBF_4 \xrightarrow{\text{MeOH}} (C_6H_6)Os\underset{Cl}{\overset{Cl}{\diagup\!\!\diagdown}}Cl-Os(C_6H_6)^+$$

$$\textbf{147} \qquad\qquad\qquad \textbf{148}$$

analogs. Treatment of the benzene osmium bispyridine chloro complex **147** with HBF_4 in methanol has been used to prepare low yields of the corresponding triply bridged complex **148** (*34*).

Reaction of $[Os(C_6H_6)Cl_2]_2$ **144a** with an excess of aqueous NaOH followed by addition of $NaBPh_4$ produces **149** in 60% yield, together with a small amount of **150**. Recrystallization of derivative **149** from acetone gives a pure sample of **150** (*35*). This compound, first found to be the triply bridged hydroxo cation **150A**, has been reformulated on the basis of X-ray structural analysis as the novel tetranuclear complex **150B** (*36*). In contrast, when the ligand is *p*-cymene instead of benzene, this reaction has

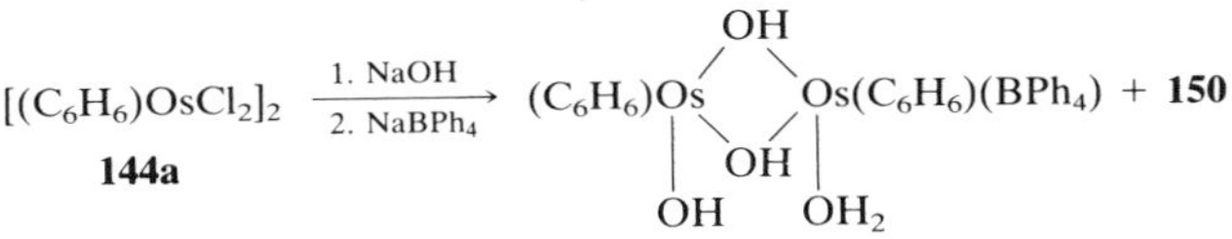

$$[(C_6H_6)OsCl_2]_2 \xrightarrow[\text{2. NaBPh}_4]{\text{1. NaOH}} (C_6H_6)Os\underset{OH\quad OH_2}{\overset{OH}{\diagup\!\!\diagdown}}\underset{OH}{Os(C_6H_6)(BPh_4)} + \textbf{150}$$

$$\textbf{144a} \qquad\qquad\qquad\qquad\qquad \textbf{149}$$

$$\left[(arene)Os\underset{OH}{\overset{OH}{\diagup\!\!\diagdown}}OH-Os(arene)\right]^+ \qquad \left[\begin{array}{c} C_6H_6\quad C_6H_6 \\ Os\quad\quad Os \\ HO\diagup\!\!\diagdown\underset{O}{}\diagup\!\!\diagdown OH \\ HO\quad\quad\quad OH \\ Os\quad\quad Os \\ C_6H_6\quad C_6H_6 \end{array}\right]^{2+}$$

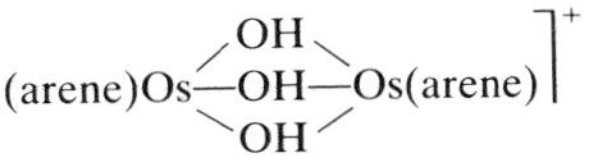

$$\textbf{150A} \qquad\qquad\qquad\qquad\qquad \textbf{150B}$$

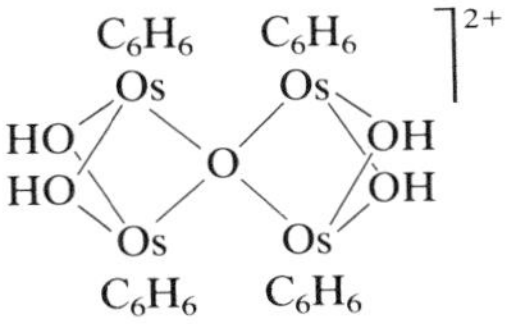

arene = benzene, *p*-cymene

been found to give exclusively the triply hydroxo-bridged cation **150A** (*109,110*). Reaction of the dimer $[Os(C_6H_6)Cl_2]_2$ (**144a**) with a solution of sodium methoxide in methanol gives the triply methoxo-bridged osmium cation **151** in 38% yield (*35*). It has also been shown by ^{1}H NMR that the mixed-metal complex $(C_6H_6)Os(\mu\text{-OMe})_3Ru(C_6H_6)^+$ (**152**) can be prepared by mixing solutions of $(C_6H_6)Ru(\mu\text{-OMe})_3Ru(C_6H_6)^+$ and cation **151** in nitromethane (*38*).

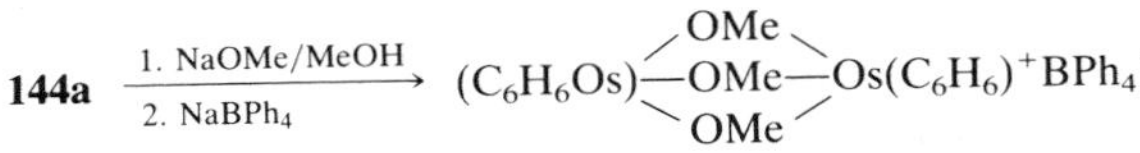

$$\textbf{144a} \xrightarrow[\text{2. NaBPh}_4]{\text{1. NaOMe/MeOH}} (C_6H_6Os)\underset{OMe}{\overset{OMe}{\diagup\!\!\diagdown}}OMe-Os(C_6H_6)^+BPh_4^-$$

$$\textbf{151}$$

Addition of aldehydes, such as formaldehyde, acetaldehyde, or propionaldehyde, to an aqueous solution of **150A** (arene = *p*-cymene) at 60°C leads to the precipitation of complexes of type **153** containing a bridging carboxylate group arising from the aldehyde. ^{1}H-NMR spectra of complexes **153** show a bridging hydride resonance as a singlet at approximately $\delta -10.4$ ppm bearing satellites owing to coupling to the ^{187}Os nucleus. The formulation of these complexes has also been confirmed by an X-ray structure of **153**. In contrast, addition of the water-insoluble cinnamaldehyde, benzaldehyde, or trimethylacetaldehyde to cation **150A** in acetone affords complexes **154**. Reaction of an excess of cinnamic, benzoic, or acetic acid to cation **150A** also gives complexes **154** (*110*).

$$\textbf{150A} \xrightarrow[\text{R = H, Me, Et}]{\text{RCHO/H}_2\text{O}} (p\text{-cymene})\text{Os}\underset{\underset{\text{R}}{|}}{\overset{\overset{\text{H}}{\diagup \diagdown}}{\underset{\diagdown \diagup}{\text{—OH—}}}}\text{Os}(p\text{-cymene})^+$$

OCO

153

$$\textbf{150A} \xrightarrow[\substack{\text{or}\\ \text{RCO}_2\text{H/acetone}}]{\text{RCHO/acetone}} (p\text{-cymene})\text{Os}\text{—OH—}\text{Os}(p\text{-cymene})^+$$

OH

O—C—O

R

154

3. Derivatives of [OsX₂(arene)]₂ and [OsX₂(L)arene] Complexes

a. Arene Os(L)(L')X⁺ Complexes. Cationic complexes of structure **155** containing different ligands can be obtained by treatment of complex **146** with phosphines (*47,108,20*) or carbon monoxide (*111*) in the pre-

$$\text{OsX}_2(\text{L})(\eta^6\text{-arene}) \xrightarrow[\text{NH}_4\text{PF}_6 \text{ or AgPF}_6/\text{MeOH}]{\text{L}'} \text{OsX}(\text{L})(\text{L}')(\eta^6\text{-arene})^+\text{PF}_6^-$$

146 (arene = C₆H₆,   **155**
Me-C₆H₄-*i*-Pr, X = Cl, I)

L = L' = PPh₃, PMe₃
L = CNR, L' = PMe₃
L = P(OMe)₃, L' = PMe₃
L = P-*i*-Pr₃, L' = CO
L = PMe₃, L' = CH₂=CH₂, CH₂=CH—CH₃

sence of $AgPF_6$ or NH_4PF_6 in methanol. Olefin osmium derivatives of type **155** are also prepared by this method (*106,111*). Treatment of complex **146** with $AgPF_6$ in acetone allows the isolation of binuclear dications **156**. Addition of 2 equiv of phosphines or phosphite L′ = PMe_3, $P(OMe)_3$, PPh_3 to cation **156** also leads to complexes **155** (*20*).

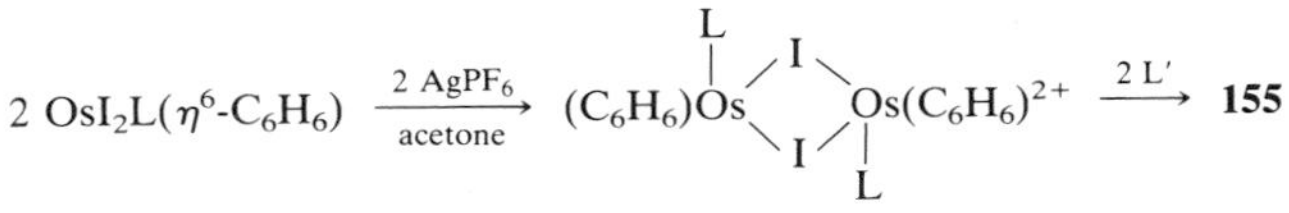

$$2\ OsI_2L(\eta^6\text{-}C_6H_6) \xrightarrow[\text{acetone}]{2\ AgPF_6} (C_6H_6)Os \overset{\overset{L}{|} \diagdown I}{\underset{\diagup I \diagdown}{\underset{\underset{L}{|}}{}}} Os(C_6H_6)^{2+} \xrightarrow{2\ L'} \textbf{155}$$

156

L = PMe_3, PPh_3, $P(OMe)_3$

b. Arene Osmium Carboxylates. Two arene osmium(II) acetates have been reported: the monoacetato and diacetato complexes **157** and **158**. They were obtained selectively by addition of silver acetate to [(*p*-cymene)OsCl$_2$]$_2$ (**144b**), depending on the quantity of the silver acetate (*112*).

$$\textbf{144b} \xrightarrow{\text{AgOAc}} (p\text{-cymene})Os(OAc)Cl \quad \text{or} \quad (p\text{-cymene})Os(OAc)_2$$

$$\textbf{157}\ (84\%) \qquad\qquad\qquad \textbf{158}\ (95\%)$$

c. Arene Osmium Allyl and Alkyl Derivatives. The benzene osmium η^3-allyl complex **159** has been obtained by cleavage of the chloro-bridged ligands in **143** by allyl mercury compounds (*45*) [Eq. (14)]. New alkyl

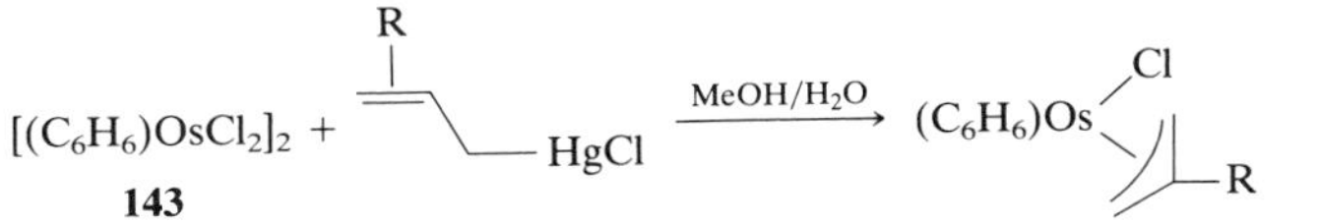

$$[(C_6H_6)OsCl_2]_2 + \overset{\overset{R}{|}}{\diagdown\diagdown} -HgCl \xrightarrow{MeOH/H_2O} (C_6H_6)Os\overset{\diagup Cl}{\diagdown} \diagdown -R \qquad (14)$$

143

159 (R = H, Me)

osmium complexes have been synthesized from **146** by reaction with Al_2Me_6 in toluene. The resulting complexes depend greatly on the nature of additional ligands L′ (*103*). Other mono- or dialkyl osmium derivatives were prepared by Werner and co-workers for study of the migration of the methyl group to methylene or carbonyl and for access to osmium(0) and hydrido osmium compounds (*104,46*). One particular method consists of the reaction catalyzed with $CuSO_4$ of complex **146** with diazomethane (*113*).

$$1 \text{ equiv Al}_2\text{Me}_6, \text{ L}' = \text{DMSO} \longrightarrow (p\text{-cymene})\text{Os(DMSO)Cl(CH}_3) \xrightarrow{\text{PPh}_3} (p\text{-cymene})\text{Os}\overset{\text{Cl}}{\underset{\text{PPh}_3}{-}}\text{CH}_3$$

$$p \qquad\qquad 160$$

146

$$\xrightarrow[\text{L}' = \text{CO, CN-}t\text{-Bu, PMe}_3]{1 \text{ equiv Al}_2\text{Me}_6 \text{ or large excess}} (p\text{-cymene})\text{OsCl(L}')(\text{CH}_3)$$

$$\mathbf{160}$$

$$\xrightarrow{\text{Al}_2\text{Me}_6, \text{ L}' = \text{PPh}_3} (p\text{-cymene})\text{Os(PPh}_3)\text{Cl(CH}_3) + (p\text{-cymene})\text{Os(PPh}_3)(\text{CH}_3)_2$$

$$\mathbf{160} \qquad\qquad\qquad\qquad \mathbf{161}$$

$$+ \text{ products of orthometallation}$$

$$(\text{C}_6\text{H}_6)\text{OsI}_2(\text{L}) \xrightarrow{2 \text{ MeLi}} (\text{C}_6\text{H}_6)\text{OsL(CH}_3)_2$$

$$\text{L} = \text{CO, PMe}_3 \qquad\qquad \mathbf{161}$$

$$\downarrow \text{Ph}_3\text{C}^+\text{PF}_6^-$$

$$\left[(\text{C}_6\text{H}_6)\text{Os}\overset{\text{L}}{\underset{\text{CH}_2}{\overset{\text{CH}_3}{-}}}\right]^+ \longrightarrow (\text{C}_6\text{H}_6)\text{Os}\overset{\text{L}}{\underset{\text{CH}_2}{\overset{\text{H}}{-}}}\overset{}{\underset{\text{CH}_2}{}}$$

$$\mathbf{162}$$

$$(\text{C}_6\text{H}_6)\text{OsI}_2(\text{P-}i\text{-Pr}_3) \xrightarrow[\text{ether}]{\text{CH}_2\text{N}_2, \text{ CuSO}_4} (\text{C}_6\text{H}_6)\text{OsI}(\text{P-}i\text{-Pr}_3)(\text{CH}_3)$$

$$\mathbf{146} \qquad\qquad\qquad\qquad \mathbf{160} \text{ (78\%)}$$

d. Hydrido-Bridged and Terminal Arene Osmium Complexes from Osmium(II) Derivatives. Complex **150A** (arene = *p*-cymene) reacts smoothly in 2-propanol at 80°C to give the tri-μ-hydrido diosmium complex **163** in 75% yield. Reaction of **144** with 2-propanol at 80°C gives the

$$\mathbf{150A} \xrightarrow[80°\text{C}]{\text{2-propanol}} (p\text{-cymene})\text{Os}\overset{\text{H}}{\underset{\text{H}}{\overset{}{-}}}\text{H}\!-\!\text{Os}(p\text{-cymene})^+$$

$$\mathbf{163}$$

mono-μ-hydrido trichloro complex **164**, but only in the presence of KPF$_6$. The other μ-hydrido osmium complexes **165** and **166** have also been made in 79 and 13% yield, respectively, under the same conditions, from the mono and diacetato osmium complexes **157** and **158**.

$$[\text{Os}(p\text{-cymene})\text{Cl}_2]_2 \xrightarrow[80°\text{C, KPF}_6]{\text{2-propanol}} (p\text{-cymene})\text{Os}\overset{\text{H}}{\underset{\text{Cl}}{\overset{\text{Cl}}{<}}}\overset{\text{Cl}}{\underset{}{}}\text{Os}(p\text{-cymene})$$

$$\mathbf{144}$$

$$\mathbf{164}$$

Os(p-cymene)Cl(OAc) $\xrightarrow[\text{80°C, KPF}_6]{\text{2-propanol}}$ (p-cymene)Os—Cl—Os(p-cymene)$^+$

157

H over Cl, OAc below

165

Os(p-cymene)(OAc)$_2$ $\xrightarrow[\text{80°C, KPF}_6]{\text{2-propanol}}$ (p-cymene)Os—H—Os(p-cymene)$^+$

158

H above, OAc below

166

Measurement of ^{187}Os chemical shifts using ^{1}H–$\{^{187}$Os$\}$ two-dimensional NMR spectroscopy for complex **163** gives $\delta = -2526 \pm 1$ ppm (from molten OsO$_4$). The multiplicity (quartet) confirms that complex **163** contains three hydride ligands. All the complexes **163–166** exhibit coupling of the hydrides to the ^{187}Os nucleus ranging from 63 to 83 Hz (*109*).

Many methods of preparing mono or dihydrido mononuclear osmium complexes have been reported. Monohydrido osmium complexes **167** can be obtained by treatment of **146** with zinc in methanol at room temperature

Os(arene)(L)(H)(X) $\xleftarrow[\text{RT}]{\text{Zn/MeOH}}$ Os(arene)X$_2$L $\xrightarrow[\text{80°C}]{\text{Et}_3\text{N/EtOH}}$ Os(arene)(L)(H)(X)

167   **146**   **167**

(arene = benzene,   (arene = p-cymene,
X = I, L = PMe$_3$,   X = Cl, L = PPh$_3$)
PMe^{t}Bu$_2$, P-i-Pr$_3$)

(*107,49*) or by heating **146** in ethanol with triethylamine (*109*). An original method of producing hydrido derivatives **167** has been found by reacting

Os(C$_6$H$_6$)(PMe$_2$H)I$_2$ $\xrightarrow[\text{MeOH}]{\text{KOtBu}}$ Os(C$_6$H$_6$)(PMe$_2$OCH$_3$)(H)(I)

146   **167**

complex **146** (L = PMe$_2$H) with a base in methanol. The mechanism possibly involves formation of an intermediate containing an Os=PMe$_2$ double bond (*105*). Monohydrido osmium complex **167** has been used to prepare azaalkylidene complexes **168** in good yields.

167 + (HO)N=CRR′ $\xrightarrow[\text{CH}_2\text{Cl}_2]{\text{AgPF}_6}$ [Os]$^+$ with H and N—OH, CRR′ $\xrightarrow{-\text{H}_2\text{O}}$ [Os]=N=CRR′$^+$

168

[Os] = (C$_6$H$_6$)(PMe-t-Bu$_2$)Os

Reduction of alkyl osmium complex **160** with $NaBH_4$ or $NaBD_4$ gives the methyl hydrido and methyl deutero osmium complexes **169**. Treatment of **169** with trityl cation gives the metallaheterocycle **170**. This reaction probably occurs via cationic intermediates **171** and **172**. The coordinatively unsaturated **172** then undergoes an intramolecular oxidative addition with

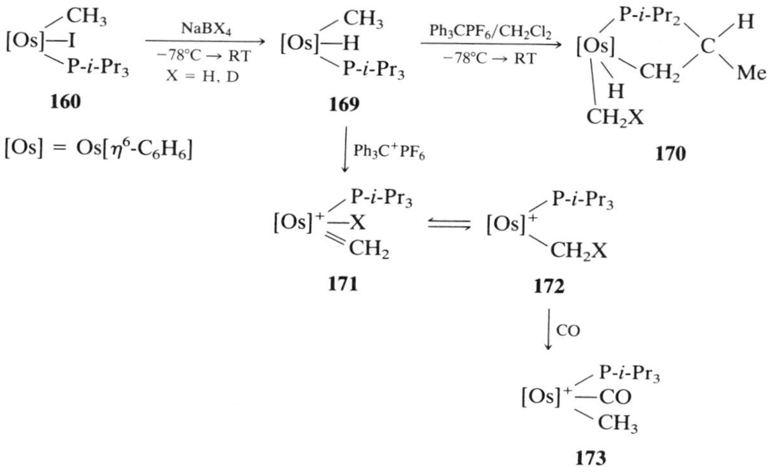

a C—H bond of an isopropyl group of the phosphine to produce **170**. In the presence of CO, the intermediate **172** can be trapped as the carbonyl methyl osmium cation **173** (*113*). One dihydrido osmium complex of structure **174** has been obtained in 80% yield by treatment of **146** with $NaBH_4$ in methanol (*111*).

$$(p\text{-cymene})OsI_2(PMe_3) \xrightarrow[\text{MeOH}]{NaBH_4} Os(p\text{-cymene})(PMe_3)H_2$$
$$\textbf{174}$$

e. Hydrido and Alkyl Arene Osmium from Osmium(0) Intermediates. Eighteen-electron arene osmium(0) complexes **175** have often been used in preparing alkyl or hydrido osmium(II) complexes by oxidative addition. Ammonium hexafluorophosphate in methanol protonates complexes of

$$Os^0(\eta^6\text{-}C_6H_6)(L)(L') \xrightarrow[]{NH_4PF_6/MeOH} Os(\eta^6\text{-}C_6H_6)(L)(L')H^+$$

175   **176**

$$\downarrow \begin{array}{l} 1.\ CH_3I \\ 2.\ NH_4PF_6 \end{array} THF$$

$$Os(\eta^6\text{-}C_6H_6)(L)(L')CH_3^+$$

173

L = L' = PPh_3, P(OMe)_3

L, L' = CO, P-i-Pr_3, CO, PMe_3, CNR, PMe_3

L, L' = PMe_3, C_2H_3R (R = H, Me)

type **175** at room temperature to afford the corresponding hydrido osmium cations **176**. By the same procedure reaction of **175** with methyl iodide gives the alkylosmium(II) cation **173** (*49,50,106,108,114*).

When ligand L is an olefin, reaction of NaI with derivatives **176** and **173** proceeds in two different ways. Complex **176** gives the neutral monoalkyl osmium(II) complex **160** by olefin insertion, whereas complex **173** gives the neutral methyl osmium(II) compound **160** by olefin displacement (*47*).

$$Os(C_6H_6)(C_2H_3R)(PMe_3)H^+ \xrightarrow[\text{acetone}/H_2O]{\text{NaI}} Os(C_6H_6)(C_2H_4R)(PMe_3)(I)$$

$$\text{\textbf{176}} \qquad\qquad\qquad\qquad \text{\textbf{160}}$$

$$Os(C_6H_6)(C_2H_3R)(PMe_3)CH_3^+ \xrightarrow[\text{acetone}/H_2O]{\text{NaI}} \overset{\displaystyle C_2H_3R}{\nearrow} Os(C_6H_6)(PMe_3)(CH_3)(I)$$

$$\text{\textbf{173}} \qquad\qquad\qquad\qquad\qquad\qquad \text{\textbf{160}}$$

With treatment of the same olefin cations **176** and **173** (L = C_2H_4 or C_3H_6) with PMe_3, olefin insertion (to produce **177**), substitution (**180**), or addition of the phosphine to the coordinated olefin (**178** and **179**) is observed (*47*).

$$Os(C_6H_6)(C_2H_3R)(PMe_3)H^+ \xrightarrow[R = H]{PMe_3} [Os]\overset{PMe_3}{\underset{C_2H_5}{\diagup\!\!\!\diagdown}}{}^+ + [Os]\overset{C_2H_4PMe_3}{\underset{H}{\diagup\!\!\!\diagdown}}{}^{\rceil+}$$

$$\text{\textbf{176}} \qquad\qquad\qquad\qquad\qquad \text{\textbf{177}} \qquad\qquad \text{\textbf{178}}$$

$$\text{\textbf{173}} \xrightarrow[R = CH_3]{PMe_3} [Os]\overset{PMe_3}{\underset{C_3H_7}{\diagup\!\!\!\diagdown}}{}^+$$

$$\text{\textbf{177}}$$

$$Os(C_6H_6)(C_2H_3R)(PMe_3)CH_3^+ \xrightarrow[R = H]{PMe_3} [Os]\overset{C_2H_4PMe_3}{\underset{CH_3}{\diagup\!\!\!\diagdown}}{}^{\rceil+}$$

$$\text{\textbf{173}} \qquad\qquad\qquad\qquad\qquad\qquad \text{\textbf{179}}$$

$$\text{\textbf{173}} \xrightarrow[R = CH_3]{PMe_3} \overset{\displaystyle C_3H_6}{\nearrow} [Os]\overset{PMe_3}{\underset{CH_3}{\diagup\!\!\!\diagdown}}{}^+$$

$$\text{\textbf{180}}$$

$$[Os] = Os(C_6H_6)(PMe_3)$$

Reduction of $Os(C_6H_6)I_2(PMe_3)$ with $NaC_{10}H_8$ in THF gives a 16-electron osmium(0) intermediate which reacts with C_6H_6 or C_6D_6 by intermolecular CH addition to form aryl hydrido osmium complexes of type **181** (*114*).

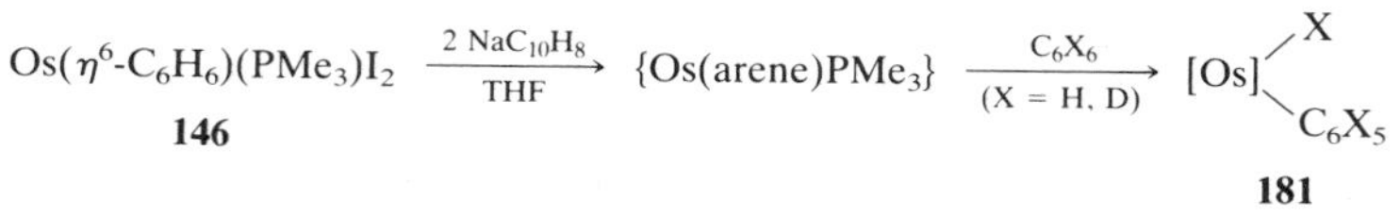

$$Os(\eta^6\text{-}C_6H_6)(PMe_3)I_2 \xrightarrow[\text{THF}]{2\ NaC_{10}H_8} \{Os(arene)PMe_3\} \xrightarrow[(X = H, D)]{C_6X_6} [Os]\langle\begin{smallmatrix}X\\C_6X_5\end{smallmatrix}$$

146   **181**

f. Phosphonate Arene Osmium(II) Complexes. Benzene osmium(II) phosphonates have been made via the initial reaction of phosphite-containing osmium cations of type **147** with NaI. Demethylation occurs, and the anionic complex **182** is obtained. Protonation with trifluoroacetic afforded adduct **183** which generated complex **184** on treatment with pyridine. The latter was shown to be a precursor of the thallium complex **185** containing the chelating osmium phosphonate (*115*).

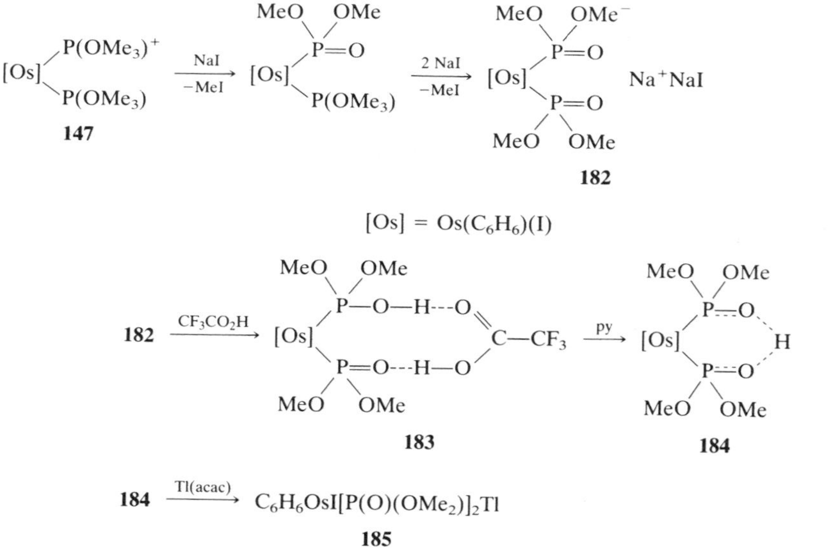

$$[Os]\langle\begin{smallmatrix}P(OMe_3)^+\\P(OMe_3)\end{smallmatrix} \xrightarrow[-MeI]{NaI} [Os]\langle\begin{smallmatrix}P(=O)(OMe)_2\\P(OMe_3)\end{smallmatrix} \xrightarrow[-MeI]{2\ NaI} [Os]\langle\begin{smallmatrix}P(=O)(OMe)(OMe)^-\\P(=O)(OMe)(OMe)\end{smallmatrix}\ Na^+NaI$$

147   **182**

$$[Os] = Os(C_6H_6)(I)$$

182 $\xrightarrow{CF_3CO_2H}$ **183** $\xrightarrow{py}$ **184**

183   **184**

184 $\xrightarrow{Tl(acac)}$ $C_6H_6OsI[P(O)(OMe_2)]_2Tl$

185

g. Acetylide and Vinyl Osmium(II) Complexes. Acetylide complex **186** has been prepared by treatment of the diiodo derivative **146** with phenyl-acetylene and AgPF$_6$ in CH$_2$Cl$_2$. Compound **186** is a useful precursor of vinylidene osmium(0) complex **187** via a vinylhydrido (**188**) and vinylha-logeno (**189**) osmium compound (*116*). Complex **146** with activated alkynes of formula RC≡C—CO$_2$Me (R = H, Me, CO$_2$Me) behaves differently and gives metallaheterocycles **190** by insertion of the alkyne into one Os—I bond. The X-ray structure of **190** has been established. An analogous metallaheterocycle of type **190** has been prepared by insertion of methyl-propiolate into the Os—H bond of complex **167** (*117*).

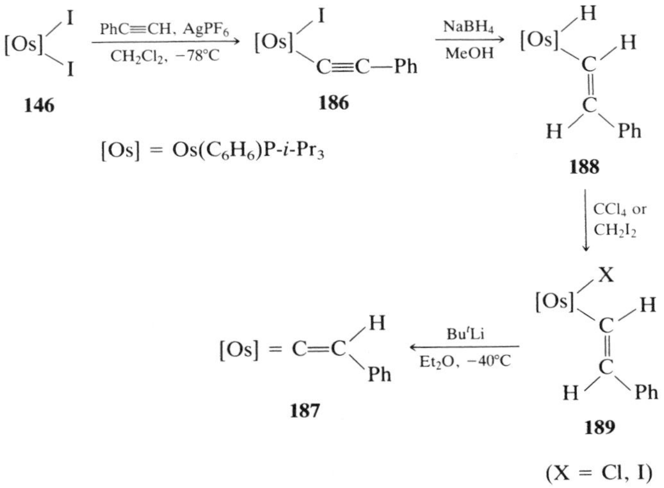

$$[Os] \overset{I}{\underset{I}{\diagdown}} \xrightarrow[\text{CH}_2\text{Cl}_2, \ -78°\text{C}]{\text{PhC} \equiv \text{CH, AgPF}_6} [Os] \overset{I}{\underset{\text{C} \equiv \text{C} - \text{Ph}}{\diagdown}} \xrightarrow[\text{MeOH}]{\text{NaBH}_4} [Os] \overset{H}{\underset{}{\diagdown}}$$

146   **186**

$$[Os] = Os(C_6H_6)P\text{-}i\text{-}Pr_3$$

188

CCl₄ or
CH₂I₂

$$[Os] = C = C \overset{H}{\underset{Ph}{\diagdown}} \xleftarrow[\text{Et}_2\text{O, } -40°\text{C}]{\text{Bu}^t\text{Li}}$$

187

189

(X = Cl, I)

$$[Os] \overset{I}{\underset{X}{\diagdown}} \xrightarrow[\text{CH}_2\text{Cl}_2, \ -78°\text{C} \rightarrow \text{RT}]{\text{AgPF}_6/\text{RC}_2\text{CO}_2\text{Me}}$$

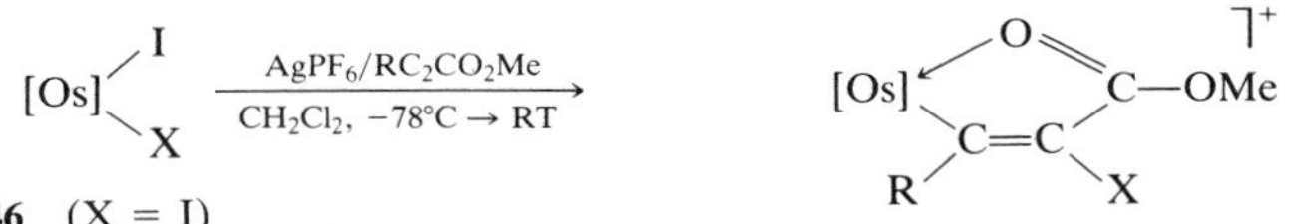

146 (X = I)
167 (X = H)

190 (X = I, R = H, Me, CO₂Me; X = H, R = H)

$$[Os] = Os(C_6H_6)P\text{-}i\text{-}Pr_3$$

4. Arene Cyclopentadienyl Osmium Complexes

Two arene cyclopentadienyl osmium complexes of type **191** have been prepared in moderate yields (17–43%) by reacting precursor **144** with thallium cylopentadienide in acetonitrile (73).

$$[\text{Os(arene)Cl}_2]_2 + \text{TlC}_5\text{H}_5 \xrightarrow[\text{NaBPh}_4]{\text{CH}_3\text{CN}} \text{Os(arene)(C}_5\text{H}_5)^+\text{BPh}_4^-$$

144   **191**

$$\text{arene} = C_6H_6, \ \text{MeC}_6H_4\text{-}i\text{-Pr}$$

5. Osmaborane and Carborane Clusters

The first reported arene osmacarborane complex has been prepared in 15% yield by reacting compound **146** with Tl(3,1,2-TlC₂B₉H₁₁) in THF

$$\text{Os(C}_6\text{H}_6)\text{Cl}_2(\text{CH}_3\text{CN}) + \text{Tl}(3,1,2\text{-TlC}_2\text{B}_9\text{H}_{11}) \rightarrow [3,1,2\text{-}(C_6H_6)\text{OsC}_2\text{B}_9\text{H}_{11}]$$

146   **192**

(*91*). Three osmaborane compounds (**193**, **194**, and **195**) have also been obtained in 13–70% yields using a route analogous to that available for the corresponding ruthenaboranes (*92,95*).

$$[Os(C_6Me_6)Cl_2]_2 +$$
144c

$$\xrightarrow{\textit{closo-}B_{10}H_{10}{}^{2-}} \textit{isocloso-}[1\text{-}(C_6Me_6)OsB_{10}H_{10}]$$
193

$$\xrightarrow{\textit{arachno-}B_3H_8{}^-} \textit{arachno-}[2\text{-}(C_6Me_6)OsB_3H_8]$$
194

$$\xrightarrow{\textit{arachno-}B_9H_{14}{}^-} \textit{nido-}[6\text{-}(C_6Me_6)OsB_9H_{13}]$$
195

III

ARENE RUTHENIUM(0) AND OSMIUM(0) COMPLEXES

A. *Arene Ruthenium(0) Complexes*

The first (η^6-arene)Ru(0) complexes to be produced were Ru(η^6-C$_6$H$_6$)-(η^4-1,3-cyclohexadiene) (**196a**), by reduction of Ru(C$_6$H$_6$)$_2{}^{2+}$ with borohydride (*118*) or of RuCl$_3$ with isopropylmagnesium bromide in the presence of 1,3-cyclohexadiene (*119*), Ru(η^6-C$_6$Me$_6$)(η^4-C$_6$Me$_6$) (**197**), by reduction of Ru(η^6-C$_6$Me$_6$)$_2{}^{2+}$ with sodium (*4*), and Ru(η^6-C$_6$H$_6$)(η^4-C$_8$H$_{12}$) (**198**), in low yield, by reaction of [RuCl$_2$(C$_6$H$_6$)]$_2$ **1a** with *i*-PrMgX in the presence of cyclooctadiene (*112*). Complex **198** is obtained in better yield (70%) by heating of the cyclooctadiene complex **199** in benzene–acetone; in addition, 16% of the derivative Ru(η^6-C$_6$H$_5$—BPh$_3$)(η^5-1-3,5,6-C$_8$H$_{11}$), identified by X-ray analysis, is obtained (*120*).

$$RuH(C_8H_{12})(NH_2NMe_2)_3(BPh_4)_2 \rightarrow Ru(C_6H_6)(C_8H_{12})$$
199   **198**

1. *Preparation from Ru(η^4-cyclooctadiene)(η^6-cyclooctatriene)*

A straightforward route to complex **196a** (70–80% yield) is by reduction of RuCl$_3$ with zinc dust in the presence of cyclohexadiene (*121,122*) [Eq. (15)]. The same reaction with cyclooctadiene leads to complex **200** (*121–123*), from which Vitulli and co-workers have developed a general

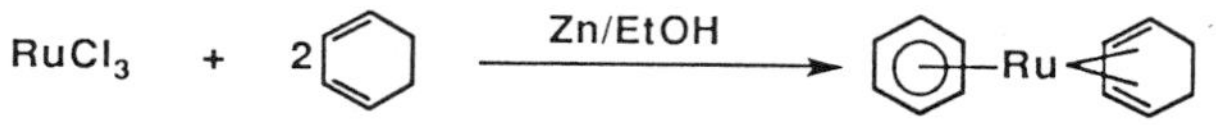

$$RuCl_3 \; + \; 2 \; \bigcirc \; \xrightarrow{\text{Zn/EtOH}} \; \bigcirc\text{—Ru}\text{—}\diamond \quad \textbf{196a} \quad (15)$$

route to complexes of type **198** by displacement of cyclooctadiene on hydrogenation (1 atm H_2) (*12,124*). Complexes **198** containing a chiral group attached to the benzene ring have also been obtained and show unusual anisochronicity of the aromatic protons (*125*). Displacement of cyclodiene from complex **200** by arene is also achieved on protonation with HPF_6 or CF_3CO_2H in aromatic solvents. (η^6-Arene)ruthenium(II) complexes of type **201** are thus obtained (*126*) [Eq. (16)].

$$(\eta^{\underline{6}}\text{arene})\,Ru \quad \bigcirc \quad \rceil^+ \quad PF_6^- \qquad 2\,0\,1$$

$$1,3\text{-}C_8H_{12} \quad \longleftarrow \quad \bigg| \quad HPF_6/\text{arene}/RT$$

$$RuCl_3,3H_2O + \bigcirc \xrightarrow[\text{Zinc}]{\text{EtOH}} \bigcirc\!\!-Ru \bigcirc \xrightarrow[\substack{H_2/\,8\,h \\ RT}]{\text{arene}} \overset{R}{\bigcirc}\!\!-Ru \bigcirc \qquad (16)$$

$$2\,0\,0 \qquad\qquad\qquad 1\,9\,8$$

arene: $C_6H_6, C_6H_{6-n}R_n$, $C_6H_5CH(CH_3)NH_2$, C_6H_5OMe, $C_6H_5COCH_3$ (121)

R : -CH(Me)Et, -CH(Me)t-Bu, -CH-(Me)OEt, -O-CH(Me)Et (125)

The formation of complexes **198** has been extended to provide access from the (η^6-naphthalene)Ru complex **202** (80%). Compound **202** is used to produce ruthenium(0) complexes of type **198** by displacement of naphthalene in acetonitrile (*127*). Complex **202** is also obtained by reaction of $[RuCl_2(COD)]_n$ with sodium naphthalene (*128*) [Eq. (17)].

$$2\,0\,0 \quad \xrightarrow[\text{H}_2/\text{pentane}]{\text{naphthalene}} \quad \bigcirc\!\!\bigcirc\!\!-Ru \bigcirc \xrightarrow[\substack{\text{arene} \\ RT}]{\text{CH}_3\text{CN}} \overset{R}{\bigcirc}\!\!-Ru \bigcirc \qquad (17)$$

$$2\,0\,2 \qquad\qquad\qquad 1\,9\,8$$

arene: C_6H_6, C_6H_5Me, $C_6H_4Me_2$, C_6Me_6, C_6H_5OMe, C_6H_5Cl, C_6H_5CHO

Reaction of the arene ruthenium precursor **200** with polystyrene and hydrogen leads to the loss of both cycloolefins and gives a polymeric ruthenium complex, used for catalytic hydrogenation, for which EXAFS indicated a Ru—C distance of 2.05 Å. A similar reaction of derivative **200** with 1,3-diphenylpropane gives complex **203** and a compound of composition (diphenylpropane)Ru_2 (**204**) (*129,130*) [Eq. (18)].

A decisive aspect of (arene)Ru(COD) complexes **198** appeared when they were found to easily give complexes of type $[RuCl_2(\text{arene})]_2$ (**1**) that are among the most useful precursors for a variety of arene ruthenium

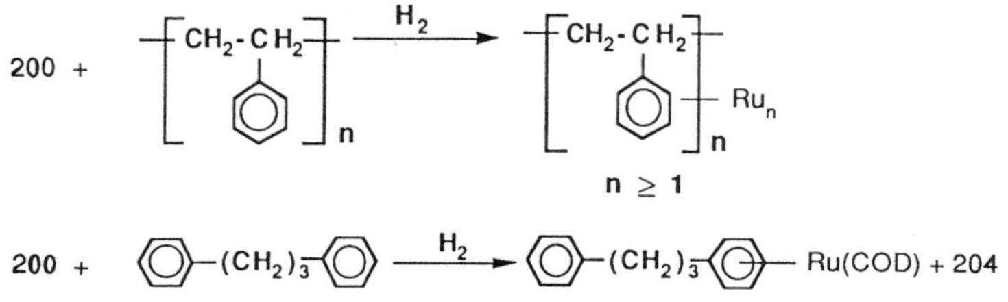

$$200 + \left[\begin{array}{c} CH_2\text{-}CH_2 \\ \bigcirc \end{array}\right]_n \xrightarrow{\ H_2\ } \left[\begin{array}{c} CH_2\text{-}CH_2 \\ \bigcirc \end{array}\right]_n Ru_n$$

$$n \geq 1 \tag{18}$$

$$200 + \bigcirc\text{-}(CH_2)_3\text{-}\bigcirc \xrightarrow{\ H_2\ } \bigcirc\text{-}(CH_2)_3\text{-}\bigcirc - Ru(COD) + 204$$

203

derivatives. By treatment of **198** with aqueous HCl, complexes of type **1** are obtained in good yield *(12,125)* [Eq. (19)].

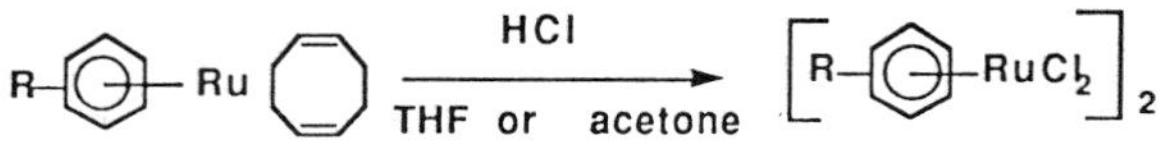

$$R\text{-}\bigcirc\text{-}Ru\,\bigcirc \xrightarrow[\text{THF or acetone}]{\text{HCl}} \left[R\text{-}\bigcirc\text{-}RuCl_2\right]_2 \tag{19}$$

198   **1**

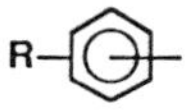

$$R\text{-}\bigcirc\text{-}\ :\ C_6H_6,\ C_6H_4Me_2,\ C_6H_3Me_3,\ C_6H_5\text{-}CH(Me)Et,\ C_6H_5iPr,$$
$$C_6H_5OMe$$

2. *Preparation from [RuCl$_2$(arene)]$_2$ Complexes*

[RuCl$_2$(arene)]$_2$ complexes (**1**) react with 1,5-cyclooctadiene and 1,3- or 1,4-cyclohexadiene in the presence of ethanol and Na$_2$CO$_3$ or zinc dust to give Ru^0(η^6-arene)(η^4-diene) compounds of type **196–198** in 60% yield [Eq. (20)]; thus, this reaction appears to be the reverse of the **198** $\rightarrow$ **1** reaction [Eq. (19)]. The same reaction with ethylene leads to the bis-ethylene ruthenium(0) complex **205** (37%) *(131,10)*. The norbornadiene complex **207** is prepared similarly from derivative **206** *(125)*. Combination of transformations **206** $\rightarrow$ **207** [Eq. (21)] or **1** $\rightarrow$ **198** [Eq. (20)] with trans-

$$\left[RuCl_2(arene)\right]_2 + cyclodiene \xrightarrow[\text{or Zinc}]{Na_2CO_3/EtOH} Ru^0(\eta^6\text{-arene})(\eta^4\text{-diene})$$

1   **196 — 198**

arene: C$_6$H$_6$, C$_6$H$_3$Me$_3$, C$_6$Me$_6$

$$\tag{20}$$

$$1c + CH_2\text{=}CH_2 \xrightarrow[\text{EtOH}]{Na_2CO_3} Ru(\eta^6\text{-}C_6Me_6)(CH_2\text{=}CH_2)_2$$

205

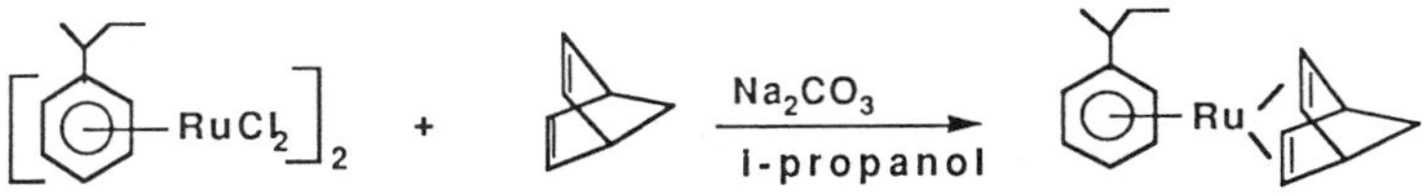

$$\text{(21)}$$

206   207

formation $198 \rightarrow 1$ [Eq. (19)] actually permits the possibility of exchanging the cyclodiene of complexes of type 198.

Reaction of $Ru(\eta^6\text{-}C_6Me_6)(\eta^4\text{-}C_6H_8)$ ($196c$) with HPF_6 leads to protonation of the η^4-diene ligand and formation of intermediate 208 which, in solution, is in equilibrium with the hydride 209. Alternatively, 208 adds carbon monoxide to afford cation 210 (131) (Scheme 14).

(Arene)(*endo*-dicyclopentadiene)M(0) complexes [M = Ru (211), Os (212)] are prepared by reduction of $1c$ or $1d$ with 2-propanol and Na_2CO_3 in the presence of dicyclopentadiene (Scheme 15). Protonation of derivatives 211 and 212 with HPF_6 leads to cations 213 and 214 which are stabilized by an agostic C—H—M bond, as shown by the X-ray structure of the osmium species (214). The cations 213 and 214 react with CN(*t*-Bu) to give complexes of type 215 in which the agostic C—H—M is no longer observed (102).

(η^6-Arene)(1-4-η-cyclooctatetraene)M(0) (216) [(M = Fe, Ru, Os; with Ru, arene = C_6H_6 (35%), $C_6H_3Me_3$ (25%), C_6Me_6 (55%) have been prepared by reduction of corresponding complexes 1 with Li_2COT or K_2COT (Scheme 16). Complex 216 (mesitylene) is obtained in similar yield by treatment of 1 (25%) or $RuCl_2(py)(C_6H_3Me_3)$ (28%) with $Li_2(COT)$. The yields increase when RuCl(acac)(arene) is used as precursor, even when prepared *in situ*. The yields of complexes of type 216 increase in the order $Os < Ru < Fe$ and $C_6H_6 < C_6H_3Me_3 < C_6Me_6$ ($13,132$). The

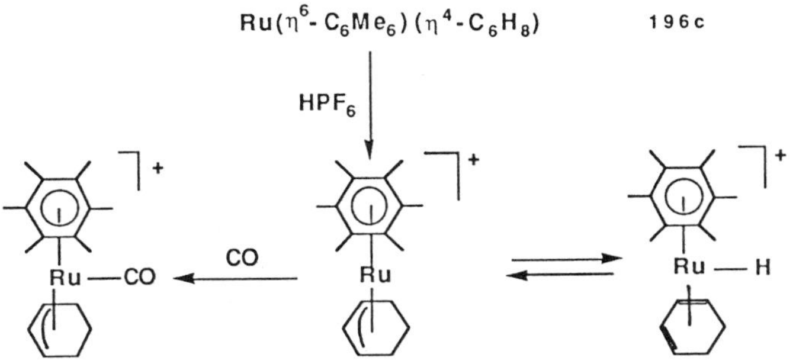

$$Ru(\eta^6\text{-}C_6Me_6)(\eta^4\text{-}C_6H_8) \qquad 196c$$

210   208   209

SCHEME 14

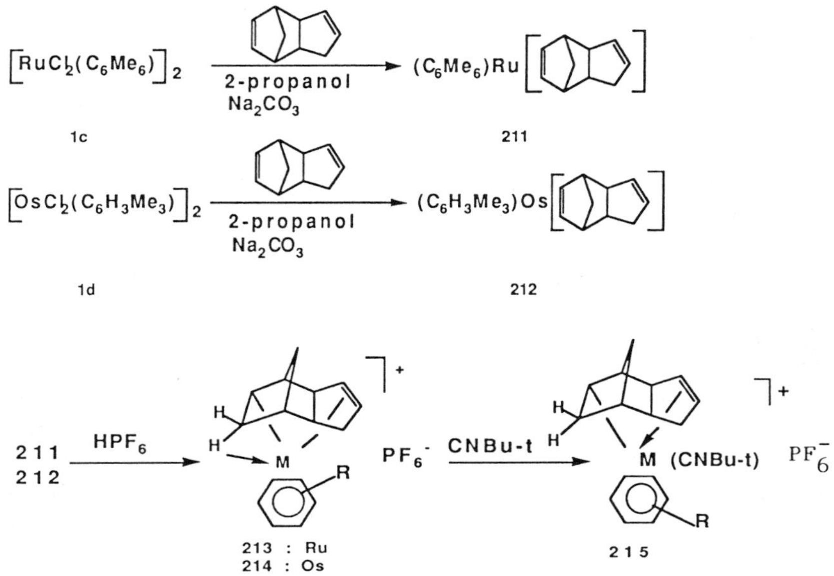

$$\left[RuCl_2(C_6Me_6) \right]_2 \xrightarrow[\substack{2\text{-propanol} \\ Na_2CO_3}]{} (C_6Me_6)Ru \left[\right]^{}$$

1c   211

$$\left[OsCl_2(C_6H_3Me_3) \right]_2 \xrightarrow[\substack{2\text{-propanol} \\ Na_2CO_3}]{} (C_6H_3Me_3)Os \left[\right]^{}$$

1d   212

$$\begin{array}{c} 211 \\ 212 \end{array} \xrightarrow{HPF_6} \quad \left[\right]^{+} PF_6^{-} \xrightarrow{CNBu\text{-}t} \quad \left[\right]^{+} PF_6^{-}$$

213 : Ru   215
214 : Os

SCHEME 15

$$\left[RuCl_2(arene) \right]_2 \xrightarrow{M_2(COT)} \text{—Ru—} R$$

1   216

Tl(acac)   Li$_2$(COT)   HX

RuCl(acac)(arene)

217

acetone   50°C

218

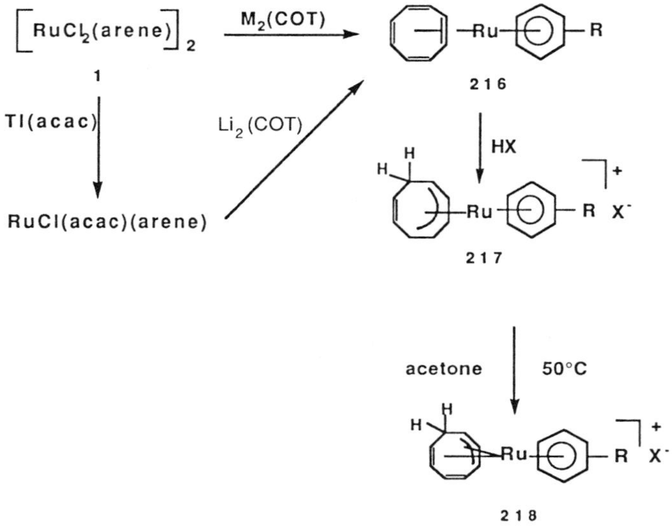

SCHEME 16

X-ray structure of **216** (C_6Me_6) shows the 1-4-η coordination type and a dihedral angle of 45.4° for the cyclooctatetraene ligand. The NMR spectrum shows that the COT ligand in **216** is more fluxional than in $M(\eta^4\text{-COT})(CO)_3$ complexes (*132*).

Addition of acids HX (HPF_6, HBF_4, CF_3CO_2H) to complexes of type **216** ($C_6H_3Me_3$, C_6Me_6, $C_6H_5\text{-}t\text{-Bu}$) leads to (1-5-$\eta$-$C_6H_9$)Ru(II) complexes of structure **217**. The isolated salts **217** (X = PF_6, BF_4) isomerize and (1-3,6-7-η-C_8H_9)Ru(II) cations of type **218** are formed. The isomerization is not complete either when CF_3CO_2H is used or when the ligand is $t\text{-BuC}_6H_5$ even by using HBF_4 and HPF_6. Complex **218** (C_6Me_6) in CD_2Cl_2 with CF_3CO_2H partially reverts to **217** to reach an equilibrium $217 \leftrightharpoons 218$ (1:4). The NMR studies are based on X-ray structures of the two key products **217** and **218** containing the mesitylene ligand (*133*).

3. Reduction of $Ru(arene)_2^{2+}$

The reduction of $Ru(C_6Me_6)_2^{2+}$ to $Ru(\eta^6\text{-}C_6Me_6)(\eta^4\text{-}C_6Me_6)$ (**197**) by sodium (*4*) has been extended to that of $Ru(arene^1)(arene^2)^{2+}$ (*27*). Reduction of cation **18a** by sodium amalgam affords complexes of type **219**, whereas similar reduction of $Ru(\eta^6\text{-}C_6Me_6)(naphthalene)^{2+}$ led to decomposition. An X-ray analysis of complex **219** (C_6Me_6) shows the η^4-coordination of the $C_{10}Me_8$ ligand. Reaction of **219** (C_6H_6) with CF_3CO_2H leads to exo protonation of the $\eta^4\text{-}C_{10}Me_8$ ligand, and reaction with $Cr(CO)_3(CH_3CN)_3$ leads to exo coordination of the $Cr(CO)_3$ moiety to the free aromatic ring and to formation of the bimetallic derivative **220** (*15*) [Eq. (22)].

$$\tag{22}$$

(C_6Me_6)_2Ru(0) (**197**) (*4*) has been extensively studied. It has the same η^4–η^6 structure in the solid state (*134*) as in solution (*4*). The molecule is fluxional, and a 16-electron intermediate $(\eta^4\text{-}C_6Me_6)_2Ru(0)$ explains the fluxional behavior (*135*), rather than the $(\eta^6\text{-}C_6Me_6)(\eta^3\text{-}CH_2\text{—}C_6Me_5)$-(H)Ru intermediate that was postulated to account for the observed H–D exchange (*136*). Addition of alane to $(C_6Me_6)_2Ru(0)$ has been shown to catalyze a new fluxional process which was suggested to be a 1,3 shift in the η^4-arene ligand (135).

It has been established that the ligands of derivative **196a** are not exchanged with arene up to 97°C (137). When **196a** is irradiated with UV light in the presence of an excess of alkyne, however, cyclotrimerization of the alkyne occurs, leading to ruthenium(0) complexes **221** or **222**, according to the nature of the alkyne: **221** [R = Ph (50%), CO_2Me (20%), CH_2OH)], **222** (R = Me, 40%). The formation of **222** probably involves the corresponding intermediate of type **221**. Phenylacetylene leads to a mixture of $Ru(C_6H_6)(C_6H_3Ph_3)$ isomers. X-Ray analysis of complex **221** (R = Ph) has established the η^4-coordination of the C_6Ph_6 ligand (138) [Eq. (23)].

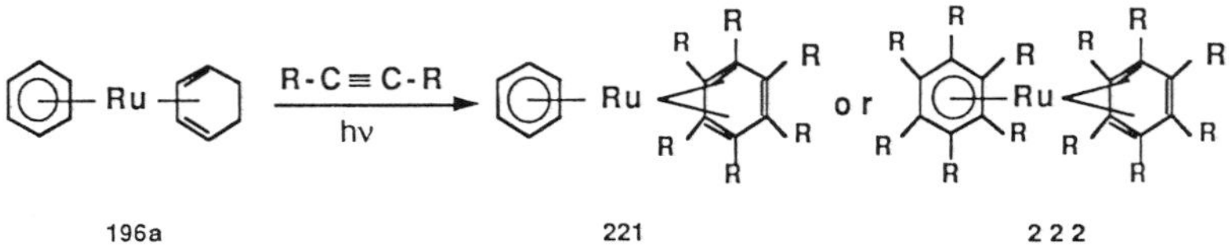

Complex **224** is also obtained by cyclotrimerization of acetylene promoted by complex **223**, with displacement of the bicyclotriene ligand. It is noteworthy that complex **223** is obtained from **200** by simple cycloaddition of alkyne to the cyclooctatriene ligand (139) [Eq. (24)].

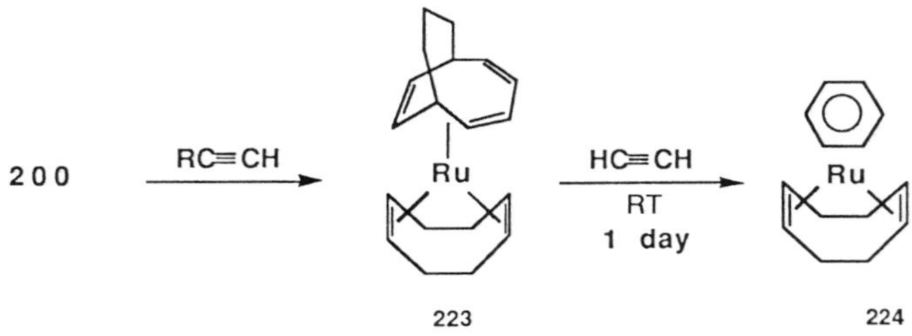

4. *Formation of $Ru^0(arene)(L)_2$ Complexes*

Sixteen-electron ruthenium(0) species of type (η^6-arene)(L)Ru(0) and containing two-electron ligands are probable intermediates for C—H bond activation and formation of metallacyclic complexes (Section II,A,3,c). A variety of 18-electron complexes of general formula (arene)$(L^1)(L^2)$Ru(0) have been prepared by H. Werner and co-workers either by deprotonation of hydride ruthenium(II) complexes or by reduction of cations $RuX(L)_2$-(arene)$^+$. Some of these Ru(0) complexes have already been discussed with the formation of alkyl or hydridoruthenium complexes (Sections

II,A,3,b II,A,3,c). Among deprotonation reactions the formation of complex $Ru(PMe_3)_2(C_6H_6)$ from $[RuCl(PMe_3)_2(C_6H_6)]PF_6$ and t-BuLi is of interest. The first step corresponds to the formation of $[RuH(PMe_3)_2$-$(C_6H_6)]PF_6$, which is then deprotonated by an excess of t-BuLi (49).

A wide variety of (arene)$Ru(PR_3)(L)X^+$ cations of type **225**, containing identical or different phosphorus groups (L = PR_3), for arenes C_6H_6, Me—C_6H_4—i-Pr, and C_6Me_6, have been reduced by sodium naphthalene in THF. Ruthenium(0) complexes **226** (L = PR_3) are isolated in good yield (19,49,106). Similarly, the ethylene complex **225** (L = C_2H_4) is reduced to complex **227** (C_6H_6) (24,50), and the analogous carbonyl derivative **225** (L = CO) affords complex **228** (17,49). All these derivatives are reactive toward electrophiles and are precursors to Ru—H and Ru—CH_2R complexes by reaction with acids or alkyl halides (Sections II,A,3,b and II,A,3,c) [Eq. (25)]. A complex of type **226**, $Ru(PMe_3)_2(C_6H_6)$, has been shown to react with halogen X_2 (Cl_2,Br_2) or Me_3SnCl in the presence of NH_4PF_6 to give back complexes of type **225**, $[RuX(PMe_3)_2(C_6H_6)]^+PF_6^-$ (X = Cl, Br, $SnMe_3$) (49).

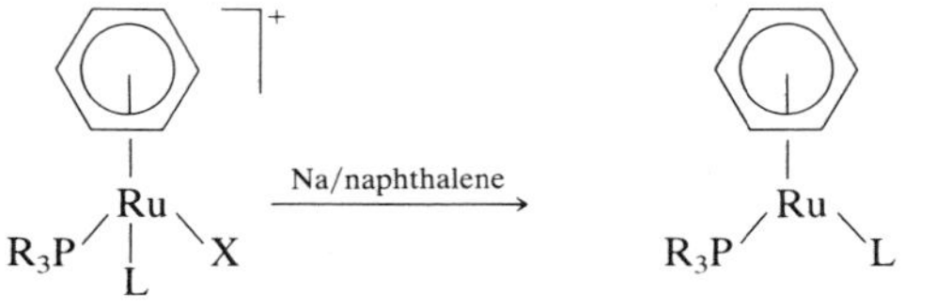

$$ (25) $$

225

226 L = PR_3 (arene)
227 L = CH_2=CH_2 (C_6H_6)
228 L = CO (C_6Me_6)

Reduction with sodium naphthalene of easily obtained diazadiene ruthenium(II) complexes of type **65** (X = Cl) affords (η^6-arene)Ru(0) complexes **64**. The latter are very reactive and undergo oxidative addition with iodine or alkyl halides to produce complexes **65** (X = I and X = CH_2R, respectively) (51) [Eq. (26)]. Other ruthenium(0) complexes have been

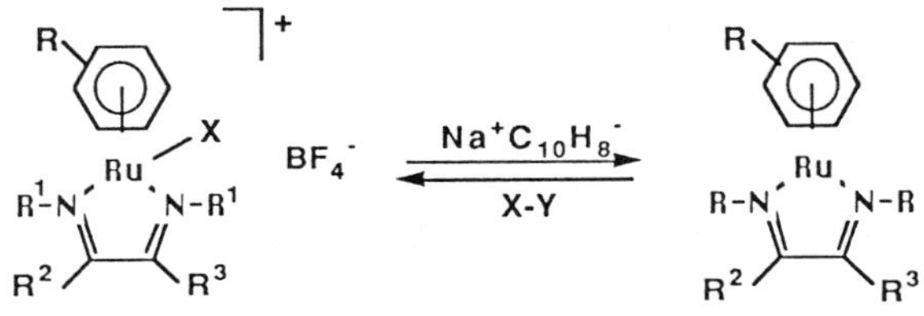

$$ (26) $$

6 5   6 4

arene: C_6H_6; 1,2-Et_2-4-MeC_6H_3
R^2 : H or Me; R^1:iPr; t-Bu; $C_6H_{5-n}R_n$

produced by reaction of the arene ligand of Ru(II) (arene), and they are described in Section IV.

B. *Arene Osmium(0) Complexes*

The first arene osmium(0) complex **229** has been obtained by treating $OsCl_3$ and 1,3-cyclohexadiene with isopropylmagnesium bromide followed by UV irradiation (119).

$$OsCl_3 + \text{1,3-cyclohexadiene} \xrightarrow[\text{2. } h\nu]{\text{1. } i\text{-PrMgBr}} Os(\eta^6\text{-}C_6H_6)(\eta^4\text{-}C_6H_8)$$
$$\textbf{229}$$

The synthesis and reactivity of arene cyclooctatetraene osmium(0) complexes of type **216** (arene = benzene, mesitylene) (13,132) and arene dicyclopentadiene osmium(0) complexes **212** (arene = mesitylene) (102) have been discussed previously (Section III,A,2). Arene osmium(0) complexes of type $Os(\eta^6\text{-}C_6H_6)(PR_3)(L)$ (**230**) have been prepared by reduction

$$OsI(C_6H_6)(PR_3)L^+PF_6^- \xrightarrow[\text{THF, } -78°C]{2 \text{ NaC}_{10}\text{H}_8} Os(C_6H_6)(PR_3)(L)$$
$$\textbf{230}$$

$$PR_3 = L = PPh_3$$
$$PR_3 = L = P(OMe_3)$$
$$PR_3 = PMe_3, L = CO, CNR, C_2H_4, C_3H_7$$

of the corresponding cationic osmium(II) complexes with sodium naphthylide in THF (*49,50,106,108*). Complex **230** (PR_3 = P-*i*-Pr_3, L = CO) has also been obtained by treatment of the hydrido complex with sodium hydride in THF (*114*).

$$Os(C_6H_6)(CO)(P\text{-}i\text{-}Pr_3)H \xrightarrow[\text{THF}]{\text{NaH}} Os(C_6H_6)(P\text{-}i\text{-}Pr_3)(CO)$$
$$\textbf{230}$$

The osmium complexes of structure **230** are effective reagents for C—H bond activation of benzene, and the nature of the arene ligand has been shown to determine intra-versus intermolecular C—H addition. Complex **230** (PR_3 = PMe_3, L = C_2H_4) reacts with an excess of PMe_3 in benzene to give the hydridophenylosmium(II) complexes **231** (*140*) (Scheme 17). This reaction occurs by *intramolecular* C—H addition of the benzene ligand as shown by reaction in deuterated benzene. In contrast, reaction of the analogous *p*-cymene complex with PMe_3 in deuterated benzene gives the deuteriophenyl complex **232** by *intermolecular* C—H addition (*111*).

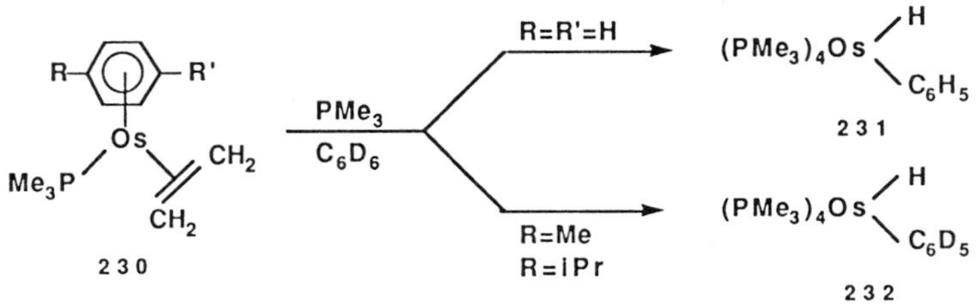

SCHEME 17

Vinylidene osmium(0) (**187**) (Section II,B,3,g) reacts with sulfur, selenium, or copper chloride to give complexes **233** via electrophilic addition to the osmium–carbon bond (*116*) (Scheme 18). Complex **187** also reacts with benzoylazide to form the five-membered metallacycle **234** (*Z*) which isomerizes **234** (*E*) on heating in benzene (*141*).

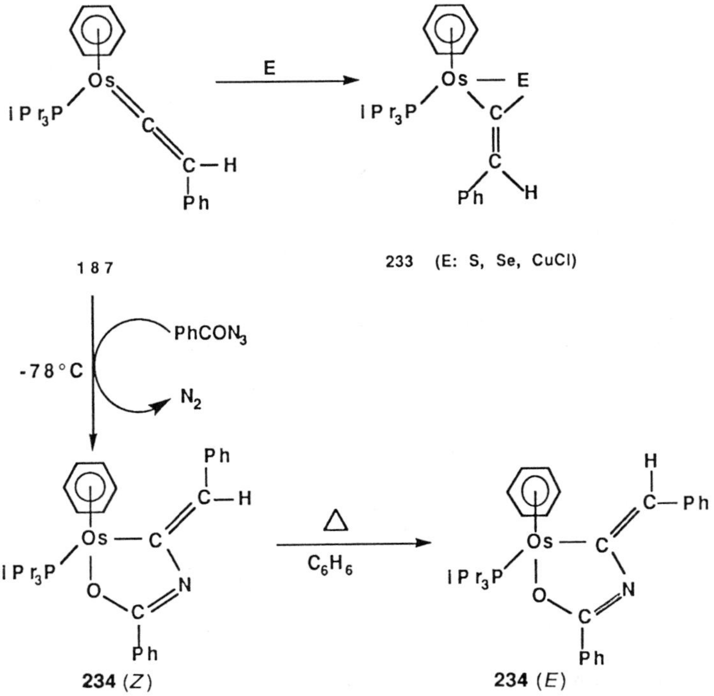

SCHEME 18

IV

REACTIVITY OF THE ARENE LIGAND

A. Access to η^5-Cyclohexadienyl Ruthenium(II) and Osmium(II) Complexes

η^5-Cyclohexadienyl ruthenium complexes have been obtained either by addition of nucleophiles to the arene ring of arene ruthenium(II) complexes or by protonation of ruthenium(0) complexes. The first complex prepared, the benzene cyclohexadienyl ruthenium cation **236a**, has been obtained together with the zero-valent arene cyclohexadiene ruthenium(0) complex **196a**, by reaction of **235a** with lithium aluminum hydride (*118*) [Eq. (27)].

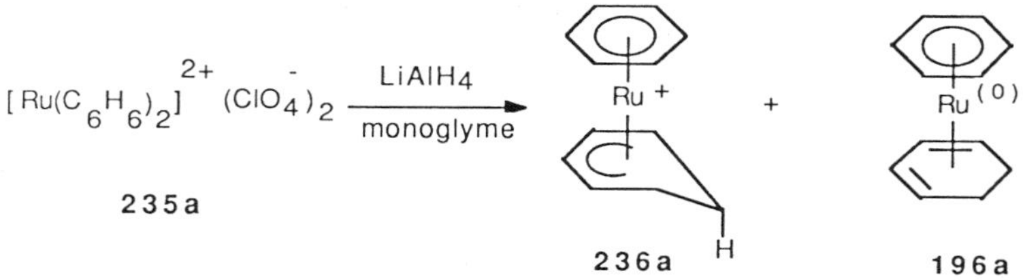

$$[Ru(C_6H_6)_2]^{2+} (ClO_4)_2^{-} \xrightarrow[\text{monoglyme}]{\text{LiAlH}_4}$$

235a

236a   **196a**   (27)

Mesitylene and hexamethylbenzene cyclohexadienyl ruthenium complexes **236** have also been prepared in good yield by reduction of their corresponding bisarene ruthenium dications **235** with sodium borohydride in water (*142,143*). Cyclohexadienyl derivative **236c** can be easily formed by treatment of bisarene ruthenium(0) complex **197** (arene = hexamethylbenzene) with a solution of hydrochloric acid in acetone. The structure of **236c** and the presence of the endo C—H bond have been clearly resolved by infrared and ^{1}H-NMR spectra (*144,145*) [Eq. (28)].

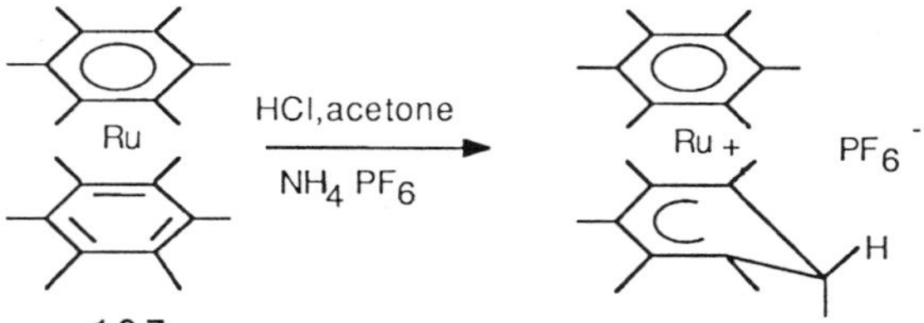

$$\xrightarrow[\text{NH}_4\text{PF}_6]{\text{HCl,acetone}}$$

197   **236c** PF_6^{-} (28)

Other nucleophiles such as phosphines and trialkylphosphites undergo nucleophilic addition to bisbenzene ruthenium and osmium dications **235a** and **142** to yield the cyclohexadienyl phosphonium adducts **237** and **238**.

Kinetic studies have shown that electrophilicity in the iron triad is strongly metal dependent with Fe $\gg$ Ru, Os, and the nucleophilic reactivity order is $PPh_3 > P(O\text{-}tBu)_3$. Adducts **237** ($PR_3$ = phosphites) react with water to give the cyclohexadienyl phosphonate complexes **239**. Complex **235** is a effective catalyst for the conversion of phosphites to $HP(O)(OR)_2$ (*99,146,147*) [Eq. (29)]. In a similar fashion, benzene ruthenium dications

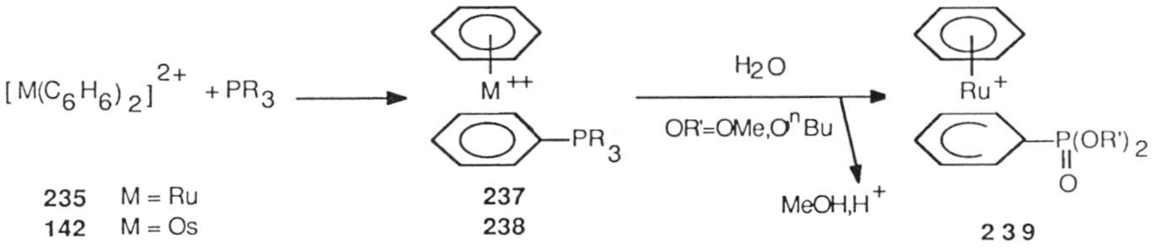

$$[M(C_6H_6)_2]^{2+} + PR_3 \longrightarrow \mathbf{237} \quad \mathbf{238} \xrightarrow[OR'=OMe,O^nBu]{H_2O} \mathbf{239}$$

235 M = Ru	**237**
142 M = Os	**238**

$$\text{(29)}$$

240 react with nucleophiles such as hydride, hydroxide, and cyanide to give the corresponding stable cyclohexadienyl ruthenium complexes **241**. No reaction with phosphines is observed (*21,148*) [Eq. (30)].

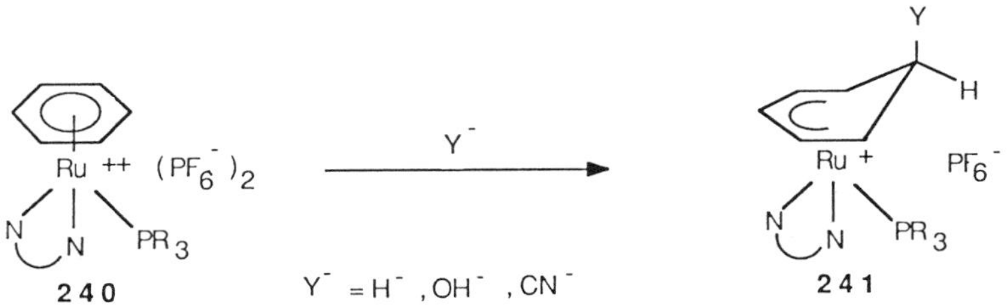

$$\mathbf{240} \quad (PF_6^-)_2 \xrightarrow{Y^-} \mathbf{241} \quad PF_6^- \qquad \text{(30)}$$

$$Y^- = H^-, OH^-, CN^-$$

Other nucleophiles such as methyllithium and triethylamine also yield cyclohexadienyl ruthenium complexes **241** and **242a** (L = PMe$_3$) (Scheme 19). When L is acetonitrile, reaction with trimethylphosphine gives the cyclohexadienyl ruthenium complex **242b**, resulting from both nucleophilic addition to the benzene ring and substitution of two ligands L. Further treatment with trifluoroacetic acid leads to elimination of the phosphonium group to produce **15** (*149,150*). When one ligand is ethylene, the formation of cyclohexadienyl ruthenium complexes **241** and **242a** is also observed with nucleophiles such as methyllithium, sodium methoxide, and triethylamine. In contrast, addition of phosphines and phosphites is observed only on the olefin ligand (**66** $\rightarrow$ **243**) (*149*).

$$[Ru(C_6H_6)(PMe_3)_2(C_2H_4)]^{2+} \xrightarrow[CH_3NO_2]{PR_3} [Ru(C_6H_6)(PMe_3)_2(C_2H_4PR_3)]^{2+}$$

66	**243**

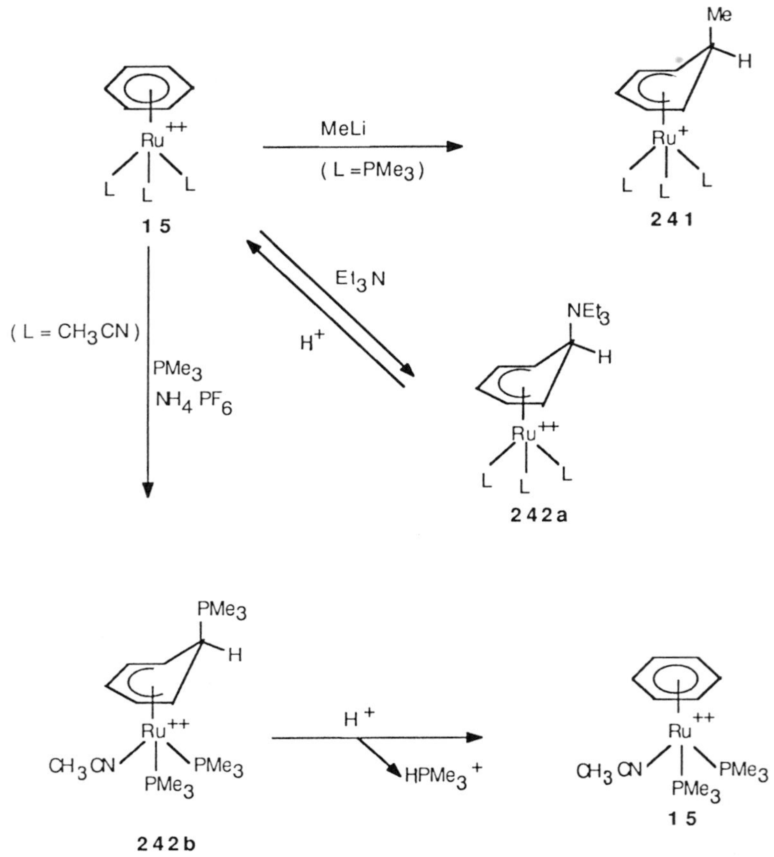

SCHEME 19

Monocations of type **5** react with 2 equiv of alkyl phosphines to give the phosphonio cyclohexadienyl ruthenium(II) compounds (**242a**). The same compounds can be obtained by addition of 3 equiv of alkylphosphines to the neutral compounds of type **3**. The IR and NMR spectra of **242a** indicate that the phosphonio group is in the exo position at the sp^3 carbon atom of the cyclohexadienyl ring. Compounds **242a** react with CF_3CO_2H via elimination of the phosphonio group to yield the hexafluorophosphate salts of the half-sandwich complexes of type **15** (*151*) [Eq. (31)].

The neutral complex **244** (arene = benzene) has been found to react with nucleophiles such as hydride, hydroxide, and cyanide to give cyclo-hexadienyl derivatives, but they were too unstable to be isolated (*7*). The complex $[Ru(C_6H_6)Cl(PMe_3)_2](PF_6)$ (**5**) reacts with methyllithium to give

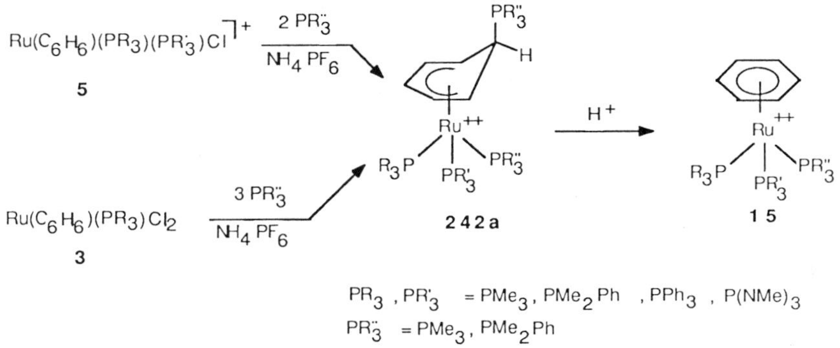

$$(31)$$

a mixture of products that contains complex **244** as the main component (Scheme 20). Reaction of **5** with LiC_6H_5 in the presence of LiBr gives first bromide compound **245** and then the complex **246** (*152*).

Similar nucleophic addition is observed with the analogous osmium complex **147** (Scheme 21). X-Ray structural analysis of **247** (R = n-C_4H_9) shows that the n-butyl group occupies the exo position (*152*). Complexes of type **247** react smoothly with trityl cation in acetone to give the arene

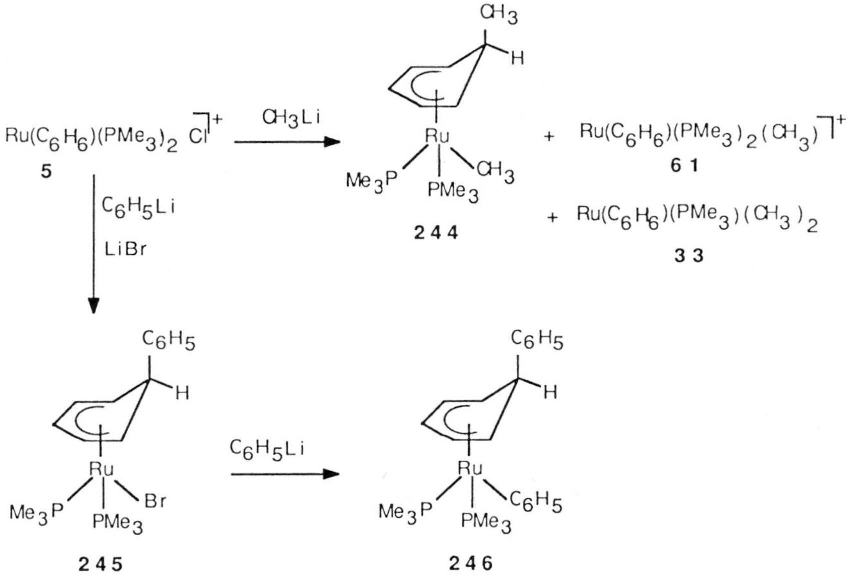

SCHEME 20

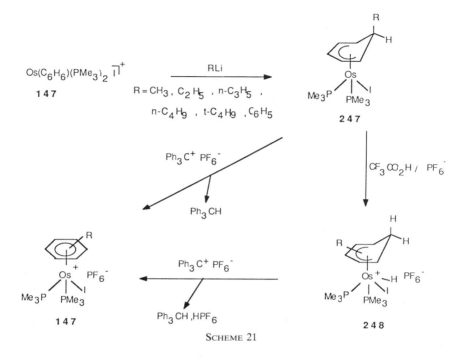

SCHEME 21

osmium(II) complexes **147** in quantitative yield. The hydride elimination proceeds via (cyclohexadienyl)hydridoosmium(IV) intermediate **248**, which can be prepared from **247** and CF_3CO_2H in the presence of NH_4PF_6 (*153*).

Benzene cyclopentadienyl ruthenium(II) complex **125** undergoes nucleophilic addition at the arene ligand via the addition of sodium borohydride or phenyllithium. Reaction with phenyllithium gives the exo-phenyl cyclohexadienyl derivative **249** in 89% yield (*154*) [Eq. (32)].

B. *Access to Diene Ruthenium(0) Complexes*

The synthesis of the zero-valent, 1, 3-cyclohexadiene benzene ruthenium complex **196a** has been mentioned as a coproduct of the cyclohexadienyl complex **236a** in the reduction of the benzene ruthenium dication **235** with lithium aluminum hydride. Reduction of **235** with sodium borohydride in THF, however, gives only the air-sensitive, yellow–green ruthenium(0) complex **196a** (*118*). This reaction has been generalized to

$$[Ru(C_6H_6)_2]^{2+}(ClO_4^-)_2 \xrightarrow[THF]{NaBH_4} Ru(\eta^6\text{-}C_6H_6)(\eta^4\text{-}C_6H_8)$$
$$\textbf{235} \qquad\qquad\qquad\qquad\qquad \textbf{196a}$$

other bisarene ruthenium complexes of type **235**. Reduction of the mixed substituted arene complexes **235** with sodium borohydride in THF leads to formation of neutral derivatives **196a** in good yield. A mixture of isomers A and B is obtained. The less alkylated benzene ring is more easily hydrogenated. Thus, reduction of **235** (R = 1,3,5-$(CH_3)_3$, R' = H) results in formation of isomer A exclusively (*142*) [Eq. 33)].

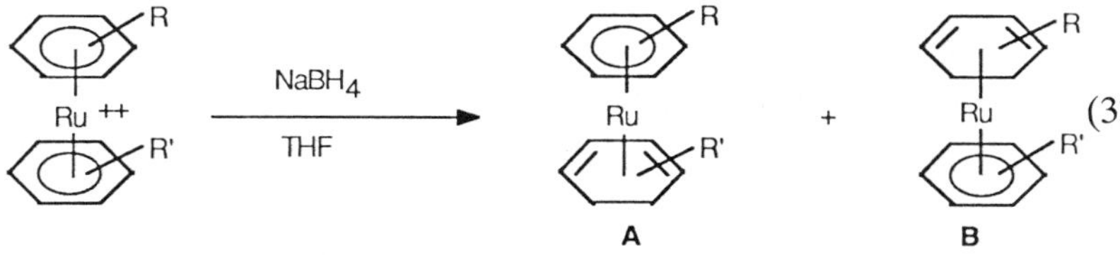

$$R = R' = 1,3,5\text{-}(CH_3)_3$$
$$R = 1,3,5\text{-}(CH_3)_3 ; R' = H$$
$$R = 1,4\text{-}(CH_3)_2 ; R' = H$$
$$R = CH_3 \quad ; R' = H$$

The analogous complex **196c** (arene = hexamethylbenzene) has been surprisingly obtained by treatment of the cyclohexadienyl complex **236c** with sodium borohydride in boiling 1,2-dimethoxyethane. In contrast, when the same complex is treated with Red-Al in THF at 0°C, the biscyclohexadienyl ruthenium(II) complex **250** is obtained (*144*) [Eq. (34)]. In a similar way, reduction of **236a** with Red-Al in toluene gives the cyclohexadiene ruthenium(0) complex in good yield. This reaction has been used to reduce selectively benzene to cyclohexene by protonation of **196a**. In the presence of benzene, bisbenzene ruthenium dication **235** can be regenerated (*143*).

Another type of diene ruthenium(0) has been obtained by deprotonation of two hexamethylbenzene ruthenium(II) complexes. Complex **235c** in

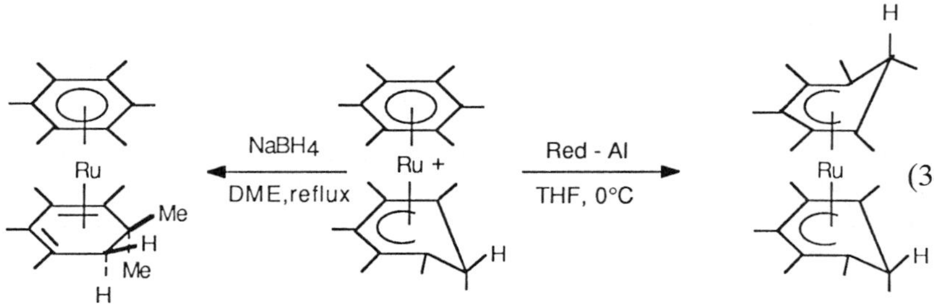

(34)

196c   **236c**   **250**

$$[Ru(\eta^6\text{-}C_6H_6)(\eta^5\text{-}C_6H_7)]^+ \xrightarrow{\text{Red-Al}} Ru(\eta^6\text{-}C_6H_6)(\eta^4\text{-}C_6H_8)$$

236a   **196a**

235

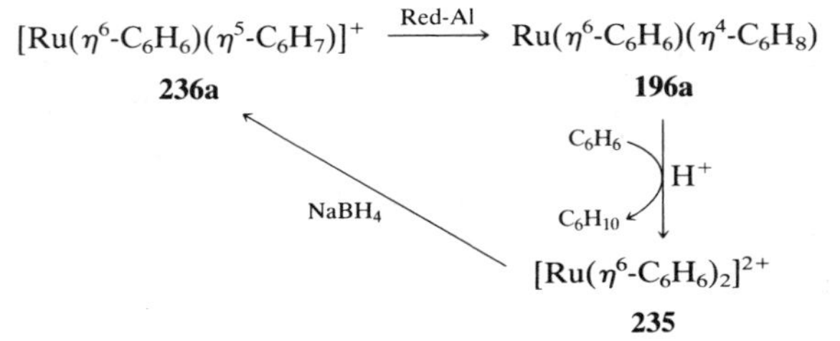

the presence of an excess of *t*-BuOK is deprotonated to give **251** in 87% yield (Scheme 22). The X-Ray structure of **251** shows that the C_6Me_4-$(CH_2)_2$ ligand is η^4-coordinated to the ruthenium(0) atom via its *intracyclic*

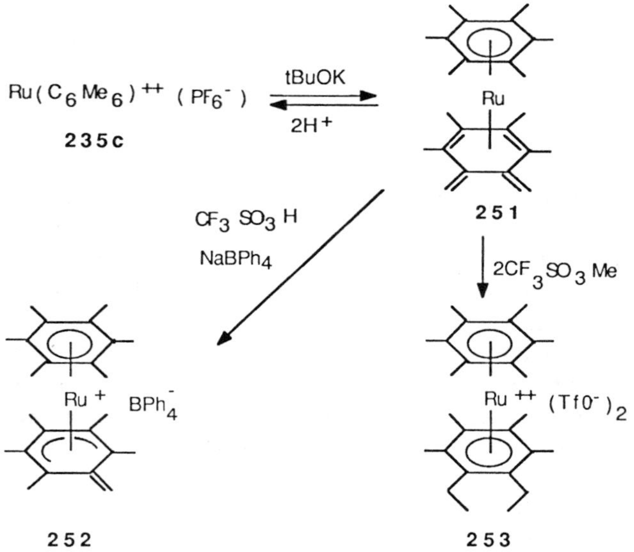

SCHEME 22

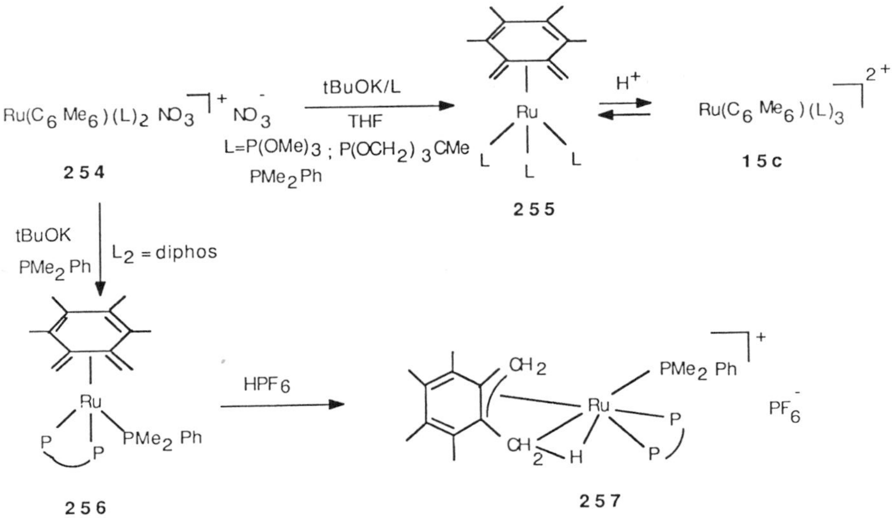

SCHEME 23

diene moiety. Treatment **251** with 1 equiv of triflic acid gives the mono-protonated species **252**, and 2 equiv of methyl triflate affords the dialkylated species **253** (155).

The cationic hexamethylbenzene ruthenium complex **254** can be deprotonated by treatment with *t*-BuOK to give the ruthenium complexes **255** and **256** in which the η^4-*o*-xylylene ligand is coordinated via its *exocyclic diene* (Scheme 23). Protonation of **255** gives the dication **15c**, whereas protonation of **256** produces the fluxional η^3-benzyl derivative **257** which is stabilized by an agostic C—H bond (156,157).

C. Nucleophilic Substitution

The halogens (Cl, F) in the arene ligand of **125** can be smoothly substituted by nucleophilic groups such as MeO, PhS, OH, CN, or $C_5H_{10}N$ (148). Nucleophilic substitution of halogens in the arene ligand of **125** has been used to prepare aromatic organoruthenium polymers such as **258** or **259** (Scheme 24). The metal-free polymer is obtained by arene displacement in acetonitrile or dimethyl sulfoxide (158). A new method for indole

$$[Ru(\eta^6\text{-}C_6H_5Cl)(\eta^5\text{-}C_5H_5)]^+ + MX \xrightarrow{\text{NaBF}_4 \text{ or NaBPh}_4} [Ru(\eta^6\text{-}C_6H_5X)(\eta^5\text{-}C_5H_5)]^+$$

125   **125**

MX = NaOH, PhSNa, NaCN

MX = MeOH/Na_2CO_3, $C_5H_{10}NH$

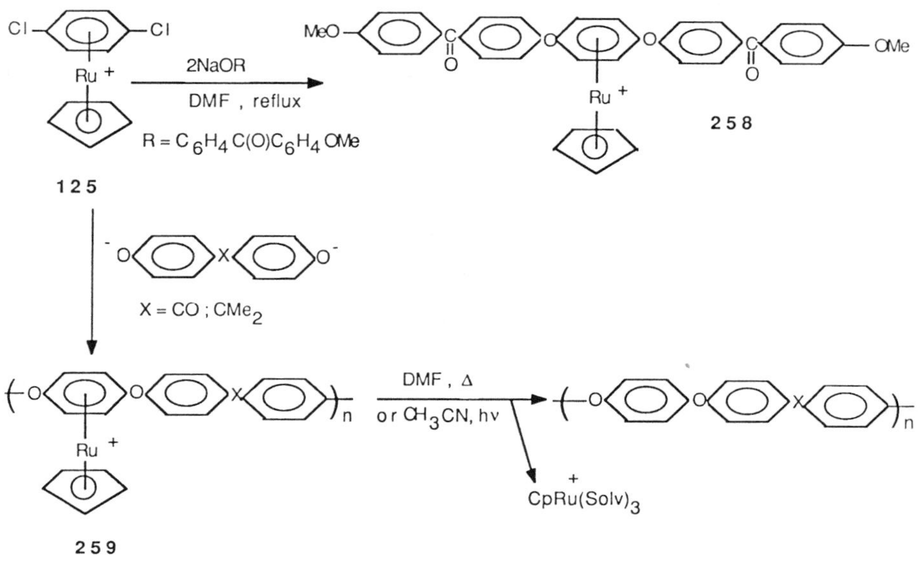

SCHEME 24

functionalization has been exemplified by the smooth nucleophilic aromatic substitution of (4- or 5-chloroindole)(cyclopentadienyl)ruthenium-
(II) derivatives (**260** → **261**) (*78*) [Eq. (35)].

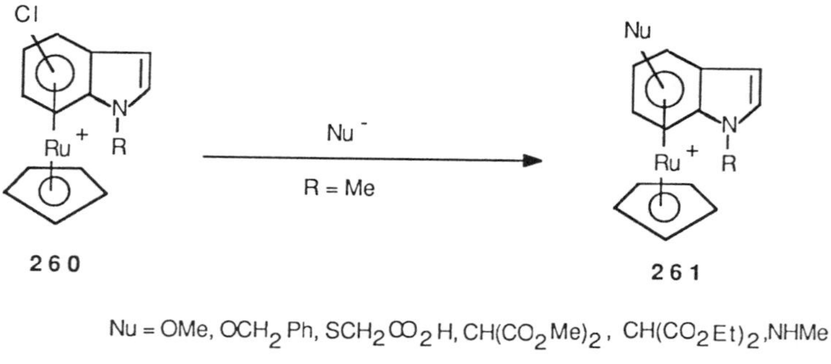

(35)

$Nu = OMe, OCH_2 Ph, SCH_2 CO_2 H, CH(CO_2 Me)_2, CH(CO_2 Et)_2, NHMe$

D. Oxidation

The central metal atom of **125** (R = CH$_3$ and R = SPh) is inert toward
potassium permanganate. Instead ligand substituents are oxidized (*148*).

$$Ru(C_6H_5R)(C_5H_5)^+ + KMnO_4$$

125

R = CH$_3$ R = SPh

$$Ru(C_6H_5CO_2H)(C_5H_5)^+ \qquad\qquad Ru(C_6H_5SO_2Ph)(C_5H_5)^+$$

V

CYCLOPHANE RUTHENIUM COMPLEXES

[2n]Cyclophane ruthenium complexes are of interest as possible mono-meric components in the construction of multilayered organometallic polymers such as **262**, which might show extended π-electron delocaliza-tion (Scheme 25).

A. Synthesis

Cyclophane ruthenium complexes have been prepared by the methods developed to make arene ruthenium complexes (Section II,A). Treatment of the solvated complexes of type **7**, obtained from the dimers **1** and silver tetrafluoroborate in acetone, with 1 equiv of cyclophanes **a–j** in the presence of trifluoroacetic acid gives the double-layered (η^6-arene)-(η^6-cyclophane)ruthenium(II) complexes **263** in good yield (Scheme 26, p. 222). When **7** is used in excess, the triple-layered complex **264** is formed (159–162).

A bis(cyclophane)ruthenium(II) complex has been prepared by using Bennett's procedure: reaction of diene **265**, obtained by Birch reduction of 4,5,7,8-tetramethyl[2$_2$](1,4)cyclophane **266**, with ruthenium chloride gives the dimeric chloride complex **267** (Scheme 27, p. 223). Treatment of the solvated complex **268** with **266** in the presence of trifluoroacetic acid leads to **269** (163). The structures of complexes **267**, **268**, and **269** are based on

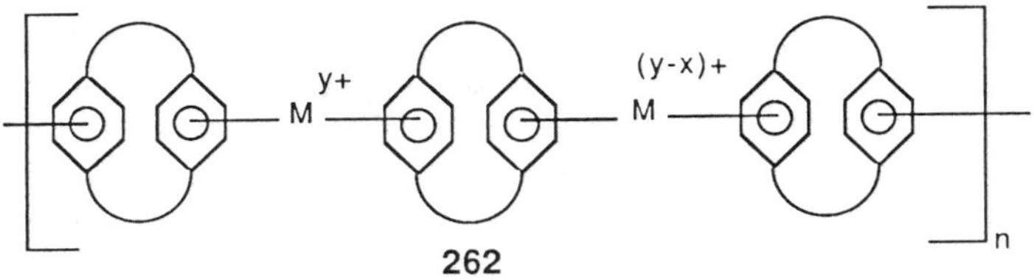

262

SCHEME 25

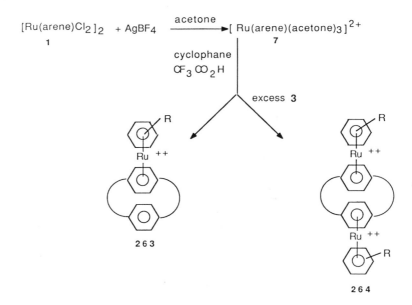

$$[Ru(arene)Cl_2]_2 \; + \; AgBF_4 \xrightarrow{\text{acetone}} [\,Ru(arene)(acetone)_3\,]^{2+}$$

1   7

cyclophane

$CF_3 CO_2 H$

excess **3**

263   **264**

[2$_n$]cyclophane : **a**, [2$_2$] (1,4) cyclophane ; **b**,anti[2$_2$] (1,3) cyclophane
c, [2$_3$] (1,3,5) cyclophane ; **d** [2$_3$] (1,2,4)(1,2,5) cyclophane
e, [2$_4$] (1,2,3,4)cyclophane ; **f**, [2$_4$] (1,2,3,5) cyclophane
g, [2$_5$] (1,2,4,5) cyclophane ; **h**, [2$_5$] (1,2,3,4,5) cyclophane
I,[2$_6$] (1,2,3,4,5,6)cyclophane ;**j**,triple layered [2$_2$]cyclophane

SCHEME 26

^{1}H- and ^{13}C-NMR analysis which rules out the expected structure **270**. The formation of **267** is surprising and could result from a ligand exchange reaction leading to the more thermodynamically stable product **267**.

A more general route to make bis(cyclophane)ruthenium(II) complexes involves a reduction of **263** (arene = benzene) with Red-Al to afford the [η^4-1,3-cyclohexadiene)(η^6-cyclophane)]ruthenium(0) derivatives **271** (Scheme 28, p. 224). Treatment of **271** with hydrochloric acid gives the dimeric chloride complexes **272**, which lead the desired bis(η^6-[2$_n$]cyclophane)ruthenium(II) complexes **274** via Bennett's procedure (*145*). Synthesis of the oligomer **275a** is also achieved in quantitative yield by heating **274** with the solvated complex **7** (arene = C_6Me_6) in neat trifluoroacetic acid.

An attempt to achieve the complexation of the cyclophane iron complex **276** by treatment with the solvated (*p*-cymene)ruthenium derivative **7** results in disruption of the arene–iron bond and formation of the (*p*-cymene)[[2$_2$](1, 4)cyclophane]ruthenium(II) salt **277** as the only-product (*164*) (Scheme 29, p. 224).

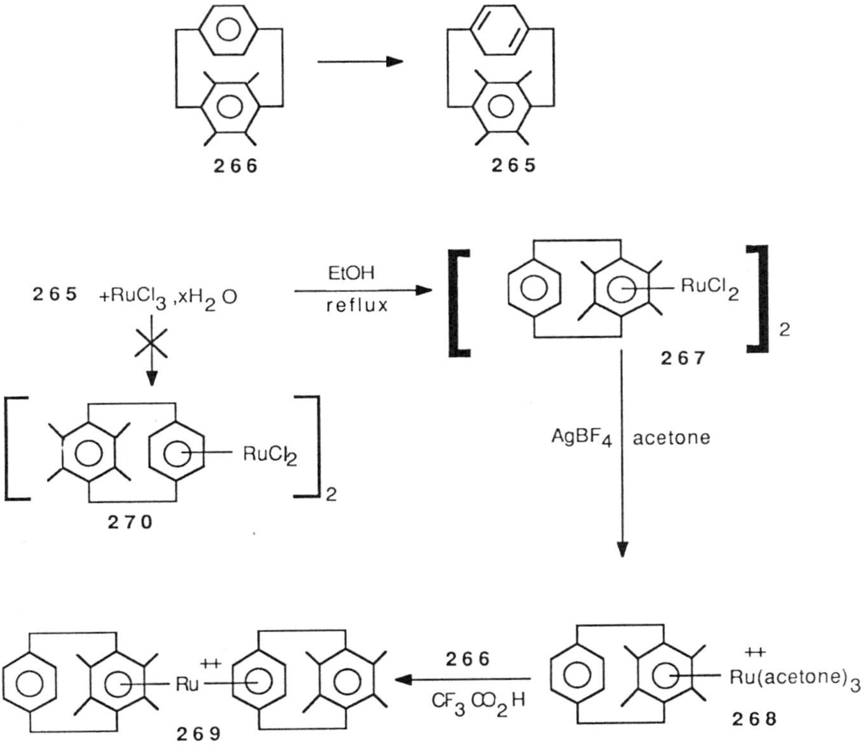

SCHEME 27

A cationic ruthenacyclophane complex **282** has been prepared by a ligand exchange reaction in trifluoroacetic acid (Scheme 30, p. 225). The dimer **281** is obtained by an indirect method in three steps starting from the dimer **1** (arene = benzene) and the arene **278**. Refluxing a solution of **281** in trifluoroacetic acid results in intramolecular exchange of chloride ligands for the arene fragment, leading to the metallocyclophane **282** in 90% yield (*165*).

B. Reduction of Cyclophane Complexes

1. Cyclic Voltammetry and Reduction of Double-Layered Cyclophane Complexes

An electrochemical study of the reduction of the double-layered (arene)-(cyclophane)ruthenium complexes **263** and of bis(arene)ruthenium complexes **235c** has been reported (*160,166*). The compounds show in

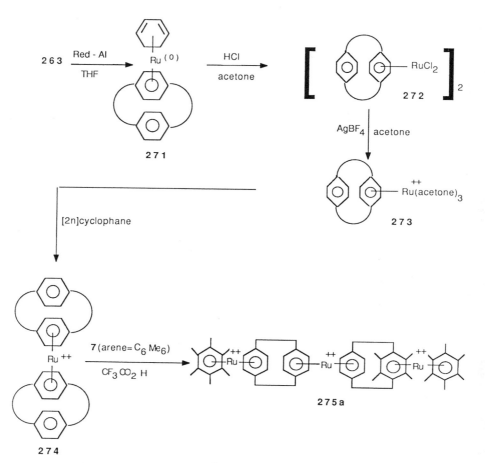

SCHEME 28

SCHEME 29

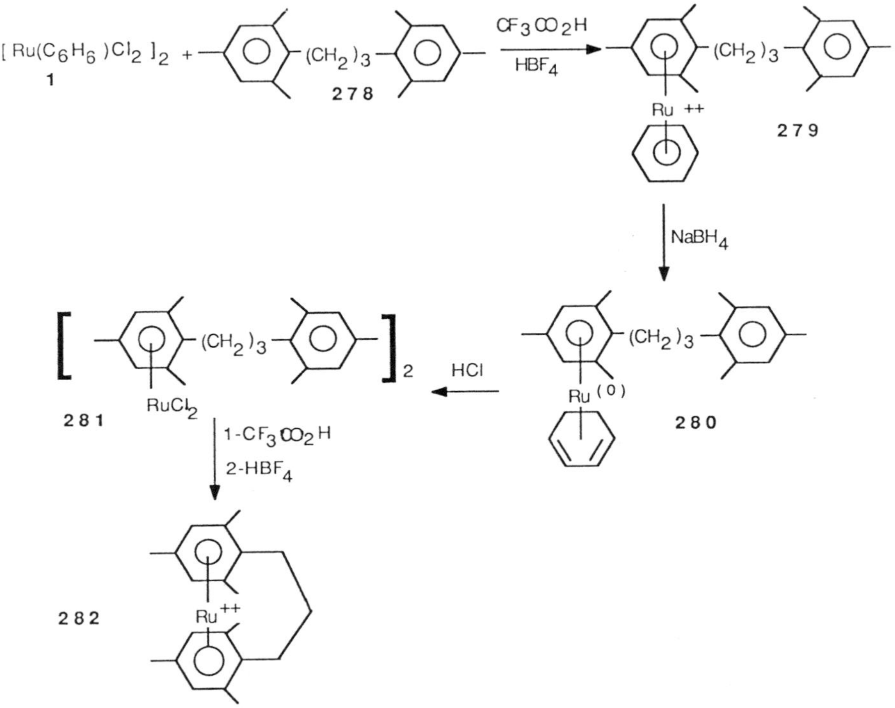

SCHEME 30

acetonitrile a two-electron wave for **235c** and either two one-electron waves or a two-electron wave for **263** which can be separated by a CH_3CN-to-CH_2Cl_2 solvent change. $(C_6Me_6)_2$ruthenium(II) complex **235c** appears to be reduced at higher potential than the cyclophane ruthenium complexes **263**. The chemical reversibility of the oxidoreduction process (arene)$_2$ $Ru^{2+} + 2e^- \rightarrow$ (arene)$_2Ru^0$(i.e., the i_a/i_c ratio) is a function of the cyclophane ligand structure (Table I, p. 226).

Chemical reduction of $[\eta^6$-$[2_2](1,4)$cyclophane][η^6hexamethylbenzene]-ruthenium(II) compound **263** is achieved with aluminum in an aqueous sodium hydroxide–hexane mixture and gives (η^4-$[2_2](14)$cyclophane)-(η^6-hexamethylbenzene)ruthenium(0) **283** (Scheme 31, p. 226). Treatment of complex **283** with concentrated hydrochloric acid gives the arene cyclo-hexadienyl derivative **284**. Reduction of **284** places an hydrogen exo on the C_6Me_6 ring to give the bis(cyclohexadienyl)ruthenium(0) complex **285**. For complex **284** it is proposed that the choice of hydrogen transfer to the

TABLE I

Cyclic Voltammetry of Some Double-Layered Ruthenium Complexes

Arene	Arene	$E_{1/2}$(V versus SCE)	i_a/i_c	Ref.
HMB[a]	C_6Me_6	-1.02^b	0.36	*160*
HMB[a]	$[2_2](1,4)$cyclophane	-0.69^b	0.94	*160*
HMB[a]	$[2_2](1,3)$cyclophane	-0.71^b	0.81	*160*
HMB[a]	$[2_3](1,3,5)$cyclophane	-0.91^b	0.27	*160*
HMB[a]	$[2_4](1,2,3,4)$cyclophane	-0.82 and -0.93^b	0.79	*160*
HMB[a]	$[2_2](1,4)$cyclophane$[2_2](1,4)$cyclophane	-0.616^c	0.95	*167*

[a] HMB, Hexamethylbenzene.
[b] In CH_3CN.
[c] In propylene carbonate.

cyclophane ligand rather than to the hexamethylbenzene ligand is determined by relief of strain on the cyclophane ring (*145,164–167*). This is supported by a similar experiment employing $[2_4](1,2,4,5)$cyclophane. Treatment of ruthenium(0) complex **286** with hydrochloric acid gives the (η^5-cyclohexadienyl)ruthenium(II)cation **287** (Scheme 32). In this case, hydrogen transfer occurs preferentially to the hexamethylbenzene ligand (*145*).

Monolayered cyclophane complexes of type **263** are also reduced by sodium bis(methoxyethoxy)aluminum hydride (Red-Al) to give (η^4-diene)-(η^6-cyclophane)ruthenium(0) complexes (Scheme 33). If the benzene ring of **263** (arene = benzene) is converted to the (1,3-cyclohexadiene)-ruthenium(0) derivative **271**, however, when the corresponding η^6-hexamethylbenzene is reduced with Red-Al, the product is the (η^4-1,4-cyclohexadiene)ruthenium(0) complex **288**. Synthesis of **271** can

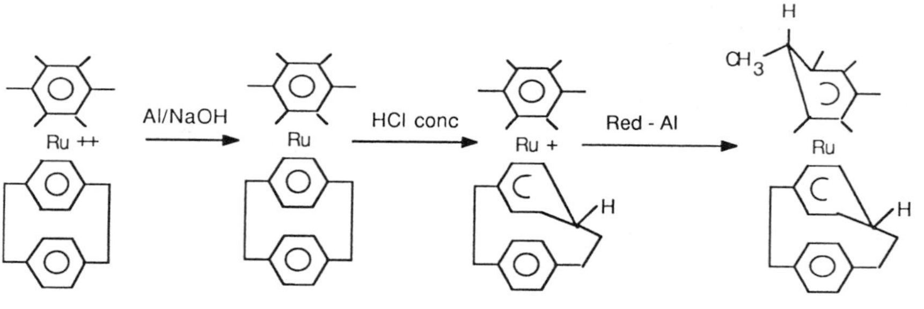

SCHEME 31

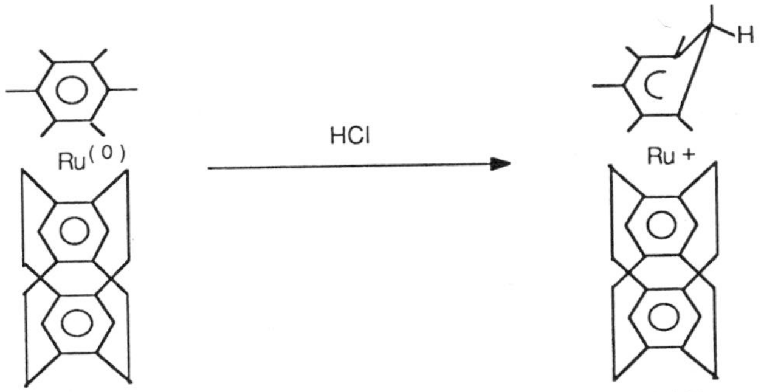

SCHEME 32

be explained by the initial formation of a 1,4-cyclohexadiene moiety which can equilibrate to the more stable 1,3-cyclohexadiene via a hydrido-ruthenium intermediate (*145*).

2. Cyclic Voltammetry and Reduction of Triple-Layered Cyclophane Complexes

The properties of model subunits of **262**, such as **264g** and **264j** (Table II), have been examined to see if they exhibit mixed-valence ion behavior, which is a requirement for complete π-electron delocalization in **262**. Cyclic voltammetry of **264g** in acetone shows a reversible net two-electron reduction wave giving the mixed valence complex **289g** and a second reversible net two-electron wave reducing **289g** to **290g** (Scheme 34, p. 228). The cyclic voltammogram of **264j** also shows two reversible, but

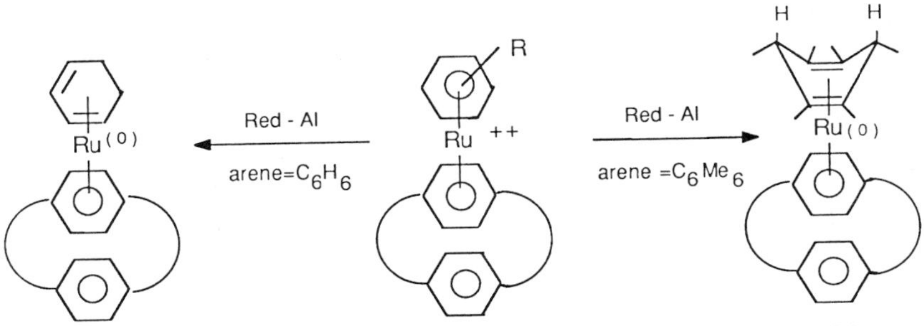

SCHEME 33

TABLE II

CYCLIC VOLTAMMETRY OF TRIPLE-LAYERED RUTHENIUM COMPLEXES **264g** AND **264j**

	$E_{1/2}^1$(V versus SCE)	$E_{1/2}^2$(V versus SCE)	$\Delta E_{1/2}$(mV)	K_c	Ref.
264g	-0.060^a	-0.593^b	533	1.07×10^{18}	*161*
264j	-0.698^a	-0.796^b	98	2.1×10^3	*162*

^{a} In acetone.
^{b} In propylene carbonate.

not completely separated, reduction waves. Comparison of the com-proportionation constant k_c for the equilibrium **264** + **280** $\rightleftharpoons$ **2 (289)** and of the $\Delta E_{1/2}$ *values between* **264g** and **264j** shows that the interaction between the two Ru atoms in **264g** is much stronger than that in **264j** (*161,162*).

Reduction of complex **264g** to **289g** is achieved either electrochemically or by use of an electron reservoir compound of the type bis(C_6Me_6)-ruthenium(0). A further two-electron reduction of **289g** gives the neutral ruthenium(0) derivative **290g**. Similarly, chemical reduction of **264j** with 2 equiv of bis(C_6Me_6)ruthenium(0) gives the corresponding ruthenium(0) derivative **290j**, but attempts to isolate the mixed-valence complex **289j** in pure form have failed. It is shown by a variable-temperature ^{1}H-NMR study that compound **290g** is a fluxional molecule. Complex **290g** is the first mixed-valence ion class II compound exhibiting a net two-electron interva-lence transfer (*161,162*).

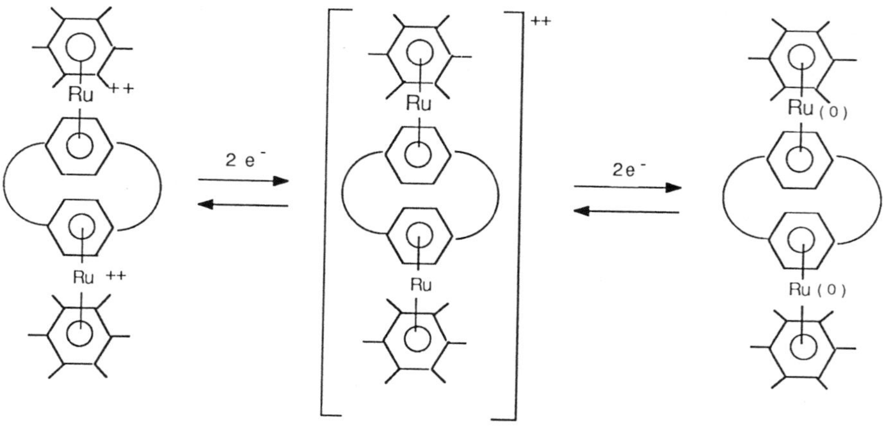

264g   **289g**   **290g**

SCHEME 34

VI

RUTHENIUM AND OSMIUM COMPLEXES AND
POLYMETALLIC DERIVATIVES

η^6-Arene ruthenium and osmium complexes have been involved recently in several aspects of cluster chemistry. (1) They have been shown to be convenient sources of a ruthenium moiety, by loss of the arene, for the preparation of heteropolymetallic complexes. (2) They have been used as an arene–metal moiety to produce (η^6-arene)Ru–or (η^6-arene)Os–metal derivatives. (3) Arene ruthenium or osmium clusters have been obtained either from arene–metal mononuclear complexes or from direct incorporation of the arene. (4) The cocondensation of arene with metal vapor has allowed access to arene-containing clusters.

A. Formation of Mixed-Metal Derivatives from
Arene Ruthenium Complexes

(η^6-Arene)ruthenium(II) complexes have been used for providing a ruthenium moiety in the building of hetero bi- or polymetallic systems. Various phosphine complexes of type **291** are partially dehalogenated by $Fe_2(CO)_9$ to afford $FeRu_2$ clusters **292**. The X-ray structure of one of these (PR_3 = Ph_2P—$C{\equiv}C$—t-Bu) shows Fe—Ru bonding and Ru$\cdots$Ru nonbonding distances of 2.808(3) and 3.185(2) Å, respectively (*168*) [Eq. (36)].

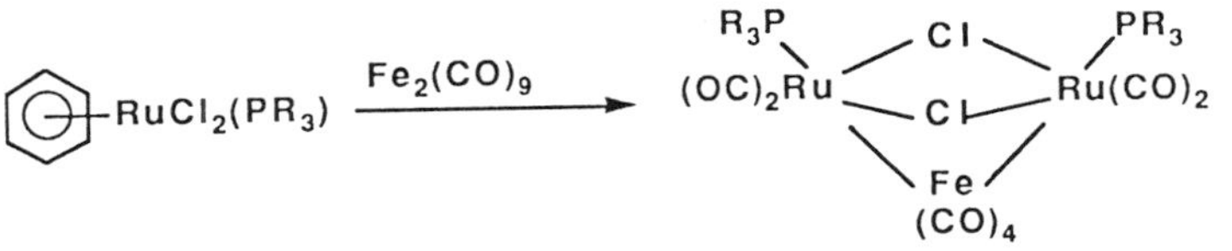

$$\text{(36)}$$

291   **292**

PR_3: $P(OMe)_3$ (19%); PPh_3 (40%); PMe_2Ph(10%); PMe_3 (33%)

$Ph_2PC{\equiv}C$-t-Bu (30%)

The dehalogenation by $Fe_2(CO)_9$ of binuclear arene ruthenium(II) derivatives **293** and **296** depends on the nature of the bridging diphosphine. With $Ph_2PCH_2CH_2PPh_2$, complexes with Ru_2Fe [**294** (20%)] and Ru_3 [**295** (12%)] are formed. With $Ph_2PCH_2PPh_2$ complete dehalogenation is favored, and complexes **297** (37%) and **298** (15%) are obtained. The X-ray structure of the latter shows that the phosphorus ligand is located almost in the Ru_3 plane (*169*) [Eqs. (37) and (38)].

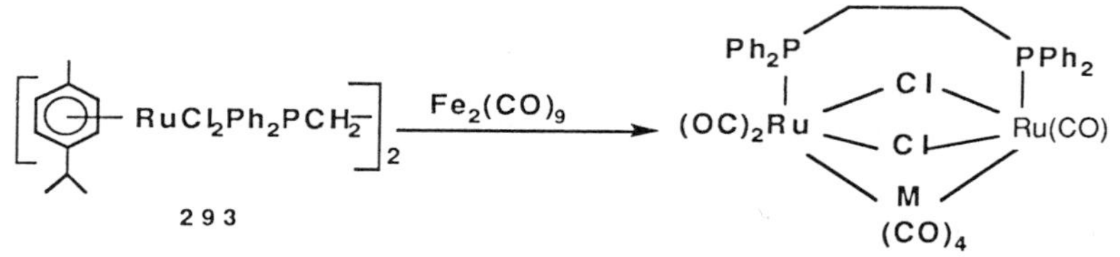

$$Ph_2P \quad\quad PPh_2$$
$$(OC)_2Ru \overset{Cl}{\underset{Cl}{\diagdown\diagup}} Ru(CO)$$

$$
\begin{bmatrix} \boxed{\bigcirc}\!\!-\!RuCl_2Ph_2PCH_2^- \end{bmatrix}_2 \xrightarrow{Fe_2(CO)_9}
$$

293

$$\qquad\qquad\qquad M \atop (CO)_4 \qquad\qquad (37)$$

294   (M=Fe)
295   (M=Ru)

$$\boxed{\bigcirc}\!\!-\!RuCl_2Ph_2PCH_2PPh_2Cl_2Ru\!-\!\boxed{\bigcirc} \xrightarrow{Fe_2(CO)_9}$$

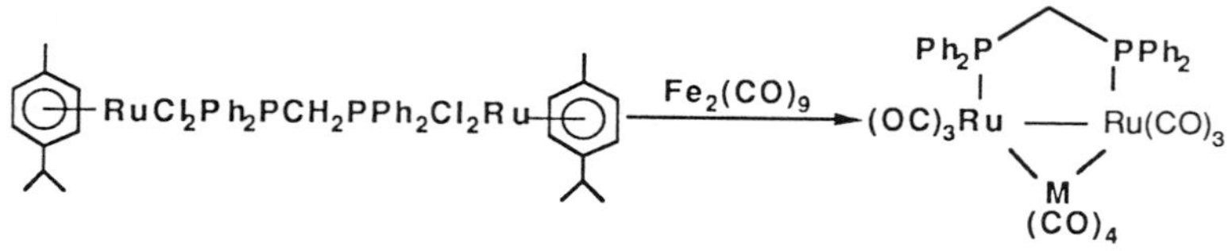

296

$$Ph_2P \quad\quad PPh_2$$
$$(OC)_3Ru \longrightarrow Ru(CO)_3$$
$$\qquad\qquad M \atop (CO)_4 \qquad (38)$$

297   M=Fe
298   M=Ru

The dehalogenation of arene ruthenium derivatives **299** and **300** with $Co_2(CO)_8$ gives access to Ru—Co complexes stabilized by a phosphido bridge. Complexes **301** (40%) and **302** (30%) are isolated in both cases. In addition, complex **300** afforded the $RuCo_2$ derivative **303** (10%) (*170,171*) [Eq. (39)].

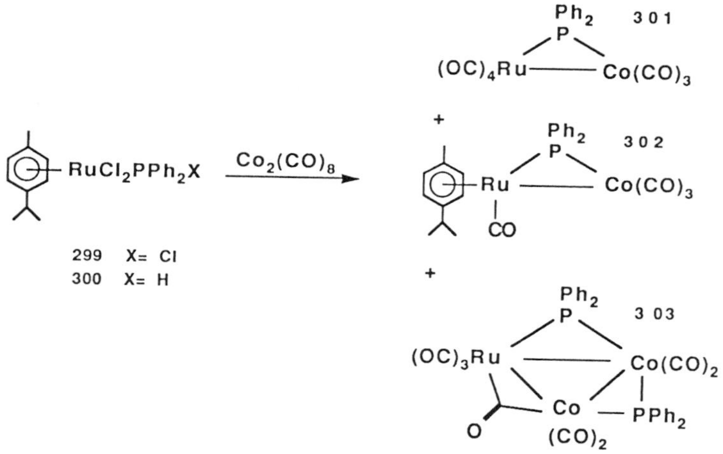

$$
\begin{array}{c}
Ph_2 \quad \mathbf{301}\\
P\\
(OC)_4Ru\!\!-\!\!-\!\!Co(CO)_3
\end{array}
$$

+

$$
\begin{array}{c}
Ph_2 \quad \mathbf{302}\\
P\\
\boxed{\bigcirc}\!\!-\!\!Ru\!\!-\!\!-\!\!Co(CO)_3\\
CO
\end{array}
$$

$$\boxed{\bigcirc}\!\!-\!RuCl_2PPh_2X \xrightarrow{Co_2(CO)_8} \qquad\qquad (39)$$

299   X= Cl
300   X= H

+

$$
\begin{array}{c}
Ph_2 \quad \mathbf{303}\\
P\\
(OC)_3Ru \!\!-\!\!-\!\! Co(CO)_2\\
Co \!\!-\!\! PPh_2\\
O \quad (CO)_2
\end{array}
$$

The arene phenylphosphine complex **304** has been used as a source of phosphinidene ruthenium moiety, as illustrated by the formation of clusters **305** (20%) and **306** (5%) (*172*) [Eq. (40)]. Complex **307**, by reaction

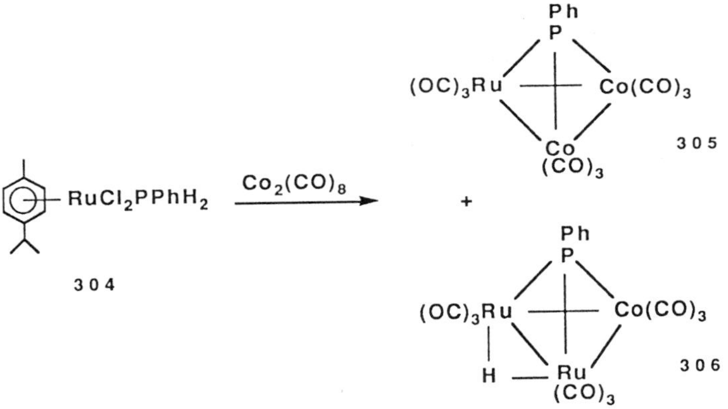

$$
\begin{array}{c}
\text{Ph} \\
\text{P} \\
(OC)_3Ru \longrightarrow Co(CO)_3 \\
\text{Co} \\
(CO)_3 \quad \mathbf{305}
\end{array}
$$

$$\mathbf{304} \xrightarrow{\text{Co}_2(\text{CO})_8} + \tag{40}$$

$$
\begin{array}{c}
\text{Ph} \\
\text{P} \\
(OC)_3Ru \longrightarrow Co(CO)_3 \\
\text{H} \longrightarrow \text{Ru} \quad \mathbf{306} \\
(CO)_3
\end{array}
$$

$$
\mathbf{307} \xrightarrow[\text{THF}]{\text{Co}_2(\text{CO})_8} \mathbf{308}
$$

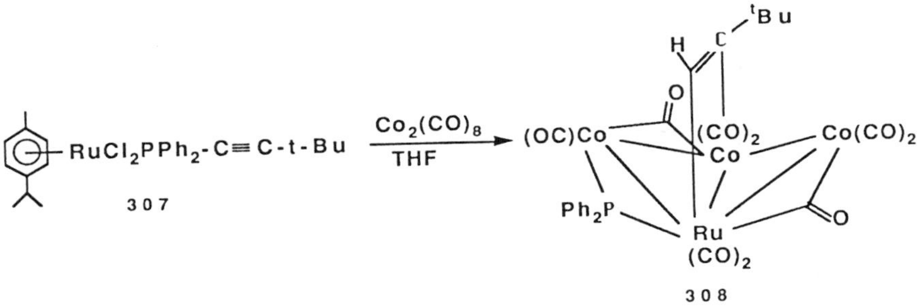

(41)

with $Co_2(CO)_8$, provides a reactive alkynephosphine ruthenium moiety for the formation of the butterfly cluster containing $RuCo_3$ **308** (30%). The X-ray structure shows that the reaction results from the cleavage of the (alkynyl)C—P bond (*173*) [Eq. (41)].

B. Arene Ruthenium Complexes as Precursors of Ruthenium/Metal Derivatives

Areneruthenium–metal complexes have been prepared by association of derivatives of type [(RuCl$_2$(arene)]$_2$ with suitable metal complexes. $(C_6H_6)Ru(\mu\text{-OMe})_3Os(C_6H_6)^+$ (**152**) is obtained by mixing analogous homonuclear Ru and Os complexes (*38*). The pyrazole complex **309** with Pd(acac)(η^3-allyl) (η^3-allyl = C_3H_5, C_4H_7, C_8H_{11}) leads to complexes of type **310** (Scheme 35, p. 232), which have been characterized by X-ray

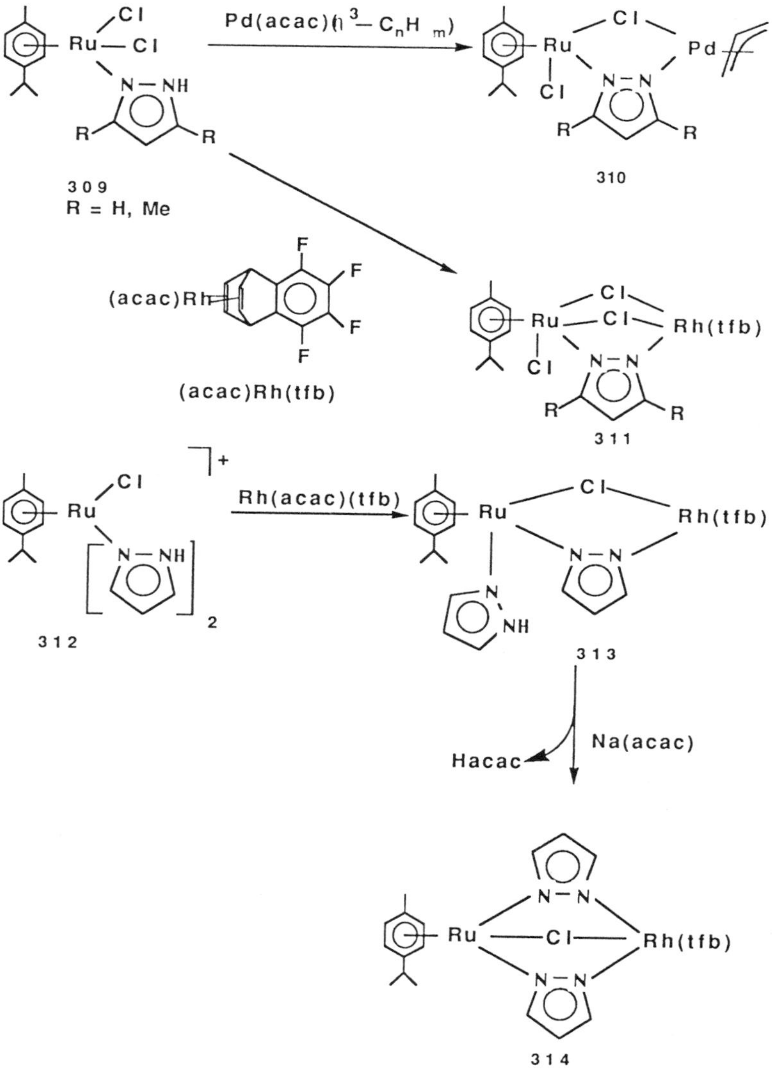

SCHEME 35

diffraction study of **310** (η^3-C_8H_{11}) (*174*). Triply bridged complexes **311** are obtained by reaction of the pyrazole **309** with Rh(acac)(tfb) and have been characterized by X-ray analysis of **311** (R = H) (*175*). Cation **312** with Rh(acac)(tfb) leads to the intermediate **313** which on deprotonation affords **314** (*174*).

Iridium complexes of type **315** act as electron-donating ligands toward [RuCl$_2$(*p*-cymene)]$_2$ (**1b**), and binuclear derivatives **316** have been obtained and characterized by X-ray analysis (*176*) [Eq. (42)].

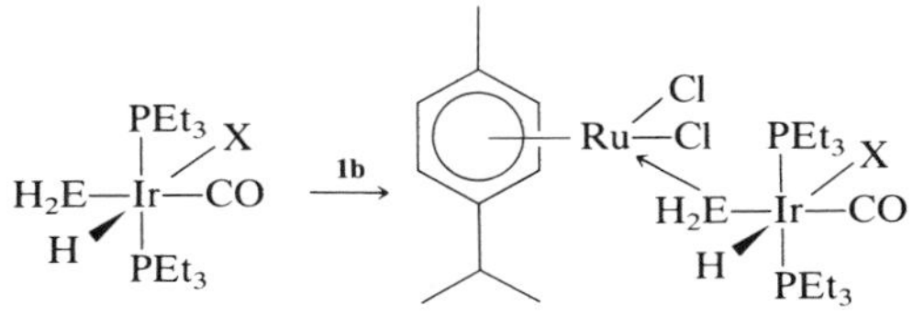

$$\text{(42)}$$

315   **316**

EH$_2$: PH$_2$; AsH$_2$

Arene–Ru/Pt or arene–Os/Pt systems of type **317** (Scheme 36) have been made by addition of the [Pt(S$_2$CNEt$_2$)(Ph$_2$PX)$_2$]$^-$ anions (X = S, O) to [RuCl$_2$(arene)]$_2$, RuCl(O$_2$CMe)(arene), or [OsCl$_2$(*p*-cymene)]$_2$. By changing the ratio of the reagents with [Pt(S$_2$CNEt$_2$)(Ph$_2$PS)$_2$]$^-$. (S—S)$^-$, (C$_6$H$_6$)(S—S)RuS—SRu(S—S)(C$_6$H$_6$)$^+$, and (C$_6$H$_6$)(S—S)Ru(η^1-S—S) are also obtained (*177*).

Of special interest is the reaction of the anionic cobalt complex **318** with [RuCl$_2$(C$_6$Me$_6$)$_2$]$_2$ in water. The triple-decker Ru/Co sandwich **319** is isolated in 65% yield (*178*) [Eq. (43)].

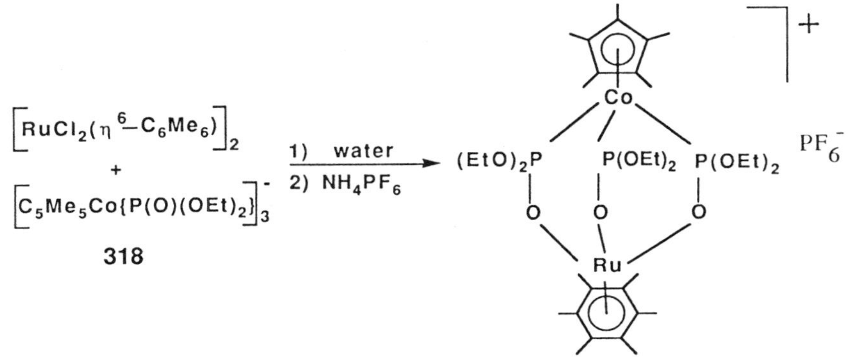

$$\text{(43)}$$

317

M : Ru, Os; X=Y=S
M : Ru; X=Y=O; X=O, Y=S
arene: C$_6$H$_6$, Me-C$_6$H$_4$-iPr

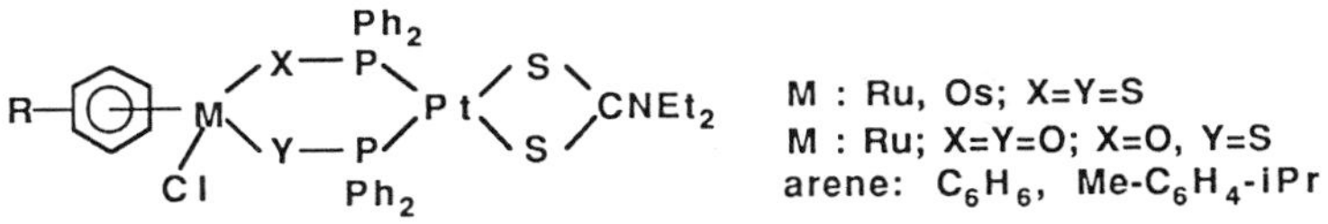

SCHEME 36

C. *Arene-Containing Clusters of Ruthenium or Osmium*

Arene-containing tetranuclear ruthenium or osmium complexes are made by treatment of $[MCl_2(arene)]_2$ derivatives with aqueous NaOH or Na_2CO_3. $[(\eta^6\text{-}C_6H_6)M_4(\mu_4\text{-}O)(\mu_2\text{-}OH)_4]^{2+}$ [M = Ru (**25B**) (*35*), Os (**150B**) (*109*)] and $[(\eta^6\text{-}C_6H_6)Ru_4(\mu\text{-}OH)_4]^{4+}$ (**26**) (*36*) are thus obtained (Section II,B,2,c).

Ruthenium and osmium clusters containing η^6-arene have been produced using high temperature reactions. $Ru_6C(CO)_{14}(\eta^6\text{-}C_6H_6)$ is obtained in low yield by heating $Ru_3(CO)_{12}$ in benzene (*179*) and in better yield by reaction of $[Ru_5C(CO)_{14}][N(PPh_3)_2]_2$ with $[Ru(NCPh)_3(\eta^6\text{-}C_6H_6)\text{-}(ClO_4)_2$ under reflux (*180*). $Ru_6(CO)_{14}(\eta^6\text{-}C_6H_5Me)$ (*181*) does not exchange its arene ligand at 150°C but produces methylcyclohexane under hydrogen at that temperature (*120*).

$Ru_6C(CO)_{14}(\eta^6\text{-}C_6H_6)^{2-}$, obtained by reduction of $Ru_6C(CO)_{14}(\eta^6\text{-}C_6H_6)$ with Na_2CO_3/MeOH, reacts with $[Ru(PhCN)_3(\eta^6\text{-}C_6H_6)](ClO_4)_2$ to afford $Ru_6C(CO)_{11}(\mu_3\text{-}\eta^2,\eta^2,\eta^2\text{-}C_6H_6)(\eta^6\text{-}C_6H_6)$ (**320**) (*180*) (Scheme 37). Dehydrogenation of 1,3-cyclohexadiene by $Ru_3(CO)_{12}$ leads to the arene cluster $Ru_4(CO)_9(\eta^6\text{-}C_6H_6)(C_6H_8)$, which has been characterized by X-ray analysis (*182*). Cyclohexadiene reacts with $Os_3H_2(CO)_{10}$ to give $Os_3H(C_6H_7)(CO)_9$ (*183*) which, on hydride abstraction with $(Ph_3C)BF_4$, leads to $[Os_3H(C_6H_6)(CO)_9]BF_4$ and further, on deprotonation, to $Os_3\text{-}(CO)_9(C_6H_6)$ (**321**). X-Ray analysis of **321** reveals the $\mu_3\text{-}\eta^2,\eta^2,\eta^2$ coordination type of the benzene ligand (*180*). Two-electron ligands such as ethylene can be coordinated to one osmium atom of cluster **321**, and the resulting derivatives show fluxionality of both the ethylene and arene ligand (*184*).

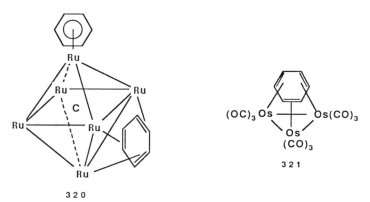

320

$(OC)_3 Os \longrightarrow Os(CO)_3$
Os
$(CO)_3$

321

SCHEME 37

Related benzyne-containing clusters have been described. For instance, heating of $Ru_3(CO)_{11}(PPh_3)$ in toluene affords three types of clusters containing the benzyne ligand coordinated in the μ_3, μ_4, or μ_5 fashion: $Ru_3(CO)_7(\mu\text{-}PPh_2)_2(\mu_3\text{-}\eta^2\text{-}C_6H_4)$ (30%), $Ru_4(CO)_{11}(\mu_4\text{-}PPh)(\mu_4\text{-}\eta^4\text{-}C_6H_4)$ (35%), and $Ru_5(CO)_{13}(\mu_4\text{-}PPh)(\mu_5\text{-}\eta^6\text{-}C_6H_4)$ (7%) (185). The former cluster has been previously made by pyrolysis of $Ru_3(CO)_9(PPh_3)_3$ (186,187). $Ru_4(CO)_{11}(\mu_4\text{-}PR)(\mu_4\text{-}\eta^4\text{-}C_6H_4)$ complexes (R = Me, CH_2-NPh_2) are also prepared from $Ru_3(CO)_{11}(Ph_2PR)$ clusters (185). Benzyne complexes of osmium of the type $Os_3(\mu\text{-}H)(\mu\text{-}A)(CO)_9(\mu_3\text{-}\eta^2\text{-}C_6H_4)$ (A = H, $AsMe_2$, SR) (188,190) or $Os_3(CO)_7\text{-}(\mu\text{-}PPh_2)_2(\mu_3\text{-}\eta^2\text{-}C_6H_4)$ (187) have also been characterized.

VII

METAL VAPORS IN THE SYNTHESIS OF ARENE RUTHENIUM AND OSMIUM COMPLEXES

A. Mononuclear Ruthenium and Osmium Complexes

The first use of ruthenium atoms in the synthesis of arene ruthenium derivatives was achieved in 1978 for the preparation of the thermally unstable bisbenzene ruthenium(0) complex **196a** by condensation of ruthenium vapor with benzene (191). The more stable bisbenzene osmium(0) complex (**322**) has also been prepared in 15% yield by cocondensation of osmium atoms with benzene (192,193).

$$Ru_{at} \quad or \quad Os_{at} + C_6H_6 \; \rightarrow \; M(\eta^6\text{-}C_6H_6)(\eta^4\text{-}C_6H_6)$$

196a (M = Ru)

322 (M = Os)

Intramolecular ring-exchange processes in **322** have been analyzed by two-dimensional (^1H–^1H) chemical exchange and ^1H magnetization transfer NMR experiments (192,193). Warming **322** in C_6D_6 to 100°C causes slow exchange of the benzene ligands with C_6D_6 to give **323** together with

$$Os(C_6H_6)_2 \xrightarrow[100°C]{C_6D_6} Os(C_6D_6)_2 + Os(C_6D_6)(C_6D_8)$$

322   **323**   **324**

$H_2 \downarrow RT$

$Os(C_6H_6)(C_6H_8)$

229

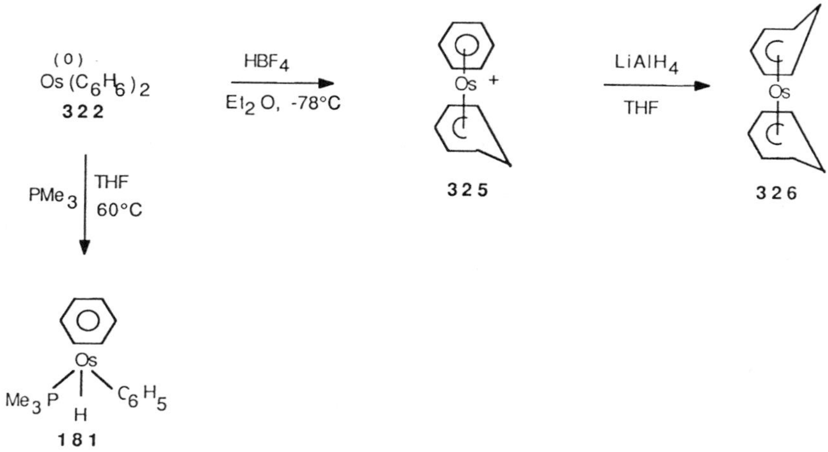

(0)
Os $(C_6H_6)_2$
322

$\xrightarrow[\text{Et}_2\text{O, -78°C}]{\text{HBF}_4}$

325

$\xrightarrow[\text{THF}]{\text{LiAlH}_4}$

326

PMe$_3$ | THF 60°C

Me$_3$P — Os — C_6H_5
H
181

SCHEME 38

the arene cyclohexadiene osmium(0) complex **324**, while treatment of **322** with dihydrogen at room temperature gives **229** in a rapid reaction (*193*).

Protonation of **322** with tetrafluoroboric acid in diethyl ether gives the cyclohexadienyl derivative **325** in 70% yield. Treatment of **325** with lithium aluminum hydride yields the biscyclohexadienyl osmium(II) complex **326**. Treatment of **322** with PMe$_3$ at 60°C gives the hydridophenyl osmium-(II) complex **181**, rather than the expected arene bistrimethylphosphine osmium(0) compound, via intramolecular C—H bond activation of the benzene ligand (*192,193*) (Scheme 38). Compound **181** as well as the analogous ruthenium complex (**92**) have also been obtained directly by co-condensation of osmium or ruthenium atoms with benzene and tri-methylphosphine (*62*) [Eq. (44)].

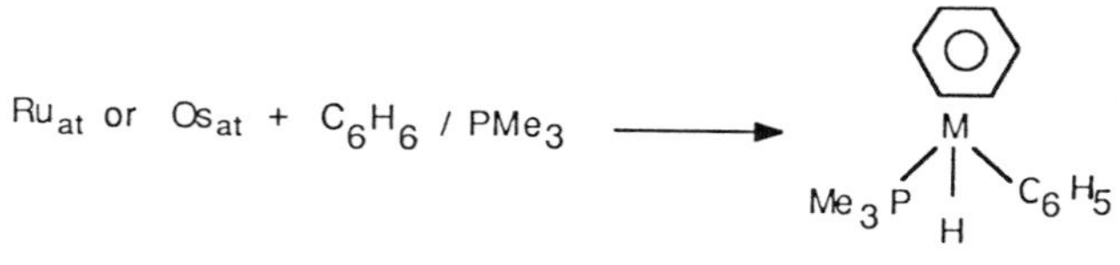

$$\text{Ru}_{at} \text{ or } \text{Os}_{at} + C_6H_6 / PMe_3 \longrightarrow \underset{H}{\overset{M}{\underset{Me_3P}{\diagup|\diagdown}}} C_6H_5 \qquad (44)$$

92 M = Ru
181 M = Os

When **92** is heated with C$_6$D$_6$ at 80°C, a ligand exchange reaction occurs. The thermal reactions of **92** with toluene, *ortho*-, *meta*-, and *para*-xylene, and mesitylene have been monitored by NMR spectroscopy. In all cases

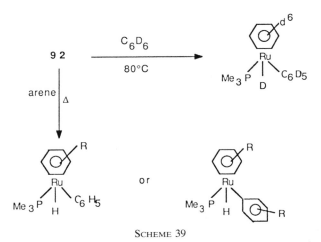

SCHEME 39

the benzene ligand is replaced by the arene. However, the σ-phenyl ligand is not replaced when the solvent arene is *p*-xylene or mesitylene (*62*) (Scheme 39).

Cocondensation of ruthenium atoms with 1,3-cyclohexadiene, and then CO, at $-196°C$ affords the cyclohexadienyl cyclohexenyl carbonyl ruthenium complex **327**, which can be transformed to the benzene cyclohexenyl carbonyl ruthenium cation **328** by hydride abstraction (*194*) [Eq. (45)].

$$Ru_{at} + 1,3-C_6H_8 \xrightarrow[-196°C]{CO} \quad \underset{\textbf{327}}{\text{Ru—CO}} \quad \underset{Ph_3C^+BF_4^-}{\rightleftharpoons} \quad \underset{\textbf{328}}{\text{Ru—CO}} \quad BF_4^-$$

(45)

B. *Bi- and Trinuclear Osmium Complexes*

In contrast to the cocondensation of osmium atoms with benzene which gives the osmium(0) derivative **322**, cocondensation of osmium atoms with mesitylene affords the binuclear complex **329** as the major product (10%) and a second minor product **330** (1%). The structure of **330** has been

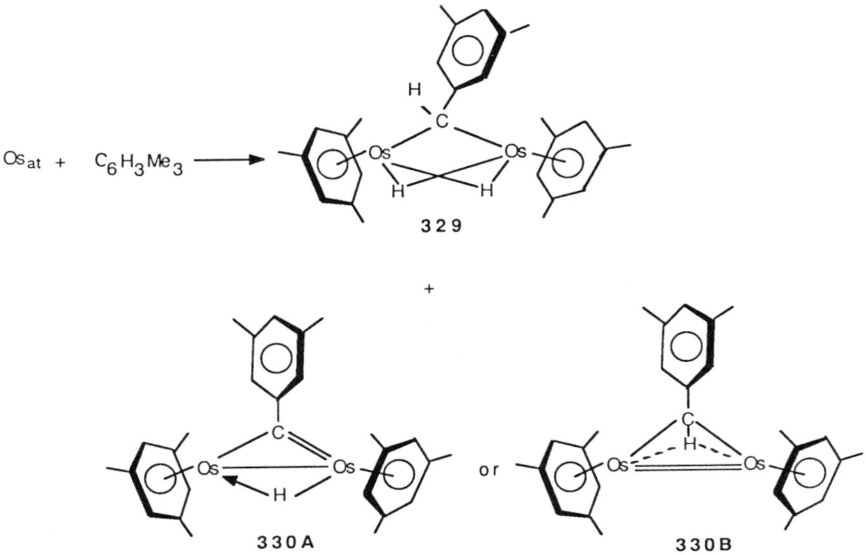

$$Os_{at} + C_6H_3Me_3 \longrightarrow$$

329

+

330A   or   **330B**

SCHEME 40

determined by X-ray crystallography and can be represented by the two forms **330A** and **330B** (*192,193*) (Scheme 40).

Cocondensation of osmium atoms with benzene–2-methylpropane (1 : 1) gives the air-sensitive triosmium compound **331** in 11% yield. ^{1}H- and ^{13}C-NMR experiments show the adamantane-like structure of **331**. Formation of **331** involves activation of the three methyl groups of 2-methylpropane. Cocondensation with C_6D_6 gives $[Os(C_6D_6)(\mu\text{-H})]_3$-$\{\mu^3\text{-}(CH_2)_3CH\}$, showing that the three bridging hydrido groups arise from 2-methylpropane (*193,195*).

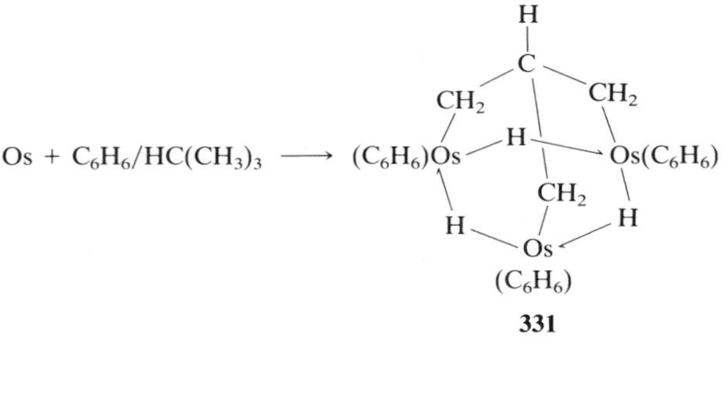

$$Os + C_6H_6/HC(CH_3)_3 \longrightarrow$$

331

VIII

ARENE RUTHENIUM AND OSMIUM DERIVATIVES AS CATALYST PRECURSORS

Arene ruthenium(II) derivatives have been shown, as have many other ruthenium complexes, to be efficient catalyst precursors for hydrogenation of olefins. Thus, $[RuCl_2(arene)]_2$ (1) complexes (arene = benzene, mesitylene, 1,3,5-triphenylbenzene) show hydrogenation activity in the presence of a base (pyrrolidine, triethylamine) or triphenylphosphine at 30°C (11,196). $[Ru_2(\mu\text{-}H)(\mu\text{-}Cl)_2(C_6Me_6)_2]Cl$, $[Ru_2(\mu\text{-}H)_3(C_6Me_6)_2]Cl$, and especially $RuH(Cl)(PPh_3)(C_6Me_6)$ catalyze the hydrogenation of 1-hexene and benzene to hexane and cyclohexane (50°C, 50 atm H_2) as well as of a variety of olefins and dienes by hydrogen transfer from 1-phenylethanol (52).

Arene ruthenium(0) derivatives also appear to be active hydrogenation catalyst precursors. $Ru(\eta^6\text{-}arene)(\eta^4\text{-}cyclooctadiene)$ hydrogenates α-olefins and cycloolefins under 1–20 atm of H_2 at room temperature (197). Polystyrene ruthenium(0) derivatives [Eq. (18)] hydrogenate olefins and aromatic hydrocarbons but also ketones, oximes, and nitrobenzene at 25–80°C under 50 atm of H_2. More drastic conditions are required (120–140°C) for the hydrogenation of propionitrile or benzonitrile to amines (129). $Ru(\eta^6\text{-}C_6Me_6)(\eta^4\text{-}C_6Me_6)$ catalyzes the hydrogenation of arene to cyclohexane at 90°C (2–3 atm H_2). In this reaction free hexamethylbenzene is not observed, and H/D exchange at the methyl groups takes place in the presence of D_2. These observations suggest a mechanism according to Eq. (46) (136). Addition of trimethylalane to the catalyst increases the rate by a factor of 4–5 and appears to stabilize the catalyst (135).

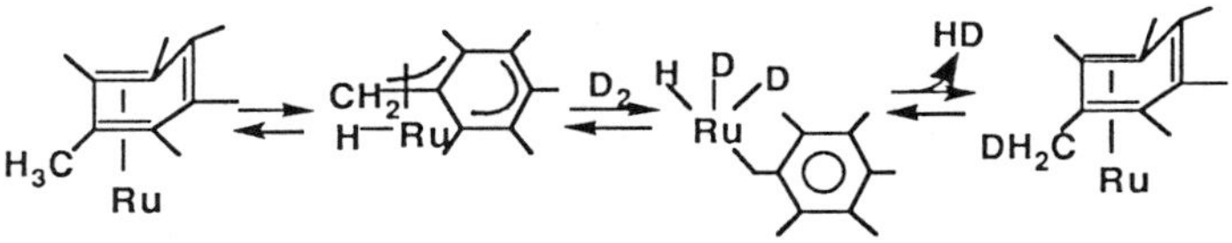

$$\tag{46}$$

$[RuCl_2(p\text{-}cymene)]_2$ and $[Ru_2(\mu\text{-}OH)_3(\eta^6\text{-}C_6Me_6)_2]Cl$ catalyze, as $[RhCl_2(C_5Me_5)]_2$ or $[Rh_2(\mu\text{-}OH)_3(C_5Me_5)_2]Cl$, the disproportionation of acetaldehyde and propionaldehyde, in the presence of water, to carboxylic acid and the corresponding alcohol (198). The corresponding osmium

$$RCHO + H_2O \rightarrow RCO_2H + RCH_2OH$$

derivative $[Os_2(\mu\text{-}OH)_3(p\text{-}cymene)_2]PF_6$, in the presence of water, catalyzes the oxidation of aldehydes to carboxylic acids with formation of
hydrogen. The osmium derivative shows a greater selectivity for oxidation
but lower activity than the ruthenium complexes. The turnover of the reaction with the osmium species is raised by the addition of a base (Na_2CO_3)
or by increase of temperature (*110*).

$$RCHO + H_2O \rightarrow RCO_2H + H_2$$

[$RuCl_2(p\text{-}cymene)$]$_2$ and $RuCl_2(PMe_3)(p\text{-}cymene)$ have been shown to
catalyze the benzylation of heterocycles such as thiophene, furan, or ben-
zofuran with benzylbromide (*199*). More interestingly, [$RuCl_2(C_6Me_6)$]$_2$
has been shown to catalyze both carbon–carbon coupling and alkylation
of furan or thiophene, using alcohol as the alkylating reagent. The first
step of the reaction is believed to be electrophilic activation of the C—H
bond of the heterocycle (*200*) [Eq. (47)].

Z: O,S

Arene ruthenium(II) complexes have been shown to catalyze the regioselective addition of ammonium carbamate to terminal alkynes, such
as phenylacetylene and 1-hexyne, to produce vinylcarbamates.

$$R_2NH + CO_2 + HC\equiv CR^1 \rightarrow R_2N\text{—}CO\text{—}O\text{—}CH\text{=}CHR^1$$

$RuCl_2(PMe_3)(C_6Me_6)$ and $RuCl_2(PMe_3)(p\text{-}cymene)$ appear to be much
more efficient catalysts than other mononuclear ruthenium complexes, or
$Ru_3(CO)_{12}$, for a variety of secondary amines and terminal alkynes, such as
diethylamine, morpholine, piperidine, and pyrrolidine (*65,201*), except for
acetylene itself (*202,203*). The regioselective addition to the terminal carbon suggests that that the reaction proceeds via an arene ruthenium vinylidene intermediate that has been characterized (*66*) (Section II,A,3,d).

$$RCO_2H + HC\equiv C\text{—}R^1 \rightarrow R\overset{\text{O}}{\underset{}{\underset{}{C}}}\text{—}O\text{—}CH\text{=}CH_2$$

This catalytic reaction is a unique synthesis involving carbon dioxide.
$RuCl_2(PMe_3)(p\text{-}cymene)$ also catalyzes the addition of a variety of carboxylic acids to alkynes to give enol esters in one step. The presence
of the basic phosphine is necessary to direct the addition of the carboxylate

group selectively to the substituted alkyne carbon (*204*). RuCl$_2$(PMe$_3$)-(*p*-cymene) catalyzes the addition of ammonium carbamates or carboxylic acids to propargylic alcohols in a different way. β-Oxopropyl derivatives are thus obtained in good yield under mild conditions (*205,206*) [Eq. (48)].

$$\text{R}_2\text{N}\underset{\text{O}}{\overset{\text{O}}{\bigvee}}\text{O}\underset{\text{R}\ \ \text{R}}{\overset{\text{O}}{\bigvee}} \xleftarrow[\text{CO}_2]{\text{R}_2\text{NH}} \quad \text{HC} \equiv \text{C} \underset{\text{R}\ \ \text{R}}{\bigvee}\text{OH} \xrightarrow[]{\text{Z-NHCHCO}_2\text{H}} \text{ZNHCH}\underset{\text{O}\ \text{R}\ \ \text{R}}{\overset{\text{R}}{\bigvee}}\text{O}\underset{}{\overset{\text{O}}{\bigvee}} \quad (48)$$

Although arene ruthenium complexes appear to be efficient catalyst precursors for the activation of alkynes, no evidence for the role or even the presence of the arene ligand in the active species has been obtained yet.

ACKNOWLEDGMENTS

We are grateful to Professors M. A. Bennett, W. L. Gladfelter, K. Itoh, E. Singleton, G. Vitulli, and H. Werner for sending us recent papers and manuscripts prior to publication. We express our appreciation to talented co-workers who have contributed to parts of this work and to D. Devanne, C. Bruneau, and N. Stantinat for their help during the preparation of the manuscript.

REFERENCES

1. M. A. Bennett, M. I. Bruce, and T. W. Matheson, *in* "Comprehensive Organometallic Chemistry" (G. Wilkinson, F. G. A. Stone, and E. W. Abel, Eds.), Vol. 4, p. 796. Pergamon, Oxford, 1982.
2. R. D. Adams and J. P. Selegue, *in* "Comprehensive Organometallic Chemistry" (G. Wilkinson, F. G. A. Stone, and E. W. Abel, Eds.), Vol. 4, p. 1020. Pergamon, Oxford, 1982.
3. E. O. Fischer and R. Böttcher, *Z. Anorg., Allg. Chem.* **291,** 305 (1957).
4. E. O. Fischer and C. Elschenbroich, *Chem. Ber.* **103,** 162 (1970).
5. E. O. Fischer, C. Elschenbroich, and C. G. Kreiter, *J. Organomet. Chem.* **7,** 481 (1967).
6. G. Winkhaus and H. Singer, *J. Organomet. Chem.* **7,** 487 (1967).
7. R. A. Zelonka and M. C. Baird, *J. Organomet. Chem.* **35,** C43 (1972); R. A. Zelonka and M. C. Baird, *Can. J. Chem.* **50,** 3063 (1972).
8. M. A. Bennett, G. B. Robertson, and A. K. Smith, *J. Organomet. Chem.* **43,** C41 (1972).
9. M. A. Bennett and A. Smith, *J. Chem. Soc., Dalton Trans.,* 233 (1974).
10. M. A. Bennett, T. N. Huang, T. W. Matheson, and A. K. Smith, *Inorg. Synth.* **21,** 74 (1982).
11. R. Iwata and I. Ogata, *Tetrahedron* **29,** 2753 (1973).
12. P. Pertici, G. Vitulli, R. Lazzaroni, P. Salvadori, and L. Barili, *J. Chem. Soc., Dalton Trans.,* 1019 (1982).
13. M. A. Bennett, T. W. Matheson, G. B. Robertson, A. K. Smith, and P. A. Tucker, *Inorg. Chem.* **19,** 1014 (1980).
14. M. A. Bennett and J. P. Ennett, *Organometallics* **3,** 1365 (1984).
15. J. W. Hull, Jr., and W. L. Gladfelter, *Organometallics* **3,** 605 (1984).

16. F. Faraone and V. Marsala, *Inorg. Chim. Acta* **27**, L109 (1978).
17. H. Werner and R. Werner, *J. Organomet. Chem.* **174**, C67 (1979).
18. F. Faraone, G. A. Loprete, and G. Tresoldi, *Inorg. Chim. Acta* **34**, L251 (1979).
19. H. Werner and R. Werner, *Angew. Chem.* **90**, 721 (1978).
20. H. Werner and R. Werner, *Chem. Ber.* **115**, 3766 (1982).
21. D. R. Robertson and T. A. Stephenson, *J. Organomet. Chem.* **142**, C31 (1977).
22. C. J. Jones, J. A. Mc Cleverty, and A. S. Rothin, *J. Chem. Soc., Dalton Trans,* 109 (1986).
23. M. A. Bennett, T. W. Matheson, G. B. Robertson, W. L. Steffen, and T. W. Turney, *J. Chem. Soc., Chem. Commun.,* 32 (1979).
24. H. Werner and R. Werner, *J. Organomet. Chem.* **174**, C63 (1979).
25. R. H. Crabtree and A. J. Pearman *J. Organomet. Chem.* **141**, 325 (1977).
26. R. J. Restivo, G. Fergusson, D. J. O'Sullivan, and F. J. Lalor, *Inorg. Chem.* **14**, 3046 (1975).
27. M. A. Bennett and T. W. Matheson, *J. Organomet. Chem.* **175**, 87 (1979).
28. M. J. H. Russel, C. White, A. Yates, and P. M. Maitlis, *Angew. Chem. Int. Ed. Engl.* **15**, 490 (1976).
29. M. J. Rybinskaya, A. R. Kudinov, and V. S. Kaganovitch, *J. Organomet. Chem.* **246**, 279 (1983).
30. H. Werner, H. Kletzin, and C. Burschka, *J. Organomet. Chem.* **276**, 231 (1984).
31. D. R. Robertson and T. A. Stephenson, *J. Organomet. Chem.* **116**, C29 (1976).
32. D. R. Robertson and T. A. Stephenson, *J. Organomet. Chem.* **162**, 121 (1978).
33. T. Arthur and T. A. Stephenson, *J. Organomet. Chem.* **168**, C39 (1979).
34. T. Arthur and T. A. Stephenson, *J. Organomet. Chem.* **208**, 369 (1981).
35. T. Arthur, D. R. Robertson, D. A. Tocher, and T. A. Stephenson, *J. Organomet. Chem.* **208**, 389 (1981).
36. R. O. Gould, T. A. Stephenson, and D. A. Tocher, *J. Organomet. Chem.* **264**, 365 (1984).
37. R. O. Gould, C. L. Jones, D. R. Robertson, and T. A. Stephenson, *J. Chem. Soc., Chem. Commun.* **222** (1977).
38. R. O. Gould, T. A. Stephenson, and D. A. Tocher, *J. Organomet. Chem.* **263**, 375 (1984).
39. D. A. Tocher, R. O. Gould, T. A. Stephenson, M. A. Bennett, J. P. Ennett, T. W. Matheson, L. Sawyer, and V. K. Shah, *J. Chem. Soc., Dalton Trans.,* 1571 (1983).
40. J. Cook, J. E. Hamlin, A. Nutton, and P.M. Maitlis, *J. Chem. Soc., Dalton Trans.,* 2342 (1981).
41. K. Isobe, P. M. Bailey, and P. M. Maitlis, *J. Chem. Soc., Dalton Trans.,* 2003 (1981).
42. H. Werner and H. Kletzin, *J. Organomet. Chem.* **228**, 289 (1982).
43. H. Kletzin and H. Werner, *J. Organomet. Chem.* **291**, 213 (1985).
44. R. .A. Zelonka and M. C. Baird, *J. Organomet. Chem.* **44**, 383 (1972).
45. A. N. Nesmeyanov and A. Z. Rubezhov, *J. Organomet. Chem.* **164**, 259 (1979).
46. H. Werner, H. Kletzin, A. Höhn, W. Paul, and W. Knaup, *J. Organomet. Chem.* **306**, 227 (1986).
47. H. Werner and H. Kletzin, *Angew. Chem. Int. Ed. Engl.* **22**, 46 (1983).
48. J. C. Hayes and N. J. Cooper, *J. Am. Chem. Soc.* **104**, 5570 (1982).
49. R. Werner and H. Werner, *Chem. Ber.* **115**, 3781 (1982).
50. R. Werner and H. Werner, *Chem. Ber.* **116**, 2074 (1983).
51. H. Tom Dieck, W. Kollvitz, and I. Kleinwächter, *Organometallics*, **5**, 1449 (1986).
52. a. M. A. Bennett, T.-N Huang, A. K. Smith, and T. W. Turney, *J. Chem. Soc., Chem. Commun.,* 582 (1978); b. M. A. Bennett. T.-N Huang, and T. W. Turney, *J. Chem. Soc., Chem. Commun.,* 312 (1979).

53. M. A. Bennett, J. P. Ennett, and K. I. Gell, *J. Organomet. Chem.* **233**, C17 (1982).
54. M. A. Bennett, T.-N Huang, and J. L. Latten, *J. Organomet, Chem.* **272**, 189 (1984); M. A. Bennett, T.-N. Huang, and J. L. Latten, *J. Organomet. Chem.* **276**, C39 (1984).
55. M. A. Bennett and J. Latten, *Aust. J. Chem.* **40**, 841 (1987).
56. R. H. Morris and M. Shiralian, *J. Organomet. Chem.* **260**, C47 (1984).
57. J. J. Hough and E. Singleton, *J. Chem. Soc., Chem. Commun.*, 371 (1972).
58. A. R. Siedle, R. A. Newmark, L. H. Pignolet, D. X. Wang, and T. A. Albright, *Organometallics* **5**, 38 (1986).
59. H. Werner and H. Kletzin, *J. Organomet. Chem.* **243**, C59 (1983).
60. H. Werner and K. Roder, *J. Organomet. Chem.* **310**, C51 (1986).
61. H. Werner and H. Kletzin, *Angew. Chem.* **95**, 916 (1983); H. Werner and H. Kletzin, *Angew. Chem. Int. Ed. Engl.* **22**, 873 (1983).
62. M. L. H. Green, D. S. Joyner, J. M. Wallis, *J. Chem. Soc., Dalton Trans.*, 2823 (1987).
63. M. I. Bruce and A. G. Swincer, *Adv. Organomet. Chem.* **22**, 59 (1983).
64. K. Ouzzine, H. Le Bozec, and P. H. Dixneuf, *J. Organomet. Chem.* **317**, C25 (1986).
65. R. Mahe, P. H. Dixneuf, and S. Lecolier, *Tetrahedron Lett.* **27**, 6333 (1986).
66. K. Ouzzine, H Le Bozec, and P. H. Dixneuf, unpublished results.
67. D. F. Dersnah and M. C. Baird, *J. Organomet. Chem.* **127**, C55 (1977).
68. H. Brunner and R. Gastinger, *J. Chem. Soc., Chem. Commun.*, 488 (1977).
69. H. Brunner and R. Gastinger, *J. Organomet. Chem.* **145**, 365 (1978).
70. J. D. Korp and I. Bernal, *Inorg. Chem.* **20**, 4065 (1981).
71. P. Pertici, P. Salvadori, A. Biasci, G. Vitulli, M. A. Bennett, L. A. P. Kane-Maguire, *J. Chem. Soc., Dalton Trans.*, in press (1988).
72. P. Pertici, P. Salvadori, A. Biasci, M. A. Bennett, and F. Marchetti, *Abstracts Euchem Conference Organometal. Chem.* Toledo, 70 (1987).
73. I. W. Robertson, T. A. Stephenson, and D. A. Tocher, *J. Organomet. Chem.* **228**, 171 (1982).
74. E. Roman and D. Astruc, *Inorg. Chim. Acta* **37**, L465 (1979).
75. A. N. Nesmeyanov, N. A. Vol'kenau, I. N. Bolesova, and L. S. Shulpina, *J. Organomet. Chem.* **182**, C36 (1979).
76. T. P. Gill and K. R. Mann, *Organometallics* **1**, 485 (1982).
77. A. Mc. Nair, J. L. Schrenk, and K. R. Mann, *Inorg. Chem.* **23**, 2633 (1984).
78. J. M. Moriarty, Y. Y. Ku, and U. S. Gill, *J. Chem. Soc., Chem. Commun.*, 1493 (1987).
79. M. O. Albers, D. J. Robinson, A. Shaver, and E. Singleton, *Organometallics* **5**, 2199 (1986).
80. R. J. Haines and A. L. Du Preez, *J. Am. Chem. Soc.* **93**, 2820 (1971).
81. R. J. Haines and A. L. Du Preez, *J. Organomet. Chem.* **84**, 357 (1975).
82. G. J. Kruger, A. L. Du Preez, and R. J. Haines. *J. Chem. Soc. Dalton Trans.*, 1302 (1974).
83. M. O. Albers, D. C. Liles, D. J. Robinson, A. Shaver, and E. Singleton, *Organometallics* **6**, 2347 (1987).
84. M. O. Albers, D. C. Liles, D. J. Robinson, and E. Singleton, *J. Chem. Soc., Chem. Commun.*, 1103 (1986).
85. M. Crocker, M. Green, A. G. Orpen, and D. M. Thomas, *J. Chem. Soc., Chem. Commun.*, 1143 (1984).
86. K. Itoh, T. Fukahori, K. Ara, H. Nagashima, H. Nishiyama, *Abstracts 5th Int. Symp. Homogeneous Catalysis,* Kobe, 87 (1986).
87. Z. L. Lutsenko, G. G. Aleksandrov, P. V. Petrovskii, E. S. Shubina, V. G. Andrianov, Yu. T. Struchkov, and A. Z. Rubezhov, *J. Organomet. Chem.* **281**, 349 (1985).

88. M. Stebler-Röthlisberger A. Salzer, H. B. Bürgi, and A. Ludi, *Organometallics* **5**, 298 (1986).
89. V. S. Kaganovich, A. R. Kudinov, and M. I. Rybinskaya, *Izv. Akad. Nauk SSSR, Ser. Khim.*, 492 (1986).
90. T. P. Hanusa, J. C. Huffman, and L. J Todd, Polyhedron **1**, 77 (1983).
91. T. P. Hanusa, J. C. Huffman, T. L. Curtis, and L. J. Todd, *Inorg. Chem.* **24**, 787 (1985).
92. M. Bown, N. N. Greenwood, and J. D. Kennedy, *J. Organomet. Chem.* **309**, C67 (1986).
93. M. Bown, X. L. R. Fontaine, N. N. Greenwood, J. D. Kennedy, and M. Thornton-Pett, *J. Organomet. Chem.* **315**, C1 (1986).
94. M. Bown, X. L. R. Fontaine, N. N. Greenwood, J. D. Kennedy, and P. Mac Kinnon, *J. Chem. Soc. Chem. Commun.*, 817 (1987).
95. M. Bown, X. L. R. Fontaine, N. N. Greenwood, and J. D. Kennedy, *J. Organomet. Chem.* **325**, 233 (1987).
96. M. Bown, X. L. R. Fontaine, N. N. Greenwood, J. D. Kennedy, and M. Thornton-Pett, *J. Chem. Soc., Dalton Trans.*, 1169 (1987).
97. M. Bown, X. L. R. Fontaine, N. N. Greenwood, P. Mac Kinnon, J. D. Kennedy, and M. Thornton-Pett, *J. Chem. Soc., Dalton Trans.*, 2781 (1987).
98. E. O. Fischer and H.. P. Fritz, *Angew. Chem.* **73**, 353 (1961).
99. P. J. Domaille, S. D. Ittel, J. P. Jesson, and D. A. Sweigart, *J. Organomet. Chem.* **202**, 191 (1980).
100. G. Winkhaus, H. Singer, and M. Kribe, *Z. Naturforsch.* **202**, 191 (1980).
101. A. Z. Rubezhov, A. S. Ivanov, and S. P. Gubin, *Bull, Acad. Sci. URSS, Div. Chem. Sci. (Engl. Transl.)* **23**, 1828 (1974).
102. M. A. Bennett, I. J. McMahon, S. Pelling, G. B. Robertson, and W. A. Wickrama-singhe, *Organometallics* **4**, 754 (1985).
103. J. A. Cabeza and P. M. Maitlis, *J. Chem. Soc. Dalton Trans.*, 573 (1985).
104. K. Roder and H. Werner, *Angew. Chem. Int. Ed. Engl.* **26**, 686 (1987).
105. R. Weinand and H. Werner, *Chem. Ber.* **119**, 2055 (1986).
106. H. Werner and R. Werner, *J. Organomet. Chem.* **194**, C7 (1980).
107. H. Werner, W. Knaup, and M. Dziallas, *Angew. Chem. Int. Ed. Engl.* **26**, 248 (1987).
108. H. Werner and R. Weinand, *Z. Naturforsch.* **38b**, 1518 (1983).
109. J. A. Cabeza, B. E. Mann, C. Brevard, and P. M. Maitlis, *J. Chem. Soc., Chem. Commun.*, 65 (1985).
110. J. A. Cabeza, A. J. Smith, H. Adams, and P. M. Maitlis, *J. Chem. Soc. Dalton Trans.*, 1155 (1986).
111. H. Werner and K. Zenkert, *J. Chem. Soc., Chem. Commun.*, 1607 (1985).
112. J. Müller, C. G. Kreiter, B. Mertschenk, and S. Schmitt, *Chem. Ber.* **108**, 273 (1975).
113. H. Werner, H. Kletzin, R. Zolk, and H. Otto, *J. Organomet. Chem.* **310**, C11 (1986).
114. H. Werner and K. Roder, *J. Organomet. Chem.* **281**, C38 (1985).
115. U. Schubert, R. Werner, L. Zinner, and H. Werner, *J. Organomet. Chem.* **253**, 363 (1983).
116. R. Weinand and H. Werner, *J. Chem. Soc., Chem. Commun.*, 1145 (1985).
117. H. Werner, R. Weinand, and H. Otto, *J. Organomet. Chem.* **307**, 49 (1986).
118. D. Jones, L. Pratt, and G. Wilkinson, *J. Chem. Soc.*, 4458 (1962).
119. E. O. Fischer and J. Müller, *Chem. Ber.* **96**, 3217 (1963).
120. T. V. Ashworth, M. G. Nolte, R. H. Reimann, and E. Singleton, *J. Chem. Soc., Chem. Commun.*, 937 (1977).
121. P. Pertici and G. Vitulli, *Inorg. Synth.* **22**, 176 (1983).

122. P. Pertici, G. Vitulli, M. Paci, and L. Porri, *J. Chem. Soc., Dalton Trans.*, 1961 (1980).
123. P. Pertici, G. Vitulli, and L. Porri, *J. Chem. Soc., Chem. Commun.*, 846 (1975).
124. P. Pertici, G. P. Simonelli, G. Vitulli, G. Deganello, P. L. Sandrini, and A. Mantovani, *J. Chem. Soc., Chem. Commun.*, 132 (1977).
125. P. Pertici, G. Vitulli, S. Bertozzi, R. Lazzaroni, and P. Salvadori, *J. Organomet. Chem.* in press (1988).
126. G. Vitulli, P. Pertici, and C. Bigelli, *Gazz. Chim. Ital.* **115,** 79 (1985).
127. G. Vitulli, P. Pertici, and P. Salvadori, *J. Chem. Soc., Dalton Trans.*, 2255 (1984).
128. M. A. Bennett, P. Pertici, P. Salvadori, and G. Vitulli, *Abstracts COMO 12 Conf., Tasmania,* P1-8 (1984).
129. P. Pertici, G. Vitulli, C. Carlini, and F. Ciardelli, *J. Mol. Catal.* **11,** 353 (1981).
130. A. Biancono, M. Dell'ariccia, A. Giovannelli, E. Burattini, N. Cavallo, P. Patteri, E. Pancini, C. Carlini, F. Ciardelli, D. Papeschi, P. Pertici, G. Vitulli, D.Dalba, P. Fornasini, S. Mobilio, and L. Palladino, *Chem. Phys. Lett.* **90,** 19 (1982).
131. M. A. Bennett and T. W. Matheson, *J. Organomet. Chem.* **153,** C25 (1978).
132. M. A. Bennett, T. W. Matheson, G. B. Robertson, A. K. Smith, and P. A. Tucker, *J. Organomet. Chem.* **121,** C18 (1976).
133. M. A. Bennett, T. W. Matheson, G. B. Robertson, A. K. Smith, and P. A. Tucker, *Inorg, Chem.* **20,** 2353 (1981).
134. G. Huttner and S. Lange, *Acta Crystallogr. Sect. B* **28,** 2049 (1972).
135. M. Y. Darensbourg and E. L. Muetterties, *J. Am. Chem. Soc.* **100,** 7425 (1978).
136. J. W. Johnson and E. L. Muetterties, *J. Am. Chem. Soc.* **99,** 7395 (1977).
137. E. L. Muetterties, J. R. Bleeke, and A. C. Sievert, *J. Organomet. Chem.* **178,** 197 (1975).
138. A. Lucherini and L. Porri, *J. Organomet. Chem.* **155,** C45 (1978).
139. K. Itoh, K. Murai, H. Nagashima, and H. Nishiyama, Chem. Lett., 499 (1983).
140. R. Werner and H. Werner, *Angew. Chem.* **93,** 826 (1981).
141. H. Werner, A. Höhn, and R. Weinand, *J. Organomet. Chem.* **299,** C15 (1986).
142. M. I. Rybinskaya, V. S. Kaganovich, and A. R. Kudinov. *J. Organomet, Chem.* **235,** 215 (1982).
143. S. L. Grundy and P. M. Maitlis. *J. Chem. Soc. Chem., Commun.*, 379 (1982).
144. R. T. Swann, A. W. Hanson, and V. Boekelheide, *J. Am. Chem. Soc.* **106,** 818 (1984).
145. R. T. Swann, A. W. Hanson, and V. Boekelheide, *J. Am. Chem. Soc.* **108,** 3324 (1986).
146. D. A. Sweigart, *J. Chem. Soc. Chem. Commun.*, 1159 (1980).
147. Y. K. Chang, E. D. Honig, and D. A. Sweigart, *J. Organomet. Chem.* **256,** 277 (1983).
148. D. R. Robertson, W. Robertson, and T. A. Stephenson, *J. Organomet. Chem.* **2021,** 309 (1980).
149. H. Werner and R. Werner, *Chem. Ber.* **118,** 4543 (1985).
150. R. Werner and H. Werner, *J. Organomet. Chem.* **210,** C11 (1981).
151. H. Werner and H. Kletzin, *J. Organomet. Chem.* **228,** 289 (1982).
152. H. Werner, R. Werner, and Ch. Burschka, *Chem. Ber.* **117,** 152 (1984).
153. R. Werner and H. Werner, *Chem. Ber.* **117,** 161 (1984).
154. N. A. Vol'kenau, I. N. Bolesova, L. S. Shul'pina, and A. N. Kitaigorodskii, *J. Organomet. Chem.* **267,** 313 (1984).
155. J. W. Hull, Jr. and W. L. Gladfelter, *Organometallics.* **1,** 1716 (1982).
156. M. A. Bennett, I. J. McMahon, and T. W. Turney, *Angew. Chem.* **94,** 373 (1982).
157. M. A. Bennett, I. J. McMahon, and T. W. Turney, *Angew. Chem. Suppl.*, 853 (1982).
158. J. A. Segal, *J. Chem. Soc., Chem. Commun.*, 1338 (1985).
159. E. D. Laganis, R. G. Finke, and V. Boekelheide, *Tetrahedron. Lett.* **21,** 4405 (1980).

160. E. D. Laganis, R. H. Voegeli, R. T. Swann, R. G. Finke, H. Hopf, and V. Boekelheide, *Organometallics* **1**, 1415 (1982).
161. R. H. Voegeli, H. C. Kang, R. G. Finke, and V. Boekelheide, *J. Am. Chem. Soc.* **108**, 7010 (1986).
162. H. C. Kang, K. D. Plitzko, V. Boekelheide, H. Higuchi, and S. Misumi, *J. Organomet. Chem.* **321**, 79 (1987).
163. W. D. Rohrbach and V. Boekelheide, *J. Organomet. Chem.* **48**, 3673 (1983).
164. J. Elzinga and M. Rosenblum, *Organometallics* **2**, 1214 (1983).
165. V. S. Kaganovich, A. R. Kudinov, and M. I. Rybinskaya, *J. Organomet. Chem.* **323**, 111 (1983).
166. R. G. Finke, R. H. Voegeli, E. D. Laganis, and V. Boekelheide, *Organometallics* **2**, 347 (1983).
167. R. T. Swann, A. W. Hanson, and V. Boekelheide, *J. Am. Chem. Soc.* **106**, 818 (1984).
168. D. F. Jones, P. H. Dixneuf, T. G. Southern, J. Y. Le Marouille, D. Grandjean, and P. Guenot, *Inorg. Chem.* **20**, 3247 (1981).
169. a. A. W. Coleman, D. F. Jones, P. H. Dixneuf, C. Brisson, J. J. Bonnet, and G. Lavigne, *Inorg. Chem.* **23**, 952 (1984); b. A. W. Coleman, H. Zhang, S. G. Bott, J. L. Atwood, and P. H. Dixneuf, *J. Coord. Chem.* **16**, 9 (1987).
170. R. Regragui and P. H. Dixneuf, *J. Organomet. Chem.* **239**, C12 (1984).
171. R. Regragui, P. H. Dixneuf, N. J. Taylor, and A. J. Carty, *Oganometallics* **3**, 1020 (1984).
172. R. Regragui, Thèse d'état, Université de Rennes, 1986.
173. D. F. Jones, P. H. Dixneuf, A. Benoit, and J. Y. Le Marouille, *J. Chem. Soc., Chem. Commun.,* 1217 (1982).
174. M. P. Garcia, A. Portilla, L. A. Oro, C. Foces-Foces, and F. H. Carro, *J. Organomet. Chem.* **322**, 111 (1987).
175. L. A. Oro, D. Carmona, M. P. Garcia, F. J. Lahoz, J. Reyes, C. Foces-Foces, and F. H. Carro, *J. Organomet. Chem.* **296**, C43 (1985).
176. E. A. V. Ebsworth, R. O. Gould, R. A. Mayo, and M. Walkinshaw, *J. Chem. Soc., Dalton Trans.,* 2831 (1987).
177. J. D. Fotheringham and T. A. Stephenson, *J. Organomet. Chem.* **284**, C12 (1985).
178. E. Roman, F. Tapia, and S. Hernandez, *Polyhedron* **5**, 917 (1986).
179. C. R. Eady, B. F. G. Johnson, and J. Lewis, *J. Chem. Soc., Dalton Trans.,* 2606 (1975).
180. M. Gomez-Sal, B. F. G. Johnson, J. Lewis, P. R. Raithby, and A. H. Wright, *J. Chem. Soc., Chem. Commun.,* 1682 (1985).
181. R. Mason and W. R. Robinson, *J. Chem. Soc., Chem. Commun.,* 468 (1968).
182. S. Aime, L. Milone, D. Osella, G. A. Vaglio, M. Valle, A. Tiripicchio, and M. Tiripicchio-Camellini, *Inorg. Chim. Acta* **34**, 49 (1979).
183. E. G. Bryan, B. F. G. Johnson, J. W. Kelland, and M. McPartlin, *J. Chem. Soc., Chem. Commun.,* 254 (1976).
184. M. A. Gallop, B. F. G. Johnson, J. Lewis, and P. Raithby, *J. Chem. Soc., Chem. Commun.,* 1809, 1831 (1987).
185. S. A. R. Knox, B. R. Lloyd, A. G. Orpen, J. M. Vinas, and M. Weber, *J. Chem. Soc., Chem. Commun.,* 1498 (1987).
186. M. I. Bruce, G. Shaw, and F. G. A. Stone, *J. Chem. Soc., Dalton Trans.,* 2094 (1972).
187. M. I. Bruce, J. M. Guss, R. Mason, B. W. Shelton, and A. H. White, *J. Organomet. Chem.* **251**, 261 (1983).
188. R. J. Goudsmit, B. F. G. Johnson, J. Lewis, P. R. Raithby, and M. J. Rosa, *J. Chem. Soc., Dalton Trans.,* 2257 (1983).

189. A. J. Deeming, I. P. Rothwell, M. B. Hursthouse, and J. D. J. Backer-Dir, *J. Chem. Soc., Dalton Trans.*, 1879 (1981).
190. R. D. Adams, D. A. Katahira, and L. W. Yang, *Organometallics* **1,** 235 (1982).
191. P. L. Timms and R. B. King, *J. Chem. Soc., Chem. Commun.*, 898 (1978).
192. J. A. Bandy, M. L. H. Green, D. O'Hare, and K. Prout, *J. Chem. Soc., Chem. Commun.*, 1402 (1984).
193. J. A. Bandy, M. L. H. Green, and D. O'Hare, *J. Chem. Soc., Dalton Trans.*, 2477 (1986).
194. D. N. Cox and R. Roulet, *Organometallics* **5,** 1886 (1986).
195. M. L. H. Green and D. O'Hare, *J. Chem. Soc., Chem. Commun.*, 355 (1985).
196. A. G. Hinze, *Recl. Trav. Chim. Pays-Bas* **92,** 542 (1973).
197. P. Pertici, G. Vitulli, S. Bigelli, and R. Lazzaroni, *J. Organomet. Chem.* **275,** 113 (1984).
198. J. Cook, J. E. Hamlin, A. Nutton, and P. M. Maitlis, *J. Chem. Soc., Chem. Commun.*, 144 (1980).
199. R. Jaouhari and P. H. Dixneuf, *Inorg. Chem. Acta* in press (1988).
200. R. Jaouhari, P. Guenot, and P. H. Dixneuf, *J. Chem. Soc., Chem. Commun.*, 1255 (1986).
201. Y. Sasaki and P. H. Dixneuf, *J. Chem. Soc., Chem. Commun.*, 790 (1986).
202. Y. Sasaki and P. H. Dixneuf, *J. Org. Chem.* **52,** 314 (1987).
203. C. Bruneau, P. H. Dixneuf, and S. Lécolier, *J. Mol. Cat.* **44,** 175 (1988).
204. C. Ruppin and P. H. Dixneuf, *Tetrahedron Lett.* **27,** 6323 (1986).
205. C. Bruneau and P. H. Dixneuf, *Tetrahedron Lett.* **28,** 2005 (1987).
206. D. Devanne, C. Ruppin, and P. H. Dixneuf, *J. Org. Chem.* **53,** 926 (1988).

Heteronuclear Cluster Chemistry of Copper, Silver, and Gold

IAN D. SALTER

Department of Chemistry
University of Exeter
Exeter EX4 4QD, England

I

INTRODUCTION

The first examples of compounds containing a bond between copper, silver, or gold and another, different transition metal were a series of dimers reported by Nyholm and co-workers (*1–3*) in 1964. Although Nyholm *et al.* (*1*) also described the first mixed-metal gold cluster, [Au$_2$-Fe(CO)$_4$(PPh$_3$)$_2$],[1] in the same year, the heteronuclear cluster chemistry of the Group IB metals was very poorly developed before 1981, despite the great interest shown in the field of heteronuclear transition metal clusters as a whole (*5*). Since the early 1980s, however, mixed-metal cluster compounds containing one or more Group IB metals have attracted a great deal of attention, and a large number of such species have now been reported, especially gold-containing clusters. This article reviews the synthesis, structures, chemical properties, and dynamic behavior of copper-, silver-, and gold-containing heteronuclear cluster compounds, and covers the literature up to the end of September 1987. For the purposes of this review, a cluster compound is considered to be a species which contains three or more framework atoms with sufficient interactions between them to define either a metal core made up of one or more trigonal planar M$_3$ units or a three-dimensional skeletal geometry based on a wide variety of polyhedra. The scope of this article has been restricted to Group IB metal cluster compounds in which the skeletal atoms consist predominantly of transition metals.

Bimetallic clusters of gold were reviewed by Braunstein and Rose (*6*) in 1985, and Hall and Mingos (*7*) surveyed both homonuclear and heteronuclear gold clusters in the previous year. In addition, Steggerda *et al.* (*8*) reviewed the preparation and properties of gold clusters in 1982, Jones (*9*) has listed gold clusters studied by single-crystal X-ray diffraction up to mid

---

[1] The existence of an interaction between the gold atoms [Au—Au 3.028(1) Å] was not recognized until later (*4*).

1985, Melnik and Parish *(10)* correlated [197]Au Mössbauer and X-ray structural data for gold clusters in 1986, and Salter *(11)* has recently published a review of the stereochemical nonrigidity exhibited in solution by the metal skeletons of some mixed-metal clusters containing copper, silver, and gold.

In the vast majority of the Group IB metal heteronuclear clusters reported so far, one two-electron donor ligand is attached to each coinage metal, and PR_3 (R = alkyl or aryl) is by far the most common ligand of this type. As the chemistry of this class of cluster compounds has been studied most intensively, it is discussed in detail in Section II. A number of species with coinage metals that do not carry any ligands or with ligands that are not simple two-electron donors bonded to the Group IB metals are also known, however, and these clusters are described more briefly in Sections III and IV.

II

HETERONUCLEAR CLUSTERS CONTAINING ONE OR MORE ML [M = Cu, Ag, OR Au; L = TWO-ELECTRON DONOR LIGAND, NORMALLY PR₃ (R = ALKYL OR ARYL)] FRAGMENTS

A. *Synthesis*

Despite the great interest shown in heteronuclear transition metal cluster chemistry over the last 20 years, widely applicable rational methods for the synthesis of a desired mixed-metal cluster remain relatively rare, and the element of chance still often plays an important role in the preparation of such species *(5)*. However, in contrast to the situation for the vast majority of heteronuclear clusters of other transition metals, a number of elegant rational synthetic strategies that afford high yields of mixed-metal clusters containing ML fragments have been developed since 1982. Indeed, the study of Group IB metal clusters has greatly benefitted from the availability of such generally applicable preparative routes, and these are outlined in Sections II,A,1 and II,A,2.

1. *Rational Preparative Routes Using Anionic Precursors*

The vast majority of the currently known heteronuclear cluster compounds containing $M(PR_3)$ (M = Cu, Ag, or Au, R = alkyl or aryl) units were synthesized by treating an anionic mono-, di-, or polynuclear precursor with the complexes $[MX(PR_3)]$ (X = Cl, Br, or I). The first example of a gold mixed-metal cluster, $[Au_2Fe(CO)_4(PPh_3)_2]$, was prepared by treating the dianion $[Fe(CO)_4]^{2-}$ with the complex $[AuCl(PPh_3)]$ [Eq. (1)] (1). Subsequently, Lewis and co-workers (12) extended this synthetic approach by utilizing a preformed cluster anion instead of a monomeric anionic complex as a precursor [Eq. (2)] (12). They also found that the yields could be increased considerably by using $TlPF_6$ to remove Cl^- from the reaction mixture and $[AuCl(PR_3)]$ (12,13). Indeed, it is now common to add either $TlPF_6$ or a similar compound which can act as a halide abstractor [e.g., $AgClO_4$ (14)] to this type of reaction.

$$Na_2[Fe(CO)_4] + 2\,[AuCl(PPh_3)] \rightarrow [Au_2Fe(CO)_4(PPh_3)_2] \qquad (1)$$

$$[N(PPh_3)_2][Os_3(\mu\text{-}H)(CO)_{11}] + [AuCl(PR_3)] \xrightarrow{TlPF_6} [AuOs_3(\mu\text{-}H)(CO)_{10}(PR_3)] \qquad (2)$$

$$R = Ph\ or\ Et$$

The preparative route was also soon adapted for the synthesis of copper- and silver-containing clusters by using the complexes $[MI(PEt_3)]$ ($M = Cu$ or Ag) instead of $[AuCl(PR_3)]$ [Eq. (3)] (15). The versatility of the

$$K[Ru_3(\mu\text{-}H)(\mu_3\text{-}PPh)(CO)_9] + [MI(PEt_3)] \rightarrow [MRu_3(\mu\text{-}H)(\mu_3\text{-}PPh)(CO)_9(PEt_3)] \qquad (3)$$

$$M = Cu\ or\ Ag$$

reaction between the complexes $[MX(PR_3)]$ ($M = Cu$, Ag, or Au; $X = Cl$, Br, or I; $R = alkyl$ or $aryl$) and an anionic precursor is illustrated as follows:

1. Two or three $Au(PR_3)$ units can be added to a monomeric anionic complex. Equations (1) and (4) show examples (1,16).

$$Cs_3[V(CO)_5] + 3[AuCl(PPh_3)] \rightarrow [Au_3V(CO)_5(PPh_3)_3] \qquad (4)$$

2. One $M(PR_3)$ fragment can be added to a dimeric anionic complex. Equation (5) shows an example (14).

$$[N(PPh_3)_2][Mn_2(\mu\text{-}PPh_2)(CO)_8] + [MX(PEt_3)] \xrightarrow{AgClO_4}$$

$$[MMn_2(\mu\text{-}PPh_2)(CO)_8(PEt_3)] \qquad (5)$$

$$M = Cu\ or\ Ag,\ X = I;\ M = Au,\ X = Cl$$

3. One $M(PR_3)$ moiety can be introduced into a preformed cluster monoanion. Equations (2) and (3) show examples (12,15).

4. Two $M(PR_3)$ groups can be introduced into a preformed cluster dianion. Equation (6) shows an example (17).

$$[N(PPh_3)_2]_2[Ru_4(\mu\text{-}H)_2(CO)_{12}] + 2\,[MX(PPh_3)] \xrightarrow{TlPF_6} [M_2Ru_4H_2(CO)_{12}(PPh_3)_2] \qquad (6)$$

$$M = Cu\ or\ Au,\ X = Cl;\ M = Ag,\ X = I$$

5. One $M_2(\mu\text{-L})$ (M = Au or Cu; μ-L = bidentate phosphine or arsine ligand) unit can be introduced into a preformed cluster dianion. Equation (7) shows an example (18).

$$K_2[Ru_3(\mu_3\text{-S})(CO)_9] + [M_2(\mu\text{-Ph}_2PCH_2PPh_2)Cl_2] \xrightarrow{TlPF_6}$$

$$[M_2Ru_3(\mu_3\text{-S})(\mu\text{-Ph}_2PCH_2PPh_2)(CO)_9] \tag{7}$$

$$M = Cu \text{ or } Au$$

6. Two $M(PR_3)$ groups containing two different coinage metals can be introduced into a preformed cluster dianion. Equation (8) gives an example (17).

$$[N(PPh_3)_2]_2[Ru_4(\mu\text{-H})_2(CO)_{12}] + [MX(PPh_3)] + [M'X(PPh_3)] \xrightarrow{TlPF_6}$$

$$[MM'Ru_4(\mu_3\text{-H})_2(CO)_{12}(PPh_3)_2] \tag{8}$$

$$M = Cu, M' = Ag \text{ or } Au; M = Ag, M' = Au; X = Cl \text{ or } I$$

The complexes $[Au(NO_3)(PR_3)]$, $[Au(PR_3)_2][PF_6]$, and $[Ag\{C_5\text{-}(CO_2Me)_5\}(PR_3)]$ have also been used instead of $[AuX(PR_3)]$ or $[AgX(PR_3)]$ (X = Cl, Br, or I) as the gold or silver precursors in the same synthetic approach. Examples are shown in Eqs. (9), (10), and (11),

$$[Co_3Fe(CO)_{12}]^- + [Au(NO_3)(PPh_3)] \rightarrow [AuCo_3Fe(CO)_{12}(PPh_3)] \tag{9}$$

$$[N(PPh_3)_2][Ru_3(\mu\text{-H})(\mu_3\text{-PPh})(CO)_9] + [Au(PR_3)_2][PF_6] \rightarrow$$

$$[AuRu_3(\mu\text{-H})(\mu_3\text{-PPh})(CO)_9(PR_3)] \tag{10}$$

$$PR_3 = PMe_2Ph \text{ or } PEt_3$$

$$K[Ru_3(\mu_3\text{-PhPCH}_2PPh_2)(CO)_9] + [Ag\{C_5(CO_2Me)_5\}(PPh_3)] \rightarrow$$

$$[AgRu_3(\mu_3\text{-PhPCH}_2PPh_2)(CO)_9(PPh_3)] \tag{11}$$

respectively (19–21). Brown and Salter (22) have found that the poor solubility of some gold halide complexes of bidentate phosphine and arsine ligands can lead to low yields of product when they are reacted with cluster anions. Their modification of the original preparative route involves initially synthesizing the gold halide complex of the desired ligand by a ligand exchange reaction with $[AuCl(SC_4H_8)]$, followed by an immediate in situ reaction between the resultant solution of the gold precursor and the appropriate cluster anion. The new procedure is also very convenient, as $[AuCl(SC_4H_8)]$ is readily prepared in very high yield (23), the ligand exchange reaction is very quick, and it is not necessary to isolate and purify the gold precursor complexes. Equation (12) shows an example of the modified approach (22).

The reaction between a cluster monoanion and the complex $[AuCl(PR_3)]$ almost always results in the introduction of just one $Au(PR_3)$

$$2[AuCl(SC_4H_8)] + Ph_2ECH_2E'Ph_2 \rightarrow [Au_2(\mu\text{-}Ph_2ECH_2E'Ph_2)Cl_2]$$

$$\xrightarrow[\text{TlPF}_6]{[N(PPh_3)_2]_2[Ru_4(\mu\text{-}H)_2(CO)_{12}]} [Au_2Ru_4(\mu_3\text{-}H)(\mu\text{-}H)(\mu\text{-}Ph_2ECH_2E'Ph_2)(CO)_{12}] \quad (12)$$

$$E = P, E' = P \text{ or } As; E = E' = As$$

fragment into a cluster, but there have been a few reports of a hydrido ligand in the cluster precursor being replaced by a $Au(PR_3)$ unit as well (e.g., 24–26). Equation (13) shows an example (24). In addition, Bruce

$$[N(PPh_3)_2][Fe_4(\mu\text{-}H)(BH)(CO)_{12}] + 2[AuCl(PPh_3)] \rightarrow$$

$$[Au_2Fe_4(BH)(CO)_{12}(PPh_3)_2] \quad (13)$$

and co-workers (27–31) have shown that the gold–oxonium reagent $[\{Au(PPh_3)\}_3O][BF_4]$ can replace a CO group or two hydrido ligands in cluster anions by two $Au(PPh_3)$ fragments as well as acting as a source of $[Au(PPh_3)]^+$. Thus, the complex can be used to incorporate up to three $Au(PPh_3)$ units into a cluster monoanion, and Eqs. (14) and (15) show examples (27,31).

$$K[Ru_4(\mu\text{-}H)_3(CO)_{12}] + [\{Au(PPh_3)\}_3O][BF_4] \rightarrow [AuRu_4(\mu_3\text{-}H)(\mu\text{-}H)_2(CO)_{12}(PPh_3)]$$

$$+ [Au_2Ru_4(\mu_3\text{-}H)(\mu\text{-}H)(CO)_{12}(PPh_3)_2] + [Au_3Ru_4(\mu_3\text{-}H)(CO)_{12}(PPh_3)_3] \quad (14)$$

$$[N(PPh_3)_2][CoRu_3(CO)_{13}] + [\{Au(PPh_3)\}_3O][BF_4] \rightarrow$$

$$[Au_3CoRu_3(CO)_{12}(PPh_3)_3] \quad (15)$$

In 1980, Albano et al. (32) showed that two $Cu(NCMe)$ units could be incorporated into a mixed-metal cluster by treating a preformed cluster dianion with the complex $[Cu(NCMe)_4][BF_4]$ [Eq. (16)]. Other workers

$$K_2[Rh_6C(CO)_{15}] + 2[Cu(NCMe)_4][BF_4] \rightarrow [Cu_2Rh_6C(CO)_{15}(NCMe)_2] \quad (16)$$

have subsequently utilized this synthetic approach to add one or two $Cu(NCMe)$ fragments to cluster mono- or dianions (e.g., 33–35), but very few clusters containing $Cu(NCMe)$ units have been reported compared to the number in which the copper atoms are ligated by PR_3 groups, presumably because the former species generally seem to be much less stable than their PR_3-containing analogs (36). Indeed, the NCMe ligands on $[Cu_2Ru_6(CO)_{18}(NCMe)_2]$ were found to be so labile that they were replaced by two C_6H_5Me ligands on recrystallization from warm toluene [Eq. (17)] (37).

$$[N(PPh_3)_2]_2[Ru_6(CO)_{18}] + 2[Cu(NCMe)_4][BF_4] \rightarrow [Cu_2Ru_6(CO)_{18}(NCMe)_2] \xrightarrow{\text{warm toluene}}$$

$$[Cu_2Ru_6(CO)_{18}(C_6H_5Me)_2] \quad (17)$$

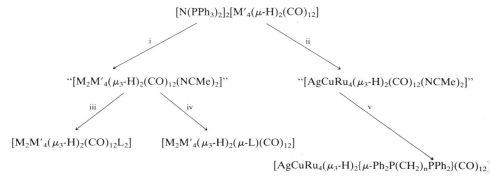

SCHEME 1. Reagents (CH$_2$Cl$_2$ solutions at -30°C): i, [M(NCMe)$_4$][PF$_6$] (M = Cu or Ag) (2 equiv); ii, [Cu(NCMe)$_4$][PF$_6$] (1 equiv) and [Ag(NCMe)$_4$][PF$_6$] (1 equiv); iii, L [M = Cu, M' = Ru, L = PPh$_3$, PMePh$_2$, PMe$_2$Ph, PEt$_3$, PMe$_3$, P(CHMe$_2$)$_3$, P(cyclo-C$_6$H$_{11}$)$_3$, or P(OEt)$_3$; M = Ag, M' = Ru, L = PPh$_3$, PMePh$_2$, P(CHMe$_2$)$_3$, or P(cyclo-C$_6$H$_{11}$)$_3$] (2 equiv); iv, L [M = Cu or Ag, M' = Ru, L = Ph$_2$P(CH$_2$)$_n$PPh$_2$ (n = 1–6) or Ph$_2$As(CH$_2$)$_n$EPh$_2$ (E = As or P; n = 1 or 2) or cis-Ph$_2$PCH=CHPPh$_2$; M = Cu or Ag, M' = Os, L$_2$ = Ph$_2$P(CH$_2$)$_2$PPh$_2$] (1 equiv); v, Ph$_2$P(CH$_2$)$_n$PPh$_2$ (n = 1 or 2, M' = Ru) (1 equiv) (36,38–42).

The lability of NCMe ligands in mixed-metal clusters has been utilized by Salter and co-workers (36) in a versatile new synthetic route to species containing one or two Cu(PR$_3$) fragments. The preparative procedure involves first incorporating one or two Cu(NCMe) units into a cluster mono- or dianion using [Cu(NCMe)$_4$][PF$_6$] and then immediately treating the relatively unstable intermediate species with the appropriate phosphine ligand to obtain the desired phosphine-containing cluster. This approach is superior to the reaction between a cluster anion and the copper halide complex of the appropriate phosphine because it frequently affords substantially better yields and it also avoids the severe problems that can occur with the latter method when the copper halide complex of the desired phosphine is not available, is unstable, or has very poor solubility in common organic solvents (36). The new synthetic route has also been extended to afford mixed-metal clusters containing one or two Ag(PR$_3$) moieties by using the complex [Ag(NCMe)$_4$][PF$_6$] instead of [Cu(NCMe)$_4$][PF$_6$], and arsine ligands can be attached to the coinage metals as well as a very wide range of mono- and bidentate phosphine ligands (36,38–44). Scheme 1 gives an example of the versatility of the procedure (36,38–42).

2. Rational Preparative Routes Using Neutral Precursors

Although most of the currently known heteronuclear clusters containing ML (M = Cu, Ag, or Au; L = two-electron donor ligand) units were synthesized from anionic precursors (Section II,A,1), there are also a number of important rational preparative procedures which utilize neutral precursors, especially for gold-containing clusters. One method for introducing $Au(PR_3)$ fragments into neutral clusters involves condensation reactions, with the elimination of methane, between preformed cluster compounds containing hydrido ligands and the complex $[AuMe(PPh_3)]$. This preparative route was first reported by Stone and co-workers (45) in 1978 [Eq. (18)]. Subsequently, a number of other mixed-metal clusters,

$$[Os_3(\mu\text{-H})_2(CO)_{10}] + [AuMe(PPh_3)] \xrightarrow{-CH_4} [AuOs_3(\mu\text{-H})(CO)_{10}(PPh_3)] \qquad (18)$$

including some containing two or three $Au(PR_3)$ fragments, have been prepared by using the complex $[AuMe(PPh_3)]$ in this type of condensation reaction (e.g., 13,14,46–51), and Eq. (19), shows an example (46).

$$[Ru_3(\mu\text{-H})_3(\mu\text{-COMe})(CO)_9] + [AuMe(PPh_3)] \xrightarrow{-CH_4}$$

$$[AuRu_3(\mu\text{-H})_2(\mu_3\text{-COMe})(CO)_9(PPh_3)] + [Au_2Ru_3(\mu\text{-H})(\mu_3\text{-COMe})(CO)_9(PPh_3)_2]$$

$$+ [Au_3Ru_3(\mu_3\text{-COMe})(CO)_9(PPh_3)_3] \qquad (19)$$

The complexes $[Au(OR)(PPh_3)]$ ($R = Bu^t$ or $2,4,6\text{-}Bu^t_3C_6H_2$) have also been used to replace two or three hydrido ligands in precursor compounds by $Au(PPh_3)$ fragments via condensation reactions (52,53). Equation (20) shows an example (52). Interestingly, when the complex $[Cu(OBu^t)(PPh_3)]$ is reacted with $[Re_2H_8(PMe_2Ph)_4]$, cleavage of Cu-P bonds occurs

$$[Re_2H_8(PMe_2Ph)_4] + 2 [Au(OBu^t)(PPh_3)] \xrightarrow{-2 Bu^tOH}$$

$$[Au_2Re_4H_6(PMe_2Ph)_4(PPh_3)_2] \qquad (20)$$

as well as elimination of Bu^tOH to afford the cluster $[Cu_2Re_4H_{14}(PMe_2Ph)_8]$, which contains no $Cu(PR_3)$ fragments (52).

Oxidative addition reactions have also been used to synthesize heteronuclear Group IB metal clusters. Clark and co-workers (54) first described this type of reaction in 1970 [Eq. (21)], and similar methods have subsequently been used to introduce one $Au(PR_3)$ unit into other neutral clusters [e.g., Eqs. (22)–(25)] (51,55–59).

$$[Os_3(CO)_{12}] + [AuX(PPh_3)] \rightarrow [AuOs_3(\mu\text{-X})(CO)_{10}(PPh_3)] \qquad (21)$$
$$X = Cl, Br, I, \text{ or } SCN$$

$$[Ru_5C(CO)_{15}] + [AuX(PPh_3)] \rightarrow [AuRu_5C(CO)_{15}X(PPh_3)] \qquad (22)^2$$
$$X = Cl, Br, \text{ or } Me$$

$$[Ru_3(CO)_{12}] + [AuCl(PPh_3)] \rightarrow [AuRu_3(\mu\text{-Cl})(CO)_{10}(PPh_3)] \qquad (23)$$

---

[2] In the case of X = Me, the Me group bonds to the O atom of the CO ligand to afford the cluster $[AuRu_5C(CO)_{14}(COMe)(PPh_3)]$.

$$[Os_3(CO)_{10}(NCMe)_2] + [AuX(PR_3)] \rightarrow [AuOs_3(CO)_{10}X(PR_3)] \qquad (24)^3$$

$$PR_3 = PPh_3 \text{ or } PMe_2Ph, X = C_2Ph; PR_3 = PEt_3, X = NCO$$

$$[Pt_3(\mu\text{-}SO_2)_3\{P(cyclo\text{-}C_6H_{11})_3\}_3] + [AuCl\{P(C_6H_4F\text{-}4)_3\}] \rightarrow$$

$$[AuPt_3(\mu\text{-}SO_2)_2(\mu\text{-}Cl)\{P(cyclo\text{-}C_6H_{11})_3\}_3\{P(C_6H_4F\text{-}4)_3\}] \qquad (25)$$

The oxidative addition reactions, however, do not always produce the expected single product. Thus, although Bruce *et al.* (*61*) have shown that the reactions between the cluster $[Os_3(\mu\text{-}H)_2(CO)_{10}]$ and the complexes $[M(C_2C_6F_5)(PPh_3)]$ (M = Cu, Ag, or Au) proceed via oxidative addition and hydrogen migration steps to afford the clusters $[MOs_3\text{-}(\mu\text{-}CH{=}CHC_6F_5)(CO)_{10}(PPh_3)]$ cleanly and in high yield [Eq. (26)],

$$[Os_3(\mu\text{-}H)_2(CO)_{10}] + [M(C_2C_6F_5)(PPh_3)] \rightarrow$$

$$[MOs_3(\mu\text{-}CH{=}CHC_6F_5)(CO)_{10}(PPh_3)] \qquad (26)$$

$$M = Cu, Ag, \text{ or } Au$$

they also found that similar reactions using $[M(C_2Ph)(PR_3)]$ (M = Cu or Ag, R = Ph; M = Au, R = Ph or Me) produce a complex mixture of products. The exact nature of the mixture is dependent on the type of Group IB metal and phosphine ligand present in the mononuclear complex.

It is possible to synthesize mixed-metal clusters containing $Au(PR_3)$ fragments by treating neutral as well as anionic precursors with either $[Au(PR_3)]^+$ or a compound that acts as a source of this cation. In many cases, a $[Au(PR_3)]^+$ unit is simply added to the cluster precursor to afford a cationic heteronuclear gold cluster (e.g., *59,62–66*). Equation (27)

$$[Pt_3(\mu\text{-}L)_3\{P(cyclo\text{-}C_6H_{11})_3\}_3] + [AuCl(PR_3)] \xrightarrow{\text{TlPF}_6}$$

$$[AuPt_3(\mu\text{-}L)_3\{P(cyclo\text{-}C_6H_{11})_3\}_3(PR_3)][PF_6] \qquad (27)$$

$$R = cyclo\text{-}C_6H_{11}, L = CO \text{ or } SO_2; R = C_6H_4F\text{-}4, L = SO_2$$

shows an example (*59,62*). Interestingly, it is possible to add either $[Au\{P(C_6H_4F\text{-}4)_3\}]^+$ or $[AuCl\{P(C_6H_4F\text{-}4)_3\}]$ to the cluster $[Pt_3(\mu\text{-}SO_2)_3\text{-}\{P(cyclo\text{-}C_6H_{11})_3\}_3]$ depending on whether the reaction with $[AuCl\text{-}\{P(C_6H_4F\text{-}4)_3\}]$ is performed in the presence of $TlPF_6$ or not [Eqs. (25) and (27)] (*59*). Two $[Au(PR_3)]^+$ fragments can also be added to a neutral precursor to form cluster dications (e.g., *66,67*). Equation (28) gives an example (*67*). In addition, one case of $[M(PR_3)]^+$ (M = Cu or Ag) units being added to a neutral complex has been reported [Eq. (29)] (*68*). Addition of $[Au(PR_3)]^+$ fragments to neutral compounds can also occur in

---

[3] The reaction between $[Os_3(CO)_{10}(NCMe)_2]$ and $[Au(C_2Ph)(PPh_3)]$ was first described by Lewis and co-workers (*60*), but they incorrectly formulated the product as $[AuOs_3(CO)_9(C_2Ph)(PPh_3)]$.

$$[RuH_2(\mu\text{-}Ph_2PCH_2PPh_2)_2] + 2\,[Au(NO_3)(PPh_3)] \rightarrow$$

$$[Au_2RuH_2(\mu\text{-}Ph_2PCH_2PPh_2)_2(PPh_3)_2][PF_6]_2 \qquad (28)$$

$$[Pt_2(\mu\text{-}S)(CO)(PPh_3)_3] + [MCl(PPh_3)] \xrightarrow{\text{TlPF}_6} [MPt_2(\mu_3\text{-}S)(CO)(PPh_3)_4][PF_6] \qquad (29)$$

$$\text{M = Cu, Ag, or Au}$$

conjunction with reactions which replace ligands attached to the precursor by Au(PR$_3$) units (e.g., *31,69–72*). Equations (30) and (31) show two examples (*69,70*).

$$\textit{trans}\text{-}[PtHCl(PEt_3)_2] + 2\,[Au(thf)(PPh_3)][CF_3SO_3] \rightarrow$$

$$[Au_2PtCl(PEt_3)_2(PPh_3)_2][CF_3SO_3] \qquad (30)$$

$$[RhH(CO)(PPh_3)_3] + [Au(NO_3)(PPh_3)] \xrightarrow{\text{KPF}_6} [Au_3Rh(\mu\text{-}H)(CO)(PPh_3)_5][PF_6] \qquad (31)$$

The Group IB metal site preferences in heteronuclear cluster compound allow coinage metal exchange reactions between clusters and mononuclear complexes to be used as a high yield synthetic route to novel trimetallic species containing two different Group IB metals. These reactions are discussed in detail in Section II,D,2.

3. Miscellaneous Syntheses

Although the vast majority of Group IB metal heteronuclear clusters have been synthesized using one of the generally applicable rational preparative procedures outlined in Sections II,A,1 and II,A,2, a number of such species have been obtained via routes for which the nature of the product cannot be predicted in advance or via procedures which have not been widely used. These other methods of synthesis are outlined in this section.

The homonuclear copper cluster [Cu$_6$(μ-H)$_6$(PPh$_3$)$_6$] has been utilized as a source of Cu(PPh$_3$) fragments for incorporation into neutral precursors [Eqs. (32) and (33)] (*73,74*).

$$[Cu_6(\mu\text{-}H)_6(PPh_3)_6] + [Os_3(\mu\text{-}H)_2(CO)_{10}] \rightarrow [CuOs_3(\mu\text{-}H)_3(CO)_{10}(PPh_3)] \qquad (32)$$

$$[Cu_6(\mu\text{-}H)_6(PPh_3)_6] + [W(\equiv CC_6H_4Me\text{-}4)(CO)_2(\eta\text{-}C_5H_5)] \rightarrow$$

$$[CuW_2(\mu_3\text{-}CC_6H_4Me\text{-}4)(CO)_4(PPh_3)(\eta\text{-}C_5H_5)_2]$$

$$+ [W(\eta^3\text{-}CH_2C_6H_4Me\text{-}4)(CO)_2(\eta\text{-}C_5H_5)] \qquad (33)$$

Mixed-metal clusters containing Ag(PR$_3$) fragments have been prepared using the reactions shown in Eqs. (34) and (35) (*75,76*). Two very high nuclearity silver–gold clusters containing M(PPh$_3$) (M = Ag or Au) units have been obtained by co-reducing gold and silver complexes or salts with NaBH$_4$ [Eqs. (36) and (37)] (*77,78*).

$$[NBu_4]_2[Pt_{12}(CO)_{24}] + [Ag(PPh_3)_4][ClO_4] + PPh_3 \rightarrow [AgPt_3(\mu\text{-}CO)_3(PPh_3)_5][ClO_4]$$

$$+ [NBu_4][AgPt_6(CO)_8(PPh_3)_4] + [NBu_4]_2[Pt_{18}(CO)_{36}] \qquad (34)$$

$$Na_2[Fe(CO)_4] + [Ag_3Cl_3\{\mu_3\text{-}(Ph_2P)_3CH\}] \rightarrow [Ag_6Fe_3\{\mu_3\text{-}(Ph_2P)_3CH\}(CO)_{12}] \qquad (35)$$

$$[AgCl(PPh_3)] + [AuCl(PPh_3)] \xrightarrow{NaBH_4} [Ag_{12}Au_{13}Cl_6(PPh_3)_{12}]^{m+} \qquad (36)$$

$$Au_2O_3 + AgAsF_6 + P(C_6H_4Me\text{-}4)_3 + HBr \xrightarrow{NaBH_4}$$

$$[Ag_{19}Au_{18}Br_{11}\{P(C_6H_4Me\text{-}4)_3\}_{12}][AsF_6]_2 \qquad (37)$$

There are a number of examples of the preparation of gold–platinum clusters from the reaction between the complexes $[PtL(PR_3)_2]$ (L = C_2H_4, R = Ph or Me; L = PPh_3, R = Ph) and various gold species [Eqs. (38)–(42)] (72,79–82).

$$[AuW_2(\mu\text{-}CC_6H_4Me\text{-}4)_2(CO)_4(\eta\text{-}C_5H_5)_2][PF_6] + [Pt(C_2H_4)(PMe_3)_2] \rightarrow$$

$$[AuPtW(\mu_3\text{-}CC_6H_4Me\text{-}4)(CO)_2(PMe_3)_3(\eta\text{-}C_5H_5)][PF_6] \qquad (38)$$

$$2\,[Au(CNC_6H_3Me_2\text{-}2,6)_2][PF_6] + 2\,[Pt(C_2H_4)(PPh_3)_2] \rightarrow$$

$$[Au_2Pt_2(PPh_3)_4(CNC_6H_3Me_2\text{-}2,6)_4][PF_6] \qquad (39)$$

$$2\,[Au(NO_3)(PPh_3)] + [Pt(C_2H_4)(PPh_3)_2] \rightarrow [Au_2Pt(NO_3)(PPh_3)_4][NO_3] \qquad (40)$$

$$[Au(NO_3)(PPh_3)] + [Pt(PPh_3)_3] \xrightarrow{CO} [AuPt_3(\mu\text{-}CO)_3(PPh_3)_5][NO_3] \qquad (41)$$

$$[Au(C_2Bu')] + [Pt(PPh_3)_3] \rightarrow [Au_6Pt(C_2Bu')(PPh_3)_7][Au(C_2Bu')_2] \qquad (42)$$

The action of hydrogen or oxygen on heteronuclear gold clusters has been used to afford species of higher nuclearity [Eqs. (43)–(45)] (72,83,84). Mixed-metal gold clusters of lower nuclearity have been

$$[Au_2Pt(NO_3)(PPh_3)_4][NO_3] \xrightarrow[\text{ii. NaBPh}_4]{\text{i. H}_2\,(1\,\text{atm})} [Au_6Pt(PPh_3)_7][BPh_4]_2 \qquad (43)$$

$$[Au_3Ir(NO_3)(PPh_3)_5][BF_4] \xrightarrow[\text{1 atm}]{H_2} [Au_4Ir(\mu\text{-}H)_2(PPh_3)_6][BF_4] \qquad (44)$$

$$[AuFe_4C(\mu\text{-}H)(CO)_{12}(PEt_3)] \xrightarrow[\text{1 atm}]{O_2} [Au_2Fe_4C(CO)_{12}(PEt_3)_2] \qquad (45)$$

prepared by treating larger species with small molecules like PPh_3, CO, I_2, HI, and O_2 [Eqs. (46)–(49)] (72,84–86).

$$[Au_5ReH_4\{P(C_6H_4Me\text{-}4)_3\}_2(PPh_3)_5][PF_6]_2 + 3\,PPh_3 \rightarrow$$

$$[Au_4ReH_4\{P(C_6H_4Me\text{-}4)_3\}_2(PPh_3)_4][PF_6] \qquad (46)$$

$$[Au_2Ru_5C(CO)_{14}L_2] + CO \rightarrow [Au_2Ru_4C(CO)_{12}L_2] \qquad (47)$$

$$L = PPh_3,\ PEt_3,\ \text{or}\ PMe_2Ph$$

$$[Au_2Ru_4C(CO)_{12}(PR_3)_2] + XI \rightarrow [AuRu_4C(\mu\text{-}X)(CO)_{12}(PR_3)] \tag{48}$$

$$X = H \text{ or } I; R = Ph \text{ or } Et$$

$$[Au_2Fe_5C(CO)_{14}(PEt_3)_2] \xrightarrow[\text{1 atm}]{O_2} [Au_2Fe_4C(CO)_{12}(PEt_3)_2]$$

$$+ \text{ unidentified cluster anion} \tag{49}$$

Mixed-metal gold clusters containing isocyanate ligands have been prepared by treating $[Os_3(CO)_{12}]$ with the complex $[AuCl(PEt_3)]$ in the presence of NaN_3 [Eq. (50)] (57). A novel diphosphane-containing gold

$$[Os_3(CO)_{12}] + [AuCl(PEt_3)] + NaN_3 \rightarrow$$

$$[AuOs_3(NCO)(CO)_{11}(PEt_3)] + [AuOs_3(\mu\text{-}NCO)(CO)_{10}(PEt_3)] \tag{50}$$

cluster has been synthesized by deprotonation of $[Ir_4(CO)_{11}(PPhH_2)]$, followed by oxidation with $AgClO_4$ in the presence of $[Au(PEt_3)]^+$ [Eq. (51)] (87).

$$[Ir_4(CO)_{11}(PPhH_2)] \xrightarrow[\text{ii. }[Au(PEt_3)][ClO_4].\ AgClO_4]{\text{i. DBU}}$$

$$[Au_2Ir_4(CO)_9(PPh_3)_2(\mu\text{-}PhPPPh)Ir_4(CO)_{11}] \tag{51}$$

$$DBU = 1,8\text{-diazabicyclo}[5.4.0]\text{undec-7-ene}$$

There are two examples of the synthesis of mixed-metal gold clusters by treating cationic complexes with $[Au(PPh_3)]^+$ [Eqs. (52) and (53)] (88,89).

$$[Ir_2(\mu\text{-}H)_3H_2(PPh_3)_4][PF_6] + [Au(NO_3)(PPh_3)] \rightarrow [Au_3Ir(NO_3)(PPh_3)_5][PF_6] \tag{52}$$

$$[IrH_2(Me_2CO)_2(PPh_3)_2][BF_4] + [Au(NO_3)(PPh_3)] \rightarrow$$

$$[Au_2IrH(NO_3)(PPh_3)_4][BF_4] \tag{53}$$

Interestingly, in a similar reaction, a $Au(NO_3)$ unit was added to the precursor instead of a $Au(PPh_3)$ fragment [Eq. (54)] (88). The nitrate ligand attached to the gold atom can subsequently be replaced by PPh_3 to afford $[AuIr_3(\mu\text{-}H)_3H_3(PPh_3)\{Ph_2P(CH_2)_2PPh_2\}_3][BF_4]$ (88).

$$[Ir_3(\mu_3\text{-}H)(\mu\text{-}H)_3H_3\{Ph_2P(CH_2)_2PPh_2\}_3][BF_4]_2 + 2\ [Au(NO_3)(PPh_3)] \rightarrow$$

$$[AuIr_3(\mu\text{-}H)_3H_3(NO_3)\{Ph_2P(CH_2)_2PPh_2\}_3][BF_4] \tag{54}$$

A homonuclear gold cluster has been used as a precursor to a heteronuclear species in which two gold atoms are bonded to $Co(CO)_4$ fragments [Eq. (55)] (90). Treatment of the cluster salt $K[Co_3Fe(CO)_{12}]$ with $[Bu^n_4N][AuI_2]$ also affords a mixed-metal cluster containing a $Au\{Co(CO)_4\}$ unit [Eq. (56)] (91).

$$[Au_8(PPh_3)_7][NO_3]_2 + 2\ Li[Co(CO)_4] \rightarrow [Au_6Co_2(CO)_8(PPh_3)_4] \tag{55}$$

$$2\ K[Co_3Fe(CO)_{12}] + [NBu^n_4][AuI_2] \rightarrow [NBu^n_4][\{(CO)_4Co\}AuCo_3Fe(CO)_{12}] \tag{56}$$

B. *Structures*

In the vast majority of heteronuclear cluster compounds which contain ML (M = Cu, Ag, or Au; L = two-electron donor ligand) units, the Group IB metals are ligated by PR_3 (R = alkyl or aryl). One of the very interesting features of the chemistry of these phosphine-containing species is the wide range of skeletal geometries that they exhibit. This flexibility in metal framework structures is thought (7,92) to arise from the bonding capabilities of the $M(PR_3)$ units. Theoretical studies by Evans and Mingos (93) have shown that, as the orbitals with a predominance of d character are filled, the bonding characteristics of $M(PR_3)$ moieties are determined primarily by the $hy(s-z)$ orbital (a_1 symmetry) and a higher lying pair of p_x and p_y orbitals (e symmetry). These frontier orbitals allow the $M(PR_3)$ groups to adopt either a bonding mode bridging a metal–metal bond or one capping a triangular three-metal face in heteronuclear clusters containing a single coinage metal (92) (Fig. 1). For species in which there is more than one $M(PR_3)$ unit, the situation is even more complicated, as the energy differences between structures with and without the Group IB metals in close contact seem to be small. However, in almost all cases, the $M(PR_3)$ groups are still either edge-bridging or capping three-metal faces. The wide range of structural types that the metal frameworks of heteronuclear clusters containing ML fragments are known to adopt are surveyed in Sections II,B,1 to II,B,4. Section II,B,5 discusses the short M---C contacts which are frequently observed between the coinage metals and CO ligands in these clusters.

1. *Mixed-Metal Clusters Containing One ML (M = Cu, Ag, or Au; L = Two-Electron Donor Ligand) Unit*

Mixed-metal clusters containing one ML fragment are very common. Table I lists the compounds of this type which have an edge-bridging ML

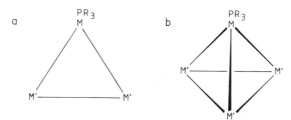

FIG. 1. Edge-bridging (a) and face-capping (b) bonding modes adopted by $M(PR_3)$ (M = Cu, Ag, or Au; R = alkyl or aryl) units in mixed-metal clusters. M' is a transition metal other than a Group IB metal.

TABLE I

MIXED-METAL CLUSTERS CONTAINING ONE EDGE-BRIDGING ML (M = Cu, Ag, OR Au; L = TWO-ELECTRON DONOR LIGAND) UNIT

Cluster	Spectroscopic data reported	Crystal structure determined?	Ref.
Copper clusters			
$[CuW_2(\mu_3-CC_6H_4Me-4)(CO)_4(PPh_3)(\eta-C_5H_5)]^a$	IR, ^1H, ^{31}P, ^{13}C NMR	Yes	74
$[CuMn_2(\mu-PPh_2)(CO)_8(PEt_3)]$	IR, ^1H NMR, Mass	No	14
$[CuMn_2\{\mu-(EtO)_2POP(OEt)_2\}_n(\mu-Br)(CO)_{8-2n}(PPh_3)]$	IR, ^1H, ^{31}P NMR	No	94
(M = Mn or Re, n = 1 or 2)			
$[N(PPh_3)_2][CuFe_4(CO)_{13}(PPh_3)]^b$	IR, ^1H, ^{31}P, ^{13}C NMR	Yes	104,105
$[N(PPh_3)_2][CuFe_4(CO)_{13}(CNC_6H_3Me_2-2,6)]^b$	IR, ^1H, ^{31}P, ^{13}C NMR	No	104,105
$[CuFe_3(\mu-CMe)(CO)_{10}(PPh_3)]$	IR, ^1H NMR, Mass	Yes	95
$[CuFe_3(\mu-H)_2(\mu_3-CMe)(CO)_9(PPh_3)]$	^1H NMR, Mass	No	95
$[CuFe_3(\mu-COMe)(CO)_{10}(PPh_3)]$	IR, ^1H, ^{31}P, ^{13}C NMR	No	43
$[CuRu_3(CO)_9(C_2Bu^t)(PPh_3)]$	IR, ^1H, ^{31}P, ^{13}C NMR	Yes	43
$[CuRu_3(CO)_9(C_2Bu^t)(NCMe)]^c$	IR	No	43
$[CuRu_3(\mu-H)(\mu_3-PPh)(CO)_9(PEt_3)]$	IR, ^1H NMR, Mass	No	20
$[CuRu_3(\mu_3-PhPCH_2PPh_2)(CO)_9(PPh_3)]$	IR, ^1H NMR	Yes	21
$[CuRu_3(\mu-H)(\mu_3-S)(CO)_8L(PPh_3)]$ (L = CO or PPh_3)	IR, ^1H, ^{31}P, ^{13}C NMR	No	18
$[CuOs_3(\mu-CH=CHR)(CO)_{10-n}(PPh_3)_{n+1}]$	IR, ^1H NMR	No	61
(n = 0, R = Ph or C_6F_5; n = 1, R = Ph)			
$[CuOs_3(\mu-H)(CO)_{10}(PPh_3)]$	IR, ^1H NMR	No	61
$[CuOs_3(\mu-H)_3(CO)_{10}(PPh_3)]$	IR, ^1H, ^{13}C NMR	Yes	73
$[CuNiOs_3(\mu-H)_2(CO)_{12}(PPh_3)(\eta-C_5H_5)]$	IR, ^1H NMR, UV	No	96
$[CuPt_2(\mu_3-S)(CO)(PPh_3)_4][PF_6]^d$	IR, ^{31}P, ^{195}Pt NMR	No	68
Silver clusters			
$[AgMn_2(\mu-PPh_2)(CO)_8(PEt_3)]$	IR, ^1H NMR, Mass	No	14
$[AgM_2\{\mu-(EtO)_2POP(OEt)_2\}_n(\mu-Br)(CO)_{8-2n}(PPh_3)]$	IR, ^1H, ^{31}P NMR	No	94
(M = Mn or Re; n = 1 or 2)			

(continued)

TABLE I (*Continued*)

Cluster	Spectroscopic data reported	Crystal structure determined?	Ref.
[AgFe$_3$(μ-COMe)(CO)$_{10}$(PPh$_3$)]^c	IR, ^{1}H NMR	No	43
[AgRu$_3$(CO)$_9$(C$_2$Bu^t)(PPh$_3$)]	IR, ^{1}H, ^{31}P, ^{13}C NMR	Yes	43,97
[AgRu$_3$(CO)$_9$(C$_2$Bu^t)(NCMe)]^c	IR	No	43
[AgRu$_3$(μ-H)(μ_3-PPh)(CO)$_9$(PEt$_3$)]	IR, ^{1}H NMR, Mass	No	20
[AgRu$_3$(μ_3-PhPCH$_2$PPh$_2$)(CO)$_9$(PPh$_3$)]	IR, ^{1}H NMR	Yes	21
[AgRu$_3$(μ-H)(μ_3-S)(CO)$_8$L(PPh$_3$)] (L = CO or PPh$_3$)	IR, ^{1}H, ^{31}P, ^{13}C NMR	No	18
[AgOs$_3$(μ-CH=CHR)(CO)$_{10-n}$(PPh$_3$)$_{n+1}$] (n = 0, R = Ph or C$_6$F$_5$; n = 1, R = Ph)	IR, ^{1}H NMR	No	61
[AgOs$_3$(μ-CH=CHPh)(CO)$_9$(PPh$_3$)]	IR, ^{1}H NMR	No	61
[AgOs$_3$(μ-H)(CO)$_{10}$(PPh$_3$)]	IR, ^{1}H NMR	No	61
[AgOs$_6$P(CO)$_{18}$(PPh$_3$)]	IR, ^{31}P NMR	No	98
[AgPt$_2$(μ_3-S)(CO)(PPh$_3$)$_4$][PF$_6$]^d	IR, ^{31}P, ^{195}Pt NMR	No	68
Gold clusters			
[AuPtW(μ_3-CC$_6$H$_4$Me-4)(CO)$_2$(PMe$_3$)$_3$(η-C$_5$H$_5$)][PF$_6$]^e	IR, ^{1}H, ^{31}P, ^{13}C NMR	Yes	79
[AuMn$_2$(μ-PPh$_2$)(CO)$_8$(PMe$_2$Ph)]	IR, ^{1}H NMR, Mass	Yes	14
[AuMn$_2$(μ-PPh$_2$)(CO)$_8$(PR$_3$)] (R = Ph or Et)	IR, ^{1}H NMR, Mass	No	14
[AuMn$_2${μ-(EtO)$_2$POP(OEt)$_2$}(μ-Br)(CO)$_6$(PPh$_3$)]	IR, ^{1}H, ^{31}P NMR	Yes	94
[AuMn$_2$(μ-H){μ-(EtO)$_2$POP(OEt)$_2$}(CO)$_6$(PPh$_3$)]	IR, ^{1}H, ^{31}P NMR	No	99
[AuRe$_2${μ-(EtO)$_2$POP(OEt)$_2$}(μ-Br)(CO)$_6$(PPh$_3$)]	IR, ^{1}H, ^{31}P NMR	No	94
[AuRe$_2$(μ-H)$_3$H$_4$(PPh$_3$)$_5$]	^{1}H, ^{31}P NMR	Yes	66
[AuRe$_2$(μ-H)$_3$H$_4$(PPh$_3$)$_4$(PEt$_3$)]		No	66
[AuRe$_2$H$_8$(PPh$_3$)$_4$(PR$_3$)][PF$_6$] (R = Ph or Et)	^{1}H, ^{31}P NMR	No	66
[NEt$_4$][[AuRe$_3$(μ-H)$_3$(CO)$_{10}$(PPh$_3$)]^c	IR, ^{1}H NMR	No	100
[AuFe$_2$(μ-C$_{11}$H$_{19}$S)(CO)$_6$(PPh$_3$)]	IR, ^{1}H, ^{13}C NMR	Yes	101
[AuFe$_3$(μ-CMe)(CO)$_{10}$(PPh$_3$)]	IR, ^{1}H NMR, Mass	No	95
[AuFe$_3$(μ-COMe)(CO)$_{10}$(PPh$_3$)]	IR, ^{1}H, ^{31}P, ^{13}C NMR	No	43,46
[AuFe$_3$(μ_3-HC=NBu^t)(CO)$_9$(PPh$_3$)]	IR, ^{1}H NMR	Yes	102

Compound	Spectroscopy	Crystal structure	References
$[AuFe_4C(\mu\text{-}H)(CO)_{12}(PPh_3)]$[f]	^1H, ^{13}C NMR	Yes	103,104
$[AuFe_4C(\mu\text{-}H)(CO)_{12}(PEt_3)]$[f]	IR, ^1H NMR, Mass	No	103
$[AuFe_4(COMe)(CO)_{12}(PEt_3)]$	IR, ^1H, ^{31}P, ^{13}C NMR	Yes	104,105
$[K(18\text{-crown-}6)][AuFe_4(CO)_{13}(PEt_3)]$[g]	IR, ^{31}P, ^{13}C NMR	Yes	104,105
$[K(18\text{-crown-}6)][AuFe_4(CO)_{13}(PPh_3)]$[g]	IR, ^{31}P, ^{13}C NMR	No	104,105
$[N(PPh_3)_2][AuFe_4(CO)_{13}(PEt_3)]$[b]	IR, ^{31}P, ^{13}C NMR	No	84,104,105
$[N(PPh_3)_2][AuFe_4(CO)_{13}(PPh_3)]$[b]	IR, ^{31}P, ^{13}C NMR	No	104,105
$[AuFe_4(\mu_4\text{-}N)(CO)_{12}(PPh_3)]$	IR, ^{15}N NMR	No	106
$[AuFeRu_3(\mu_4\text{-}N)(CO)_{12}(PPh_3)]$	IR, ^{15}N NMR	Yes	106
$[AuCoFeRu(\mu_3\text{-}X)(CO)_9(PPh_3)]$[h] (X = S or PMe)	IR, ^1H NMR	Yes	107
$[AuFeOs_3(\mu\text{-}H)(CO)_{13}(PR_3)]$ (R = Ph or Et)	IR, ^1H NMR	No	108
$[AuCoFe_2(\mu_3\text{-}COMe)(\mu_3\text{-}CO)(CO)_6(PPh_3)(\eta\text{-}C_5H_5)]$	IR, ^1H, ^{31}P, ^{13}C NMR	Yes	50
$[AuCo_2Fe(\mu_3\text{-}CPh)(CO)_9(PPh_3)]$	IR	Yes	109
$[AuRu_3(CO)_9(C_2\text{-}Bu^t)(PPh_3)]$	IR, ^1H, ^{31}P, ^{13}C NMR, ^{197}Au Mössbauer	Yes	29,43 110,111
$[AuRu_3(\mu\text{-}H)(\mu_3\text{-}PPh)(CO)_9(PMe_2Ph)]$	IR, ^1H NMR, Mass	Yes	20
$[AuRu_3(\mu\text{-}H)(\mu_3\text{-}PPh)(CO)_9(PEt_3)]$	IR, ^1H NMR, Mass	No	20
$[AuRu_3(\mu_3\text{-}PhPCH_2PPh_2)(CO)_9(PPh_3)]$	IR, ^1H NMR	Yes	21
$[AuRu_3(\mu_3\text{-}PhAsCH_2AsPPh_2)(CO)_9(PPh_3)]$	IR, ^1H NMR	No	21
$[AuRu_3\{\mu_3\text{-}PhP(CH_2)_2PPh_3\}(CO)_9(PPh_3)]$	IR, ^1H NMR	No	21
$[AuRu_3(\mu\text{-}H)(\mu_3\text{-}S)(CO)_9(PPh_3)]$	IR, ^1H, ^{31}P, ^{13}C NMR, Mass	Yes	47,112,113
$[AuRu_3(\mu\text{-}H)(\mu_3\text{-}S)(CO)_8(PPh_3)_2]$	IR, ^1H, ^{31}P, ^{13}C NMR	No	47
$[AuRu_3(\mu_3\text{-}SBu^t)(CO)_9(PPh_3)]$	IR, ^1H NMR	Yes	112
$[AuRu_3(\mu_3\text{-}Cl)(CO)_{10}(PPh_3)]$	IR, ^1H NMR	Yes	56
$[AuRu_3(\mu\text{-}COMe)(CO)_{10}(PPh_3)]$	IR, ^1H, ^{31}P, ^{13}C NMR	Yes	46
$[AuRu_3(\mu\text{-}COMe)(CO)_9(PPh_3)_2]$	IR, ^1H, ^{31}P, ^{13}C NMR	No	114
$[AuRu_3(\mu\text{-}H)_2(\mu_3\text{-}COMe)(CO)_9(PPh_3)]$	IR, ^1H, ^{31}P, ^{13}C NMR, ^{197}Au Mössbauer	Yes	46,111
$[AuRu_4(\mu_3\text{-}H)(\mu\text{-}H)_2(CO)_{12}(PPh_3)]$	IR, ^1H, ^{31}P, ^{13}C NMR, ^{197}Au Mössbauer	Yes	27,43 111,115
$[AuRu_4(\mu_3\text{-}H)(\mu\text{-}H)_2(CO)_{12}(PEt_3)]$	^{197}Au Mössbauer	No	108

(continued)

263

TABLE I (Continued)

Cluster	Spectroscopic data reported	Crystal structure determined?	Ref.
[{AuRu₄(μ₃-H)(μ-H)₂(CO)₁₂}₂(μ-Ph₂PCH₂PPh₂)]	IR, ¹H, ³¹P NMR	No	43
[AuRu₄C(μ-H)(CO)₁₂(PPh₃)]ⁱ	IR	Yes	86
[AuRu₄C(μ-H)(CO)₁₂(PEt₃)]ⁱ	IR, ¹H, ³¹P NMR, Mass	No	86
[AuRu₄C(μ-I)(CO)₁₂(PPh₃)]ⁱ		No	86
[AuRu₄C(μ-I)(CO)₁₂(PEt₃)]ⁱ	IR, ³¹P NMR, Mass	Yes	86
[AuRu₄(μ₄-N)(CO)₁₂(PPh₃)]	IR, ¹⁵N NMR	No	106
[AuRu₅C(CO)₁₃(NO)(PEt₃)]ʲ	IR	Yes	116
[AuRu₅C(CO)₁₃(PPh₃)(η-C₅H₅)]	IR, Mass	Yes	117
[AuRu₅C(CO)₁₃(PEt₃)(η-C₅H₅)]	IR, Mass	No	117
[AuRu₅C(CO)₁₃X(PPh₃)] (X = SEt or μ-I)	IR	No	51
[AuRu₅C(CO)₁₃(μ-I)(PPh₃)₂]	IR	Yes	51
[AuRu₅C(CO)₁₄(μ-Cl)(PPh₃)]		No	55
[AuRu₅C(CO)₁₄(μ-Br)(PPh₃)]		Yes	55
[AuRu₅C(CO)₁₄(μ-I)(PPh₃)]		No	51
[AuRu₅C(CO)₁₄(μ-η-MeCO)(PPh₃)]	IR, ¹H NMR	Yes	51,117
[AuRu₅C(CO)₁₄(SEt)(PPh₃)]	IR, ¹H NMR, Mass	No	51
[AuRu₅C(CO)₁₅Cl(PPh₃)]	IR	Yes	55
[AuRu₅C(CO)₁₅Br(PPh₃)]	IR	No	55
[AuOs₃(CO)₁₁(NCO)(PEt₃)]	IR, ¹H, ¹³C NMR, Mass	No	57
[AuOs₃H(CO)₁₁(PPh₃)]	IR, ¹H NMR	No	25
[AuOs₃H(CO)₁₁(PEt₃)]	¹H NMR, Mass	No	25
[NEt₃H][AuOs₃(CO)₁₁(PPh₃)]	IR	No	25
[NEt₃H][AuOs₃(CO)₁₁(PEt₃)]		No	25
[AuOs₃(μ-CH=CHR)(CO)₁₀(PPh₃)] (R = C₆F₅ or Ph)	IR, ¹H NMR	Yes	61
[AuOs₃(μ-CH=CHPh)(CO)₉L(PMe₃)] (L = CO or PMe₃)	IR, ¹H NMR, Mass	No	61
[AuOs₃(μ-CH=CHPh)(CO)₉(PPh₃)₂]	IR, ¹H NMR	No	61
[AuOs₃(μ-CH=CHPh)(CO)₉(PMe₃)]	IR, ¹H NMR, Mass	No	61
[AuOs₃(μ-CH=CHPh)(CO)₉(PPh₃)]	IR, ¹H NMR	No	61

Compound	Characterization	Structure	Ref.
[AuOs$_3$(μ-H)(CO)$_{10}$(PPh$_3$)]	IR, ^{1}H NMR, Mass	Yes	12,45,61
[AuOs$_3$(μ-H)(CO)$_{10}$(PMe$_3$)]	IR, ^{1}H NMR, Mass	No	61
[AuOs$_3$(μ-H)(CO)$_{10}$(PEt$_3$)]	IR	No	12
[AuOs$_3$(μ-Cl)(CO)$_{10}$(PPh$_3$)]	IR, R, Mass	Yes	54
[AuOs$_3$(μ-Br)(CO)$_{10}$(PPh$_3$)]	IR	Yes	54
[AuOs$_3$(μ-SCN)(CO)$_{10}$(PPh$_3$)]	IR, R	Yes	12,54
[AuOs$_3$(μ-X)(CO)$_{10}$(PR$_3$)] (X = Cl, R = C$_6$H$_4$Me; X = I, R = Ph)	IR	No	54
[AuOs$_3$(μ-NCO)(CO)$_{10}$(PEt$_3$)]	IR, ^{1}H, ^{13}C NMR, Mass	No	57
[AuOs$_3$(μ-NCO)(CO)$_9$(PPh$_3$)(PEt$_3$)]	IR, ^{1}H NMR, Mass	No	57
[AuOs$_3$(μ-NO)(CO)$_{10}$(PEt$_3$)]	IR, ^{1}H NMR, Mass	No	57
[AuOs$_3$(μ-X)(CO)$_{10}$(PEt$_3$)] (X = NCNPh, NHCNPhH, or NHCNPhNHCH$_2$Ph)	IR, ^{1}H NMR, Mass	No	57
[AuOs$_3$(μ-NHCOY)(CO)$_{10}$(PEt$_3$)] [Y = H, OH, OMe, PhCH$_2$NH, or PhCH$_2$(CO$_2$Me)CHNH]	IR, ^{1}H NMR, Mass	No	57
[AuOs$_3$(μ-X)(CO)$_{10}$(PPh$_3$)] (X = OH or NHSO$_2$C$_6$H$_4$Me-4)	IR, ^{1}H NMR, Mass	No	60
[AuOs$_3$(μ-OEt)(CO)$_{10}$(PEt$_3$)]	IR, ^{1}H NMR, Mass	No	25
[AuOs$_3$(CO)$_{10}$(C$_2$Ph)(PMe$_2$Ph)]	IR, ^{1}H NMR	Yes	58
[AuOs$_3$(CO)$_{10}$(C$_2$Ph)(PPh$_3$)]	IR, ^{1}H NMR, Mass	No	58
[AuOs$_3$(CN)(CO)$_{10}$(PPh$_3$)][l]	IR	No	58
[AuOs$_3$(CO)$_9$(C$_2$Ph)(PMe$_2$Ph)][k]	IR	No	58
[AuOs$_3$(CO)$_9$(C$_2$Ph)(PPh$_3$)]	IR	No	58
[AuOs$_3$(CO)$_9$(2-NHC$_5$H$_4$N)(PPh$_3$)]	IR, ^{1}H NMR	No	60
[AuOs$_3$(CO)$_8$(2-NHC$_5$H$_4$N)(PPh$_3$)$_2$]	IR, ^{1}H NMR	Yes	60
[AuOs$_4$(μ-H)$_3$(CO)$_{12}$(PEt$_3$)]		Yes	108
[AuOs$_4$(μ-H)$_3$(CO)$_{12}$(PPh$_3$)]		No	108
[AuOs$_4$(μ-H)(CO)$_{13}$(PEt$_3$)]		Yes	108
[AuOs$_4$(μ-H)(CO)$_{13}$(PPh$_3$)]		No	108
[AuOs$_5$(μ-H)(CO)$_{15}$(PR$_3$)] (R = Et or Ph)		No	108
[AuOs$_6$P(CO)$_{18}$(PPh$_3$)]	IR, ^{31}P NMR	Yes	98

(continued)

265

TABLE I (Continued)

Cluster	Spectroscopic data reported	Crystal structure determined?	Ref.
[PMePh$_3$][AuOs$_{10}$C(CO)$_{24}$(PPh$_3$)]	IR	Yes	35
[AuRh$_2$(μ-CO)(μ-Ph$_2$PCH$_2$PPh$_2$)(PPh$_3$)(η-C$_5$H$_5$)$_2$][BF$_4$]^m	IR, ^{1}H, ^{31}P NMR	Yes	63
AuPt$_2$(μ-Ph$_2$PCH$_2$PPh$_2$)$_2$Cl$_2$(PPh$_3$)][NO$_3$]	^{1}H, ^{31}P NMR	No	64
AuPt$_2$(μ-Ph$_2$PCH$_2$PPh$_2$)$_2$Cl$_2$L][NO$_3$]		No	64
[L = Ph$_2$PCH$_2$PPh$_2$AuCl or Ph$_2$P(CH$_2$)$_2$PPh$_2$AuCl]			
[AuPt$_2$(μ_3-S)(CO)(PPh$_3$)$_4$][PF$_6$]^{d,n}	IR, ^{31}P, ^{195}Pt NMR	Yes	68

^{a} As well as bridging a W—W edge, the Cu(PPh$_3$) fragment is also bonded to the carbyne C atom in the CC$_6$H$_4$Me-4 unit.

^{b} In the solid state, the cluster anion adopts the structure of isomer A (Fig. 9), but skeletal isomers A and B (Fig. 9) are both present in equilibrium in solution.

^{c} The cluster is too unstable to be isolated as a pure compound.

^{d} As well as bridging a Pt—Pt edge, the M(PPh$_3$) (M = Cu, Ag, or Au) unit is also bonded to the sulfur atom.

^{e} As well as bridging a Pt—W edge, the Au(PMe$_3$) group is also bonded to the carbyne C atom of the CC$_6$H$_4$Me-4 unit.

^{f} The edge-bridging Au(PR$_3$) (R = Ph or Et) unit interacts with the carbido C atom as well as two Fe atoms.

^{g} In the solid state, the cluster anion adopts the structure of isomer B (Fig. 9), but skeletal isomers A and B (Fig. 9) are both present in equilibrium in solution.

^{h} Only the heavy atom framework of [AuCoFeRu(μ_3-S)(CO)$_9$(PPh$_3$)] could be determined unequivocally, because of disorder.

^{i} The edge-bridging Au(PR$_3$) (R = Ph or Et) unit interacts with the carbido C atom as well as two Ru atoms.

^{j} The cluster exists as two skeletal isomers in the solid state (Fig. 8).

^{k} The cluster was first synthesized by Lewis and co-workers (60), but they incorrectly formulated it as [AuOs$_3$(CO)$_9$(C$_2$Ph)(PPh$_3$)] (58).

^{l} The cluster could only be obtained as an oil.

^{m} The cluster cation has also been isolated as the [PF$_6$]$^-$ salt.

^{n} The two Au—Pt separations in this cluster are long [3.312(2) and 3.016(2) Å]. The larger separation is well out of the range normally associated with Au—Pt bonding (68).

unit, and Table II presents those with a face-capping ML moiety.[4] For the vast majority of both structural types, L = PR_3 (R = alkyl or aryl).

According to the calculations of Evans and Mingos (93), the $Au(PR_3)$ fragment effectively functions as $[Au(PR_3)]^+$ in heteronuclear clusters, and both the edge-bridging and the face-capping bonding modes adopted by this moiety result in an increment of 12 to the polyhedral electron count of the cluster. When the $Au(PR_3)$ group bridges a metal–metal bond, it utilizes its a_1 hybrid (s–z) orbital in a three-center, two-electron bonding interaction, and the e set of p_x and p_y orbitals are too high lying in energy to make an effective contribution. Alternatively, the a_1 and e frontier orbitals of a capping $Au(PR_3)$ unit are destabilized by interactions with a triangular three-metal face so that they are not available for skeletal bonding. The polyhedral skeletal electron pair theory assumes that each edge-bridging transition metal fragment will donate 14 electrons to the cluster and each face-capping transition metal unit will donate 12 electrons (92). Thus, an edge-bridging $Au(PR_3)$ moiety in a mixed-metal cluster leads to a polyhedral electron count of two fewer than that predicted by the above theory, whereas a face-bridging $Au(PR_3)$ group will give the correct number of electrons.

As the p_x and p_y orbitals of the $Au(PR_3)$ fragment play only a secondary bonding role, Evans and Mingos (93) have suggested that the $Au(PR_3)$ fragment can be considered to be isolobal with a $Mn(CO)_5$ unit or a hydrido ligand. Indeed, the experimental evidence for mixed-metal clusters containing just one $Au(PR_3)$ group supports the idea of an isolobal relationship between the gold–phosphine unit and the hydrido ligand. In almost all compounds where structural comparisons are possible, the $Au(PR_3)$ fragment occupies a similar position to that of the isolobal hydrido ligand in the related hydrido–metal cluster. Figs. 2 and 3 show edge-bridging (12,129) and face-capping (19,130) examples, respectively. There are, however, a few exceptions to this isolobal relationship, and two examples are shown in Figs. 4 and 5. The $Au(PR_3)$ group in $[AuFe_3$-$(\mu_3$-HC=NBu^t)(CO)$_9$(PPh$_3$)] bridges a different edge of a trigonal planar Fe_3 unit to that bridged by the hydrido ligand in $[Fe_3(\mu$-H)(μ_3-MeC=NH) (CO)$_9$] (Fig. 4) (102,131). The cluster dianion $[Re_7C(\mu_3$-H)-(CO)$_{21}]^{2-}$ exists as two isomers in solution. Although the predominant isomer has the hydrido ligand capping the same Re_3 face as the $Au(PPh_3)$ moiety in $[AuRe_7C(CO)_{21}(PPh_3)]^{2-}$, the minor isomer has the hydrido ligand capping a different Re_3 face (Fig. 5) (125).

---

[4] Abbreviations used for the spectroscopic data listed in the tables: IR, infrared; R, Raman; UV, ultraviolet; NMR, nuclear magnetic resonance; Mass, mass spectrometry; ESR, electron spin resonance.

TABLE II

MIXED-METAL CLUSTERS CONTAINING ONE FACE-CAPPING ML (M = Cu, Ag, or Au; L = TWO-ELECTRON DONOR LIGAND) UNIT

Cluster	Spectroscopic data reported	Crystal structure determined?	Ref.
Copper clusters			
$[NEt_4]_2[CuRe_7C(CO)_{21}(NCMe)]$	IR, ^1H NMR	No	118
$[N(PPh_3)_2][CuFe_4(CO)_{13}(PPh_3)]$[a]	IR, ^1H, ^{31}P, ^{13}C NMR	Yes	104,105
$[N(PPh_3)_2][CuFe_4(CO)_{13}(CNC_6H_3Me_2\text{-}2,6)]$[a]	IR, ^1H, ^{13}C NMR	No	104,105
$[CuCo_3Fe(CO)_{12}(PPh_3)]$	IR, ^{31}P NMR, UV	No	119,120
$[CuCo_3Fe(CO)_{12}L]$	IR, UV	No	119,120
(L = AsPh₃, NCMe, or ⟨P–Ph ring⟩)			
$[CuRu_4(\mu_3\text{-}H)_3(CO)_{12}(PMePh_2)]$	IR, ^1H, ^{31}P, ^{13}C NMR	Yes	43
$[CuRu_4(\mu_3\text{-}H)_3(CO)_{12}(PPh_3)]$	IR, ^1H, ^{31}P, ^{13}C NMR	No	43,115
$[CuRu_4(\mu_3\text{-}H)_3(CO)_{12}(PR_3)]$	IR, ^1H, ^{31}P NMR	No	115
(R = $C_6H_4Me\text{-}2$ or $C_6H_4Me\text{-}4$)			
$[\{CuRu_4(\mu_3\text{-}H)_3(CO)_{12}\}_2\{\mu\text{-}Ph_2P(CH_2)_2PPh_2\}]$[b]			
$[CuRu_4(\mu_3\text{-}H)_3(CO)_{12}(NCMe)]$[b]	IR, ^1H, ^{31}P, ^{13}C NMR	No	43
$[CuCo_3Ru(CO)_{12}(PPh_3)]$	IR	No	43
$[CuCo_3Ru(CO)_{12}(PPh_3)]$	IR, ^{31}P NMR, UV	Yes	119,120
$[CuCo_3Ru(CO)_{12}(NCMe)]$	IR, UV, Mass	No	120
$[CuCo_3Ru(CO)_{12}L]$	IR, UV	No	120
(L = ⟨P–Ph ring⟩)			
$[CuOs_4\{\mu_3\text{-}N(CO)Me\}(CO)_{12}(PPh_3)]$	IR, ^1H NMR	Yes	121
$[PMePh_3][CuOs_{10}C(CO)_{24}(NCMe)]$	IR	Yes	35

Compound	Methods	X-ray	Ref.
[PMePh$_3$][CuOs$_{11}$C(CO)$_{27}$(NCMe)]^c	IR	Yes	122
[NMe$_3$(CH$_2$Ph)][CuRh$_6$C(CO)$_{15}$(NCMe)]^a	IR, ^{13}C, ^{103}Rh NMR	No	32,123
Silver clusters			
[NEt$_4$]$_2$[AgRe$_7$C(CO)$_{21}$(PPh$_3$)]	IR, ^{1}H NMR	No	118
[AgCo$_3$M(CO)$_{12}$L]	IR, UV	No	120
(M = Fe or Ru; L = $\overset{\displaystyle P}{\underset{\displaystyle Ph}{}}$)			
[AgRu$_4$(μ_3-H)$_3$(CO)$_{12}$(PPh$_3$)]	IR, ^{1}H, ^{31}P, ^{13}C, ^{109}Ag INEPT NMR	No	43,124
[(AgRu$_4$(μ_3-H)$_3$(CO)$_{12}$)$_2${μ-Ph$_2$P(CH$_2$)$_2$PPh$_2$}]	IR, ^{1}H, ^{31}P NMR	No	43
[AgRu$_4$(μ_3-H)$_3$(CO)$_{12}$(NCMe)]^b		No	43
[NMe$_3$(CH$_2$Ph)][AgRh$_6$C(CO)$_{15}$(PEt$_3$)]^{d,e}	^{31}P, ^{13}C, ^{103}Rh NMR	No	123
[AgPt$_3$(μ_3-CO)$_3$(PPh$_3$)$_5$][ClO$_4$]	IR, ^{1}H NMR	Yes	75
Gold clusters			
[PPh$_4$][AuRe$_3$(μ-H)$_3$(CO)$_9$(PPh$_3$)]	IR, ^{1}H NMR	Yes	100
[AsPh$_4$]$_2$[AuRe$_7$C(CO)$_{21}$(PPh$_3$)]	^{13}C NMR, Mass	Yes	125
[N(PPh$_3$)$_2$]$_2$[AuRe$_7$C(CO)$_{21}$(PPh$_3$)]	IR, ^{1}H, ^{31}P NMR	No	125
[NEt$_4$]$_2$[AuRe$_7$C(CO)$_{21}$(PPh$_3$)]	IR, ^{31}P, ^{13}C NMR	No	118
[K(18-crown-6)][AuFe$_4$(CO)$_{13}$(PEt$_3$)]^f	IR, ^{31}P, ^{13}C NMR	Yes	104,105
[K(18-crown-6)][AuFe$_4$(CO)$_{13}$(PPh$_3$)]^f	IR, ^{31}P, ^{13}C NMR	No	104,105
[N(PPh$_3$)$_2$][AuFe$_4$(CO)$_{13}$(PEt$_3$)]^a	IR, ^{31}P, ^{13}C NMR	No	84,104,105
[N(PPh$_3$)$_2$][AuFe$_4$(CO)$_{13}$(PPh$_3$)]^a	IR, ^{31}P, ^{13}C NMR	No	104,105
AuCo$_3$Fe(CO)$_{12}$(PPh$_3$)	IR	Yes	19
[NBu^4][((CO)$_4$Co}AuCo$_3$Fe(CO)$_{12}$]	IR, UV	No	91
AuRu$_5$C(CO)$_{13}$(NO)(PEt$_3$)^g	IR	Yes	116
AuRu$_6$C(CO)$_{15}$(NO)(PPh$_3$)^h	IR	Yes	126
AuCoRu$_3$(μ-CO)$_3$(CO)$_{10}$(PPh$_3$)	IR, ^{1}H NMR	Yes	31
[AuCo$_3$Ru(CO)$_{12}$(PPh$_3$)]	IR, UV	Yes	119

(continued)

TABLE II (Continued)

Cluster	Spectroscopic data reported	Crystal structure determined?	Ref.
[AuCo$_{3-x}$Rh$_x$Ru(CO)$_{12}$]^i (x = 1–3)	IR, ^{31}P NMR	Yes	127
[AuOs$_4${μ_3-N(CO)Me}(CO)$_{12}$(PPh$_3$)]	IR, ^{1}H NMR	Yes	121
[PMePh$_3$][AuOs$_{11}$C(CO)$_{27}$(PPh$_3$)]^j	IR	No	122
[PPh$_4$][AuRh$_6$C(CO)$_{15}$(PPh$_3$)]	IR	Yes	128
[NMe$_3$(CH$_2$Ph)][AuRh$_6$C(CO)$_{15}$(PEt$_3$)]^d	IR, ^{31}P, ^{13}C, ^{103}Rh NMR	No	123
[AuIr$_3$(μ-H)$_3$H$_3$(NO$_3$){Ph$_2$P(CH$_2$)$_2$PPh$_2$}$_3$][BF$_4$]	^{1}H, ^{31}P NMR, Mass	Yes	88
[AuIr$_3$(μ-H)$_3$H$_3${Ph$_2$P(CH$_2$)$_2$PPh$_2$}$_3$(PPh$_3$)][BF$_4$]	^{1}H, ^{31}P NMR	No	88
[AuPt$_3$(μ-CO)$_3$(PPh$_3$)$_5$][NO$_3$]	IR, ^{31}P, ^{195}Pt NMR	Yes	81
[AuPt$_3$(μ-CO)$_3${P(cyclo-C$_6$H$_{11}$)$_3$}$_4$][PF$_6$]	IR, ^{31}P, ^{195}Pt NMR	Yes	59,62
[AuPt$_3$(μ-CO)$_2$(μ-SO$_2$){P(cyclo-C$_6$H$_{11}$)$_3$}$_4$][PF$_6$]	IR, ^{31}P, ^{195}Pt NMR	Yes	59
[AuPt$_3$(μ-SO$_2$)$_2$(μ-Cl){P(cyclo-C$_6$H$_{11}$)$_3$}$_3${P(C$_6$H$_4$F-4)$_3$}]	IR, ^{31}P, ^{195}Pt NMR	Yes	59
[AuPt$_3$(μ-SO$_2$)$_3${P(cyclo-C$_6$H$_{11}$)$_3$}$_4$][PF$_6$]	IR, ^{31}P, ^{195}Pt NMR	No	59

^{a} In the solid state, the cluster anion adopts the structure of isomer A (Fig. 9), but skeletal isomers A and B (Fig. 9) are both present in equilibrium in solution.

^{b} The cluster is too unstable to be isolated as a pure compound.

^{c} The Cu(NCMe) unit caps a triangular Os$_3$ face, and the Cu atom is also involved in a very long interaction with another Os atom [3.156(4) Å].

^{d} The cluster anion has not been isolated in the solid state.

^{e} The cluster is stable in solution only at low temperatures.

^{f} In the solid state, the cluster anion adopts the structure of isomer B (Fig. 9), but skeletal isomers A and B (Fig. 9) are both present in equilibrium in solution.

^{g} The cluster exists as two skeletal isomers in the solid state (Fig. 8).

^{h} The Au(PPh$_3$) unit adopts an unusual asymmetrical bonding mode with two short Au—Ru distances (mean 2.782 Å) one long Au—Ru separation (3.19 Å).

^{i} The mixture of cluster compounds was not separated.

^{j} The cluster anion decomposes on standing in CH$_2$Cl$_2$ solution, losing Au(PPh$_3$) and Os(CO)$_3$ fragments to form the dianion [Os$_{10}$C(CO)$_{24}$]$^{2-}$.

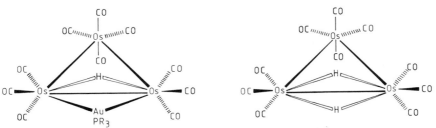

FIG. 2.   Structures of the clusters $[AuOs_3(\mu\text{-H})(CO)_{10}(PR_3)]$ (R = Ph or Et) and $[Os_3\text{-}(\mu\text{-H})_2(CO)_{10}]$, showing the isolobal relationship between the edge-bridging $Au(PR_3)$ unit and the hydrido ligand.

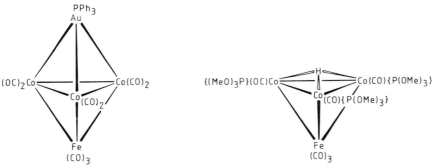

FIG. 3.   Structures of the clusters $[AuCo_3Fe(\mu\text{-CO})_3(CO)_9(PPh_3)]$ and $[Co_3Fe(\mu_3\text{-H})\text{-}(\mu\text{-CO})_3(CO)_6\{P(OMe)_3\}_3]$, showing the isolobal relationship between the face-capping $Au(PPh_3)$ unit and the hydrido ligand. The CO ligands bridging the three Co—Co vectors in each cluster have been omitted for clarity.

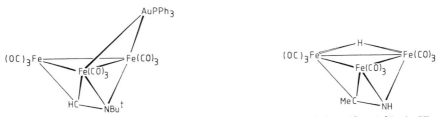

FIG. 4.   Structures of the clusters $[AuFe_3(\mu_3\text{-HC}{=}NBu^t)(CO)_9(PPh_3)]$ and $[Fe_3(\mu\text{-H})\text{-}(\mu_3\text{-MeC}{=}NH)(CO)_9]$.

The bonding situation is more confused, however, for $M(PR_3)$ (M = Cu or Ag) moieties. According to Evans and Mingos (93), the degenerate p_x and p_y orbitals of the $Cu(PR_3)$ fragment are much lower in energy than those of the $Au(PR_3)$ unit, with the e set of orbitals of the $Ag(PR_3)$ group

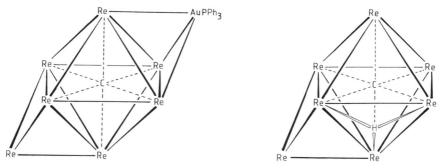

Fig. 5. Structures of the cluster anion $[AuRe_7C(CO)_{21}(PPh_3)]^{2-}$ and the less abundant isomer of $[Re_7C(\mu_3-H)(CO)_{21}]^{2-}$, with the three terminal CO ligands attached to each Re atom omitted for clarity.

lying at an intermediate energy. Thus, the $Cu(PR_3)$ moiety can be considered to be isolobal with a $Mn(CO)_3$ unit, rather than with a hydrido ligand. Although a transition metal tricarbonyl fragment has suitable frontier molecular orbitals to act as a face-capping group, this moiety does not function effectively as an edge-bridging unit in cluster compounds, as one component of its e set of molecular orbitals is not effectively utilized in bonding when it adopts such a position. Therefore, an edge-bridging bonding mode for a $Cu(PR_3)$ moiety in a mixed-metal cluster would be expected to be less favorable than that for a $Au(PR_3)$ fragment in the analogous gold species (43). Indeed, there is some experimental evidence for such site preferences in Group IB metal heteronuclear clusters (35,43), and Fig. 6 shows an example (43,115). Braunstein et al. (119), however,

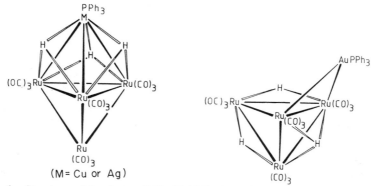

Fig. 6. Structures of the clusters $[MRu_4H_3(CO)_{12}(PPh_3)]$ (M = Cu, Ag, or Au), showing the site preferences of the $M(PPh_3)$ units.

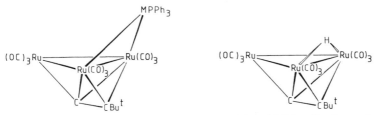

FIG. 7.   Structures of the clusters [MRu$_3$(CO)$_9$(C$_2$Bu^t)(PPh$_3$)] (M = Cu, Ag, or Au) and [Ru$_3$(μ-H)(CO)$_9$(C$_2$Bu^t)].

have recently reported that they found no significant difference in energy for the p_x and p_y orbitals of Cu(PH$_3$) and Au(PH$_3$) units. This theoretical result is supported by the experimental observation that, taking the different covalent radii of Cu and Au into account, the ranges of the Cu—Co and Au—Co distances for the face-capping M(PPh$_3$) fragments in [MCo$_3$-Ru(CO)$_{12}$(PPh$_3$)] (M = Cu or Au) are not significantly different (119). Either way, Table I clearly shows that a number of stable mixed-metal cluster compounds containing edge-bridging Cu(PR$_3$) and Ag(PR$_3$) fragments have been isolated and fully characterized. In a number of cases, the M(PR$_3$) (M = Cu or Ag) unit occupies the same position as that of the hydrido ligand in the related hydrido metal cluster, as is frequently observed for the clusters containing a single Au(PR$_3$) moiety. Fig. 7 shows an example (43,97,132).

There is a considerable amount of experimental evidence available to suggest that the energy differences between face-capping and edge-bridging bonding modes for M(PR$_3$) (M = Cu, Ag, or Au) fragments can be very small. Two recent reports of skeletal isomerism in heteronuclear clusters containing a single M(PR$_3$) moiety provide some of this evidence (104,105,116) (see also Sections II,B,2, II,B,3, II,B,4, and II,E,1). A single-crystal X-ray diffraction study on the hexanuclear cluster [AuRu$_5$-C(CO)$_{13}$(NO)(PEt$_3$)] revealed two independent molecules with different metal core geometries (116). Both of these molecules contained square-based pyramidal Ru$_5$ units, but in one case the Au(PEt$_3$) fragment capped a triangular Ru$_3$ face of the unit and in the other it bridged a Ru—Ru edge (Fig. 8). Two isomeric forms of the cluster anions [MFe$_4$(CO)$_{13}$L]$^-$ (M = Cu, L = PPh$_3$ or CNC$_6$H$_3$Me$_2$-2,6; M = Au, L = PPh$_3$ or PEt$_3$) exist in equilibrium in solution (104,105). One isomer (A in Fig. 9) has a metal framework structure consisting of a Fe$_4$ tetrahedron with one triangular face capped by the ML fragment, whereas the other (B in Fig. 9) exhibits a skeletal geometry consisting of a Fe$_4$ "butterfly" with the ML fragment edge-bridging the "hinge" of the "butterfly." Interestingly, the

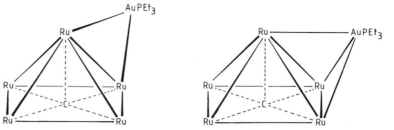

FIG. 8. Structures of the two skeletal isomers of the clusters [AuRu$_5$C(CO)$_{13}$(NO)-(PEt$_3$)], with the CO and NO ligands omitted for clarity.

two copper–iron and the two gold–iron cluster anions exhibit the tetra-hedral structure of isomer A in the solid state when they are isolated as the [N(PPh$_3$)$_2$]$^+$ salts. In the case of both gold–iron species, however, a change of cation to [K(18-crown-6)]$^+$ alters the solid state cluster geometry to the "butterfly" structure of isomer B (104,105).

Detailed structural investigations of all three members of a series of Group IB metal congeners are rare. However, two comparative studies of series of analogous clusters containing edge-bridging M(PPh$_3$) (M = Cu, Ag, or Au) units have been performed. Single-crystal X-ray diffraction studies on the clusters [MRu$_3$(μ-PhPCH$_2$PPh$_2$)(CO)$_9$(PPh$_3$)] and [MRu$_3$-(CO)$_9$(C$_2$Bu^t)(PPh$_3$)], which all exhibit "butterfly" metal core structures with the coinage metals occupying "wing-tip" sites, reveal some interesting structural trends (21,43,97,110). As can be seen from Table III, the Ag—P bond lengths in the two silver–ruthenium clusters are very much longer than either the Cu—P or the Au—P separations in the analogous copper–ruthenium and gold–ruthenium species. In addition, the M—Ru bond

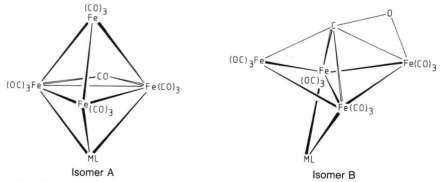

FIG. 9. Structures of the two skeletal isomers of the cluster anions [MFe$_4$(CO)$_{13}$L]$^-$ (M = Cu, L = PPh$_3$ or CNC$_6$H$_3$Me$_2$-2,6; M = Au, L = PPh$_3$ or PEt$_3$).

TABLE III

M—P and M—Ru Separations and Interplanar Angles in the Two Series of Group IB
Metal Congeners [MRu$_3$(μ_3-PhPCH$_2$PPh$_2$)(CO)$_9$(PPh$_3$)] and [MRu$_3$(CO)$_9$(C$_2$Bu^t)(PPh$_3$)]
(M = Cu, Ag, or Au)

Cluster	M—P separation (Å)	M—Ru separations (Å)	Interplanar (MRu$_2$/Ru$_3$) angle (°)
[CuRu$_3$(μ_3-PhPCH$_2$PPh$_2$)(CO)$_9$(PPh$_3$)][a]	2.228(2)	2.607(1), 2.622(1)	133.6
[AgRu$_3$(μ_3-PhPCH$_2$PPh$_2$)(CO)$_9$(PPh$_3$)][a]	2.422(3)	2.767(1), 2.806(1)	143.9
[AuRu$_3$(μ-PhPCH$_2$PPh$_2$)(CO)$_9$(PPh$_3$)][a]	2.297(2)	2.751(1), 2.786(1)	144.2
[CuRu$_3$(CO)$_9$(C$_2$Bu^t)(PPh$_3$)][b]	2.217(2)	2.603(1), 2.603(1)	115.7
[AgRu$_3$(CO)$_9$(C$_2$Bu^t)(PPh$_3$)][c]	2.405(9)	2.785(3), 2.788(3)	120.6
[AuRu$_3$(CO)$_9$(C$_2$Bu^t)(PPh$_3$)][d]	2.276(3)	2.757(1), 2.763(1)	129.3

[a] Ref. 21.
[b] Ref. 43.
[c] Ref. 97.
[d] Ref. 110.

lengths of the edge-bridging M(PPh$_3$) units decrease with the changing M in the order Ag > Au > Cu in both series of clusters, and the dihedral angles between the "wings" of the "butterfly" (MRu$_2$/Ru$_3$) are smallest for the copper–ruthenium species in each case.

2. Mixed-Metal Clusters Containing Two ML (M = Cu, Ag, or Au; L = Two-Electron Donor Ligand) Units

A considerable number of mixed-metal cluster compounds containing two ML fragments have been reported and, again, in the vast majority of cases, L = PR$_3$ (R = alkyl or aryl). The ML fragments can adopt either edge-bridging or face-capping bonding modes and the energy differences between metal core structures with and without the Group IB metals in close contact often seem to be small, so the clusters exhibit a wide range of skeletal geometries. These can be classified into four structural types (Tables IV–VII), depending on whether the two ML moieties are edge-bridging or face-capping and whether or not there is close contact between the Group IB metals. Interestingly, there are only two examples of copper- or silver-containing species with two edge-bridging ML groups (99), which is consistent with the theoretical prediction that an edge-bridging bonding mode for a Cu(PR$_3$) unit in a mixed-metal cluster will be less favorable than that for a Au(PR$_3$) fragment in the analogous gold species (Section II,B,1). In addition to the above structural classes, there are a number of examples of clusters which exhibit a trigonal planar Au$_2$M (M = transition metal) skeletal geometry, and these are listed in Table VIII. Table IX

TABLE IV

Mixed-Metal Clusters Containing Two Edge-Bridging $M(PR_3)$ (M = Cu, Ag, or Au; R = Alkyl or Aryl) Units, with No Close Contact between the Group IB Metals

Cluster	Spectroscopic data reported	Crystal structure determined?	Ref.
Copper clusters			
$[Cu_2Mn_2\{\mu\text{-}(EtO)_2POP(OEt)_2\}(CO)_6(PPh_3)_2]$	IR	No	99
Silver clusters			
$[Ag_2Mn_2\{\mu\text{-}(EtO)_2POP(OEt)_2\}(CO)_6(PPh_3)_2]$	IR	No	99
Gold clusters			
$[Au_2Mn_2\{\mu\text{-}(EtO)_2POP(OEt)_2\}(CO)_6(PPh_3)_2]$	IR	No	99
$[Au_2Re_2H_6(PMe_2Ph)_4(PPh_3)_2]^a$	^{1}H, ^{31}P NMR	Yes	52
$[Au_2Re_2H_8(PPh_3)_4(PR_3)_2][PF_6]_2^b$ (R = Ph or Et)	^{1}H, ^{31}P NMR	No	66
$[Au_2Re_2H_7(PPh_3)_6][PF_6]^b$	^{1}H, ^{31}P NMR, Mass	No	66,72
$[Au_2Re_2H_7(PPh_3)_4(PEt_3)_2][PF_6]^c$	^{1}H, ^{31}P NMR	No	66
$[Au_2Ru_4C(CO)_{12}(PMe_2Ph)_2]^d$	IR, Mass	Yes	86
$[Au_2Ru_4C(CO)_{12}(PEt_3)_2]^d$	IR, ^{31}P NMR	No	86
$[Au_2Ru_4C(CO)_{12}(PPh_3)_2]^d$	IR	No	86
$[Au_2Ru_6C(CO)_{16}(PMePh_2)_2]$		Yes	133
$[Au_2Ru_6C(CO)_{16}(PEt_3)_2]^e$	IR, ^{31}P NMR	No	133
$[Au_2Ru_5WC(CO)_{17}(PEt_3)_2]^f$	IR, ^{31}P NMR	Yes	133
$[Au_2Ru_5M(CO)_{17}(PEt_3)_2]^f$ (M = Cr or Mo)	IR, ^{31}P NMR	No	133
$[Au_2Os_3(CO)_{10}(PPh_3)_2]$	IR, ^{1}H, NMR	No	25
$[Au_2Os_3(CO)_{10}(PEt_3)_2]$	^{1}H NMR, Mass	Yes	25
$[Au_2Os_5C(CO)_{14}(PPh_3)_2]$	IR	Yes	134

^{a} The cluster exists as two skeletal isomers in the solid state and in solution (Fig. 11).

^{b} The ^{31}P–$\{^1$H$\}$ NMR spectrum of the cluster shows a single resonance for the $Au(PR_3)$ (R = Ph or Et) units. Therefore, it seems likely that the cluster adopts a metal core structure similar to that established by X-ray diffraction for the minor isomer of the very closely related species $[Au_2Re_2H_6(PMe_2Ph)_4(PPh_3)_2]$ (Fig. 11) (52).

^{c} The NMR spectra of the cluster suggest that two skeletal isomers are present in solution at ambient temperature. It seems likely that these two isomers exhibit metal core structures similar to those which the two isomeric forms of the very closely related cluster $[Au_2Re_2H_6(PMe_2Ph)_4(PPh_3)_2]$ are thought to adopt (Fig. 11) (52).

^{d} One of the two edge-bridging $Au(PR_3)$ (R = Ph or Et) units interacts with the carbido C atom as well as the two Ru atoms.

^{e} A preliminary photographic study suggests that, in the solid state, the cluster adopts a metal core geometry similar to that of the analogous $PMePh_2$-containing species (isomer A in Fig. 14), but skeletal isomers A and B (Fig. 14) are both thought to be present in solution at low temperatures.

^{f} In the solid state, the cluster adopts the structure of isomer B (Fig. 14), but skeletal isomers A and B (Fig. 14) are both thought to be present in solution at low temperatures.

TABLE V

Mixed-Metal Clusters Containing Two Edge-Bridging $Au(PR_3)$ (R = Alkyl or Aryl) Units, with the Gold Atoms in Close Contact

Cluster	Spectroscopic data reported	Crystal structure determined?	Ref.
$[Au_2Re_2H_6(PMe_2Ph)_4(PPh_3)]^a$	^1H, ^{31}P NMR	No	52
$[Au_2Re_2H_7(PPh_3)_4(PEt_3)_2][PF_6]^b$	^1H, ^{31}P NMR	No	66
$[Au_2Ru_3(\mu_3\text{-}S)(\mu\text{-}Ph_2PCH_2PPh_2)(CO)_9]$	IR, ^1H, ^{31}P, ^{13}C NMR, ^{197}Au Mössbauer	Yes	18,111
$[Au_2Ru_3(\mu\text{-}H)(\mu_3\text{-}COMe)(CO)_9(PPh_3)_2]$	IR, ^1H, ^{31}P, ^{13}C NMR, ^{197}Au Mössbauer	Yes	46,47,111
$[Au_2Ru_4(\mu_3\text{-}H)(\mu\text{-}H)(\mu\text{-}Ph_2PCH_2EPh_2)(CO)_{12}]$ (E = P or As)	IR, ^1H, ^{31}P NMR, ^{197}Au Mössbauer	Yes	111,135–137
$[Au_2Ru_4(\mu_3\text{-}H)(\mu\text{-}H)(\mu\text{-}Ph_2AsCH_2AsPh_2)(CO)_{12}]$	IR, ^1H NMR, ^{197}Au Mössbauer	No	136
$[Au_2Ru_4(\mu_3\text{-}H)(\mu\text{-}H)\{\mu\text{-}Ph_2P(CH_2)_2PPh_2\}(CO)_{12}]^c$	IR, ^1H, ^{31}P NMR, ^{197}Au Mössbauer	Yes	136,137
$[Au_2Ru_4(\mu_3\text{-}H)(\mu\text{-}H)(\mu\text{-}cis\text{-}Ph_2PCH=CHPPh_2)(CO)_{12}]$	IR, ^1H, ^{31}P NMR	Yes	41,138
$[Au_2Os_3(CO)_{11}(PPh_3)]$	IR	Yes	25,139
$[Au_2Os_3(CO)_{11}(PEt_3)_2]^d$	Mass	No	25
$[Au_2Os_4(CO)_{13}(PEt_3)_2]$	IR, ^{31}P NMR	Yes	26
$[Au_2Os_4(CO)_{13}\text{-}L_2]$ (L = PPh$_3$ or PMePh$_2$)	IR	No	26

[a] The cluster exists as two skeletal isomers in the solid state and in solution (Fig. 11).

[b] The NMR spectra of the cluster suggest that two skeletal isomers are present. It seems likely that these two isomers exhibit metal core structures similar to those which the two isomeric forms of the very closely related cluster $[Au_2Re_2H_6(PMe_2Ph)_4(PPh_3)_2]$ are thought to adopt (Fig. 11) (52).

[c] Although the ^{197}Au Mössbauer spectrum suggests (136) that the cluster exhibits a metal core structure similar to that adopted by $[Au_2Ru_4(\mu_3\text{-}H)(\mu\text{-}H)(CO)_{12}(PPh_3)_2]$, in which there are two face-capping $Au(PPh_3)$ units (Table VII), an X-ray diffraction study shows that one of the Au—Ru distances [3.446(4) Å] is too long for any significant bonding interaction between the two atoms (Fig. 19) (137).

[d] The cluster reacts with NEt$_3$ and PPh$_3$ to afford species of formula $[AuOs_3(CO)_{11}L(PEt_3)_2]$ (L = NEt$_3$ or PPh$_3$), the structures of which are unknown (25).

TABLE VI

Mixed-Metal Clusters Containing Two Face-Capping ML (M = Cu, Ag, or Au; L = Two-Electron Donor Ligand) Units, with No Close Contact between the Group IB Metals

Cluster	Spectroscopic data reported	Crystal structure determined?	Ref.
Copper clusters			
$[Cu_2Ru_4(\mu\text{-}CO)_3(CO)_{10}(PPh_3)_2]$	IR, ^1H, ^{31}P NMR	Yes	*44*
$[Cu_2Ru_4(\mu\text{-}CO)_3(CO)_{10}(NCMe)_2]^a$		No	*44*
$[Cu_2Ru_4(\mu_3\text{-}H)_2(CO)_{12}\{P(CHMe_2)_3\}_2]^b$	IR, ^1H, ^{31}P NMR	Yes	*140*
$[Cu_2Ru_6(CO)_{18}(C_6H_5Me)_2]$		Yes	*37*
$[Cu_2Ru_6(CO)_{18}(NCMe)_2]^c$		No	*37*
$[Cu_2Rh_6C(CO)_{15}(PPh_3)_2]$	IR	Yes	*128*
$[Cu_2Rh_6C(CO)_{15}(NCMe)_2]$	IR, ^{13}C, ^{103}Rh NMR	Yes	*32,123,128*
Silver clusters			
$[Ag_2Ru_4(\mu\text{-}CO)_3(CO)_{10}(PPh_3)_2]$	IR, ^1H, ^{31}P NMR	Yes	*44*
$[Ag_2Ru_4(\mu\text{-}CO)_3(CO)_{10}(NCMe)_2]^a$		No	*44*
$[Ag_2Rh_6C(CO)_{15}(PPh_3)_2]$	IR	Yes	*128*
$[Ag_2Rh_6C(CO)_{15}(NCMe)_2]$	IR	Yes	*128*
Gold clusters			
$[Au_2Ru_4(\mu\text{-}CO)_3(CO)_{10}(PPh_3)_2]$	IR, ^1H, ^{31}P NMR	No	*44*
$[Au_2Os_4(CO)_{12}(PMePh_2)_2]^d$	IR	Yes	*26*
$[Au_2Os_4(CO)_{12}(PEt_3)_2]$	IR, ^{31}P NMR	No	*26*
$[Au_2Os_4(CO)_{12}(PPh_3)_2]$	IR	No	*26*
$[Au_2Os_8(CO)_{22}(PPh_3)_2]^e$	IR, Mass	Yes	*141*
$[Au_2Rh_6C(CO)_{15}(PPh_3)_2]$	IR	Yes	*128*
$[Au_2Pt_3(\mu\text{-}SO_2)_2(\mu\text{-}Cl)\{P(cyclo\text{-}C_6H_{11})_3\}_3\text{-}\{P(C_6H_4F\text{-}4)_3\}_2][PF_6]$	IR, ^{31}P, ^{195}Pt NMR	Yes	*65*

^{a} The cluster is too unstable to be isolated as a pure compound.

^{b} In the solid state, the cluster adopts the structure of isomer A (Fig. 13), but skeletal isomers A and B (Fig. 13) are both thought to be present in solution at low temperatures.

^{c} The cluster has not been isolated as a pure compound.

^{d} The common Os—Os edge of the two Os_3 faces capped by the $Au(PMePh_2)$ moieties has an Os—Os distance of 3.250(2) Å, which is probably too long for any significant bonding interaction between the two atoms.

^{e} The common Os—Os edge of the two Os_3 faces capped by the $Au(PPh_3)$ moieties has an Os—Os distance of 3.373(3) Å, which is too long for any significant bonding interaction between the two atoms.

presents the clusters containing two $M(PR_3)$ (M = Cu or Au; R = alkyl or aryl) moieties which do not appear in Tables IV–VIII.

Unlike the situation for species containing a single $Au(PR_3)$ unit (Section II,B,1), cases of gold–phosphine fragments occupying similar positions to those of the isolobal hydrido ligands in the related hydrido–metal

TABLE VII

Mixed-Metal Clusters Containing Two Face-Capping ML (M = Cu, Ag, or Au; L = Two-Electron Donor Ligand) Units, with the Group IB Metals in Close Contact

Cluster	Spectroscopic data reported	Crystal structure determined?	Ref.
Copper clusters			
[Cu$_2$Ru$_3$(μ_3-S)(μ-Ph$_2$PCH$_2$PPh$_2$)(CO)$_9$]	IR, ^{1}H, ^{31}P, ^{13}C NMR	Yes	18
[Cu$_2$Ru$_3$(μ_3-S)(CO)$_9$(PPh$_3$)$_2$]	IR, ^{1}H, ^{31}P NMR	No	18
[Cu$_2$Ru$_4$(μ_3-H)$_2$(CO)$_{12}$(PPh$_3$)$_2$]	IR, ^{1}H, ^{31}P, ^{13}C NMR	Yes	142
[Cu$_2$Ru$_4$(μ_3-H)$_2$(CO)$_{12}${P(CHMe$_2$)$_3$}$_2$][a]	IR, ^{1}H, ^{31}P NMR	Yes	140
[Cu$_2$Ru$_4$(μ_3-H)$_2${μ-Ph$_2$P(CH$_2$)$_n$EPh$_2$}(CO)$_{12}$] (E = P, n = 2, 3, or 5; E = As, n = 2)	IR, ^{1}H, ^{31}P NMR	Yes	39,40,143
[Cu$_2$Ru$_4$(μ_3-H)$_2$(CO)$_{12}$-L$_2$] [L$_2$ = μ-Ph$_2$P(CH$_2$)$_n$PPh$_2$ (n = 1, 4, or 6), μ-cis-Ph$_2$PCH=CHPPh$_2$, or μ-Ph$_2$AsCH$_2$PPh$_2$; L = PMePh$_2$, PMe$_2$Ph, PEt$_3$, PMe$_3$, P(OPh)$_3$, P(OEt)$_3$, or P(OMe)$_3$]	IR, ^{1}H, ^{31}P NMR	No	39–42,143
[Cu$_2$Ru$_4$(μ_3-H)$_2${μ-Ph$_2$As(CH$_2$)$_n$AsPh$_2$}(CO)$_{12}$] (n = 1 or 2)	IR, ^{1}H, ^{13}C NMR	No	40
[Cu$_2$Ru$_4$(μ_3-H)$_2$(CO)$_{12}$(NCMe)$_2$][b]	IR	No	142
[CuAgRu$_4$(μ_3-H)$_2$(CO)$_{12}$(PPh$_3$)]	IR, ^{1}H, ^{31}P, ^{13}C NMR	Yes	144
[CuAgRu$_4$(μ_3-H)$_2${μ-Ph$_2$P(CH$_2$)$_n$PPh$_2$}(CO)$_{12}$] (n = 1 or 2)	IR, ^{1}H, ^{31}P NMR	No	36,145
[CuAgRu$_4$(μ_3-H)$_2$(CO)$_{12}$(NCMe)$_2$][b]	IR, ^{1}H, ^{31}P, ^{13}C NMR	No	36
[CuAuRu$_4$(μ_3-H)$_2$(CO)$_{12}$(PPh$_3$)$_2$]	IR, ^{1}H, ^{31}P NMR	No	144
[CuAuRu$_4$(μ_3-H)$_2${μ-Ph$_2$P(CH$_2$)$_2$PPh$_3$}(CO)$_{12}$]	IR, ^{1}H, ^{31}P NMR	Yes	138
[Cu$_2$Ru$_6$C(CO)$_{16}$(NCMe)$_2$]	IR, ^{1}H, ^{13}C NMR	Yes	33
[Cu$_2$Os$_4$(μ_3-H)$_2${μ-Ph$_2$P(CH$_2$)$_2$PPh$_2$}(CO)$_{12}$]	IR, ^{1}H, ^{31}P NMR	Yes	36,146
[Cu$_2$Os$_4$(μ_3-H)$_2$(CO)$_{12}$(NCMe)$_2$][b]	IR, ^{1}H, ^{31}P NMR	No	36

(continued)

TABLE VII (*Continued*)

Cluster	Spectroscopic data reported	Crystal structure determined?	Ref.
Silver clusters			
$[Ag_2Ru_3(\mu_3\text{-}S)(CO)_9(PPh_3)_2]$	IR, ^1H, ^{31}P NMR	No	18
$[Ag_2Ru_4(\mu_3\text{-}H)_2(CO)_{12}(PPh_3)_2]$	IR, ^1H, ^{31}P, ^{13}C NMR	Yes	142
$[Ag_2Ru_4(\mu_3\text{-}H)_2(\mu\text{-}Ph_2PCH_2PPh_2)(CO)_{12}]$	IR, ^1H, ^{31}P, ^{109}Ag INEPT NMR	Yes	124,143,147
$[Ag_2Ru_4(\mu_3\text{-}H)_2\{\mu\text{-}Ph_2P(CH_2)_nPPh_2\}(CO)_{12}]$ (n = 2 or 4)	IR, ^1H, ^{31}P, ^{109}Ag INEPT NMR	No	124,143,147
$[Ag_2Ru_4(\mu_3\text{-}H)_2(CO)_{12}L_2]$ [$L_2 = \mu\text{-}Ph_2P(CH_2)_nPPh_2$ (n = 3, 5, or 6), $\mu\text{-}cis\text{-}Ph_2PCH=CHPPh_2$, or $\mu\text{-}Ph_2As(CH_2)_nPPh_2$ (n = 1 or 2); L = PMePh_2, P(CHMe_2)_3, or P(cyclo-C_6H_{11})_3]	IR, ^1H, ^{31}P NMR	No	36,39–41, 124,143,147, 148
$[Ag_2Ru_4(\mu_3\text{-}H)_2(\mu\text{-}Ph_2AsCH_2AsPh_2)(CO)_{12}]$	IR, ^1H, ^{109}Ag INEPT NMR	No	40
$[Ag_2Ru_4(\mu_3\text{-}H)_2\{\mu\text{-}Ph_2As(CH_2)_2AsPh_2\}(CO)_{12}]$	IR, ^1H NMR	No	40
$[Ag_2Ru_4(\mu_3\text{-}H)_2(CO)_{12}(NCMe)_2]^b$		No	142
$[AgAuRu_4(\mu_3\text{-}H)_2(CO)_{12}(PPh_3)_2]$	IR, ^1H, ^{31}P, ^{13}C NMR	No	144
$[AgAuRu_4(\mu_3\text{-}H)_2\{\mu\text{-}Ph_2P(CH_2)_2PPh_2\}(CO)_{12}]$	IR, ^1H, ^{31}P NMR	No	145

See also clusters $[CuAgRu_4(\mu_3\text{-}H)_2(CO)_{12}L_2]$ [$L = PPh_3$ or NCMe; $L_2 = \mu\text{-}Ph_2P(CH_2)_nPPh_2$ (n = 1 or 2)] listed with copper clusters above.

Cluster	Spectroscopic data reported	Crystal structure determined?	Ref.
Gold clusters			
$[Au_2Fe_3(\mu_3\text{-}S)(CO)_9(PPh_3)_2]$		Yes	149
$[Au_2Ru_3(\mu_3\text{-}S)(CO)_9(PPh_3)_2]$	IR, ^1H, ^{31}P, ^{13}C NMR, Mass	Yes	47,112,113
$[Au_2Ru_3(\mu_3\text{-}S)(CO)_8(PPh_3)_3]$	IR, ^1H, ^{31}P, ^{13}C NMR	Yes	47
$[Au_2Ru_3(\mu_3\text{-}C=CHBu^t)(CO)_9(PPh_3)_2]$	IR, ^1H NMR	Yes	29

Compound	Techniques		References
[Au$_2$Ru$_4$(μ_3-H)(μ-H)(CO)$_{12}$(PPh$_3$)$_2$]	IR, ^{1}H, ^{31}P, ^{13}C NMR, ^{197}Au Mössbauer	Yes	27,111,142
[Au$_2$Ru$_4$H$_2$(CO)$_{12}$L$_2$] [L$_2$ = μ-Ph$_2$P(CH$_2$)$_n$PPh$_2$ (n = 3–6); L = P(CHMe$_2$)$_3$ or P(cyclo-C$_6$H$_{11}$)$_3$]	IR, ^{1}H, ^{31}P NMR	No	137,150
[Au$_2$Ru$_4$(μ_3-H)(μ-H){μ-Ph$_2$As(CH$_2$)$_2$PPh$_2$}(CO)$_{12}$]^c	IR, ^{1}H, ^{31}P NMR, ^{197}Au Mössbauer	No	136
[Au$_2$Ru$_4$(μ_3-H)(μ-H){μ-Ph$_2$As(CH$_2$)$_2$AsPh$_2$}(CO)$_{12}$]	IR, ^{1}H NMR, ^{197}Au Mössbauer	No	136
[Au$_2$CoRu$_3$(μ-H)(CO)$_{12}$(PPh$_3$)$_2$]	IR, ^{1}H NMR	Yes	31
[Au$_2$Co$_2$Ru$_2$(μ-CO)$_2$(CO)$_{10}$(PPh$_3$)$_2$]		Yes	149
[Au$_2$Ru$_6$C(CO)$_{16}$(PEt$_3$)$_2$]^c	^{31}P NMR	No	133
[Au$_2$Ru$_5$WC(CO)$_{17}$(PEt$_3$)$_2$]^d	^{31}P NMR	Yes	133
[Au$_2$Ru$_5$MC(CO)$_{17}$(PEt$_3$)$_2$]^d (M = Cr or Mo)	^{31}P NMR	No	133
[Au$_2$Os$_3$S$_2$(CO)$_{10}$(PPh$_3$)$_2$]^e	IR	Yes	54
[Au$_2$Os$_4$(μ-H)$_2$(CO)$_{12}$(PPh$_3$)$_2$]^{f,g}	IR	Yes	13
[Au$_2$Os$_4$(μ-H)$_2$(CO)$_{12}$(PEt$_3$)$_2$]^g	IR, Mass	No	13
[Au$_2$Ir$_4$(CO)$_9$(PEt$_3$)$_2$(μ-PhPPPh)Ir$_4$(CO)$_{11}$]	IR	Yes	87

See also clusters [MAuRu$_4$(μ_3-H)$_2$(CO)$_{12}$L$_2$] [M = Cu or Ag, L = PPh$_3$ or L$_2$ = μ-Ph$_2$P(CH$_2$)$_2$PPh$_2$] listed with copper or silver clusters above.

^{a} In the solid state, the cluster adopts the structure of isomer A (Fig. 13), but skeletal isomers A and B (Fig. 13) are both thought to be present in solution at low temperatures.

^{b} The cluster is too unstable to be isolated as a pure compound.

^{c} A preliminary photographic study suggests that the cluster adopts the structure of isomer A (Fig. 14) in the solid state, but skeletal isomers A and B (Fig. 14) are both thought to be present in solution at low temperatures.

^{d} In the solid state, the cluster adopts the structure of isomer B (Fig. 14), but skeletal isomers A and B (Fig. 14) are both thought to be present in solution at low temperatures.

^{e} Only the results of a preliminary X-ray diffraction study have been reported for the cluster.

^{f} One of the Au—Os distances for one of the capping Au(PPh$_3$) fragments [3.159(4) Å] may be too long to indicate any significant bonding interaction between the two atoms.

^{g} The cluster exists as two isomers in the solid state, but the structure of the second isomeric form is not known.

TABLE VIII

MIXED-METAL CLUSTERS CONTAINING A TRIGONAL PLANAR Au_2M (M = TRANSITION METAL) UNIT

Cluster	Spectroscopic data reported	Crystal structure determined?	Ref.
$[Au_2Nb(PPh_3)_2(\eta\text{-}C_5H_4SiMe_3)_2][PF_6]$	IR, ^1H, ^{13}C NMR	Yes	71
$[Au_2Nb(PPh_3)_2(\eta\text{-}C_5H_4R)_2]X$ (X = ClO_4, R = H or $SiMe_3$; X = PF_6, R = H)	IR, ^1H, ^{13}C NMR	No	71
$[Au_2Fe(CO)_4(PPh_3)_2]$	IR, ^{31}P NMR	Yes	1,4,114
$[\{Au_2Fe(CO)_4\}_2\{\mu\text{-}Ph_2P(CH_2)_2PPh_2\}_2]^a$	IR	Yes	151
$[Au_2RuH_2(Ph_2PCH_2PPh_2)_2(PPh_3)_2][NO_3]_2$	IR, ^1H, ^{31}P NMR	Yes	67
$[Au_2RuH_2(Ph_2PCH_2PPh_2)_2(PPh_3)_2][PF_6]_2$	Mass	No	67,72
$[Au_2Ru(CO)_4(PPh_3)_2]$		No	152
$[Au_2Os(CO)_4(PPh_3)_2]$	IR	Yes	153,154
$[Au_2IrH(NO_3)(PPh_3)_4][BF_4]$	IR, ^1H, ^{31}P NMR	Yes	89
$[Au_2Ir(CO)_2(PPh_3)_4][PF_6]$	IR, ^1H, ^{31}P NMR, Mass	No	85
$[Au_2PtCl(PPh_3)_2(PEt_3)_2][CF_3SO_3]$	^{31}P NMR	Yes	69
$[Au_2Pt(NO_3)(PPh_3)_4][PF_6]$	IR, ^{31}P NMR	No	72
$[Au_2Pt(NO_3)(PPh_3)_4][NO_3]$	Mass	No	72

[a] Although there are four $Au(PR_3)$ fragments in this cluster and it is, therefore, listed in Table XIV, it has also been included here in Table VIII as it contains two trigonal planar Au_2Fe units which are very similar to that in $[Au_2Fe(CO)_4(PPh_3)_2]$ (1,4).

TABLE IX

MIXED-METAL CLUSTERS CONTAINING TWO M(PR$_3$) (M = Cu OR Au; R = ALKYL OR ARYL) UNITS WHICH DO NOT BELONG TO THE STRUCTURAL CLASSES LISTED IN TABLES IV–VIII

Cluster	Spectroscopic data reported	Crystal structure determined?	Skeletal geometry adopted	Ref.
Copper clusters				
[Cu$_2$Ru$_4$(μ_3-H)$_2$(CO)$_{12}$[P(cyclo-C$_6$H$_{11}$)$_3$]$_2$]	IR, ^{1}H, ^{31}P NMR	Yes	Ru$_4$ tetrahedron with one μ- and one μ_3-Cu[P(cyclo-C$_6$H$_{11}$)$_3$]. No Cu—Cu close contact	140
Gold clusters				
[Au$_2$Re$_2$H$_6$(PPh$_3$)$_6$][PF$_6$][a]	UV, Mass	No	Unknown	72
[Au$_2$Fe$_4$(BH)(CO)$_{12}$(PPh$_3$)$_2$]	IR, ^{1}H, ^{31}P, ^{11}B NMR, Mass	Yes	Fe$_4$ "butterfly" with B interacting with all four Fe. One Au(PPh$_3$) caps a Fe$_2$B face and the other a AuFeB face. Au—Au close contact	24
[Au$_2$Fe$_4$C(CO)$_{12}$(PEt$_3$)$_2$]	IR, Mass	Yes	Au$_2$Fe$_4$ octahedron with interstitial C. Au—Au close contact	103
[Au$_2$Fe$_4$C(CO)$_{12}$(PPh$_3$)$_2$]		No	As PEt$_3$-containing analog	103
[Au$_2$Fe$_5$C(μ-CO)$_3$(CO)$_{11}$(PEt$_3$)$_2$]	IR	Yes	Fe$_5$ square-based pyramid with interstitial C. One μ- and one μ_4-Au(PEt$_3$). No Au—Au close contact.	84
[Au$_2$Pt$_2$(CNC$_6$H$_3$Me$_2$-2,6)$_4$(PPh$_3$)$_4$][PF$_6$]$_2$[b]	IR, ^{31}P, ^{195}Pt NMR, UV	Yes	Au$_2$Pt$_2$ "butterfly" with Pt in the "wing-tip" sites. Au—Au close contact	80

[a] The cluster cation has also been isolated as the [BPh$_4$]$^-$ salt.
[b] The cluster cation has also been isolated as the [BF$_4$]$^-$ salt.

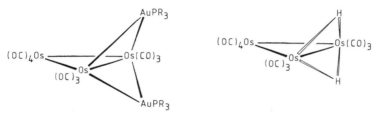

FIG. 10. Structures of the clusters [Au$_2$Os$_3$(CO)$_{10}$(PR$_3$)$_2$] (R = Ph or Et) and [Os$_3$-(μ-H)$_2$(CO)$_{10}$], showing the isolobal relationship between the edge-bridging Au(PR$_3$) units and the hydrido ligands.

cluster are the exceptions rather than the normal rule for clusters containing more than one gold atom. Although some examples of the structural relationship do exist, and one is shown in Fig. 10 (25,129), the fact that the hydrido ligand cannot be involved in bonding interactions similar to those which can occur between adjacent Au(PR$_3$) groups means that the isolobal analogy does not normally apply when there is more than one gold–phosphine fragment in a cluster.

There is a considerable amount of experimental evidence that the energy differences between the various types of skeletal geometry adopted by mixed-metal clusters containing two ML fragments can be small (see also Section II,E,1). A number of examples of skeletal isomerism have been reported (13,52,66,133,140). It is actually possible to separate and isolate the two isomeric forms of the clusters [Au$_2$Re$_2$H$_6$(PMe$_2$Ph)$_4$(PPh$_3$)$_2$] and [Au$_2$Os$_4$(μ-H)$_2$(CO)$_{12}$(PR$_3$)$_2$] (R = Ph or Et) in the solid state. The major isomer (A) of the gold–rhenium species is thought to exhibit a skeletal geometry in which one Au(PPh$_3$) fragment bridges a Re—Re bond and one edge of the trigonal planar AuRe$_2$ unit so formed is further bridged by the second Au(PPh$_3$) group so that the two Au atoms are in close contact, whereas in the minor isomer (B), the two Au(PPh$_3$) moieties bridge a Re—Re bond with no bonding interaction between the Au atoms (Fig. 11) (52). The NMR spectra of the closely related cluster [Au$_2$Re$_2$H$_7$(PPh$_3$)$_4$-(PEt$_3$)$_2$][PF$_6$] suggest that two skeletal isomers are present in solution at ambient temperature (66). It seems likely that these two isomeric forms exhibit metal core structures similar to those of [Au$_2$Re$_2$H$_6$(PMe$_2$Ph)$_4$-(PPh$_3$)$_2$] (Fig. 11). One skeletal isomer of [Au$_2$Os$_4$(μ-H)$_2$(CO)$_{12}$(PR$_3$)$_2$] (R = Ph or Et) adopts the metal core geometry shown in Fig. 12, but, unfortunately, the structure of the second isomeric form is not known (13).

The clusters [Cu$_2$Ru$_4$(μ-H)$_2$(CO)$_{12}${P(CHMe$_2$)$_3$}$_2$], [Au$_2$Ru$_6$C(CO)$_{16}$-(PEt$_3$)$_2$], and [Au$_2$Ru$_5$M(CO)$_{17}$(PEt$_3$)$_2$] (M = Cr, Mo, or W) exhibit only one metal core structure in the solid state, but two skeletal isomers exist in solution at low temperatures. Interestingly, at ambient temperature in

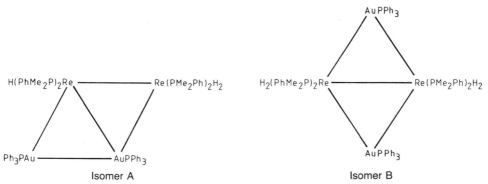

FIG. 11.   Structures of the skeletal isomers of the cluster [Au₂Re₂H₆(PMe₂Ph)₄(PPh₃)₂], with the three and two hydrido ligands which bridge the Re—Re vectors of isomers A and B, respectively, omitted for clarity.

solution, each pair of skeletal isomers interconvert via intramolecular metal core rearrangements (see also Ref. *11* and Section II,E,1). In the solid state, the copper–ruthenium cluster adopts a metal framework geometry in which one Cu{P(CHMe₂)₃} moiety caps a face of a Ru₄ tetrahedron and one CuRu₂ face of the CuRu₃ tetrahedron so formed is further capped by the second Cu{P(CHMe₂)₃} fragment so that the two Cu atoms are in close contact (isomer A in Fig. 13). However, at −90°C in solution, ¹H and ³¹P–{¹H} NMR spectra show the presence of two skeletal isomers. The predominant isomer (77% abundance) is thought to exhibit the same metal core structure as that adopted in the solid state, whereas the other isomeric form (B in Fig. 13; 23% abundance) probably has a skeletal geometry with no bonding interaction between the Cu atoms (*140*). The two skeletal isomers of the clusters [Au₂Ru₆C(CO)₁₆(PEt₃)₂] and [Au₂Ru₅MC(CO)₁₇(PEt₃)₂] (M = Cr, Mo, or W) both have metal

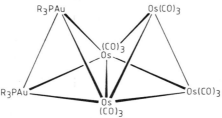

FIG. 12.   Structure of one skeletal isomer of the cluster [Au₂Os₄(μ-H)₂(CO)₁₂(PR₃)₂] (R = Ph or Et), with the hydrido ligands, which are thought to bridge two Os—Os vectors, omitted for clarity.

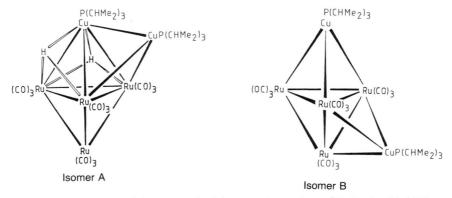

FIG. 13. Structures of the two skeletal isomers of the cluster $[Cu_2Ru_4(\mu_3\text{-}H)_2(CO)_{12}$-$\{P(CHMe_2)_3\}_2]$. The positions of the hydrido ligands in isomer B are not known. [Reprinted with permission of the Royal Society of Chemistry (140).]

core structures based on a Ru_5M (M = Ru, Cr, Mo, or W) octahedral unit containing a central interstitial carbon atom (Fig. 14) (133). However, for isomer A, the two $Au(PEt_3)$ groups bridge opposite edges of this unit, whereas in isomer B, a Ru_3 face is capped by a $Au(PEt_3)$ fragment and one $AuRu_2$ face of the $AuRu_3$ tetrahedron so formed is further capped by the second $Au(PEt_3)$ moiety so that the Au atoms are in close contact.

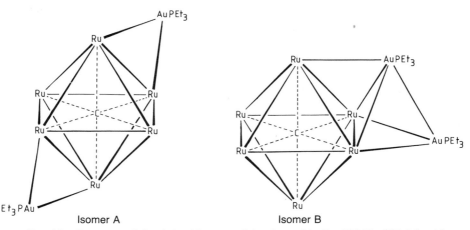

FIG. 14. Structures of the skeletal isomers of the cluster $[Au_2Ru_6C(CO)_{16}(PEt_3)_2]$, with the CO ligands omitted for clarity. [Reprinted with permission of Freund Publishing House Ltd. (11).] The skeletal isomers of $[Au_2Ru_5MC(CO)_{17}(PEt_3)_2]$ (M = Cr, Mo, or W) adopt similar structures with M replacing one Ru atom.

Interestingly, in the solid state, $[Au_2Ru_6C(CO)_{16}(PEt_3)_2]$ is thought to adopt the same structure as that of isomer A, whereas $[Au_2Ru_5M(CO)_{17}$-$(PEt_3)_2]$ (M = Cr, Mo, or W) exhibit skeletal geometries similar to that of isomer B (133).

The fact that relatively small changes in the attached ligands can have a remarkable effect on the skeletal geometries adopted by Group IB metal heteronuclear clusters provides further evidence that the energy differences between the various structural types are small in many cases for these species. The formal replacement of two hydrido ligands in the clusters $[M_2Ru_4H_2(CO)_{12}(PPh_3)_2]$ (M = Cu, Ag, or Au) by a CO ligand in $[M_2$-$Ru_4(\mu\text{-}CO)_3(CO)_{10}(PPh_3)_2]$ causes a fundamental change in the positions that the Group IB metals adopt on the Ru_4 tetrahedra of these compounds (Fig. 15) (44,142). A similar alteration in coinage metal arrangement has also been observed for two Cu_2Ru_6 clusters when an interstitial

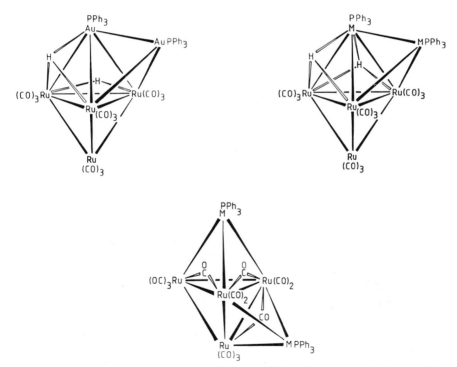

FIG. 15. Structures of the clusters $[M_2Ru_4H_2(CO)_{12}(PPh_3)_2]$ and $[M_2Ru_4(\mu\text{-}CO)_3$-$(CO)_{10}(PPh_3)_2]$ (M = Cu, Ag, or Au), showing the change in skeletal geometry when two hydrido ligands are formally replaced by a CO group. [Reprinted with permission of the Royal Society of Chemistry (142) and Elsevier Sequoia S.A. (44).]

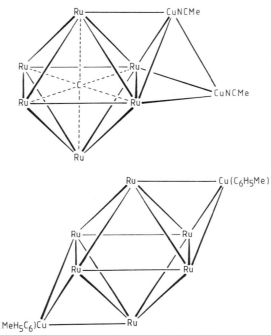

FIG. 16. Structures of the clusters [Cu₂Ru₆C(CO)₁₆(NCMe)₂] and [Cu₂Ru₆(CO)₁₈-(C₆H₅Me)₂], with the CO ligands omitted for clarity, showing the change in skeletal geometry when an interstitial carbido ligand is formally replaced by two CO groups. [Reprinted with permission of Freund Publishing House Ltd. (11).]

carbido ligand is formally replaced by two CO groups. Whereas in [Cu₂-Ru₆C(CO)₁₆(NCMe)₂], the Cu atoms are in close contact (33), the two Cu(C₆H₅Me) units cap opposite faces of a Ru₆ octahedron in [Cu₂-Ru₆(CO)₁₈(C₆H₅Me)₂] (37) (Fig. 16).

An increase in the cone angle of the phosphine ligands attached to the copper atoms in [Cu₂Ru₄(μ₃-H)₂(CO)₁₂L₂] [L = PR₃ or P(OR)₃; R = alkyl or aryl] also causes a change from a skeletal geometry in which the Cu atoms are in close contact to a framework structure with no Cu—Cu bonding interaction. For L = PPh₃, PMePh₂, PMe₂Ph, PEt₃, PMe₃, P(OPh)₃, P(OEt)₃, or P(OMe)₃, the copper–ruthenium clusters all adopt the metal core structure shown in Fig. 15 for the PPh₃-containing analog (42,142). When the Cu atoms are ligated by the bulky phosphine P(cyclo-C₆H₁₁)₃, however, the cluster adopts a skeletal geometry consisting of a Ru₄ tetrahedron, with one Ru—Ru edge bridged by a Cu{P(cyclo-C₆H₁₁)₃} unit and an adjacent Ru₃ face capped by the second Cu{P(cyclo-C₆H₁₁)₃} fragment (Fig. 17) (140).

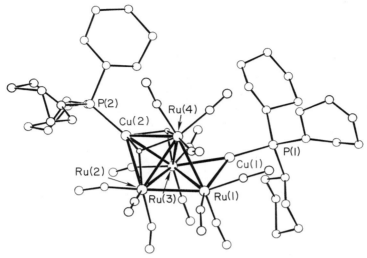

FIG. 17.   Structure of the cluster $[Cu_2Ru_4(\mu_3\text{-}H)_2(CO)_{12}\{P(cyclo\text{-}C_6H_{11})_3\}_2]$. [Reprinted with permission of the Royal Society of Chemistry (140).]

Linking two gold atoms together with the bidentate ligands Ph_2-$ECH_2E'Ph_2$ (E = P, E' = As or P; E = E' = As), $Ph_2P(CH_2)_2PPh_2$, or $cis\text{-}Ph_2PCH{=}CHPPh_2$ can also alter the metal core geometry of a cluster. The formal replacement of the two PPh_3 groups attached to the Au atoms in $[Au_2Ru_4(\mu_3\text{-}H)(\mu\text{-}H)(CO)_{12}$ $(PPh_3)_2]$ (142) by a $Ph_2ECH_2E'Ph_2$ or $cis\text{-}Ph_2PCH{=}CHPPh_2$ ligand causes the capped trigonal bipyramidal metal framework to change to a capped square-based pyramidal structure (Fig. 18) (41,135–138). When the Au atoms are ligated by the bidentate diphosphine $Ph_2P(CH_2)_2PPh_2$, the skeletal geometry is reasonably similar to the capped trigonal bipyramidal structure of the analogous PPh_3-containing cluster, but one of the Au—Ru distances [Au(2)—Ru(4) in Fig. 19] is 3.446(4) Å, which is too long for any significant bonding interaction between the two atoms (137). Thus, the metal framework structure of the cluster is somewhat distorted toward a square-based pyramidal geometry. The formal replacement of the PPh_3 groups attached to the Au atoms in $[Au_2Ru_3(\mu_3\text{-}S)(CO)_9(PPh_3)_2]$ by a $Ph_2PCH_2PPh_2$ ligand causes a similar alteration in metal framework structure. Whereas the PPh_3-containing species adopts a trigonal bipyramidal skeletal geometry (47,112), one of the Au—Ru distances in $[Au_2Ru_3(\mu_3\text{-}S)$-$(\mu\text{-}Ph_2PCH_2PPh_2)(CO)_9]$ is 3.335(1) Å, which is again too long for any significant bonding Au—Ru interaction (Fig. 20) (18). Interestingly, no significant changes in metal core geometry for clusters containing Cu

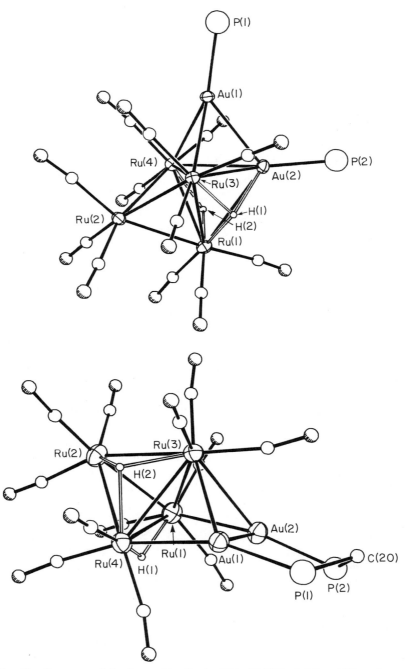

FIG. 18. Structures of the clusters $[Au_2Ru_4(\mu_3\text{-}H)(\mu\text{-}H)(CO)_{12}L_2]$ (L = PPh_3 or L_2 = μ-$Ph_2PCH_2PPh_2$), with the phenyl groups omitted for clarity. [Reprinted with permission of the Royal Society of Chemistry (135,142).] A similar change in skeletal geometry occurs when the two PPh_3 groups are formally replaced by $Ph_2AsCH_2EPh_2$ (E = As or P) or cis-Ph_2PCH=$CHPPh_2$ instead of by a $Ph_2PCH_2PPh_2$ ligand.

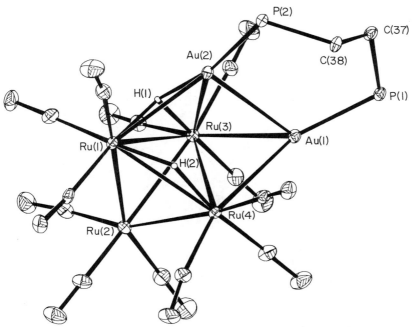

FIG. 19. Structure of the cluster [Au$_2$Ru$_4$(μ_3-H)(μ-H){μ-Ph$_2$P(CH$_2$)$_2$PPh$_2$}(CO)$_{12}$], with the phenyl groups omitted for clarity. [Reprinted with permission of the Royal Society of Chemistry (137).]

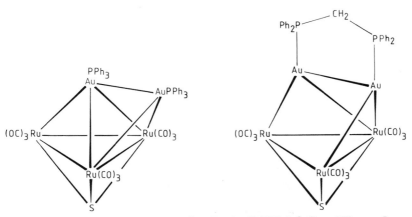

FIG. 20. Structures of the clusters [Au$_2$Ru$_3$(μ_3-S)(CO)$_9$L$_2$] (L = PPh$_3$ or L$_2$ = μ-Ph$_2$PCH$_2$PPh$_2$), showing the change in skeletal geometry when two PPh$_3$ groups are formally replaced by a Ph$_2$PCH$_2$PPh$_2$ ligand. [Reprinted with permission of the Royal Society of Chemistry (18).]

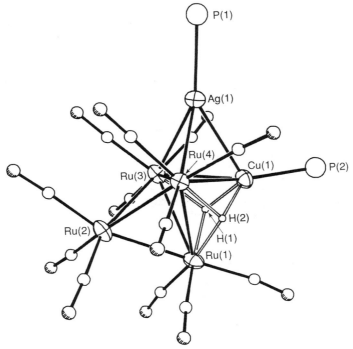

FIG. 21. Structure of the cluster $[AgCuRu_4(\mu_3\text{-H})_2(CO)_{12}(PPh_3)_2]$, with the phenyl groups omitted for clarity. The Cu(1)—Ru(1) contact is obscured by the hydrido ligand H(2). [Reprinted with permission of the Royal Society of Chemistry (144).] The analogous clusters $[MM'Ru_4(\mu_3\text{-H})_2(CO)_{12}L_2]$ [M = Cu, M' = Ag, $L_2 = \mu\text{-Ph}_2P(CH_2)_nPPh_2$ (n = 1 or 2); M = Cu or Ag, M' = Au, L = PPh_3 or $L_2 = \mu\text{-Ph}_2P(CH_2)_2PPh_2$] all adopt similar structures, with the lighter of the pair of Group IB metals occupying the copper site and the heavier in the position of the silver atom.

or Ag atoms linked by bidentate ligands have been observed (e.g., 18,40,124,135,143).

The trimetallic clusters $[MM'Ru_4(\mu_3\text{-H})_2(CO)_{12}L_2]$ [M = Cu, M' = Ag, L = PPh_3 or $L_2 = \mu\text{-Ph}_2P(CH_2)_nPPh_2$ (n = 1 or 2); M = Cu or Ag, M' = Au, L = PPh_3 or $L_2 = \mu\text{-Ph}_2P(CH_2)_2PPh_2$] all adopt capped trigonal bipyramidal metal core structures similar to those of their bimetallic analogs (36,124,138,142–145). It is interesting, however, that the lighter of the pair of Group IB metals occupies the site of higher coordination number in all of the trimetallic species (Fig. 21). Evidence that these coinage metal site preferences are thermodynamic is provided by the fact that no dynamic behavior involving Group IB metal site exchange is observed in solution for the trimetallic clusters (144,145), even though an

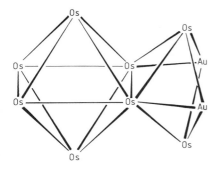 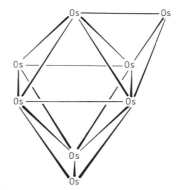

FIG. 22. Metal framework structures of the cluster $[Au_2Os_8(CO)_{22}(PPh_3)_2]$ and the cluster anion $[Os_8(CO)_{22}]^{2-}$, with the CO and PPh$_3$ ligands omitted for clarity.

intramolecular pathway for such a process is known to exist for the analogous bimetallic species (124,142,143 and Section II,E,1), and by the coinage metal exchange reactions discussed in Section II,D,2.

Since the formal addition of $[ML]^+$ (M = Cu, Ag, or Au; L = two-electron donor ligand) fragments to cluster anions contributes no extra skeletal cluster bonding electrons to the system (92,93), then according to the polyhedral skeletal electron pair theory (155), the original metal framework structure of the anion should not be altered in the Group IB metal heteronuclear cluster. In almost all cases, the experimental evidence is in agreement with the theoretical prediction, but one interesting exception is $[Au_2Os_8(CO)_{22}(PPh_3)_2]$, which exhibits an Os$_8$ framework geometry very different from that of its anionic precursor $[Os_8(CO)_{22}]^{2-}$ (Fig. 22) (141,156). This structural change seems likely to be a consequence of the small energy diferences which exist between alternative skeletal geometries for Os$_8$ clusters with no interstitial C atoms (141,157).

3. Mixed-Metal Clusters Containing Three ML (M = Cu, Ag, or Au; L = Two-Electron Donor Ligand) Units

Heteronuclear clusters containing three ML fragments are much rarer than those with just one or two. Apart from $[Cu_3Ir_2H_6(NCMe)_3(PMe_2Ph)_6][PF_6]_3$, which adopts a trigonal bipyramidal Cu$_3$Ir$_2$ metal core structure with three equatorial Cu(NCMe) moieties (158), all of the known examples contain three Au(PR$_3$) (R = alkyl or aryl) groups. With one exception, these gold species can be classified into three different structural types. Table X lists clusters with three face-capping Au(PR$_3$) groups and close contact between the Au atoms. Species which contain tetrahedral (e.g., Fig. 23) (70) or planar (or nearly planar) (e.g., Fig. 24) (88) Au$_3$M

TABLE X

MIXED-METAL CLUSTERS CONTAINING THREE FACE-CAPPING Au(PR₃) (R = ALKYL OR ARYL) UNITS,
WITH THE GOLD ATOMS IN CLOSE CONTACT

Cluster	Spectroscopic data reported	Crystal structure determined?	Ref.
$[Au_3Ru_3(\mu_3\text{-COMe})(CO)_9(PPh_3)_3]$	IR, ¹H, ³¹P, ¹³C NMR, ¹⁹⁷Au Mössbauer	Yes	46,111
$[Au_3Ru_3(\mu_3\text{-}C_{12}H_{15})(CO)_8(PPh_3)_3]$	IR, ¹H NMR, Mass	Yes	28,113
$[Au_3Ru_3(CO)_8(C_2Bu^t)(PPh_3)_3]^a$	IR, ¹H NMR	No	29
$[Au_3Ru_4(\mu_3\text{-H})(CO)_{12}(PPh_3)_3]$	IR, ¹H, ³¹P, ¹³C NMR, Mass, ¹⁹⁷Au Mössbauer	Yes	27,48,111
$[Au_3CoRu_3(CO)_{12}(PPh_3)_3]$	IR, ¹H NMR	Yes	31

ᵃ The cluster is thought to exhibit a similar metal core structure to that adopted by $[Au_3Ru_3\text{-}(\mu_3\text{-}C_{12}H_{15})(CO)_8(PPh_3)_3](28)$, although the IR spectra of the two compounds differ significantly.

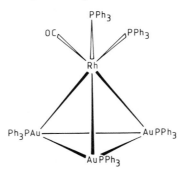

FIG. 23. Structure of the cluster cation $[Au_3Rh(\mu\text{-H})(CO)(PPh_3)_5]^+$, with the hydrido ligand, which is thought to bridge the Au—Rh vector trans to the CO ligand, omitted for clarity. [Reprinted with permission of Freund Publishing House Ltd. (11).]

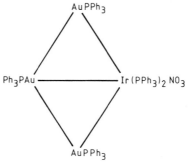

FIG. 24. Structure of the cluster cation $[Au_3Ir(NO_3)(PPh_3)_5]^+$. [Reprinted with permission of Freund Publishing House Ltd. (11).]

TABLE XI

MIXED-METAL CLUSTERS CONTAINING A TETRAHEDRAL Au₃M (M = TRANSITION METAL) UNIT

Cluster	Spectroscopic data reported	Crystal structure determined?	Ref.
[Au$_3$V(CO)$_5$(PPh$_3$)$_3$]	IR	Yes	16
[Au$_3$Re(μ-H)$_2$(PMe$_2$Ph)$_3$(PPh$_3$)$_3$][a]	^{1}H, ^{31}P NMR	Yes	53
[Au$_3$Re(μ-H)$_3$(PMe$_2$Ph)$_3$(PPh$_3$)$_3$][BF$_4$][b]	^{1}H, ^{31}P NMR	Yes	53
[Au$_3$Rh(μ-H)(CO)(PPh$_3$)$_5$][PF$_6$]	IR, ^{1}H, ^{31}P NMR, Mass	Yes	70,72

[a] In the solid state, the cluster adopts the structure of isomer A (Fig. 27), but skeletal isomers A and B (Fig. 27) are thought to be present in solution at low temperatures.

[b] The cluster cation has also been isolated as the [OC$_6$H$_2$Bu$_3^t$-2,4,6]$^-$ salt, which slowly decomposes to [Au$_3$Re(μ-H)$_2$(PMe$_2$Ph)$_3$(PPh$_3$)$_3$] in benzene solution at 25°C over a period of 80 hours.

(M = transition metal) units are presented in Tables XI and XII, respectively. Table XIII lists the clusters containing three ML (M = Cu or Au; L = two-electron donor ligand) that do not appear in Tables X–XII.

The experimental evidence shows that the energy differences between the various types of skeletal geometry adopted by mixed-metal clusters containing three ML fragments can also be small (see also Section II,E,1). The clusters [Au$_3$Ru$_3$(μ_3-COMe)(CO)$_9$(PPh$_3$)$_3$], [Au$_3$Ru$_4$(μ_3-H)(CO)$_{12}$-(PPh$_3$)$_3$], and [Au$_3$CoRu$_3$(CO)$_{12}$(PPh$_3$)$_3$] all exhibit similar Au atom arrangements, with one Au(PPh$_3$) fragment capping a MRu$_2$ (M = Co or Ru) face and two AuMRu faces of the AuMRu$_2$ tetrahedron so formed further capped by the other two Au(PPh$_3$) moieties (Au arrangement A in Fig. 25) (27,31,46,48). Interestingly, however, the third Au(PPh$_3$) group in [Au$_3$Ru$_3$(μ_3-C$_{12}$H$_{15}$)(CO)$_8$(PPh$_3$)$_3$] adopts a different position, capping a

TABLE XII

MIXED-METAL CLUSTERS CONTAINING A PLANAR OR NEARLY PLANAR Au₃M
(M = TRANSITION METAL) UNIT

Cluster	Spectroscopic data reported	Crystal structure determined?	Ref.
[Au$_3$Mn(CO)$_4$(PPh$_3$)$_3$][a]	IR	Yes	159,160
[Au$_3$Re(CO)$_4$(PPh$_3$)$_3$]	IR	No	159
[Au$_3$Re(μ-H)$_2$(PMe$_2$Ph)$_3$(PPh$_3$)$_3$][b]	^{1}H, ^{31}P NMR	Yes	53
[Au$_3$Ir(NO$_3$)(PPh$_3$)$_5$][PF$_6$][c]	^{1}H, ^{31}P NMR	Yes	88

[a] Preliminary X-ray diffraction results show the presence of a [Au$_3$M(PPh$_3$)$_3$] structural unit very similar to that in [Au$_3$Ir(NO$_3$)(PPh$_3$)$_5$][PF$_6$] (160).

[b] In the solid state, the cluster adopts the structure of isomer A (Fig. 27), but skeletal isomers A and B (Fig. 27) are thought to be present in solution at low temperatures.

[c] The cluster cation can be prepared in better yield as the [BF$_4$]$^-$ salt (89).

TABLE XIII

Mixed-Metal Clusters Containing Three ML (M = Cu or Au; L = Two-Electron Donor Ligand) Units Which Do Not Belong to the Structural Classes Listed in Tables X–XII

Cluster	Spectroscopic data reported	Crystal structure determined?	Skeletal geometry adopted	Ref.
Copper clusters				
$[Cu_3Ir_2H_6(NCMe)_3(PMe_2Ph)_6]$ $[PF_6]_3$	IR, ^{1}H, ^{31}P NMR	Yes	Cu_3Ir_2 trigonal bipyramid with three equatorial Cu atoms	158
Gold clusters				
$[Au_3Ru_4(\mu\text{-}H)(\mu\text{-}Ph_2PCH_2PPh_2)\text{-}$ $(CO)_{12}(PPh_3)]$	IR, ^{1}H, ^{31}P NMR	Yes	Distorted Au_2Ru_3 square-based pyramid with adjacent $AuRu_2$ and Ru_3 faces capped by $Au(PPh_3)$ and $Ru(CO)_3$, respectively	48,161

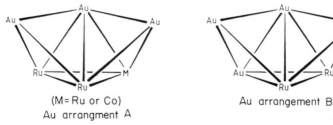

(M = Ru or Co)
Au arrangment A

Au arrangement B

Fig. 25. Arrangements of the Au atoms in the clusters $[Au_3Ru_3(\mu_3\text{-}COMe)(CO)_9\text{-}(PPh_3)_3]$, $[Au_3Ru_4(\mu_3\text{-}H)(CO)_{12}(PPh_3)_3]$, $[Au_3CoRu_3(CO)_{12}(PPh_3)_3]$, and $[Au_3Ru_3\text{-}(\mu_3\text{-}C_{12}H_{15})(CO)_8(PPh_3)_3]$.

Au_2Ru face of the Au_2Ru_2 tetrahedron formed by the other two capping $Au(PPh_3)$ moieties (Au arrangement B in Fig. 25) (28).[5] The reasons for the two different Au atom arrangements are not clear, but Bruce et al. (28) have suggested that the less common Au arrangement B is, in fact, electro-

[5] The cluster $[Au_3Ru_3(CO)_8(C_2Bu^t)(PPh_3)_3]$ is thought (29) to adopt a similar metal framework structure, although its IR spectrum differs significantly from that reported (28) for $[Au_3Ru_3(\mu_3\text{-}C_{12}H_{15})(CO)_8(PPh_3)_3]$.

nically favored, because theoretical calculations (93) indicate that closo-Au$_3$ units are particularly stable. They propose that the other three clusters adopt Au arrangement A for steric reasons. The Ru$_3$ face of [Au$_3$-Ru$_3$(μ_3-C$_{12}$H$_{15}$)(CO)$_8$(PPh$_3$)$_3$] is less sterically constrained than the MRu$_2$ (M = Co or Ru) faces in the species which adopt Au arrangement A, as only eight instead of nine CO ligands are associated with it.

The metal framework of [Au$_3$Ru$_4$(μ_3-H)(CO)$_{12}$(PPh$_3$)$_3$] (27,48) contains a trigonal bipyramidal Au$_2$Ru$_3$ unit similar to those in [Au$_2$-Ru$_4$(μ_3-H)(μ-H)(CO)$_{12}$(PPh$_3$)$_2$] (142) and [Au$_2$Ru$_3$(μ_3-S)(CO)$_9$(PPh$_3$)$_2$] (47,112). When two of the PPh$_3$ groups attached to the Au atoms in the heptanuclear cluster are formally replaced by a Ph$_2$PCH$_2$PPh$_2$ ligand, this Au$_2$Ru$_3$ unit undergoes a distortion toward a square-based pyramidal geometry (Fig. 26, p. 298) (161) somewhat similar to that observed for the latter two species (Section II,B,2) (18,135,137). There also seems to be little difference in energy between tetrahedral and planar (or nearly planar) geometries for Au$_3$M units. Although [Au$_3$Re(μ-H)$_2$(PMe$_2$Ph)$_3$-(PPh$_3$)$_3$] adopts a planar metal framework structure (isomer A in Fig. 27, p. 299) in the solid state, ^{1}H and ^{31}P–{^{1}H} NMR spectroscopy suggests that skeletal isomers A and B (Fig. 27) are present in solution at low temperatures (53). In addition, formal protonation of the neutral cluster changes the metal core geometry to a tetrahedral structure very similar to that of isomer B (Fig. 27) (53).

4. Mixed-Metal Clusters Containing More Than Three ML (M = Cu, Ag, or Au; L = Two-Electron Donor Ligand) Units

Heteronuclear clusters containing more than three ML fragments are relatively rare, and they adopt a wide variety of metal framework structures (Table XIV, pp. 300–301). Many of the lower nuclearity species exhibit metal core geometries containing one or two tetrahedral Au$_3$M (M = transition metal) or Au$_4$ units (e.g., 70,72,83,85,90), and the three high nuclearity silver–gold clusters have metal framework structures consisting of icosahedral fragments fused together by sharing common vertices and faces (e.g., 77,78,162). The clusters [Au$_4$Fe$_2${μ-Ph$_2$P(CH$_2$)$_n$PPh$_2$}$_2$-(CO)$_8$] (n = 1 or 2) provide further evidence that the energy differences between the various skeletal geometries can be small for mixed-metal clusters containing ML units. When n is 2, the two bidentate diphosphine ligands link together two well-separated trigonal planar Au$_2$Fe units, each one similar to the metal framework observed for [Au$_2$Fe(CO)$_4$(PPh$_3$)$_2$] (1,4), whereas a metal core structure consisting of a Au$_4$ "butterfly" with two Fe(CO)$_4$ units bridging opposite Au—Au edges of the two "wings" is adopted when n = 1 (Fig. 28, p. 302) (151).

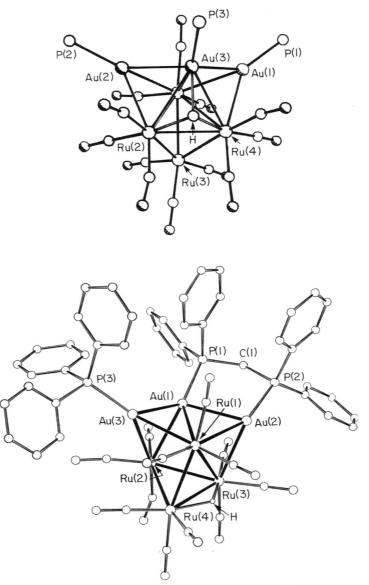

FIG. 26. Structures of the clusters $[Au_3Ru_4(\mu_3\text{-}H)(CO)_{12}(PPh_3)_3]$ and $[Au_3Ru_4(\mu\text{-}H)\text{-}(\mu\text{-}Ph_2PCH_2PPh_2)(CO)_{12}(PPh_3)]$, with the phenyl groups in the former species omitted for clarity, showing the change in skeletal geometry when two PPh_3 groups are formally replaced by a $Ph_2PCH_2PPh_2$ ligand. [Reprinted with permission of Pergamon Journals Ltd. (48) and the Royal Society of Chemistry (161).]

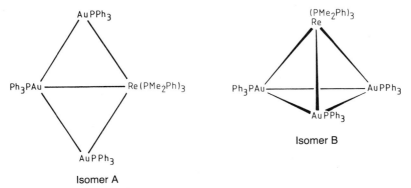

FIG. 27. Structures of the skeletal isomers of the cluster $[Au_3Re(\mu\text{-}H)_2(PMe_2Ph)_3$-$(PPh_3)_3]$, with the hydrido ligands, which are thought to bridge two Au—Re vectors, omitted for clarity.

5. Short M---C Contacts between Group IB Metals and Carbonyl Ligands Bonded to Adjacent Metals

An interesting structural feature of heteronuclear clusters containing ML (M = Cu, Ag, or Au; L = two-electron donor ligand) units is the short M---C contacts (<3 Å) that often occur between the coinage metals and essentially linear carbonyl ligands which are bonded to adjacent metals. These short contacts have been observed for copper- (e.g., *18,32,33,35,43,73,74,119,143*), silver- (e.g., *97,124*), and gold-containing species (e.g., *14,18,35,46,47,108,109,119,136,161*). The actual M---C distances vary considerably, but sometimes they can be very short, for example, 2.274(23) Å for M = Cu (*18*), 2.687(33) Å for M = Ag (*124*), and 2.51(3) Å for M = Au (*161*). It is not clear whether the short M---C contacts represent some degree of long-range interaction, either attractive (e.g., *35,74*) or repulsive (e.g., *33*), between the coinage metals and the carbonyl ligands or result from steric effects in the solid (e.g., *119*).

C. Characterization

As is the case for transition metal cluster compounds in general, single-crystal X-ray diffraction is normally the only technique available for the unambiguous structural characterization of heteronuclear Group IB metal clusters. Tables I, II, and IV–XIV indicate the mixed-metal clusters containing one or more ML (M = Cu, Ag, or Au; L = two-electron donor ligand) units which have been studied by X-ray crystallography. Other

TABLE XIV

MIXED-METAL CLUSTERS CONTAINING MORE THAN THREE M(PR₃) (M = Ag or Au; R = Alkyl or Aryl) Units

$$\text{MIXED-METAL CLUSTERS CONTAINING MORE THAN THREE } M(PR_3) \text{ } (M = Ag \text{ or } Au; R = \text{ALKYL or ARYL}) \text{ UNITS}$$

Cluster	Spectroscopic data reported	Crystal structure determined?	Skeletal geometry adopted	Ref.
Clusters containing four M(PR₃) units				
$[Au_4ReH_4\{P(C_6H_4Me-4)_3\}_2(PPh_3)_4][PF_6]$	^1H, ^{31}P NMR, Mass	No	Au_3Re tetrahedron with $Au(PPh_3)$ bridging one Au—Re edge	72
$[Au_4ReH_4\{P(C_6H_4Me-4)_3\}_2(PPh_3)_4][BPh_4]$	^1H, ^{31}P NMR	Yes	Au_3Re tetrahedron with $Au(PPh_3)$ bridging one Au—Re edge	85
$[Au_4Fe_2(\mu\text{-}Ph_2PCH_2PPh_2)_2(CO)_8]$	IR	Yes	Au_4 "butterfly" with two $Fe(CO)_4$ units bridging opposite Au—Au edges of the two "wings"	151
$[Au_4Fe_2\{\mu\text{-}Ph_2P(CH_2)_2PPh_2\}_2(CO)_8]$	IR	Yes	Two trigonal planar Au_2Fe units linked by two $Ph_2P(CH_2)_2PPh_2$ ligands	83
$[Au_4Ir(\mu\text{-}H)_2(PPh_3)_6][BF_4]$	IR, ^1H, ^{31}P NMR	Yes	Au_4Ir trigonal bipyramid with equatorial Ir atom	83
Clusters containing five M(PR₃) units				
$[Au_5ReH_4(PPh_3)_7][PF_6]_2$	^1H, ^{31}P NMR	Yes	Two Au_3Re tetrahedra sharing a common Au—Re edge	70
$[Au_5ReH_4\{P(C_6H_4Me-4)_3\}_2(PPh_3)_5][PF_6]_2$	Mass	No	Two Au_3Re tetrahedra sharing a common Au—Re edge	72

Clusters containing six M(PR$_3$) units

Cluster	Method	Ag–Ag	Description	Ref.
[Ag$_6$Fe$_3${μ_3-(Ph$_2$P)$_3$CH}(CO)$_{12}$]^a	IR, ^{31}P NMR	Yes	Ag$_6$ octahedron with Fe(CO)$_4$ units capping three Ag$_3$ faces	76
[Au$_6$Co$_2$(CO)$_8$(PPh$_3$)$_4$]		Yes	Au$_6$ edge-sharing bitetrahedron with two Au atoms ligated by Co(CO)$_4$ units	90
[Au$_6$Pt(C$_2$Bu^t)(PPh$_3$)$_7$][Au(C$_2$Bu^t)$_2$]	^{1}H, ^{31}P NMR	Yes	Two Au$_4$Pt square-based pyramidal units sharing a common Au$_2$Pt face; apical Pt atom	82
[Au$_6$Pt(PPh$_3$)$_7$][BPh$_4$]$_2$	^{1}H, ^{31}P NMR, Mass	No	Unknown	72

Clusters containing more than six M(PR$_3$) units

Cluster	Method	Ag–Ag	Description	Ref.
[Ag$_{12}$Au$_{13}$Cl$_6$(PPh$_3$)$_{12}$]$^{m+}$		Yes	Three interpenetrating icosahedral units sharing common pentagonal faces so that an interstitial Au atom is encapsulated within each	77
[Ag$_{19}$Au$_{18}$Br$_{11}${P(C$_6$H$_4$Me-4)$_3$}$_{12}$][AsF$_6$]$_2$		Yes	Three icosahedra sharing three vertices in a cyclic manner with one capping atom	78
[Ag$_{20}$Au$_{18}$Cl$_{14}$(PPh$_3$)$_{12}$]		Yes	Three icosahedra sharing three vertices in a cyclic manner with two capping atoms	162

^{a}Three of the Ag—Ag contacts (3.288–3.415 Å) in the octahedral Ag$_6$ unit within the metal framework of the cluster are too long for there to be any significant bonding interaction between the Ag atoms.

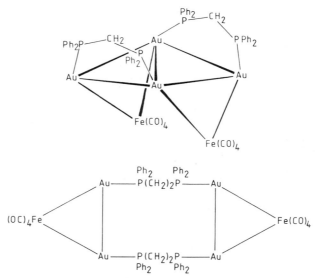

FIG. 28.   Structures of the clusters $[Au_4Fe_2\{\mu\text{-}Ph_2P(CH_2)_nPPh_2\}_2(CO)_8]$ (n = 1 or 2).

conventional spectroscopic and analytical techniques, such as multinuclear NMR, IR, and UV/visible spectroscopy, mass spectrometry, and microanalysis, are generally of much more limited value for the elucidation of structures, but they can all facilitate characterization in some cases. The spectroscopic data available for each cluster are also listed in Tables I, II, and IV–XIV.

In principle, multinuclear NMR spectroscopy, especially $^{31}P-\{^1H\}$, should provide valuable structural information for mixed-metal clusters containing $M(PR_3)$ (R = alkyl or aryl) units. In practice, however, the stereochemical nonrigidity exhibited in solution by the metal frameworks of many such species (Section II,E,1) means that the observed spectra are often simpler than would be anticipated (e.g., *11,48,142,143*), and, in many cases, it is not possible to obtain spectra consistent with the ground state structures of clusters even when solutions are cooled to −90 or −100°C (e.g., *18,46,47*). Thus, it is particularly important to find other techniques, apart from X-ray diffraction, that can provide information about the ground state skeletal geometries of heteronuclear Group IB metal clusters. High-resolution $^{31}P-\{^1H\}$ NMR spectroscopy in the solid state and ^{197}Au Mössbauer spectroscopy are two such possibilities that have been investigated recently.

High-resolution $^{31}P-\{^1H\}$ NMR spectra have been reported (*163,164*) for a number of homonuclear gold clusters, and it has proved possible to obtain spectra consistent with the ground state structures even in cases where the stereochemical nonrigidity of the metal frameworks of the clusters have prevented this at low temperatures in solution. A recent preliminary study by Dobson, Salter, and co-workers (165) has demonstrated that high-resolution $^{31}P-\{^1H\}$ NMR spectra consistent with the ground state skeletal geometries can also be observed for mixed-metal clusters containing $M(PR_3)$ units when only "averaged" phosphorus signals are visible in the solution spectra. For example, the $^{31}P-\{^1H\}$ NMR spectra of $[M_2Ru_4H_2(CO)_{12}(PPh_3)_2]$ (M = Ag or Au) in solution show a single phosphorus resonance for each cluster at ambient temperature (142), but two signals, consistent with the capped trigonal bipyramidal ground state metal core structures, are visible in the solid state spectra (Figs. 15 and 29) (165). Although a similar spectrum can be obtained at −90°C in solution for the silver-containing species, the signal for the gold cluster is still severely broadened by the dynamic process at this temperature (142).

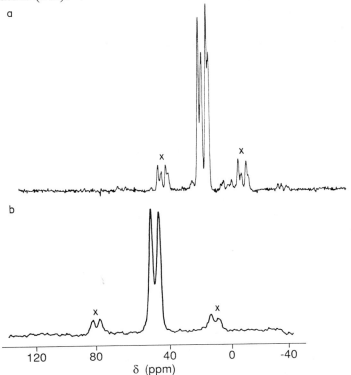

FIG. 29. Solid-state $^{31}P-\{^1H\}$ NMR spectra of the clusters $[M_2Ru_4H_2(CO)_{12}(PPh_3)_2]$ [M = Ag (a) or Au (b)]. The two resonances in (a) are each split into doublets by ^{107,109}Ag–^{31}P coupling. Groups of peaks marked by x are spinning sidebands.

Attempts to use ^{197}Au Mössbauer spectroscopy to investigate the skeletal geometries of homonuclear gold clusters have proved to be somewhat disappointing (7,8,10). Although it is easy to distinguish the signals for peripheral Au atoms bearing different exo ligands (e.g., halides and tertiary phosphines), there is only one report of resolved subspectra being observed for structurally nonequivalent Au atoms coordinated to the same ligand but to different arrangements of metal atoms (166).[6]

Parish, Salter, and co-workers (111,136,167) have recently demonstrated, however, that the more marked differences between the gold sites in heteronuclear clusters make their ^{197}Au Mössbauer spectra more sensitive to structural effects. First, the technique can often resolve the signals arising from nonequivalent $Au(PR_3)$ units in the metal cores of such species. For example, the change in metal framework geometry that occurs when two PPh$_3$ groups attached to the gold atoms in $[Au_2Ru_4(\mu_3\text{-H})(\mu\text{-H})(CO)_{12}(PPh_3)_2]$ (142) are formally replaced by a $Ph_2PCH_2PPh_2$ ligand in $[Au_2Ru_4(\mu_3\text{-H})(\mu\text{-H})(\mu\text{-}Ph_2PCH_2PPh_2)(CO)_{12}]$ (135,137) is easily detected. The ^{197}Au Mössbauer spectrum of the PPh$_3$-containing cluster clearly shows two overlapping signals, one for each of the two geometrically distinct gold sites within the capped trigonal bipyramidal cluster skeleton (Figs. 18 and 30) (111,167). However, only one signal is observed for the bidentate diphosphine-containing species, which adopts a capped square-based pyramidal metal framework structure in which the two gold atoms are equivalent except for the position of one of the hydrido ligands (Figs. 18 and 30) (111). Interestingly, when the signals that are due to more than one Au environment can be resolved, the intensity ratios reflect the effective masses of the recoiling units (i.e., the Au atoms together with their immediate neighbors) (111). Second, the isomer shift and quadrupole splitting values seem to be strongly correlated with the bonding modes of the $Au(PR_3)$ fragments. In a series of gold–ruthenium clusters, these parameters have been found to show marked decreases as the connectivity of the Au atoms (Table XV) and average Au—Ru bond length increase (111,136).

Another technique that has recently proved to be very useful for facilitating the characterization of heteronuclear Group IB metal cluster compounds is fast atom bombardment (FAB) mass spectrometry. Although the mass spectra of some mixed-metal clusters containing ML units have

---

[6] The spectrum of $[Au_7(PPh_3)_7][OH]$ shows two doublets thought to correspond to the axial and equatorial Au atoms of a pentagonal bipyramidal cluster. However, the area ratio of the two subspectra is approximately 1.4:1 instead of the expected 0.4:1. Even when the expected relative intensities are treated to take the effective masses of the recoiling units into account, as described by Parish, Salter, and co-workers (111), the expected ratio increases to only 0.48:1.

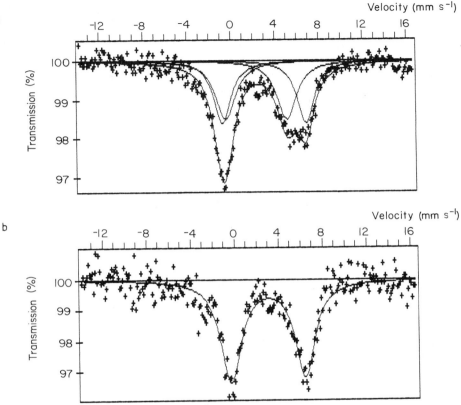

FIG. 30. ^{197}Au Mössbauer spectra of the clusters $[Au_2Ru_4(\mu_3\text{-}H)(\mu\text{-}H)(CO)_{12}L_2]$ [L = PPh$_3$ (a) or L$_2$ = μ-Ph$_2$PCH$_2$PPh$_2$ (b)]. [Reprinted with permission of the Royal Society of Chemistry (*167*).]

been recorded using conventional electron-impact techniques (e.g., *13,14,20,25,21,57,60,61,95*), the high molecular weight and relatively low volatility of these species generally mean that it is very difficult to obtain signals due to the molecular ions in this manner. Recently, however, a number of workers have used the FAB technique to obtain mass spectra, which include peaks due to the molecular ions, of heteronuclear Group IB clusters with molecular weights of over 2000 (*48,113*) and of cationic (*72,85*) and anionic species (*125*). The molecular weight data afforded by FAB mass spectrometry are particularly useful for the characterization of clusters containing hydrido ligands, because of the difficulty of detecting these ligands using X-ray crystallography (*72*).

TABLE XV

GOLD-197 MÖSSBAUER DATA FOR THE GOLD SITES IN GOLD–RUTHENIUM CLUSTER COMPOUNDS, IN ORDER OF ISOMER SHIFT

Cluster[a]	Coordination of Au atom[b]	Isomer shift (mm second^{-1})	Quadrupole splitting (mm second^{-1})	Linewidth (mm second^{-1})
Parameters for Au atoms ligated by PR$_3$ (R = alkyl or aryl)				
[AuRu$_3$(CO)$_9$(C$_2$Bu^t)(PPh$_3$)]	Ru$_2$P	3.51	7.33	2.13 1.97
[AuRu$_3$(μ-H)$_2$(μ_3-COMe)(CO)$_9$(PPh$_3$)]	Ru$_2$P	3.38	7.38	1.99 2.05
[AuRu$_4$(μ_3-H)(μ-H)$_2$(CO)$_{12}$(PPh$_3$)]	Ru$_2$P	3.22	7.39	2.21 1.85
[Au$_2$Ru$_3$(μ-H)(μ_3-COMe)(CO)$_9$(PPh$_3$)$_2$]	AuRu$_2$P	3.11	6.93	1.92 2.27
[Au$_2$Ru$_4$(μ_3-H)(μ-H){μ-Ph$_2$P(CH$_2$)$_2$PPh$_2$}(CO)$_{12}$][c] (gold site A)	AuRu$_2$P	3.1	7.2	2.0
[Au$_2$Ru$_4$(μ_3-H)(μ-H)(μ-Ph$_2$PCH$_2$PPh$_2$)(CO)$_{12}$]	AuRu$_2$P	3.02	6.71	1.90 1.98
[Au$_2$Ru$_4$(μ_3-H)(μ-H)(CO)$_{12}$(PPh$_3$)$_2$] (gold site A)	AuRu$_2$P	3.0	7.5	1.8
[Au$_3$Ru$_4$(μ_3-H)(CO)$_{12}$(PPh$_3$)$_3$] (gold site A)	AuRu$_2$P	2.9	7.4	2.2
[Au$_2$Ru$_4$(μ_3-H)(μ-H)(μ-Ph$_2$AsCH$_2$PPh$_2$)(CO)$_{12}$][c]	AuRu$_2$P	2.9	7.1	1.9
[Au$_2$Ru$_3$(μ_3-COMe)(CO)$_9$(PPh$_3$)$_3$] (gold site A)	AuRu$_2$P	2.8	7.4	2.0
[Au$_2$Ru$_3$(μ_3-S)(μ-Ph$_2$PCH$_2$PPh$_2$)(CO)$_9$]	AuRu$_2$P	2.68	6.45	2.03 2.22
[Au$_2$Ru$_4$(μ_3-H)(μ-H){μ-Ph$_2$As(CH$_2$)$_2$PPh$_2$}(CO)$_{12}$][c] (gold site A)	AuRu$_2$P	2.6	7.0	2.0
[Au$_2$Ru$_4$(μ_3-H)(μ-H){μ-Ph$_2$P(CH$_2$)$_2$PPh$_2$}(CO)$_{12}$][c] (gold site B)	AuRu$_2$HP[d]	2.5	5.8	1.9

[Au$_3$Ru$_3$(μ_3-COMe)(CO)$_9$(PPh$_3$)$_3$] (gold site B)	Au$_2$Ru$_3$P	2.5	5.2	2.2
[Au$_2$Ru$_4$(μ_3-H)(μ-H)(CO)$_{12}$(PPh$_3$)$_2$] (gold site B)	AuRu$_3$HP	2.3	5.6	2.2
[Au$_3$Ru$_4$(μ_3-H)(CO)$_{12}$(PPh$_3$)$_3$] (gold site B)	Au$_2$Ru$_3$HP	1.9	4.9	1.9

Parameters for Au atoms ligated by AsR$_3$ (R = alkyl or aryl)

[Au$_2$Ru$_4$(μ_3-H)(μ-H)(μ-Ph$_2$AsCH$_2$AsPh$_2$)(CO)$_{12}$][c]	AuRu$_2$As	2.53	5.93	1.94 2.06
[Au$_2$Ru$_4$(μ_3-H)(μ-H)(μ-Ph$_2$AsCH$_2$PPh$_2$)(CO)$_{12}$][c]	AuRu$_2$As	2.5	5.5	1.6
[Au$_2$Ru$_4$(μ_3-H)(μ-H){μ-Ph$_2$As(CH$_2$)$_2$AsPh$_2$}(CO)$_{12}$][c] (gold site A)	AuRu$_2$As	2.4	6.6	1.6
[Au$_2$Ru$_4$(μ_3-H)(μ-H)[μ-Ph$_2$As(CH$_2$)$_2$AsPh$_2$}(CO)$_{12}$][c] (gold site B)	AuRu$_3$HAs	2.3	5.1	2.2
[Au$_2$Ru$_4$(μ_3-H)(μ-H)[μ-Ph$_2$As(CH$_2$)$_2$PPh$_2$}(CO)$_{12}$][c] (gold site B)	AuRu$_3$HAs	2.2	5.3	2.1

[a] Data from Ref. *111*, unless otherwise stated.
[b] Where signals arising from two inequivalent gold atoms can be resolved in the Mössbauer spectrum, the site of lower connectivity has been labeled A and that of higher connectivity, B.
[c] Data from Ref. *136*.
[d] An X-ray diffraction study shows (*137*) that the cluster adopts a fairly similar metal core structure to that exhibited by [Au$_2$Ru$_4$(μ_3-H)-(μ-H)(CO)$_{12}$(PPh$_3$)$_2$], but one of the Au—Ru distances [3.446(4) Å] for the gold atom in site B is too long for any significant bonding interaction between the two atoms. Therefore, the coordination of that gold atom is AuRu$_2$HP rather than AuRu$_3$HP.

D. Chemical Properties

This section outlines the main types of reaction which heteronuclear Group IB metal clusters undergo. However, investigations of the chemical properties of these species tend to be rather hampered by the relatively weak bonding of the ML units (*92,93*). This means that many reactions result in cleavage of the coinage metal fragment(s) from the cluster (Section II,D,1), the production of an unexpected heteronuclear Group IB metal cluster of higher or lower nuclearity (e.g., Section II,A,3), or simply the formation of a complex mixture of products.

1. Cleavage of ML (M = Cu, Ag, or Au; L = Two-Electron Donor Ligand) Units

One notable feature of the chemistry of mixed-metal clusters containing ML units is the ease with which the Group IB metal moiety can often be removed. In many cases, this reaction can be achieved simply by the addition of PR_3 (R = alkyl or aryl) to the cluster (e.g., *14,25,51,57,66, 115,119,133*), and Eqs. (57) and (58) show two examples (*119*). Treatment with PR_3 does not always cleave the coinage metal fragments from heteronuclear clusters (e.g., *16,25,47*), however, and selective removal of some of the $Au(PR_3)$ groups from higher nuclearity species using PPh_3 has been utilized as a synthetic route to gold-containing mixed-metal clusters and dimers [e.g., *67,72,85* and Eq. (46) in Section II, A,3]. Other reagents, such as $[N(PPh_3)_2]X$ (X = NO_2 or Cl) (e.g., *106*), $[NR_4]X$ (R = Me or Bu^n; X = Cl or Br) (e.g., *65,66*), X^- (X = Cl, Br, or I) (e.g., *25,35*), X_2 (X = Cl, Br, or I) (e.g., *35,86*), KOH (e.g., *20*), HI, and $[BH_4]^-$ (e.g., *86*) have also been used to cleave ML units from heteronuclear clusters. In some cases, the ML fragments are so weakly bonded that the mixed-metal cluster dissociates into a cluster anion and a Group IB metal cation when it is dissolved in coordinating solvents, such as tetrahydrofuran, acetone, or hexamethylphosphoramide (e.g., *35,119,159*).

$$[CuCo_3M(CO)_{12}(PPh_3)] + 2 \ PPh_3 \ \rightarrow \ [Cu(PPh_3)_3][Co_3M(CO)_{12}] \qquad (57)$$

$$M = Fe \ or \ Ru$$

$$[AuCo_3Ru(CO)_{12}(PPh_3)] + PPh_3 \ \rightarrow \ [Au(PPh_3)_2][Co_3Ru(CO)_{12}] \qquad (58)$$

2. Group IB Metal Exchange Reactions

The thermodynamic site preferences of the coinage metals in the metal cores of $[MM'Ru_4(\mu_3\text{-}H)_2(CO)_{12}(PPh_3)_2]$ (M = Cu, M' = Ag or Au; M = Ag, M' = Au) (Section II,B,2) mean that the clusters $[M_2\text{-}$

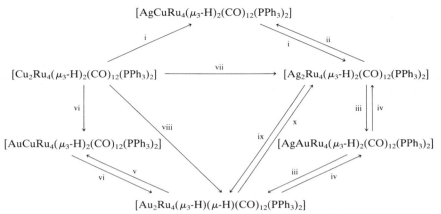

SCHEME 2. Reagents: i, [AgI(PPh$_3$)] (2 equiv) and TlPF$_6$ (3 equiv); ii, [CuCl(PPh$_3$)] (5 equiv) and TlPF$_6$ (7 equiv); iii, [AuCl(PPh$_3$)] (1.5 equiv) and TlPF$_6$ (2 equiv); iv, [AgI(PPh$_3$)] (3.5 equiv) and TlPF$_6$ (5 equiv); v, [CuCl(PPh$_3$)] (1.5 equiv) and TlPF$_6$ (2 equiv); vi, [AuCl(PPh$_3$)] (1.2 equiv) and TlPF$_6$ (2 equiv); vii, [AgI(PPh$_3$)] (4 equiv) and TlPF$_6$ (6 equiv); viii, [AuCl(PPh$_3$)] (2.4 equiv) and TlPF$_6$ (4 equiv); ix, [AuCl(PPh$_3$)] (3 equiv) and TlPF$_6$ (4 equiv); x, [AgI(PPh$_3$)] (7 equiv) and TlPF$_6$ (10 equiv) (*144,168*).

Ru$_4$H$_2$(CO)$_{12}$] (M = Cu, Ag, or Au) undergo novel intermolecular M-(PPh$_3$) unit exchange reactions in solution [Eq. (59)] (*144*). In addition,

$$[M_2Ru_4H_2(CO)_{12}(PPh_3)_2] + [M'_2Ru_4H_2(CO)_{12}(PPh_3)_2] \xrightarrow[\text{24 hours}]{CH_2Cl_2}$$

$$2 \ [MM'Ru_4(\mu_3\text{-}H)_2(CO)_{12}(PPh_3)_2] \qquad (59)$$

$$M = Cu, \ M' = Ag \ or \ Au; \ M = Ag, \ M' = Au$$

treatment of the bimetallic species with appropriate amounts of the mononuclear complexes [M'X(PPh$_3$)] (M' = Cu or Au, X = Cl; M' = Ag, X = I), in the presence of TlPF$_6$, results in the replacement of either one or both of the original M(PPh$_3$) units present in the cluster by similar fragments containing the different Group IB metal M', as shown in Scheme 2 (*144,168*). Although it is possible to exchange a heavier coinage metal, and in the case of [Au$_2$Ru$_4$(μ_3-H)(μ-H)(CO)$_{12}$(PPh$_3$)$_2$] both heavier coinage metals, present in the hexanuclear clusters with lighter Group IB metals using these reactions, it can be seen from Scheme 2 that large excesses of the mononuclear complexes are always required. The procedure is much more economical when the lighter of the pair or pairs of coinage metals to be exchanged is initially present in the cluster, as these reactions require only relatively small excesses of the mononuclear complexes (Scheme 2).

Recently, Brown and Salter (145) have found that one or two of the Group IB metals in the clusters $[M_2Ru_4(\mu_3\text{-}H)_2\{\mu\text{-}Ph_2P(CH_2)_2PPh_2\}(CO)_{12}]$ (M = Cu or Ag), which adopt very similar metal framework structures to those of the analogous PPh$_3$-containing species (124,143), can be replaced by utilizing the complexes $[Ag(NCMe)_4][PF_6]$ and $[AuCl(SC_4H_8)]$ as sources of silver and gold atoms, respectively [e.g., Eq. (60)].

$$[M_2Ru_4(\mu_3\text{-}H)_2\{\mu\text{-}Ph_2P(CH_2)_2PPh_2\}(CO)_{12}] + [AuCl(SC_4H_8)] \rightarrow$$

$$[AuMRu_4(\mu_3\text{-}H)_2\{\mu\text{-}Ph_2P(CH_2)_2PPh_2\}(CO)_{12}] \qquad (60)$$

$$M = Cu \text{ or } Ag$$

As mentioned in Section II,A,2, the above types of Group IB metal exchange reactions are a potentially very useful synthetic route to cluster compounds containing two or more different coinage metals, which are at present very rare.

3. Loss of Carbon Monoxide

A number of Group IB metal heteronuclear clusters undergo reactions involving the loss of a CO ligand (e.g., 25,26,51,55,57,58,100), and Eq. (61) shows an example (58). In some cases, the reaction is known to be

$$[AuOs_3(CO)_{10}(C_2Ph)L] \xrightarrow[-CO]{96°C} [AuOs_3(CO)_9(C_2Ph)L] \qquad (61)$$

$$L = PPh_3 \text{ or } PMe_2Ph$$

reversible (e.g. 25,26,55). Although the analogous clusters containing an isolobal hydrido ligand (93) instead of a Au(PR$_3$) moiety sometimes undergo similar CO elimination reactions to those of the Group IB metal species (e.g., 25,55), there can also be significant differences in chemical behavior between the two classes of compound. For example, the cluster anion $[AuRe_3(\mu\text{-}H)_3(CO)_{10}(PPh_3)]^-$ is stable only at low temperatures, even under a CO atmosphere, and a CO ligand is lost to form $[AuRe_3(\mu\text{-}H)_3(CO)_9(PPh_3)]^-$ on raising the temperature, whereas the analogous species $[Re_3(\mu\text{-}H)_4(CO)_{10}]^-$ is stable with respect to CO elimination up to its decomposition temperature (100). Also, the clusters $[Au_2Os_4(CO)_{13}L_2]$ (L = PPh$_3$, PEt$_3$, or PMePh$_2$) are thermally unstable in the solid state and solution and they lose a CO ligand to produce $[Au_2Os_4(CO)_{12}L_2]$ (26) in contrast to $[AuOs_4(\mu\text{-}H)(CO)_{13}(PR_3)]$ and $[Os_4H_2(CO)_{13}]$, which do not lose CO readily and which undergo cluster fragmentation reactions when refluxed in xylene for prolonged time periods (169).

4. *Ligand Addition and Substitution*

Although treatment of heteronuclear Group IB metal clusters with nucleophilic ligands often results in cleavage of ML fragments (Section II,D,1), in some cases ligand addition (e.g., *25,26,51*) and ligand substitution (e.g., *18,26,57*) reactions are known to occur instead, especially for phosphine ligands. Equations (62) and (63) show an example of each type of reaction (*18,51*).

$$[AuRu_5C(\mu\text{-}I)(CO)_{13}(PPh_3)] + PPh_3 \rightarrow [AuRu_5C(\mu\text{-}I)(CO)_{13}(PPh_3)_2] \quad (62)$$

$$[MRu_3(\mu\text{-}H)(\mu_3\text{-}S)(CO)_9(PPh_3)] + PPh_3 \rightarrow [MRu_3(\mu\text{-}H)(\mu_3\text{-}S)(CO)_8(PPh_3)_2] \quad (63)$$

$$M = Cu, Ag, \text{ or } Au$$

5. *Cleavage of M—P (M = Cu, Ag, or Au) Bonds*

In some cases, clusters containing $M(PR_3)$ units undergo reactions involving the cleavage of M—P bonds to produce other Group IB metal heteronuclear clusters (e.g., *170,171*). Equation (64) shows an example (*170*).

$$[AuOs_3(\mu\text{-}H)(CO)_{10}(PR_3)] \xrightarrow[\text{CH}_2\text{Cl}_2,\text{ reflux}]{2 \text{ [N(PPh}_3)_2]\text{Cl}} [N(PPh_3)_2][Au\{Os_3(\mu\text{-}H)(CO)_{10}\}_2] \quad (64)$$

$$R = Ph \text{ or } Et$$

6. *Protonation and Deprotonation*

Protonation and deprotonation reactions have been reported for Group IB metal heteronuclear clusters (e.g., *66,103*), but cleavage of the coinage metal moiety from the cluster often occurs instead (e.g., *20*).

7. *Homogeneous Catalysis*

Only a few Group IB metal heteronuclear clusters have been investigated as potential homogeneous catalysts (*6*), but some examples have been reported. The clusters $[AuRu_4(\mu_3\text{-}H)(\mu\text{-}H)_2(CO)_{12}(PPh_3)]$ and $[Au_2Ru_4(\mu_3\text{-}H)(\mu\text{-}H)(CO)_{12}(PPh_3)_2]$ are more effective catalysts for pent-1-ene isomerization than $[Ru_4(\mu\text{-}H)_4(CO)_{12}]$, although the analogous copper species are not (*172*). The preparation of methanol from CO and H_2 is catalyzed by the clusters $[M_2Ru_6C(CO)_{16}(NCR)_2]$ (M = Cu, Ag, or Au; R = alkyl or aryl) (*173*), and CO conversion to methanol, ethylene glycol, glycerol, and 1,2-propylene glycol is catalyzed by $[Au_2Rh_{12}(CO)_{30}]$ (*174*).

8. Supported Group IB Metal Heteronuclear Clusters

The clusters $[AuOs_3(\mu-X)(CO)_{10}(PPh_3)]$ have been attached to phosphine-functionalized silica for X = H or Cl (175,176) or polymer (styrene–divinylbenzene) for X = H (176). On both supports, the immobilized hydrido cluster was found to be inactive for alkene hydrogenation and isomerization, whereas the supported Cl-containing species catalyzed alkene hydrogenation. The different behavior was initially incorrectly attributed to different metal framework structures for the two clusters, but, in fact, both species adopt similar "butterfly" skeletal geometries (12,54).

E. Dynamic Behavior

Group IB metal heteronuclear clusters undergo a variety of interesting dynamic processes in solution, and these are surveyed in this section. Only processes which are rapid on the NMR time scale and which can, therefore, be observed by NMR spectroscopy have been included.

1. Metal Core Rearrangements

Although the vast majority of transition metal clusters have polyhedral frameworks which do not undergo dynamic behavior in solution (11), the fact that the differences in energy between alternative skeletal geometries can be very small for mixed-metal clusters containing ML (M = Cu, Ag, or Au; L = two-electron donor ligand) fragments (Sections II,B,1 to II,B,4) means that the metal cores of these species often exhibit novel stereochemical nonrigidity. When two or more ML units adopt structurally inequivalent positions in the ground state geometries of such clusters, dynamic behavior involving coinage metal exchange between the distinct sites is frequently observed in solution at ambient temperatures. Interestingly, this site exchange takes place even when the Group IB metals are linked together by bidentate ligands. In the majority of cases, the occurrence of these metal core rearrangements is inferred from $^{31}P-\{^1H\}$ or ^1H NMR studies, but recently $^{109}Ag-\{^1H\}$ INEPT NMR spectroscopy has been utilized to observe this dynamic behavior directly in the clusters $[Ag_2Ru_4(\mu_3-H)_2\{\mu-Ph_2E(CH_2)_nEPh_2\}(CO)_{12}]$ (E = P, n = 1, 2, or 4; E = As, n = 1) (40,124,147). Although these clusters all have two inequivalent silver atoms in their capped trigonal bipyramidal metal core structures, only a single silver resonance is observed at ambient temperature in each case (e.g., Fig. 31). Skeletal rearrangements in Group IB metal

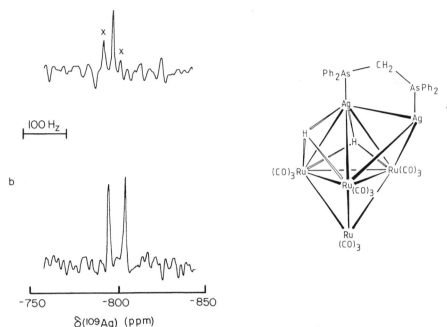

FIG. 31. $^{109}Ag-\{^{1}H\}$ INEPT NMR spectra and structure of the cluster $[Ag_2Ru_4(\mu_3\text{-}H)_2\text{-}(\mu\text{-}Ph_2AsCH_2AsPh_2)(CO)_{12}]$ (a) with the INEPT timings optimized to enhance the signals from the $^{109}Ag^{109}Ag$ isotopomer, and (b) with the INEPT timings optimized to enhance the signals from the $^{107}Ag^{109}Ag$ isotopomer $[J(^{107}Ag^{109}Ag) = 40$ Hz]. In (a), the main peak is flanked by vestigial peaks due to the $^{107}Ag^{109}Ag$ isotopomer (marked by x). [Reprinted with permission of the Royal Society of Chemistry (40).]

clusters have been the subject of a recent review (11) which covers the literature up to the middle of 1986. Therefore, heteronuclear clusters with stereochemically nonrigid metal frameworks which were reported before that date have been summarized in Table XVI, and the reader is referred to Ref. 11 for a detailed discussion of the dynamic behavior of these species.

A number of mixed-metal clusters which undergo metal core rearrangements have been reported since 1986, and these species are listed in Table XVII. The effect of the monodentate phosphine or phosphite ligands

TABLE XVI

SUMMARY OF THE HETERONUCLEAR CLUSTERS WITH STEREOCHEMICALLY NONRIGID METAL FRAMEWORKS WHICH ARE DISCUSSED IN DETAIL IN REF. [11]

Cluster	Ground state metal core structure[a]	Number of structurally inequivalent CuL (L = two-electron donor ligand) units	NMR evidence for dynamic behavior at ambient temperature	Low temperature NMR spectrum consistent with ground state structure observed?	Ref.
Copper clusters					
$[Cu_2Ru_3(\mu_3\text{-}S)(CO)_9(PPh_3)_2]$	Trigonal bipyramid	2	$^{31}P\text{-}\{^1H\}$ spectrum shows equivalent P atoms	No	18
$[Cu_2Ru_3(\mu_3\text{-}S)(\mu\text{-}Ph_2PCH_2PPh_2)(CO)_9]$	Trigonal bipyramid	2	$^{31}P\text{-}\{^1H\}$ spectrum shows equivalent P atoms	No	18
$[Cu_2Ru_4(\mu_3\text{-}H)_2(CO)_{12}\text{-}L_2]$ [L = PPh$_3$ or L$_2$ = μ-Ph$_2$P(CH$_2$)$_n$PPh$_2$ (n = 1–6)]	Capped trigonal bipyramid	2	^1H and $^{31}P\text{-}\{^1H\}$ spectra show equivalent P atoms	Yes	142,143
$[Cu_2Ru_4(\mu_3\text{-}H)_2\{\mu\text{-}Ph_2As(CH_2)_n\text{-}PPh_2\}(CO)_{12}]^b$ (n = 1 or 2)	Capped trigonal bipyramid	2	^1H and $^{31}P\text{-}\{^1H\}$ spectra suggest Cu atom site exchange	Yes	39,40
$[Cu_2Ru_6C(CO)_{16}(NCMe)_2]$	Ru$_6$ octahedron and Cu$_2$Ru$_3$ trigonal bipyramid sharing common Ru$_3$ face	2	^1H spectrum shows equivalent NCMe ligands	No	33
$[Cu_2Os_4(\mu_3\text{-}H)_2\{\mu\text{-}Ph_2P(CH_2)_2PPh_2\}\text{-}(CO)_{12}]$	Capped trigonal bipyramid	2	^1H and $^{31}P\text{-}\{^1H\}$ spectra show equivalent P atoms	Yes	36,146

Silver clusters

$[Ag_2Ru_3(\mu_3\text{-S})(CO)_9(PPh_3)_2]$	Trigonal bipyramid	2	No	$^{31}P-\{^1H\}$ spectrum shows equivalent P atoms	18
$[Ag_2Ru_4(\mu_3\text{-H})_2(CO)_{12}L_2]$ [L = PPh_3 or $L_2 = \mu\text{-}Ph_2P(CH_2)_nPPh_2$ (n = 3, 5, or 6)]	Capped trigonal bipyramid	2	Yes	^1H and $^{31}P-\{^1H\}$ spectra[c] show equivalent P atoms	124,142, 143,147
$[Ag_2Ru_4(\mu_3\text{-H})_2(\mu\text{-}Ph_2PCH_2PPh_2)(CO)_{12}]$	Capped trigonal bipyramid	2	No	$^{109}Ag-\{^1H\}$ INEPT spectrum shows equivalent Ag atoms. ^1H and $^{31}P-\{^1H\}$ spectra show equivalent P atoms	124,135, 143,147
$[Ag_2Ru_4(\mu_3\text{-H})_2\{\mu\text{-}Ph_2P(CH_2)_nPPh_2\}$-$(CO)_{12}]$ (n = 2 or 4)	Capped trigonal bipyramid	2	Yes	$^{109}Ag-\{^1H\}$ INEPT spectrum shows equivalent Ag atoms. ^1H and $^{31}P-\{^1H\}$ spectra show equivalent P atoms	124,143, 147

Gold clusters

$[Au_5ReH_4(PPh_3)_7][PF_6]_2$	Two Au_3Re tetrahedra sharing a common Au—Re edge	3	Yes	$^{31}P-\{^1H\}$ spectrum shows single resonance for $Au(PPh_3)$ moieties	70
$[Au_2Ru_3(\mu_3\text{-S})(CO)_8L(PPh_3)_2]$ (L = CO or PPh_3)	Trigonal bipyramid	2	No	$^{31}P-\{^1H\}$ spectrum shows single resonance for $Au(PPh_3)$ moieties	47,112

(continued)

TABLE XVI (*Continued*)

Cluster	Ground state metal core structure[a]	Number of structurally inequivalent CuL (L = two-electron donor ligand) units	NMR evidence for dynamic behavior at ambient temperature	Low temperature NMR spectrum consistent with ground state structure observed?	Ref.
[Au$_2$Ru$_3$(μ_3-S)(μ-Ph$_2$PCH$_2$PPh$_2$)-(CO)$_9$]	Trigonal bipyramid distorted toward square-based pyramid	2	^{31}P-{^{1}H} spectrum shows equivalent P atoms	No	18
[Au$_2$Ru$_4$(μ_3-H)(μ-H)(CO)$_{12}$-(PPh$_3$)$_2$]	Capped trigonal bipyramid	2	^{1}H and ^{31}P-{^{1}H} spectra show equivalent P atoms	No	142
[Au$_3$Ru$_3$(μ_3-COMe)(CO)$_9$-(PPh$_3$)$_3$]	Capped trigonal bipyramid	2	^{31}P-{^{1}H} spectrum shows equivalent P atoms	No	46
[Au$_3$Ru$_4$(μ_3-H)(CO)$_{12}$(PPh$_3$)$_3$]	Bicapped trigonal bipyramid	2	^{31}P-{^{1}H} spectrum shows equivalent P atoms	Yes	48
[Au$_3$Ru$_4$(μ-H)(μ-Ph$_2$PCH$_2$PPh$_2$)-(CO)$_{12}$(PPh$_3$)]	Distorted bicapped square-based bipyramid	3	^{31}P-{^{1}H} spectrum shows equivalent P atoms for Ph$_2$PCH$_2$PPh$_2$ ligand	No	48,161

Compound	Structure		NMR	Fluxional	Ref.
[Au₂Ru₆C(CO)₁₆(PEt₃)₂]	Two skeletal isomers based on Ru₆ octahedron, one with μ-Au(PEt₃) groups and no Au—Au interaction and one with μ₃-Au(PEt₃) units and Au atoms in close contact, exist in solution at low temperature[d]	2 and 1	³¹P–{¹H} spectrum shows a single P environment[e]	Yes[f]	133
[Au₃Rh(μ-H)(CO)(PPh₃)₅][PF₆]	Tetrahedron	2	³¹P–{¹H} spectrum shows one resonance for Au(PPh₃) moieties	No	70
[Au₄Ir(μ-H)₂(PPh₃)₆][BF₄]	Trigonal bipyramid	3	³¹P–{¹H} spectrum shows one resonance for Au(PPh₃) moieties	Yes	83

[a] The solid state metal core structure is given, unless otherwise stated.
[b] The analogous silver-containing clusters are mentioned in Ref. 11, but detailed interpretation of their NMR spectra was not reported until later (40).
[c] As the ambient temperature spectra are complicated by intermolecular PPh₃ ligand exchange between clusters, the ¹H and ³¹P–{¹H} NMR spectra were recorded at −50°C.
[d] See Fig. 14 (Section II,B,2).
[e] ³¹P–{¹H} NMR studies on the analogous clusters [Au₂Ru₅M(CO)₁₇(PEt₃)₂] (M = Cr, Mo, or W) revealed similar behavior, although the limiting low temperature spectra were more complex in these cases (133).
[f] Although signals arising from the two skeletal isomers can be observed at low temperatures, the Au atoms in the isomer with two inequivalent Au(PEt₃) moieties are still undergoing site exchange.

TABLE XVII
Heteronuclear Clusters with Stereochemically Nonrigid Metal Frameworks Which Are Not Included in Ref. 11

Cluster	Ground state metal core structure[a]	Number of structurally inequivalent CuL (L = two-electron donor ligand) units	NMR evidence for dynamic behavior at ambient temperature	Low temperature NMR spectrum consistent with ground state structure observed?	Ref.
Copper clusters					
[Cu₂Ru₄(μ₃-H)₂(CO)₁₂L₂] [L = PMePh₂, PMe₂Ph, PEt₃, PMe₃, P(OEt)₃, P(OPh)₃, and (POMe)₃]	Capped trigonal bipyramid	2	^1H and $^{31}P-\{^1H\}$ spectra show equivalent P atoms	Yes	42
[Cu₂Ru₄(μ₃-H)₂(μ-cis-Ph₂PCH=CHPPh₂)(CO)₁₂]	Capped trigonal bipyramid	2	^1H and $^{31}P-\{^1H\}$ spectra show equivalent P atoms	Yes	41
[Cu₂Ru₄(μ₃-H)₂{μ-Ph₂As(CH₂)₂AsPh₂}-(CO)₁₂][b]	Capped trigonal bipyramid	2	^1H and $^{13}C-\{^1H\}$ spectra show equivalent CH₂ groups in ligand	Yes	40
[Cu₂Ru₄(μ₃-H)₂(CO)₁₂{P(CHMe₂)₃}₂]	Two capped trigonal bipyramidal skeletal isomers, one with and one without the Cu atoms in close contact, at −90°C in solution[c]	2 and 1	^1H and $^{31}P-\{^1H\}$ spectra show equivalent P atoms	Yes	140
[Cu₂Ru₄(μ₃-H)₂(CO)₁₂{P(cyclo-C₆H₁₁)₃}₂]	Ru₄ tetrahedron with one μ- and one μ₃-Cu{P(cyclo-C₆H₁₁)₃} unit	2	^1H and $^{31}P-\{^1H\}$ spectra show equivalent P atoms	Yes	140
Silver clusters					
[Ag₂Ru₄(μ₃-H)₂(CO)₁₂L₂] [L = PMePh₂, P(CHMe₂)₃ or P(cyclo-C₆H₁₁)₃]	Capped trigonal bipyramid	2	^1H and $^{31}P-\{^1H\}$ spectra show equivalent P atoms	Yes	36,148

Compound	Structure		NMR	Fluxional	Ref
[Ag$_2$Ru$_4$(μ_3-H)$_2$(μ-cis-Ph$_2$PCH=CHPPh$_2$)(CO)$_{12}$]	Capped trigonal bipyramid	2	^{1}H and ^{31}P–{^{1}H}[d] spectra show equivalent P atoms	No	41
[Ag$_2$Ru$_4$(μ_3-H)$_2${μ-Ph$_2$As(CH$_2$)$_n$-PPh$_2$}(CO)$_{12}$][e] (n = 1 or 2)	Capped trigonal bipyramid	2	^{1}H and ^{31}P–{^{1}H} spectra suggest Ag atom site exchange	No	40
[Ag$_2$Ru$_4$(μ_3-H)$_2$(μ-Ph$_2$AsCH$_2$AsPh$_2$)(CO)$_{12}$]	Capped trigonal bipyramid	2	^{1}H and ^{109}Ag–{^{1}H} INEPT spectra show equivalent Ag atoms	No	40
[Ag$_2$Ru$_4$(μ_3-H)$_2${μ-Ph$_2$As(CH$_2$)$_2$AsPh$_2$}-(CO)$_{12}$]	Capped trigonal bipyramid	2	^{1}H spectrum shows equivalent Ag atoms	No	40
Gold clusters					
[Au$_3$Re(μ-H)$_2$(PMe$_2$Ph)$_3$(PPh$_3$)$_3$]	Two skeletal isomers, one with a tetrahedral and one with a planar Au$_3$Re unit, in solution at low temperatures[f]	2 and 2	^{31}P–{^{1}H} spectrum shows a single resonance for the Au(PPh$_3$) moieties	Yes	53
[Au$_4$ReH$_4${P(C$_6$H$_4$Me-4)$_3$}$_2$(PPh$_3$)$_4$]X (X = BPh$_4$ or PF$_6$)	Au$_3$Re tetrahedron with Au(PPh$_3$) bridging one Au—Re edge	4	^{31}P–{^{1}H} spectrum shows a single resonance for the Au(PPh$_3$) moieties	No	72,85
[Au$_2$Fe$_4$(BH)(CO)$_{12}$(PPh$_3$)$_2$]	Fe$_4$ "butterfly" with B interacting with all four Fe. One Au(PPh$_3$) caps a Fe$_2$B face and the other a AuFeB face	2	^{31}P–{^{1}H} spectrum shows equivalent P atoms	No	24
[Au$_2$Ru$_4$H$_2$(CO)$_{12}$L$_2$] [L = P(CHMe$_2$)$_3$ or P(cyclo-C$_6$H$_{11}$)$_3$]	Capped trigonal bipyramd	2	^{1}H and ^{31}P–{^{1}H} spectra show equivalent P atoms	Yes	150
[Au$_2$Ru$_4$(μ_3-H)(μ-H)-{μ-Ph$_2$P(CH$_2$)$_2$PPh$_2$}(CO)$_{12}$]	Capped trigonal bipyramid distorted toward capped square-based pyramid	2	^{1}H and ^{31}P–{^{1}H} spectra show equivalent P atoms	No	136,137

(continued)

TABLE XVII (*Continued*)

Cluster	Ground state metal core structure[a]	Number of structurally inequivalent CuL (L = two-electron donor ligand) units	NMR evidence for dynamic behavior at ambient temperature	Low temperature NMR spectrum consistent with ground state structure observed?	Ref.
$[Au_2Ru_4(\mu_3\text{-}H)(\mu\text{-}H)\text{-}\{\mu\text{-}Ph_2P(CH_2)_nPPh_2\}(CO)_{12}]$ ($n = 3$ or 6)	Capped trigonal bipyramid	2	^1H and $^{31}P\text{-}\{^1H\}$ spectra show equivalent P atoms	Yes	137
$[Au_2Ru_4(\mu_3\text{-}H)(\mu\text{-}H)\text{-}\{\mu\text{-}Ph_2P(CH_2)_nPPh_2\}(CO)_{12}]$ ($n = 4$ or 5)	Capped trigonal bipyramid	2	^1H and $^{31}P\text{-}\{^1H\}$ spectra show equivalent P atoms	No	137
$[Au_2Ru_4(\mu_3\text{-}H)(\mu\text{-}H)\text{-}\{\mu\text{-}Ph_2As(CH_2)_2PPh_2\}(CO)_{12}]$	Capped trigonal bipyramid	2	^1H and $^{31}P\text{-}\{^1H\}$ spectra suggest Au atom site exchange	No	136
$[Au_2Ru_4(\mu_3\text{-}H)(\mu\text{-}H)\text{-}\{\mu\text{-}Ph_2As(CH_2)_2AsPPh_2\}(CO)_{12}]$	Capped trigonal bipyramid	2	^1H spectrum shows equivalent CH_2 groups in ligand	No	136
$[Au_6Pt(C_2Bu^t)(PPh_3)_7][Au(C_2Bu^t)_2]$	Two Au_4Pt square-based pyramidal units sharing a common Au_2Pt face. Apical Pt atom	4	$^{31}P\text{-}\{^1H\}$ spectrum shows a single resonance for the $Au(PPh_3)$ moieties	No	82

[a] The solid state metal core structure is given, unless otherwise stated.
[b] The analogous cluster which contains $Ph_2AsCH_2AsPh_2$ probably also undergoes an intramolecular metal core rearrangement in solution, but it is not possible to detect such a process using NMR spectroscopy (40).
[c] See Fig. 13 (Section II,B,2).
[d] As the ambient temperature spectrum is complicated by the dynamic behavior of the *cis*-$Ph_2PCH=CHPPh_2$ ligand, the $^{31}P\text{-}\{^1H\}$ NMR spectrum was recorded at −10°C.
[e] The clusters are mentioned in Ref. 11, but detailed interpretation of their NMR spectra was not reported until later (40).
[f] See Fig. 27 (Section II,B,3).

attached to the copper atoms in $[Cu_2Ru_4(\mu_3\text{-}H)_2(CO)_{12}L_2]$ [L = PPh_3, $PMePh_2$, PMe_2Ph, PEt_3, PMe_3, $P(OEt)_3$, $P(OPh)_3$, or $P(OMe)_3$] on the free energies of activation $(\Delta G^{\ddagger})$ for the metal framework rearrangements that these species undergo in solution has recently been investigated (42). Interestingly, the magnitude of $\Delta G^{\ddagger}$ for the fluxional process varies by only about 3 kJ mol^{-1} within the series of clusters, in contrast to the much larger differences of up to around 12 kJ mol^{-1} which were observed for the analogous bidentate diphosphine-containing clusters $[Cu_2Ru_4(\mu_3\text{-}H)_2\text{-}\{\mu\text{-}Ph_2P(CH_2)_nPPh_2\}(CO)_{12}]$ (n = 1–6) (11,135). The monodentate phosphine-containing clusters showed an increase in the magnitude of $\Delta G^{\ddagger}$ with increasing phosphine cone angle, but no correlation between $\Delta G^{\ddagger}$ values and the steric or electronic properties of the ligands was found for the species containing the phosphite groups (42). Examples of the interconversion of skeletal isomers of Group IB metal heteronuclear clusters at ambient temperature in solution via metal core rearrangements have been reported previously (11). Similar dynamic behavior has also been observed for the pairs of skeletal isomers that exist at low temperatures in solution for $[Cu_2Ru_4(\mu_3\text{-}H)_2(CO)_{12}\{P(CHMe_2)_3\}_2]$ (140) (Fig. 13 in Section II,B,2) and $[Au_3ReH_2(PMe_2Ph)_3(PPh_3)_3]$ (53) (Fig. 27 in Section II,B,3). In contrast, it is interesting that no interconversion of the pairs of skeletal isomers that are present in the solid state for $[Au_2Re_2H_7(PPh_3)_4(PEt_3)_2][PF_6]$ (66) and $[Au_2Re_2H_6(PMe_2Ph)_4(PPh_3)_2]$ (52) (Fig. 11 in Section II,B,2) occurs in solution, on the NMR time scale at ambient temperature for the former species and over a 24-hour period at 90°C for the latter.

2. Fluxional Decapping of M(PR₃) (M = Cu, Ag, or Au; R = Alkyl or Aryl) Moieties

Interestingly, the PR_3-promoted (R = Ph or $C_6H_4Me\text{-}4$) dissociation of the copper–phosphine moiety from $[CuRu_4(\mu_3\text{-}H)_3(CO)_{12}(PR_3)]$ to form $[Cu(PR_3)_3][Ru_4(\mu\text{-}H)_3(CO)_{12}]$ [Eq. 65] (cf. Section II,D,1) occurs as a novel fluxional process (115). An increase in the steric bulk of the phosphine ligand to that of $P(C_6H_4Me\text{-}2)_3$ or a change in the coinage metal from copper to gold retards this intermolecular exchange process. In addition, the loss of the splittings arising from $^{107,109}Ag\text{-}^1H$ and $^{31}P\text{-}^1H$ couplings on the high field ^1H NMR hydrido ligand signals of the analogous silver-containing species $[AgRu_4(\mu_3\text{-}H)_3(CO)_{12}(PPh_3)]$ and $[\{AgRu_4\text{-}(\mu_3\text{-}H)_3(CO)_{12}\}_2\{\mu\text{-}Ph_2P(CH_2)_2PPh_2\}]$ at ambient temperature suggests that the $Ag(PR_3)$ (R = alkyl or aryl) units undergo intermolecular exchange between clusters in solution (43). The severe broadening observed in the ambient temperature ^{31}P–{^{1}H} NMR spectra of these clusters implies that the PR$_3$ ligands also exchange between the silver atoms.

$$[CuRu_4(\mu_3\text{-}H)_3(CO)_{12}(PR_3)] + 2\ PR_3 \rightleftharpoons [Cu(PR_3)_3][Ru_4(\mu\text{-}H)_3(CO)_{12}] \qquad (65)$$

$$R = Ph\ or\ C_6H_4Me\text{-}4$$

3. *Processes Involving Ligands Attached to the Group IB Metals*

Dynamic behavior involving intermolecular exchange of phosphine ligands is often observed at ambient temperature in solution for mixed-metal clusters containing one or two Ag(PR$_3$) (R = alkyl or aryl) moieties (e.g., *18,43,142*). Phosphine groups bonded to silver will also exchange with the free ligand under the same conditions (e.g., *43,142*). These types of dynamic process are also well established for silver-containing mononuclear complexes (e.g., *177*) and dimers (e.g., *178,179*). Interestingly, however, not all heteronuclear silver clusters undergo intermolecular phosphine ligand exchange in solution. For example, although the PPh$_3$ groups in [Ag$_2$Ru$_4(\mu_3\text{-}H)_2(CO)_{12}(PPh_3)_2$] do exchange between clusters (*142*), the fluxional process is not observed at room temperature for the closely related species [Ag$_2$Ru$_4(\mu_3\text{-}H)_2(CO)_{12}L_2$] [L = P(CHMe$_2$)$_3$ or P(*cyclo*-C$_6$H$_{11}$)$_3$ or L$_2$ = μ-Ph$_2$P(CH$_2$)$_n$PPh$_2$ (*n* = 1–6)] (*124,148*), [AgMRu$_4(\mu_3\text{-}H)_2(CO)_{12}(PPh_3)_2$] (M = Cu or Au) (*144*), and [Ag$_2$-Ru$_4(\mu\text{-}CO)_3(CO)_{10}(PPh_3)_2$] (*44*). In addition, although the phosphine groups in the vast majority of cluster compounds containing M(PR)$_3$ (M = Cu or Au; R = alkyl or aryl) units do not undergo intermolecular exchange between clusters or with free ligand at ambient temperature in solution (e.g., *18,43,47,142*). Some examples of this type of dynamic behavior have been reported for copper- (e.g., *43*) and gold-containing (e.g., *85*) species.

Variable-temperature ^{1}H and ^{31}P–{^{1}H} NMR studies show that the arsenic and phosphorus atoms of the bidentate ligands in the clusters [Ag$_2$Ru$_4(\mu_3\text{-}H)_2(\mu\text{-}L)(CO)_{12}$] [L = Ph$_2$As(CH$_2$)$_nPPh_2$ (*n* = 1 or 2) or *cis*-Ph$_2$PCH=CHPPh$_2$] undergo novel intramolecular site exchange between the two silver atoms at ambient temperatures in solution (*40,41*). At higher temperatures, the Ph$_2$As(CH$_2$)$_n$PPh$_2$ ligands undergo intermolecular exchange between clusters, but this dynamic process does not occur for the *cis*-Ph$_2$PCH=CHPPh$_2$ ligand, even at 70°C.

In the ground state structures, the hydrido ligands in [Cu$_2$Ru$_4(\mu_3\text{-}H)_2$-{μ-Ph$_2$P(CH$_2$)$_n$PPh$_2$}(CO)$_{12}$] (*n* = 1–6) are not equivalent (e.g., Fig. 32) (*143*). However, only one hydrido ligand resonance is observed in the ^{1}H NMR spectrum of each cluster at temperatures low enough to prevent the metal framework rearrangement process from operating, although for the

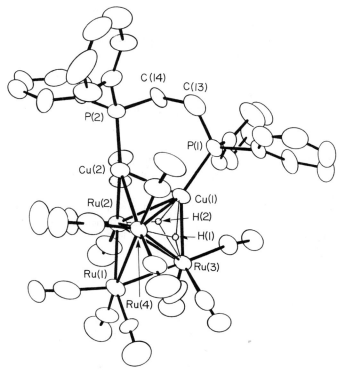

FIG. 32. Structure of the cluster $[Cu_2Ru_4(\mu_3\text{-}H)_2\{\mu\text{-}Ph_2P(CH_2)_2PPh_2\}(CO)_{12}]$. [Reprinted with permission of the Royal Society of Chemistry (143).]

species in which $n = 2$, two hydrido ligand resonances are visible at $-100°C$ (41). These results can be explained by a process involving hydrido ligand site exchange or by one in which the methylene groups in the backbones of the diphosphine ligands exchange rapidly between the various conformations so that a plane of symmetry is created through the cluster core [e.g., through Cu(1), Cu(2), Ru(1), and Ru(3) in Fig. 32]. As the free energy of activation for the fluxional process is considerably altered by changes in the number of methylene groups in the diphosphine backbone, it seems more likely that the dynamic behavior involves the diphosphine ligand rather than the hydrido ligands (41).

4. Processes Involving Ligands Attached to Other Transition Metals

Dynamic behavior involving carbonyl (e.g., 18,43,46,73,104,105,125, 142,144) and hydrido ligand (e.g., 73,135–137,142) site exchange is often observed for Group IB metal heteronuclear clusters. As these processes are very common for transition metal cluster compounds in general (e.g., *180–184*), they are not discussed in detail here.

III

HETERONUCLEAR CLUSTERS IN WHICH NO LIGANDS ARE ATTACHED TO THE GROUP IB METALS

Heteronuclear cluster compounds in which no ligands are attached to the Group IB metals are much rarer than those containing ML (M = Cu, Ag, or Au; L = two-electron donor ligand) units, and Table XVIII lists the known examples. Interestingly, copper- and silver-containing compounds are much more numerous than gold species, in contrast the clusters containing ML fragments for which M = Au is by far the most common (Section II).

A. *Synthesis*

Most heteronuclear clusters in which no ligands are attached to the Group IB metals were prepared using one of the few generally applicable rational methods that are currently available for the synthesis of these species. These preparative routes all involve the addition of Ag^+, either to preformed cluster anions (*98,118,123,190,192*) or to neutral polynuclear (*188*) or dimeric (*189,191*) compounds. Equations (66) and (67) show examples (*189,190*).

$$2 \text{[N(PPh}_3)_2][\text{Os}_3(\mu\text{-H})(\text{CO})_{11}] + \text{AgPF}_6 \rightarrow [\text{N(PPh}_3)_2][\text{Ag}\{\text{Os}_3(\mu\text{-H})(\text{CO})_{10}\}_2] \quad (66)$$

$$[\text{Fe}_2\{\mu\text{-CHCPh(NHMe)}\}(\mu\text{-PPh}_2)(\text{CO})_6] + \text{AgClO}_4 \rightarrow$$
$$[\text{AgFe}_2\{\mu\text{-CHCPh(NHMe)}\}(\mu\text{-PPh}_2)(\text{CO})_6][\text{ClO}_4] \quad (67)$$

B. *Structures*

Most of the heteronuclear clusters in which no ligands are attached to the Group IB metals have structural features in common with those containing ML (M = Cu, Ag, or Au; L = two-electron donor ligand) units (Sections II,B,1 to II,B,4). For example, "naked" silver atoms can edge-bridge a metal–metal bond (*189,191*) or cap a triangular three-metal face (*118,123*) in mixed-metal clusters. In addition, two cluster units can actually be linked together by sharing a common edge-bridging (*98,170,190*) or

TABLE XVIII

HETERONUCLEAR CLUSTERS IN WHICH NO LIGANDS ARE ATTACHED TO THE GROUP IB METALS

Cluster	Spectroscopic data reported	Crystal structure determined?	Skeletal geometry adopted	Ref.
Copper clusters				
[NEt$_4$]$_3$[Cu$_3$Fe$_3$(CO)$_{12}$]	IR	Yes	Trigonal planar Cu$_3$ unit with all three edges bridged by Fe atoms in the same plane	186
Na$_3$[Cu$_3$Fe$_3$(CO)$_{12}$]	^{13}C, ^{17}O, ^{63}Cu NMR	No	Trigonal planar Cu$_3$ unit with all three edges bridged by Fe atoms in the same plane	185, 186
[NEt$_4$]$_3$[Cu$_5$Fe$_4$(CO)$_{16}$]	IR	Yes	Cu$_4$Fe$_2$ hexagon, with a central Cu atom, and two opposite Cu—Cu edges bridged by Fe atoms in the same plane	185
Na$_3$[Cu$_5$Fe$_4$(CO)$_{16}$]	IR, ^{13}C, ^{17}O, ^{63}Cu NMR	No	Cu$_4$Fe$_2$ hexagon, with a central Cu atom, and two opposite Cu—Cu edges bridged by Fe atoms in the same plane	185, 186
[Cu{Ph$_2$P(CH$_2$)$_2$PPh$_2$}$_2$]$_2$[Cu$_6$Fe$_4$(CO)$_{16}$]^a	IR	Yes	Cu$_6$ octahedron with four faces capped by Fe atoms	187
Na$_2$[Cu$_6$Fe$_4$(CO)$_{16}$]	^{13}C, ^{17}O, ^{63}Cu NMR	No	Cu$_6$ octahedron with four faces capped by Fe atoms	185, 186
[Cu{Pt$_3$(μ-CO)$_3$(P(CHMe$_2$)$_3$)$_3$}$_2$]$^{+b}$		No	Two trigonal planar Pt$_3$ units linked by sharing a common face-capping Cu atom	188

(continued)

TABLE XVIII (Continued)

Cluster	Crystal structure determined?	Spectroscopic data reported	Skeletal geometry adopted	Ref.
Silver clusters				
$[NEt_4]_2[AgRe_6C(CO)_{21}]$	No	IR, ^1H NMR	Re_6 octahedron, with an interstitial C atom, capped on opposite faces by Ag and Re	118
$[AgFe_2\{\mu\text{-CHCPh(NHMe)}\}(\mu\text{-PPh}_2)(CO)_6][ClO_4]$	Yes	IR, ^1H, ^{31}P NMR, ^{57}Fe Mössbauer	Trigonal planar $AgFe_2$ unit	189
$[AgFe_2\{\mu\text{-CHCPh(NHR)}\}(\mu\text{-PPh}_2)(CO)_6]X$ ($X = ClO_4$, $R = Et$, $CHMe_2$, or $cyclo\text{-}C_6H_{11}$; $X = PF_6$, $R = Me$)	No	IR, ^1H, ^{31}P NMR, ^{57}Fe Mössbauer	Trigonal planar $AgFe_2$ unit	189
$[AgFe_2\{\mu\text{-CHCPh(NEt}_2)\}(\mu\text{-PPh}_2)(CO)_6][ClO_4]$	No	IR, ^{57}Fe Mössbauer	Trigonal planar $AgFe_2$ unit	189
$[N(PPh_3)_2][Ag\{Os_3(\mu\text{-H})(CO)_{10}\}_2]$	Yes	IR, ^1H, ^{13}C NMR	Two trigonal planar Os_3 units linked by sharing a common edge-bridging Ag atom	190
$[N(PPh_3)_2][Ag\{Os_6P(CO)_{18}\}_2]$	No	IR, ^{31}P NMR	Two trigonal prismatic Os_6 units, with an interstitial P atom in each, linked by sharing a common edge-bridging Ag atom	98
$[AgRh_2(\mu\text{-CO})(\mu\text{-Ph}_2PCH_2PPh_2)\text{-}(\eta\text{-}C_5H_5)_2]X^c$ ($X = BF_4$, PF_6, or ClO_4)	No	IR	Trigonal planar $AgRh_2$ unit	191
$[Ag_2Rh_6C(CO)_{15}]^d$	No	IR, ^{13}C, ^{103}Rh NMR	Rh_6 trigonal prism, with an interstitial C atom, and the two Rh_3 faces capped by Ag atoms	123

Compound	Method	Structure	Description	Ref.
$[PPh_4]_3[Ag\{Rh_6C(CO)_{15}\}_2]$	IR, ^{13}C, ^{103}Rh NMR	Yes	Two trigonal prismatic Rh_6 units, with an interstitial C atom in each, linked by sharing a common face-capping Ag atom	123
$[NMe_3(CH_2Ph)][Ag_3\{Rh_6C(CO)_{15}\}_2]^d$	IR, ^{13}C, ^{103}Rh NMR	No	Two trigonal prismatic Rh_6 units, with an interstitial C atom in each, linked by sharing a common face-capping Ag atom, with the two vacant Rh_3 faces capped by Ag atoms	123
$[NMe_3(CH_2Ph)]_4[Ag_2\{Rh_6C(CO)_{15}\}_3]^d$	IR, ^{13}C, ^{103}Rh NMR	No	Three trigonal prismatic Rh_6 units, with an interstitial C atom in each, linked by sharing two Ag atoms which each cap two Rh_3 faces	123
$[NMe_3(CH_2Ph)]_2[Ag_4\{Rh_6C(CO)_{15}\}_3]^d$	IR, ^{13}C, ^{103}Rh NMR	No	Three trigonal prismatic Rh_6 units, with an interstitial C atom in each, linked by sharing two Ag atoms which each cap two Rh_3 faces, with the two vacant Rh_3 faces capped by Ag atoms	123
$[NMe_3(CH_2Ph)]_n[Ag_n\{Rh_6C(CO)_{15}\}_n]^{d,e}$	IR, ^{13}C, ^{103}Rh NMR	No	n trigonal prismatic Rh_6 units, with an interstitial C atom in each, linked by sharing $(n-1)$ Ag atoms which each cap two Rh_3 faces. The other Ag atom caps one vacant Rh_3 face	123

(continued)

TABLE XVIII (Continued)

Cluster	Spectroscopic data reported	Crystal structure determined?	Skeletal geometry adopted	Ref.
$[AsPh_4][AgNi_9(CO)_{18}]$	IR	No	Unknown	192
$Ag\{Pt_3(\mu\text{-}CO)_3(P(CHMe_2)_3)_3\}_2]\text{-}[CF_3SO_3]$	IR, ^{31}P, ^{19}F, ^{195}Pt NMR	Yes	Two trigonal planar Pt_3 units linked by sharing a common face-capping Ag atom	188
Gold clusters				
$[N(PPh_3)_2][Au\{Os_3(\mu\text{-}H)(CO)_{10}\}_2]$	IR, ^1H NMR	Yes	Two trigonal planar Os_3 units linked by sharing a common edge-bridging Au atom	170
$[N(PPh_3)_2]_2[AuOs_{20}C_2(CO)_{48}]^{f,g}$	IR, Mass, ESR	Yes	Staggered stacked metal array $(6:3:3:6)$ of trigonal planar Os_3 fragments, trigonal planar Os_3 units with all three edges bridged by Os atoms, and a central trigonal planar $AuOs_2$ moiety	193

[a] The cluster anion has also been isolated as the $[Cu\{Me_2P(CH_2)_2PMe_2\}_2]^+$ and $[Cu(PPh_3)_3]^+$ salts.

[b] The nature of the anion was not reported.

[c] The clusters are too unstable in solution for full spectroscopic characterization.

[d] The cluster anion has not been isolated in the solid state.

[e] A mixture of cluster anions of different nuclearity exists in solution.

[f] The cluster anion has also been isolated as the $[PMePh_3]^+$, $[AsPh_4]^+$, and $[K(18\text{-crown-6})]^+$ salts.

[g] Only a preliminary X-ray diffraction study has been reported for the cluster.

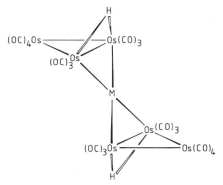

FIG. 33.   Structure of the cluster anion $[M\{Os_3(\mu\text{-}H)(CO)_{10}\}_2]^-$ (M = Ag or Au).

face-capping *(123,188)* copper, silver, or gold atom (e.g., Figs. 33 and 34). Interestingly, it is possible to prepare oligomers which consist of a whole series of trigonal prismatic $[Rh_6C(CO)_{15}]$ units fused together via shared Ag atoms capping Rh_3 faces (cf. Fig. 34) *(123)*.

A novel series of clusters containing Cu atoms and $Fe(CO)_4$ units has been reported *(185–187)*. The anions $[Cu_3Fe_3(CO)_{12}]^{3-}$ and $[Cu_5Fe_4(CO)_{16}]^{3-}$ adopt unusual planar "raft" structures, whereas the metal core of $[Cu_6Fe_4(CO)_{16}]^{2-}$ exhibits a three-dimensional structure (Fig. 35). The

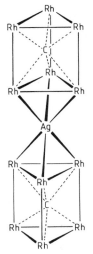

FIG. 34.   Structure of the cluster anion $[Ag\{Rh_6C(CO)_{15}\}_2]^{3-}$, with the CO ligands omitted for clarity.

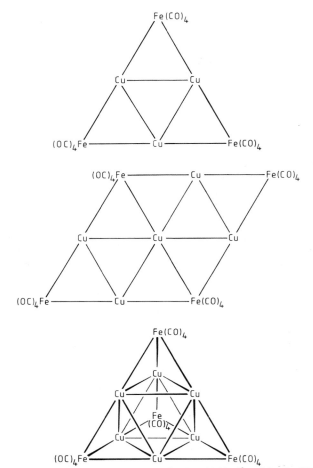

Fig. 35. Structures of the cluster anions $[Cu_3Fe_3(CO)_{12}]^{3-}$, $[Cu_5Fe_4(CO)_{16}]^{3-}$, and $[Cu_6Fe_4(CO)_{16}]^{2-}$.

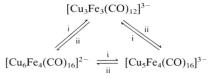

Scheme 3. Interconversion reactions for the cluster anions $[Cu_3Fe_3(CO)_{12}]^{3-}$, $[Cu_5Fe_4(CO)_{16}]^{3-}$, and $[Cu_6Fe_4(CO)_{16}]^{2-}$. Reagents: i, $Na_2[Fe(CO)_4]$; ii, CuBr (185).

open nature of the structures and the relatively weak bonding holding the clusters together have been suggested (*185*) as the reasons why the three cluster anions undergo interesting interconversion reactions under mild conditions (Scheme 3).

IV

HETERONUCLEAR CLUSTERS IN WHICH LIGANDS OTHER THAN SIMPLE TWO-ELECTRON DONORS ARE ATTACHED TO THE GROUP IB METALS

Although the vast majority of Group IB metal heteronuclear clusters have simple two-electron donor ligands attached to the coinage metals, a small number of species with other types of ligand bonded to the Group IB metals have also been reported. Examples are listed in Table XIX.

A. *Synthesis*

In contrast to the heteronuclear Group IB metal clusters discussed previously (Sections II,A and III,A), most of the clusters of the type listed in Table XIX were synthesized via reactions in which the nature of the product could not be predicted in advance. However, the compounds [Cu(thf)(η-C$_5$Me$_5$)] (*194*), AgX (X = ClO$_4$, OPF$_2$O, O$_2$CMe, or S$_2$CNEt$_2$) (*191,205,208*), [Au(C$_2$CBu^t)] (*64,210*), and [AuCl(SMe$_2$)] (*64*) have been utilized to introduce Cu(η-C$_5$Me$_5$) units, AgX fragments, Au-(C$_2$Bu^t) moieties, and AuCl groups, respectively, into monomeric or dimeric precursors.

B. *Structures*

The clusters listed in Table XIX display a wide variety of skeletal geometries. The coinage metal moieties are edge-bridging or face-capping in many, as observed for the majority of other heteronuclear Group IB metal clusters (Sections II,B and II,B). However, a number of the copper- and silver-containing species in Table XIX adopt unusual planar or nearly planar "raft" structures made up of trigonal planar three-metal units fused together by sharing common edges (e.g., *52,99,171,195,196,206*). Figure 36 shows an example (*196,206*).

TABLE XIX

Heteronuclear Clusters in Which Ligands Other Than Simple Two-Electron Donors Are Attached to the Group IB Metals

Cluster	Spectroscopic data reported	Crystal structure determined?	Skeletal geometry reported	Ref.
Copper clusters				
$[Cu_2W(\mu_3-C_6H_4Me-4)(CO)_2(\eta-C_5H_5)(\eta-C_5Me_5)_2]$	IR, ^1H, ^{13}C NMR	No	Trigonal planar Cu_2W unit	194
$[CuWPt(\mu_3-C_6H_4Me-4)(CO)_2(PMe_3)_2(\eta-C_5H_5)(\eta-C_5Me_5)][a]$	IR, ^1H, ^{13}C NMR	Yes	Trigonal planar CuWPt unit	194
$[Cu_2Mn_4H_6\{\mu-(EtO)_2POP(OEt)_2\}_2(CO)_{12}]$	^1H NMR	Yes	Cu_2Mn_2 rhombus with two Mn atoms, in the same plane, "spiked" to the Cu atoms	99
$[Cu_2Re_4H_{14}(PMe_2Ph)_8]$	^1H, ^{31}P NMR	Yes	Cu_2Re_2 rhombus, with two Cu—Re edges bridged by Re atoms almost in the same plane	52
$[Cu_2Re_4H_{16}L_8][PF_6]_2$ (L = $PMePh_2$ or PMe_2Ph)	^1H NMR	Yes	Cu_2Re_2 rhombus, with two Cu—Re edges bridged by Re atoms almost in the same plane	52,195
$[CuRu_5H(CO)_{18}(PPh_3)]$	IR, ^1H, ^{31}P NMR, Mass	Yes	Four trigonal planar M_3 (M = Cu or Ru) units edge-connected in a puckered ladderlike manner, with a Cu—Ru bond forming the central "rung"	171
$[Cu_3Os_3(\mu-H)_9(PMe_2Ph)_9]$	^1H, ^{31}P NMR	Yes	Trigonal planar Cu_3 unit with all three edges bridged by Os atoms in the same plane	196
$[CuRh_2(\mu-CO)_2(\eta-C_5Me_5)_3]$	IR, ^1H, ^{13}C NMR	Yes	Trigonal planar $CuRh_2$ unit	194

Compound	Method		Structure	Ref.
[Cu₄Rh₂(C₂Ar)₈(PRPh₂)₂] (R = Ph or Me; Ar = Ph, C₆H₄Me-4, C₆H₂F-4, or C₆F₅)	IR, ¹H NMR	No	Octahedral Cu₄Rh₂ unit with axial Rh atoms	197
[Cu₄Ir₂(C₂Ph)₈(PPh₃)₂]	IR, ¹H NMR	Yes	Octahedral Cu₄Ir₂ unit with axial Ir atoms	197,198
[Cu₄Ir₂(C₂Ar)₈(PRPh₂)₂] (R = Ph, Ar = C₆H₄Me-4, C₆H₄F-4, or C₆F₅; R = Me, Ar = Ph or C₆F₅)	IR, ¹H NMR	No	Octahedral Cu₄Ir₂ unit with axial Ir atoms	197
[Cu₄Ir₂Fe₂(C₂Ar)₈(CO)₈(PPh₃)₂] (Ar = Ph or C₆H₄Me-4)	IR	No	Octahedral Cu₄Ir₂ unit with axial Ir atoms and two Fe(CO)₄ units coordinated to two C≡CAr ligands	197
[NBu₄ⁿ][Cu₂Au₃(C₂Ph)₆]	¹H NMR	Yes	Cu₂Au₃ trigonal bipyramid with no bonding interaction between the three equatorial Au atoms	199
[Cu₂Ag₄(C₆H₄NMe₂-2)₄Br₂]ᵇ	¹H NMR	No	Cu₂Ag₄ octahedron	200,201
[Cu₄Ag₂(C₆H₄NMe₂-2)₄Br₂]ᵇ	¹H NMR	No	Cu₄Ag₂ octahedron	201
[Cu₄Ag₂(C₆H₄NMe₂-2)₄(O₃SCF₃)₂]ᶜ	¹H NMR	No	Cu₄Ag₂ octahedron	202
[Cu₄Au₂(C₆H₄NMe₂-2)₄X₂] (X = I or O₃SCF₃)	IR, ¹H NMR	No	Cu₄Au₂ octahedron with apical Au atoms	202
[Cu₄Au₂(C₆H₄NMe₂-2)₄Br₂]	¹H NMR	No	Cu₄Au₂ octahedron with apical Au atoms	202
[Li₄Cl₂(OEt₂)₁₀][Cu₃Li₂Ph₆]₂		Yes	Cu₃Li₂ trigonal bipyramid with no bonding interaction between the three equatorial Cu atoms	203
[Li(OEt₂)₄][Cu₄LiPh₆]ᵈ		Yes	Cu₄Li trigonal bipyramid with no bonding interaction between the three equatorial Cu atoms	204

(continued)

TABLE XIX (*Continued*)

Cluster	Spectroscopic data reported	Crystal structure determined?	Skeletal geometry reported	Ref.
[Cu$_4$MgPh$_6$]		Yes	Cu$_4$Mg trigonal bipyramid with no bonding interaction between the three equatorial Cu atoms	204
Silver clusters				
[NBu$_4$]$_5$[{AgRe$_7$C(CO)$_{21}$}$_2$(μ-Br)]	IR	Yes	Two Re$_6$ octahedra, each with an interstitial C atom and two opposite faces capped by Ag and Re, linked together by sharing a common Br atom bonded to both Ag atoms	118
[AgRh$_2$(μ-CO)(μ-Ph$_2$PCH$_2$PPh$_2$)(OPF$_2$O)(η-C$_5$H$_5$)$_2$]	IR, ^{1}H, ^{31}P NMR	Yes	Trigonal planar AgRh$_2$ unit	191
[AgRh$_2$(μ-CO)(μ-Ph$_2$PCH$_2$PPh$_2$)(O$_2$CMe)(η-C$_5$H$_5$)$_2$]	IR, ^{1}H, ^{31}P NMR	Yes	Trigonal planar AgRh$_2$ unit	205
[AgRh$_2$(μ-CO)(μ-Ph$_2$PCH$_2$PPh$_2$)L(η-C$_5$H$_5$)$_2$] (L = O$_2$CCF$_3$ or S$_2$CNEt$_2$)	IR, ^{1}H, ^{31}P NMR	No	Trigonal planar AgRh$_2$ unit	205
[{AgRh$_2$(μ-CO)(μ-Ph$_2$PCH$_2$PPh$_2$)(η-C$_5$H$_5$)$_2$}$_2$-{μ-(O$_2$C)$_2$CH$_2$}]	IR, ^{1}H, ^{31}P NMR	No	Two trigonal planar AgRh$_2$ units linked together by (O$_2$C)$_2$CH ligand	205
[Ag$_3$Rh$_3$(μ-H)$_9${CH$_3$C(CH$_2$PPh$_3$)$_3$}$_3$]-[O$_3$SCF$_3$]$_3$	IR, ^{1}H, ^{31}P NMR	Yes	Trigonal planar Ag$_3$ unit with all three edges bridged by Rh atoms in the same plane	206
[NBu$_4$]$_2$[Ag$_2$Pt$_2$Cl$_4$(C$_6$F$_5$)$_4$]	IR	Yes	Ag$_2$Pt$_2$ rhombus	207
[Ag$_2$Au$_2$(CH$_2$PPh$_3$)$_4$(OClO$_3$)$_4$]	IR	Yes	Ag$_2$Au$_2$ rhombus with no Ag—Ag or Au—Au close contact	208

[Ag$_2$Au$_2$(CH$_2$PMePhR)$_4$(OClO$_3$)$_4$] (R = Ph or Me)	IR	No	Ag$_2$Au$_2$ rhombus with no Ag—Ag or Au—Au close contact	208
[Ag$_4$Au$_2$(C$_6$H$_4$NMe$_2$-2)$_4$(O$_3$SCF$_3$)$_2$]	IR, ^{1}H NMR	No	Ag$_4$Au$_2$ octahedron with apical Au atoms	202
[Li$_6$Br$_4$(OEt$_2$)$_{10}$][Ag$_3$Li$_2$Ph$_6$]$_2^{c,e}$		Yes	Ag$_3$Li$_2$ trigonal bipyramid with no bonding interaction between the three equatorial Ag atoms.	

See also clusters [Cu$_2$Ag$_4$(C$_6$H$_4$NMe$_2$-2)$_4$Br$_2$] and [Cu$_4$Ag$_2$(C$_6$H$_4$Me$_2$-2)$_4$X$_2$] (X = Br or O$_3$SCF$_3$) listed with copper clusters above.

Gold clusters

[N(PPh$_3$)$_2$][[AuOs$_{10}$C(CO)$_{24}$Br]	IR	Yes	Os$_{10}$ tetracapped octahedron with an interstitial C atom and Au triply bridging a Os$_3$ face of one cap	193
[AuPt$_2$(μ-Ph$_2$PCH$_2$PPh$_2$)$_2$(C$_2$Bu^t)$_2$I]	^{1}H, ^{31}P NMR	Yes	Trigonal planar AuPt$_2$ unit	64, 210
[AuPt$_2$(μ-Ph$_2$PCH$_2$PPh$_2$)$_2$(C$_2$Bu^t)X] (X = Cl or C$_2$Bu^t)	^{1}H, ^{31}P NMR	No	Trigonal planar AuPt$_2$ unit	64, 210
[AuPt$_2$(μ-Ph$_2$PCH$_2$PPh$_2$)$_2$Cl$_3$]	^{1}H, ^{31}P NMR	No	Trigonal planar AuPt$_2$ unit	64
[Au{Pt$_2$(μ_3-S)(CO)(PPh$_3$)$_3$}$_2$][PF$_6$]	IR, ^{31}P, ^{195}Pt NMR	No	Two trigonal planar AuPt$_2$ units fused by sharing a common Au atom	68

See also clusters [NBu$^t_4$][Cu$_2$Au$_3$(C$_2$Ph)$_6$], [M$_4$Au$_2$(C$_6$H$_4$NMe$_2$-2)$_4$X$_2$] (M = Cu, X = Br, I, or O$_3$SCF$_3$; M = Ag, X = O$_3$SCF$_3$), and [Ag$_2$Au$_2$(CH$_2$PPhRR′)$_4$(OClO$_3$)$_4$] (R = R′ = Ph; R = Me, R′ = Ph or Me) listed with copper or silver clusters above.

---

^{a} The Cu—Pt distance of 2.807(3) Å is probably too long for any significant bonding interaction between the two atoms.

^{b} The cluster is not stable in solution and undergoes interaggregate exchange reactions to produce an equilibrium mixture of the various possible hexanuclear species [Cu$_{(6-n)}$Ag$_n$(C$_6$H$_4$NMe$_2$-2)$_4$Br$_2$] (n = 1–5).

^{c} The cluster is not stable at ambient temperatures.

^{d} The Li site in the trigonal bipyramidal Cu$_4$Li core is contaminated by 11% Cu.

^{e} One of the Li sites in the trigonal bipyramidal Ag$_3$Li$_2$ core is contaminated by 2% Ag.

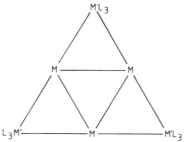

Fig. 36.   Structures of the clusters $[M_3M'_3(\mu\text{-}H)_9L_9]$ [M = Cu, M' = Os, L = PMe_2Ph; M = Ag, M' = Rh, L_3 = $CH_3C(CH_2PPh_2)_3$], with the pairs of and single hydrido ligands which alternately bridge the M—M' vectors omitted for clarity.

V

SUMMARY

From the data presented herein, it is clear that the chemistry of Group IB metal heteronuclear clusters has attracted a great deal of recent interest. Species containing $M(PR_3)$ or M(NCR) (M = Cu, Ag, or Au; R = alkyl or aryl) units, which have been investigated most extensively, have been found to exhibit novel structural features, chemical properties, and dynamic behavior that are in marked contrast to those displayed by most heteronuclear clusters of other transition metals. Mixed-metal clusters in which no ligand or ligands other than simple two-electron donors are attached to the coinage metals have not been studied to the same extent, but initial results suggest that the chemistry of these species will also have unusual features.

ACKNOWLEDGMENTS

I thank Drs. S. S. D. Brown and P. J. McCarthy for helpful discussions, Mrs. L. J. Salter for drawing the diagrams, and Mrs. B. R. Hurren for typing the manuscript.

REFERENCES

1. C. E. Coffey, J. Lewis, and R. S. Nyholm, *J. Chem. Soc.*, 1741 (1964).
2. A. S. Kasenally, R. S. Nyholm, R. J. O'Brien, and M. H. B. Stiddard, *Nature (London)* **204,** 871 (1964).
3. A. S. Kasenally, R. S. Nyholm, and M. H. B. Stiddard, *J. Am. Chem. Soc.* **86,** 1884 (1964).
4. J. W. Lauher, unpublished results, see Ref. 7 for details.
5. D. A. Roberts and G. L. Geoffroy, *in* "Comprehensive Organometallic Chemistry" (G. Wilkinson, F. G. A. Stone, and E. W. Abel, Eds.), Vol. 6, p. 763. Pergamon, Oxford, 1982.

6. P. Braunstein and J. Rose, *Gold Bull.* **18,** 17 (1985).
7. K. P. Hall and D. M. P. Mingos, *Prog. Inorg. Chem.* **32,** 237 (1984).
8. J. J. Steggerda, J. J. Bour, and J. W. A. van der Velden, *Recl. Trav. Chim. Pays-Bas* **101,** 164 (1982).
9. P. G. Jones, *Gold Bull.* **19,** 46 (1986); P. G. Jones, *Gold Bull.* **16,** 114 (1983); P. G. Jones, *Gold Bull.* **14,** 102 (1981).
10. M. Melnik and R. V. Parish, *Coord. Chem. Rev.* **70,** 157 (1986).
11. I. D. Salter, *Adv. Dynamic Stereochem.* **2,** in press (1988).
12. B. F.G. Johnson, D. A. Kaner, J. Lewis, and P. R. Raithby, *J. Organomet. Chem.* **215,** C33 (1981).
13. B. F. G. Johnson, D. A. Kaner, J. Lewis, P. R. Raithby, and M. J. Taylor, *Polyhedron* **1,** 105 (1982).
14. J. A. Iggo, M. J. Mays, P. R. Raithby, and K. Henrick, *J. Chem. Soc., Dalton Trans.,* 633 (1984).
15. M. J. Mays, P. R. Raithby, P. L. Taylor, and K. Henrick, *J. Organomet. Chem.* **224,** C45 (1982).
16. J. E. Ellis, *J. Am. Chem. Soc.* **103,** 6106 (1981).
17. M. J. Freeman, M. Green, A. G. Orpen, I. D. Salter, and F. G. A. Stone, *J. Chem. Soc., Chem. Commun.,* 1332 (1983).
18. S. S. D. Brown, S. Hudson, I. D. Salter, and M. McPartlin, *J. Chem. Soc., Dalton Trans.,* 1967 (1987).
19. J. W. Lauher and K. Wald, *J. Am. Chem. Soc.* **103,** 7648 (1981).
20. M. J. Mays, P. R. Raithby, P. L. Taylor, and K. Henrick, *J. Chem. Soc., Dalton Trans.,* 959 (1984).
21. M. I. Bruce, M. L. Williams, J. M. Patrick, B. W. Skelton, and A. H. White, *J. Chem. Soc., Dalton Trans.,* 2557 (1986).
22. S. S. D. Brown and I. D. Salter, *Organomet. Synth.* **4,** 247 (1988).
23. R. Uson and A. Laguna, *Organomet. Synth.* **3,** 324 (1986).
24. C. E. Housecroft and A. L. Rheingold, *Organometallics* **6,** 1332 (1987).
25. K. Burgess, B. F. G. Johnson, D. A. Kaner, J. Lewis, P. R. Raithby, and S. N. A. B. Syed-Mustaffa, *J. Chem. Soc., Chem. Commun.,* 455 (1983).
26. C. M. Hay, B. F. G. Johnson, J. Lewis, R. C. S. McQueen, P. R. Raithby, R. M. Sorrell, and M. J. Taylor, *Organometallics* **4,** 202 (1985).
27. M. I. Bruce and B. K. Nicholson, *J. Organomet. Chem.* **252,** 243 (1983).
28. M. I. Bruce, O. bin Shawkataly, and B. K. Nicholson, *J. Organomet. Chem.* **275,** 223 (1984).
29. M. I. Bruce, E. Horn, O. bin Shawkataly, and M. R. Snow, *J. Organomet. Chem.* **280,** 289 (1985).
30. M. I. Bruce, O. bin Shawkataly, and B. K. Nicholson, *J. Organomet. Chem.* **286,** 427 (1985).
31. M. I. Bruce and B. K. Nicholson, *Organometallics* **3,** 101 (1984).
32. V. G. Albano, D. Braga, S. Martinengo, P. Chini, M. Sansoni, and D. Strumolo, *J. Chem. Soc., Dalton Trans.,* 52 (1980).
33. J. S. Bradley, R. L. Pruett, E. Hill, G. B. Ansell, M. E. Leonowicz, and M. A. Modrick, *Organometallics* **1,** 748 (1982).
34. D. Braga, K. Henrick, B. F. G. Johnson, J. Lewis, M. McPartlin, W. J. H. Nelson, A. Sironi, and M. D. Vargas, *J. Chem. Soc., Chem. Commun.,* 1131 (1983).
35. B. F. G. Johnson, J. Lewis, W. J. H. Nelson, M. D. Vargas, D. Braga, K. Henrick, and M. McPartlin, *J. Chem. Soc., Dalton Trans.,* 975 (1986).
36. S. S. D. Brown, I. D. Salter, and B. M. Smith, *J. Chem. Soc., Chem. Commun.,* 1439 (1985).

37. G. B. Ansell, M. A. Modrick, and J. S. Bradley, *Acta Crystallogr.* **C40,** 365 (1984); P. G. Jones, *Acta Crystallogr.* **C42,** 1099 (1986).
38. S. S. D. Brown and I. D. Salter, *Organomet. Synth.* **3,** 315 (1986).
39. S. S. D. Brown, P. J. McCarthy, and I. D. Salter, *J. Organomet. Chem.* **306,** C27 (1986).
40. S. S. D. Brown, P. J. McCarthy, I. D. Salter, P. A. Bates, M. B. Hursthouse, I. J. Colquhoun, W. McFarlane, and M. Murray, *J. Chem. Soc., Dalton Trans.,* in press (1988).
41. S. S. D. Brown, Ph.D. Thesis, University of Exeter, 1987.
42. P. J. McCarthy, I. D. Salter, and V. Šik, *J. Organomet. Chem.* **344,** 411 (1988).
43. R. A. Brice, S. C. Pearse, I. D. Salter, and K. Henrick, *J. Chem. Soc., Dalton Trans.,* 2181 (1986).
44. S. S. D. Brown, I. D. Salter, T. Adatia, and M. McPartlin, *J. Organomet. Chem.* **332,** C6 (1987).
45. L. J. Farrugia, J. A. K. Howard, P. Mitrprachachon, J. L. Spencer, F. G. A. Stone, and P. Woodward, *J. Chem. Soc., Chem. Commun.,* 260 (1978).
46. L. W. Bateman, M. Green, K. A. Mead, R. M. Mills, I. D. Salter, F. G. A. Stone, and P. Woodward, *J. Chem. Soc., Dalton Trans.,* 2599 (1983).
47. L. J. Farrugia, M. J. Freeman, M. Green, A. G. Orpen, F. G. A. Stone, and I. D. Salter, *J. Organomet. Chem.* **249,** 273 (1983).
48. J. A. K. Howard, I. D. Salter, and F. G. A. Stone, *Polyhedron* **3,** 567 (1984).
49. I. D. Salter, *Organomet. Synth.* **3,** 312 (1986).
50. A. A. Aitchison and L. J. Farrugia, *Organometallics* **5,** 1103 (1986).
51. A. G. Cowie, B. F. G. Johnson, J. Lewis, J. N. Nicholls, P. R. Raithby, and M. J. Rosales, *J. Chem. Soc., Dalton Trans.,* 2311 (1983).
52. B. R. Sutherland, D. M. Ho, J. C. Huffman, and K. G. Caulton, *Angew. Chem. Int. Ed. Engl.* **26,** 135 (1987).
53. B. R. Sutherland, K. Folting, W. E. Steib, D. M. Ho, J. C. Huffman, and K. G. Caulton, *J. Am. Chem. Soc.* **109,** 3489 (1987).
54. C. W. Bradford, W. van Bronswijk, R. J. H. Clark, and R. S. Nyholm, *J. Chem. Soc. A,* 2889 (1970).
55. B. F .G. Johnson, J. Lewis, J. N. Nicholls, J. Puga, and K. H. Whitmire, *J. Chem. Soc., Dalton Trans.,* 787 (1983).
56. G. Lavigne, F. Papageorgiou, and J.-J. Bonnet, *Inorg. Chem.* **23,** 609 (1984).
57. K. Burgess, B. F. G. Johnson, and J. Lewis, *J. Chem. Soc., Dalton Trans.,* 1179 (1983); K. Burgess, B. F. G. Johnson, and J. Lewis, *J. Organomet. Chem.* **247,** C42 (1983).
58. A. J. Deeming, S. Donovan-Mtunzi, and K. Hardcastle, *J. Chem. Soc., Dalton Trans.,* 543 (1986).
59. D. M. P. Mingos and R. W. M. Wardle, *J. Chem. Soc., Dalton Trans.,* 73 (1986).
60. K. Burgess, B. F. G. Johnson, J. Lewis, and P. R. Raithby, *J. Chem. Soc., Dalton Trans.,* 1661 (1983).
61. M. I. Bruce, E. Horn, J. G. Matisons, and M. R. Snow, *J. Organomet. Chem.* **286,** 271 (1985).
62. C. E. Briant, R. W. M. Wardle, and D. M. P. Mingos, *J. Organomet. Chem.* **267,** C49 (1984).
63. S. Lo Schiavo, G. Bruno, F. Nicholo, P. Piraino, F. Faraone, *Organometallics* **4,** 2091 (1985).
64. G. J. Arsenault, L. Manojlović-Muir, K. W. Muir, R. J. Puddephatt, and I. Treurnicht, *Angew. Chem. Int. Ed. Engl.* **26,** 86 (1987).
65. D. M. P. Mingos, P. Oster, and D. J. Sherman, *J. Organomet. Chem.* **320,** 257 (1987).

66. G. A. Moehring, P. E. Fanwick, and R. A. Walton, *Inorg. Chem.* **26**, 1861 (1987).
67. B. D. Alexander, B. J. Johnson, S. M. Johnson, A. L. Casalnuovo, and L. H. Pignolet, *J. Am. Chem. Soc.* **108**, 4409 (1986).
68. M. F. Hallam, M. A. Luke, D. M. P. Mingos, and I. D. Williams, *J. Organomet. Chem.* **325**, 271 (1987).
69. P. Braunstein, H. Lehner, D. Matt, A. Tiripicchio, and M. Tiripicchio-Camellini, *Angew. Chem. Int. Ed. Engl.* **23**, 304 (1984).
70. P. D. Boyle, B. J. Johnson, A. Buehler, and L. H. Pignolet, *Inorg. Chem.* **25**, 5 (1986).
71. M. Fajardo, M. P. G.-S. P. Royo, S. M. Carrera, and S. G. Blanco, *J. Organomet. Chem.* **312**, C44 (1986).
72. P. D. Boyle, B. J. Johnson, B. D. Alexander, J. A. Casalnuovo, P. R. Gannon, S. M. Johnson, E. A. Larka, A. M. Mueting, and L. H. Pignolet, *Inorg. Chem.* **26**, 1346 (1987).
73. B. F. G. Johnson, J. Lewis, P. R. Raithby, S. N. A. B. Syed-Mustaffa, M. J. Taylor, K. H. Whitmire, and W. Clegg, *J. Chem. Soc., Dalton Trans.*, 2111 (1984).
74. M. Müller-Glieman, S. V. Hoskins, A. G. Orpen, A. L. Ratermann, and F. G. A. Stone, *Polyhedron* **5**, 791 (1986).
75. S. Bhaduri, K. Sharma, P. G. Jones, and C. Freire Erdbrügger, *J. Organomet. Chem.* **326**, C46 (1987).
76. C. E. Briant, R. G. Smith, and D. M. P. Mingos, *J. Chem. Soc., Chem. Commun.*, 586 (1984).
77. B. K. Teo and K. Keating, *J. Am. Chem. Soc.* **106**, 2224 (1984).
78. B. K. Teo, M. C. Hong, H. Zhang, and D. B. Huang, *Angew. Chem. Int. Ed. Engl.* **26**, 897 (1987).
79. G. A. Carriedo, J. A. K. Howard, F. G. A. Stone, and M. J. Went, *J. Chem. Soc., Dalton Trans.*, 2545 (1984).
80. C. E. Briant, D. I. Gilmour, and D. M. P. Mingos, *J. Chem. Soc., Dalton Trans.*, 835 (1986).
81. J. J. Bour, R. P. F. Kanters, P. P. J. Schlebos, W. Bos, W. P. Bosman, H. Behm, P. T. Beurkers, and J. J. Steggerda, *J. Organomet. Chem.* **329**, 405 (1987).
82. D. E. Smith, A. J. Welch, I. Treurnicht, and R. J. Puddephatt, *Inorg. Chem.* **25**, 4616 (1986).
83. A. L. Casalnuovo, J. A. Casalnuovo, P. V. Nilsson, and L. H. Pignolet, *Inorg. Chem.* **24**, 2554 (1985).
84. B. F. G. Johnson, D. A. Kaner, J. Lewis, and M. J. Rosales, *J. Organomet. Chem.* **238**, C73 (1982).
85. B. D. Alexander, P. D. Boyle, B. J. Johnson, J. A. Casalnuovo, S. M. Johnson, A. M. Mueting, and L. H. Pignolet, *Inorg. Chem.* **26**, 2547 (1987).
86. A. G. Cowie, B. F. G. Johnson, J. Lewis, and P. R. Raithby, *J. Chem. Soc., Chem. Commun.*, 1710 (1984).
87. J. N. Nicholls, P. R. Raithby, and M. D. Vargas, *J. Chem. Soc., Chem. Commun.*, 1617 (1986).
88. A. L. Casalnuovo, L. H. Pignolet, J. W. A. van der Velden, J. J. Bour, and J. J. Steggerda, *J. Am. Chem. Soc.* **105**, 5957 (1983).
89. A. L. Casalnuovo, T. Laska, P. V. Nilsson, J. Olofson, L. H. Pignolet, W. Bos, J. J. Bour, and J. J. Steggerda, *Inorg. Chem.* **24**, 182 (1985).
90. J. W. A. van der Velden, J. J. Bour, B. F. Otterloo, W. P. Bosman, and J. H. Noordik, *J. Chem. Soc., Chem. Commun.*, 583 (1981).
91. P. Braunstein, J. Rosé, A. Tiripicchio, and M. Tiripicchio-Camellini, *Angew. Chem. Int. Ed. Engl.* **24**, 767 (1985).

92. D. M. P. Mingos, *Polyhedron* **3**, 1289 (1984).
93. D. G. Evans and D. M. P. Mingos, *J. Organomet. Chem.* **232**, 171 (1982).
94. V. Riera, M. A. Ruiz, A. Tiripicchio, and M. Tiripicchio-Camellini, *J. Chem. Soc., Dalton Trans.*, 1551 (1987).
95. S. Attali, F. Dahan, and R. Mathieu, *J. Chem. Soc., Dalton Trans.*, 2521 (1985).
96. P. Braunstein, J. Rosé, A. M. Manotti-Lanfredi, A. Tiripicchio, and E. Sappa, *J. Chem. Soc., Dalton Trans.*, 1843 (1984).
97. T. Adatia, M. McPartlin, and I. D. Salter, *J. Chem. Soc., Dalton Trans.*, in press (1988).
98. S. B. Colbran, C. M. Hay, B. F. G. Johnson, F. J. Lahoz, J. Lewis, and P. R. Raithby, *J. Chem. Soc., Chem. Commun.*, 1766 (1986).
99. V. Riera, M. A. Ruiz, A. Tiripicchio, and M. Tiripicchio-Camellini, *J. Chem. Soc., Chem. Commun.*, 1505 (1985).
100. T. Beringhelli, G. Ciani, G. D'Alfonso, V. De Maldé, and M. Freni, *J. Chem. Soc., Chem. Commun.*, 735 (1986).
101. H. Umland and U. Behrens, *J. Organomet. Chem.* **287**, 109 (1985).
102. M. I. Bruce and B. K Nicholson, *J. Organomet. Chem.* **250**, 627 (1983).
103. B. F. G. Johnson, D. A. Kaner, J. Lewis, P. R. Raithby, and M. J. Rosales, *J. Organomet. Chem.* **231**, C59 (1982).
104. C. P. Horwitz and D. F. Shriver, *J. Am. Chem. Soc.* **107**, 8147 (1985).
105. C. P. Horwitz, E. M. Holt, C. P. Brock, and D. F. Shriver, *J. Am. Chem. Soc.* **107**, 8136 (1985).
106. M. L. Blohm and W. L. Gladfelter, *Inorg. Chem.* **26**, 459 (1987).
107. K. Fischer, M. Müller, and H. Vahrenkamp, *Angew. Chem. Int. Ed. Engl.* **23**, 140 (1984).
108. B. F. G. Johnson, D. A. Kaner, J. Lewis, P. R. Raithby, and M. J. Taylor, *J. Chem. Soc., Chem. Commun.*, 314 (1982).
109. M. Ahlgren, T. T. Pakkanen, and I. Tahvanainen, *J. Organomet. Chem.* **323**, 91 (1987).
110. P. Braunstein, G. Predieri, A. Tiripicchio, and E. Sappa, *Inorg. Chim. Acta* **63**, 113 (1982).
111. L. S. Moore, R. V. Parish, S. S. D. Brown, and I. D. Salter, *J. Chem. Soc., Dalton Trans.*, 2333 (1987).
112. M. I. Bruce, O. bin Shawkataly, and B. K. Nicholson, *J. Organomet. Chem.* **286**, 427 (1985).
113. T. Blumenthal, M. I. Bruce, O. bin Shawkataly, B. N. Green, and I. Lewis, *J. Organomet. Chem.* **269**, C10 (1984).
114. I. D. Salter, Ph.D. Thesis, University of Bristol, 1983.
115. J. Evans, A. C. Street, and M. Webster, *Organometallics* **6**, 794 (1987).
116. K. Henrick, B. F. G. Johnson, J. Lewis, J. Mace, M. McPartlin, and J. Morris, *J. Chem. Soc., Chem. Commun.*, 1617 (1985).
117. A. G. Cowie, B. F. G. Johnson, J. Lewis, J. N. Nicholls, P. R. Raithby, and A. G. Swanson, *J. Chem. Soc., Chem. Commun.*, 637 (1984).
118. T. Beringhelli, G. D'Alfonso, M. Freni, G. Ciani, and A. Sironi, *J. Organomet. Chem.* **295**, C7 (1985).
119. P. Braunstein, J. Rosé, A. Dedieu, Y. Dusausoy, J.-P. Mangeot, A. Tiripicchio, and M. Tiripicchio-Camellini, *J. Chem. Soc., Dalton Trans.*, 225 (1986).
120. P. Braunstein and J. Rosé, *J. Organomet Chem.* **262**, 223 (1984).
121. J. Puga, R. A. Sanchez-Delgado, J. Ascanio, and D. Braga, *J. Chem. Soc., Chem. Commun.*, 1631 (1986).

122. D. Braga, K. Henrick, B. F. G. Johnson, J. Lewis, M. McPartlin, W. J. H. Nelson, A. Sironi, and M. D. Vargas, *J. Chem. Soc., Chem. Commun.,* 1131 (1983).

123. B. T. Heaton, L. Strona, S. Martinengo, D. Strumolo, V. G. Albano, and D. Braga, *J. Chem. Soc., Dalton Trans.,* 2175 (1983).

124. S. S. D. Brown, I. D. Salter, V. Šik, I. J. Colquhoun, W. McFarlane, P. A. Bates, M. B. Hursthouse, and M. Murray, *J. Chem. Soc., Dalton Trans.,* 2177 (1988).

125. T. J. Henly, J. R. Shapley, and A. L. Rheingold, *J. Organomet. Chem.* **310,** 55 (1986).

126. B. F. G. Johnson, J. Lewis, W. J. H. Nelson, J. Puga, P. R. Raithby, D. Braga, M. McPartlin, and W. Clegg, *J. Organomet. Chem.* **243,** C13 (1983); B. F. G. Johnson, J. Lewis, W. J. H. Nelson, J. Puga, M. McPartlin, and A. Sironi, *J. Organomet. Chem.* **253,** C5 (1983).

127. J. Pursiainen, M. Ahlgren, and T. A. Pakkanen, *J. Organomet. Chem.* **297,** 391 (1985).

128. A. Fumagalli, S. Martinengo, V. G. Albano, and D. Braga, *J. Chem. Soc., Dalton Trans.,* 1237 (1988).

129. A. G. Orpen, A. V. Rivera, E. G. Bryan, D. Pippard, G. M. Sheldrick, and K. D. Rouse, *J. Chem. Soc., Chem. Commun.,* 723 (1978); R. W. Broach, and J. M. Williams, *Inorg. Chem.* **18,** 314 (1979); and references cited therein.

130. B. T. Huie, C. B. Knobler, and H. D. Kaesz, *J. Am. Chem. Soc.* **100,** 3059 (1978); R. G. Teller, R. D. Wilson, R. K. McMullan, T. F. Koetzle, and R. Bau, *J. Am. Chem. Soc.* **100,** 3071 (1978).

131. M. A. Andrews, G. van Buskirk, C. B. Knobler, and H. D. Kaesz, *J. Am. Chem. Soc.* **101,** 7245 (1979).

132. M. Catti, G. Gervasio, and S. A. Mason, *J. Chem. Soc., Dalton Trans.,* 2260 (1977).

133. S. R. Bunkhall, H. D. Holden, B. F. G. Johnson, J. Lewis, G. N. Pain, P. R. Raithby, and M. J. Taylor, *J. Chem. Soc., Chem. Commun.,* 25 (1984).

134. B. F. G. Johnson, J. Lewis, W. J. H. Nelson, J. N. Nicholls, J. Puga, P. R. Raithby, M. J. Rosales, M. Schroder, and M. D. Vargas, *J. Chem. Soc., Dalton Trans.,* 2447 (1983).

135. P. A. Bates, S. S. D. Brown, A. J. Dent, M. B. Hursthouse, G. F. M. Kitchen, A. G. Orpen, I. D. Salter, and V. Šik, *J. Chem. Soc., Chem. Commun.,* 600 (1986).

136. S. S. D. Brown, I. D. Salter, D. B. Dyson, R. V. Parish, P. A. Bates, and M. B. Hursthouse, *J. Chem. Soc., Dalton Trans.,* 1795 (1988).

137. S. S. D. Brown, I. D. Salter, A. J. Dent, G. F. M. Kitchen, A. G. Orpen, P. A. Bates, and M. B. Hursthouse, *J. Chem. Soc., Dalton Trans.,* in press (1989).

138. T. Adatia, S. S. D. Brown. M. McPartlin and I. D. Salter, unpublished results.

139. J. A. K. Howard, L. J. Farrugia, C. L. Foster, F. G. A. Stone, and P. Woodward, *Eur. Cryst. Meeting* **6,** 73 (1980); L. J. Farrugia, Ph.D. Thesis, University of Bristol, 1979; C. L. Foster, Part II Thesis, University of Bristol, 1980.

140. T. Adatia, P. J. McCarthy, M. McPartlin, M. Rizza, and I. D. Salter, *J. Chem. Soc., Chem. Commun.,* 1106 (1988).

141. B. F. G. Johnson, J. Lewis, W. J. H. Nelson, P. R. Raithby, and M. D. Vargas, *J. Chem. Soc., Chem. Commun.,* 608 (1983).

142. M. J. Freeman, A. G. Orpen, and I. D. Salter, *J. Chem. Soc., Dalton Trans.,* 379 (1987).

143. S. S. D. Brown, I. D. Salter, and L. Toupet, *J. Chem. Soc., Dalton Trans.,* 757 (1988).

144. M. J. Freeman, A. G. Orpen, and I. D. Salter, *J. Chem. Soc., Dalton Trans.,* 1001 (1987).

145. S. S. D. Brown and I. D. Salter, unpublished results.

146. P. A. Bates, M. B. Hursthouse, and I. D. Salter, unpublished results.

147. S. S. D. Brown, I. J. Colquhoun, W. McFarlane, M. Murray, I. D. Salter, and V. Šik, *J. Chem. Soc., Chem. Commun.*, 53 (1986).

148. C. J. Brown, P. J. McCarthy, and I. D. Salter, unpublished results.

149. E. Roland, K. Fischer, and H. Vahrenkamp, *Angew. Chem. Int. Ed. Engl.* **22**, 326 (1983).

150. P. J. McCarthy and I. D. Salter, unpublished results.

151. C. E. Briant, K. P. Hall, and D. M. P. Mingos, *J. Chem. Soc., Chem. Commun.*, 843 (1983).

152. C. T. Sears and F. G. A. Stone, unpublished results; see S. A. R. Knox and F. G. A. Stone, *J. Chem. Soc. A*, 2559 (1969) for details.

153. B. F .G. Johnson, J. Lewis, P. R. Raithby, and A. Sanders, *J. Organomet. Chem.* **260**, C29 (1984).

154. R. D. George, S. A. R. Knox, and F. G. A. Stone, *J. Chem. Soc., Dalton Trans.*, 972 (1973).

155. M. E. O'Neill and K. Wade, *in* "Comprehensive Organometallic Chemistry" (G. Wilkinson, F. G. A. Stone, and E. W. Abel, Eds.) Vol. 1, p. 1. Pergamon, Oxford, 1982; K. Wade, *Adv. Inorg. Chem. Radiochem.* **18**, 1 (1976); D. M. P. Mingos, *Adv. Organomet. Chem.* **15**, 1 (1977); and references cited therein.

156. P. F. Jackson, B. F. G. Johnson, J. Lewis, and P. R. Raithby, *J. Chem. Soc., Chem. Commun.*, 60 (1980).

157. D. Braga, K. Henrick, B. F. G. Johnson, J. Lewis, M. McPartlin, W. J. H. Nelson, and M. D. Vargas, *J. Chem. Soc., Chem. Commun.*, 419 (1982); and references cited therein.

158. L. F. Rhodes, J. C. Huffman, and K. G. Caulton, *J. Am. Chem. Soc.* **107**, 1759 (1985).

159. J. E. Ellis and R. A. Faltynek, *J. Am. Chem. Soc.* **99**, 1801 (1977).

160. J. E. Ellis, unpublished results; see Refs. *16* and *88* for details.

161. T. Adatia, M. McPartlin, and I. D. Salter, *J. Chem. Soc., Dalton Trans.*, 751 (1988).

162. B. K. Teo, K. Keating, and Y.-H. Kao, *J. Am. Chem. Soc.* **109**, 3494 (1987); and references cited therein.

163. J. W. Diesveld, E. M. Menger, H. T. Edzes, and W. S. Veeman, *J. Am. Chem. Soc.* **102**, 7935 (1980).

164. N. J. Clayden, C. M. Dobson, K. P. Hall, D. M. P. Mingos, and D. J. Smith, *J. Chem. Soc., Dalton Trans.*, 1811 (1985).

165. S. S. D. Brown, C. M. Dobson, I. D. Salter, and D. J. Smith, unpublished results.

166. J. W. A. van der Velden, P. T. Buerskens, J. J. Bour, W. P. Bosman, J. H. Noordik, M. Kolenbrander, and J. A. K. M. Buskes, *Inorg. Chem.* **23**, 146 (1984).

167. S. S. D. Brown, L. S. Moore, R. V. Parish, and I. D. Salter, *J. Chem. Soc., Chem. Commun.*, 1453 (1986).

168. I. D. Salter, *J. Organomet. Chem.* **295**, C17 (1985).

169. B. F. G. Johnson, J. Lewis, and M.J. Taylor, unpublished results; see Ref. *26* for details.

170. B. F. G. Johnson, D. A. Kaner, J. Lewis, and P. R. Raithby, *J. Chem. Soc., Chem. Commun.*, 753 (1981).

171. J. Evans, A. C. Street, and M. Webster, *J. Chem. Soc. Chem. Commun.*, 637 (1987).

172. J. Evans and G. Jingxing, *J. Chem. Soc., Chem. Commun.*, 39 (1985).

173. R. L. Pruett and J. S. Bradley, Eur. Pat. Appl. No. 81301398.4 and U.S. Pat. Nos. 4301086 and 4342838.

174. Union Carbide Corp., U.S. Pat. No. 3878 292; see Ref. *6* for details.

175. M. Wolf, H. Knözinger, and B. Tesche, *J. Mol. Catal.* **25**, 273 (1984).

176. R. Pierantozzi, K. J. McQuade, B. C. Gates, M. Wolf, H. Knözinger, and

W. Ruhmann, *J. Am. Chem. Soc.* **101,** 5436 (1979); R. Pierantozzi, K. J. McQuade, and B. C. Gates, *Stud. Surf. Sci. Catal.* **7,** 941 (1981).

177. E. L. Muetterties and C. W. Alegranti, *J. Am. Chem. Soc.* **94,** 6386 (1972).
178. A. F. M. J. van der Ploeg and G. van Koten, *Inorg. Chim. Acta* **51,** 225 (1981).
179. M. Green, A. G. Orpen, I. D. Salter, and F. G. A. Stone, *J. Chem. Soc., Dalton Trans.,* 2497 (1984).
180. B. F. G. Johnson, *Adv. Dynamic Stereochem.* **2,** in press (1988).
181. B. F. G. Johnson and R. E. Benfield, *in* "Transition Metal Clusters" (B. F. G. Johnson, Ed.), p. 471. Wiley, New York, 1980.
182. B. E. Mann, *in* "Comprehensive Organometallic Chemistry" (G. Wilkinson, F. G. A. Stone, and E. W. Abel, Eds.), Vol. 3, p. 89. Pergamon, Oxford, 1982.
183. E. Band and E. L. Muetterties, *Chem. Rev.* **78,** 639 (1978).
184. J. Evans, *Adv. Organomet. Chem.* **16,** 319 (1977).
185. G. Doyle, K. A. Eriksen, and D. Van Engen, *J. Am. Chem. Soc.,* **108,** 445 (1986).
186. G. Doyle, B. T. Heaton, and E. Occhiello, *Organometallics* **4,** 1224 (1985).
187. G. Doyle, K. A. Eriksen, and D. Van Engen, *J. Am. Chem. Soc.* **107,** 7914 (1985).
188. A. Albinati, K.-H. Dahmen, A. Togni, and L. M. Venanzi, *Angew. Chem. Int. Ed. Engl.* **24,** 766 (1985).
189. G. M. Mott, N. J. Taylor, and A. J. Carty, *Organometallics* **2,** 447 (1983).
190. M. Fajardo, M. P. Gómez-Sal, H. D. Holden, B. F. G. Johnson, J. Lewis, R. C. S. McQueen, and P. R. Raithby, *J. Organomet. Chem.* **267,** C25 (1984).
191. G. Bruno, S. Lo Schiavo, P. Piraino, and F. Faraone, *Organometallics* **4,** 1098 (1985).
192. G. Longoni and P. Chini, *Inorg. Chem.* **15,** 3029 (1976).
193. S. R. Drake, K. Henrick, B. F. G. Johnson, J. Lewis, M. McPartlin, and J. Morris, *J. Chem. Soc., Chem. Commun.,* 928 (1986).
194. G. A. Carriedo, J. A. K. Howard, and F. G. A. Stone, *J. Chem. Soc., Dalton Trans.,* 1555 (1984).
195. L. F. Rhodes, J. C. Huffman, and K. G. Caulton, *J. Am. Chem. Soc.* **105,** 5137 (1983).
196. T. H. Lemmen, J. C. Huffman, and K. G. Caulton, *Angew. Chem. Int. Ed. Engl.* **25,** 262 (1986).
197. O. M. Abu Salah and M. I. Bruce, *Aust. J. Chem.* **29,** 531 (1976).
198. M. R. Churchill and S. A. Bezman, *Inorg. Chem.* **13,** 1418 (1974).
199. O. M. Abu-Salah, A.-R. A. Al-Ohaly, and C. B. Knobler, *J. Chem. Soc., Chem. Commun.,* 1502 (1985).
200. A. J. Leusink, G. van Koten, and J. G. Noltes, *J. Organomet, Chem.* **56,** 379 (1973).
201. G. van Koten and J. G. Noltes, *J. Organomet. Chem.* **102,** 551 (1975).
202. G. van Koten, J. T. B. H. Jastrzebski, and J. G. Noltes, *Inorg. Chem.* **16,** 1782 (1977).
203. H. Hope, D. Oram, and P. Power, *J. Am. Chem. Soc.* **106,** 1149 (1984).
204. S. I. Khan, P. G. Edwards, H. S. H. Yuan, and R. Bau, *J. Am. Chem. Soc.* **107,** 1682 (1985).
205. S. Lo Schiavo, G. Bruno, P. Piraino, and F. Faraone, *Organometallics* **5,** 1400 (1986).
206. F. Bachechi, J. Ott, and L. M. Venanzi, *J. Am. Chem. Soc.* **107,** 1760 (1985).
207. R. Uson, J. Fomiés, B. Menjón, F. A. Cotton, L. R. Falvello, and M. Tomás, *Inorg. Chem.* **24,** 4651 (1985).
208. R. Usón, A. Laguna, M. Laguna, A. Usón, P. G. Jones, and C. Friere Erdbrügger, *Organometallics* **6,** 1778 (1987).
209. M. Y. Chiang, E. Böhlen, and R. Bau, *J. Am. Chem. Soc.* **107,** 1679 (1985).
210. L. Manojlović-Muir, K. W. Muir, I. Treurnicht, and R. J. Puddephatt, *Inorg. Chem.* **26,** 2418 (1987).

Index

Cumulative List of Contributors

Abel, E. W., **5**, 1; **8**, 117
Aguilo, A., **5**, 321
Albano, V. G., **14**, 285
Alper, H., **19**, 183
Anderson, G. K., **20**, 39
Angelici, R. J., **27**, 51
Armitage, D. A., **5**, 1
Armor, J. N., **19**, 1
Ash, C. E., **27**, 1
Atwell, W. H., **4**, 1
Baines, K. M., **25**, 1
Barone, R., **26**, 165
Bassner, S. L., **28**, 1
Behrens, H., **18**, 1
Bennett, M. A., **4**, 353
Birmingham, J., **2**, 365
Blinka, T. A., **23**, 193
Bogdanović, B., **17**, 105
Bottomley, F., **28**, 339
Bradley, J. S., **22**, 1
Brinckman, F. E., **20**, 313
Brook, A. G., **7**, 95; **25**, 1
Brown, H. C., **11**, 1
Brown, T. L., **3**, 365
Bruce, M. I., **6**, 273; **10**, 273; **11**, 447; **12**, 379; **22**, 59
Brunner, H., **18**, 151
Buhro, W. E., **27**, 311
Cais, M., **8**, 211
Calderon, N., **17**, 449
Callahan, K. P., **14**, 145
Cartledge, F. K., **4**, 1
Chalk, A. J., **6**, 119
Chanon, M., **26**, 165
Chatt, J., **12**, 1
Chini, P., **14**, 285
Chisholm, M. H., **26**, 97; **27**, 311
Chiusoli, G. P., **17**, 195
Churchill, M. R., **5**, 93
Coates, G. E., **9**, 195
Collman, J. P., **7**, 53
Connelly, N. G., **23**, 1; **24**, 87
Connolly, J. W., **19**, 123
Corey, J. Y., **13**, 139
Corriu, R. J. P., **20**, 265

Courtney, A., **16**, 241
Coutts, R. S. P., **9**, 135
Coyle, T. D., **10**, 237
Crabtree, R. H., **28**, 299
Craig, P. J., **11**, 331
Csuk, R., **28**, 85
Cullen, W. R., **4**, 145
Cundy, C. S., **11**, 253
Curtis, M. D., **19**, 213
Darensbourg, D. J., **21**, 113; **22**, 129
Darensbourg, M. Y., **27**, 1
Deacon, G. B., **25**, 237
de Boer, E., **2**, 115
Deeming, A. J., **26**, 1
Dessy, R. E., **4**, 267
Dickson, R. S., **12**, 323
Dixneuf, P. H., **29**, 163
Eisch, J. J., **16**, 67
Emerson, G. F., **1**, 1
Epstein, P. S., **19**, 213
Erker, G., **24**, 1
Ernst, C. R., **10**, 79
Evans, J., **16**, 319
Evans, W. J., **24**, 131
Faller, J. W., **16**, 211
Faulks, S. J., **25**, 237
Fehlner, T. P., **21**, 57
Fessenden, J. S., **18**, 275
Fessenden, R. J., **18**, 275
Fischer, E. O., **14**, 1
Ford, P. C., **28**, 139
Forniés, J., **28**, 219
Forster, D., **17**, 255
Fraser, P. J., **12**, 323
Fritz, H. P., **1**, 239
Fürstner, A., **28**, 85
Furukawa, J., **12**, 83
Fuson, R. C., **1**, 221
Gallop, M. A., **25**, 121
Garrou, P. E., **23**, 95
Geiger, W. E., **23**, 1; **24**, 87
Geoffroy, G. L., **18**, 207; **24**, 249; **28**, 1
Gilman, H., **1**, 89; **4**, 1; **7**, 1
Gladfelter, W. L., **18**, 207; **24**, 41
Gladysz, J. A., **20**, 1